AF360785

Advances in Modelling and Prediction on the Impact of Human Activities and Extreme Events on Environments

Advances in Modelling and Prediction on the Impact of Human Activities and Extreme Events on Environments

Editors

Songdong Shao
Min Luo
Matteo Rubinato
Xing Zheng
Jaan H. Pu

MDPI • Basel • Beijing • Wuhan • Barcelona • Belgrade • Manchester • Tokyo • Cluj • Tianjin

Editors
Songdong Shao
The University of Sheffield
UK

Min Luo
Zhejiang University
China

Matteo Rubinato
Coventry University
UK

Xing Zheng
Harbin Engineering University
China

Jaan H. Pu
University of Bradford
UK

Editorial Office
MDPI
St. Alban-Anlage 66
4052 Basel, Switzerland

This is a reprint of articles from the Special Issue published online in the open access journal *Water* (ISSN 2073-4441) (available at: https://www.mdpi.com/journal/water/special_issues/river_coastal_environments?authAll=true).

For citation purposes, cite each article independently as indicated on the article page online and as indicated below:

LastName, A.A.; LastName, B.B.; LastName, C.C. Article Title. *Journal Name* **Year**, *Article Number*, Page Range.

ISBN 978-3-03936-802-0 (Hbk)
ISBN 978-3-03936-803-7 (PDF)

Cover image courtesy of Jaan H. Pu.

Contents

About the Editors

Songdong Shao (Senior Lecturer): Dr Shao attained his BSc, MSc, and Ph.D. degrees at the Department of Hydraulic Engineering at Tsinghua University. He has worked at the University of Bradford, University of Hong Kong, University of Plymouth, Kyoto University, and Nanyang Technological University, before joining the Department of Civil and Structural Engineering at the University of Sheffield. Dr Shao's research concentrates on the influence of global warming on coastal disasters. He develops numerical models to solve the hydrodynamic equations to predict natural hazards, involving river and coastal environments. He is well-known for work in the area of Smoothed Particle Hydrodynamics, a robust mesh-free numerical modeling approach for various free surface flows, with a high citation record of 4000+ in Google Scholar. Dr Shao has served as an Associate Editor for the Coastal Engineering Journal (CEJ) from 2009 to 2019.

Min Luo (Professor): Professor Luo received his BEng degree (2010) in Civil Engineering from Harbin Institute of Technology (HIT) and PhD degree (2015) in Ocean Engineering from the National University of Singapore (NUS). He worked as a Post-doctor (2015–2017) at NUS and a Lecturer in the Zienkiewicz Centre for Computational Engineering at Swansea University (2017–2020), before joining the Ocean College at Zhejiang University in 2020. Professor Luo's research concentrates on the computational modelling and laboratory study of wave hydrodynamics, with an emphasis on the extreme wave interactions with the ocean/coastal structures. He is the core developer of the mesh-free Consistent Particle Method (CPM).

Matteo Rubinato (Lecturer): Dr Rubinato works in the Department of Civil and Environmental Engineering in the Faculty of Engineering, Environment & Computing at Coventry University, and he is an associate member of the Centre for Agroecology, Water and Resilience (CAWR). He completed his bachelor's and master's degrees in Environmental Engineering at the University of Padova and obtained his Ph.D. degree in 2015 at the University of Sheffield. Since then, throughout his academic career, he has conducted research with members at over 25 institutions across Europe, USA, and China with longer periods spent at the University of Colorado Boulder, Sichuan University, Beijing Normal University, and Xi'an University of Technology. His outputs have led to multiple high-quality journal articles (25+), conference proceedings (15+), and a book-chapter. Dr Rubinato is one of the leaders in large-scale physical modeling of urban floods in the UK and the main focus of his research involves novel quantification and explanation of flow processes in sewers and linked urban surfaces. Furthermore, Dr Rubinato's research also explores engineering in the field of environmental fluid mechanics typical of river and coastal environments to develop innovative and useful approaches to help reduce the vulnerability, as well as enhance the resilience, adaptive capacity, and sustainability of environmental systems around the world in the face of increasing challenges, uncertainty, and climate change variability.

Xing Zheng (Professor): Prof Zheng received his BEng degree (2003) in Naval Architecture and Ocean Engineering from Harbin Engineering University (HEU), and Ph.D. degree (2010) in Fluid Mechanics from HEU as well. His research concentrates on the numerical simulation and laboratory study of breaking waves, marine hydrodynamics, renewable engineering, and ice–ship–water interactions. He has established very influential works in the development of novel higher-order numerical schemes for the mesh-free Smoothed Particle Hydrodynamics (SPH) method, especially the incompressible SPH (ISPH) approaches.

Jaan H. Pu (Associate Professor): Dr Pu received both his BEng (1st Class Honours) and Ph.D. degrees from the University of Bradford in 2003 and 2008, respectively. Since 2008, he has taken up several research and faculty positions at various universities/institutions around the world. He was appointed as a Lecturer in Civil Engineering at the University of Bradford in 2014; as Senior Lecturer in 2017; and as Associate Professor since 2020. Dr Pu's research concentrates on numerical and laboratory approaches to representing various water engineering applications that include the naturally compound riverine flow, sediment transport, scouring, water quality, and vegetated flow. His research outputs have led to several high-quality journal articles (30+), conference proceedings (10+), book-chapters, and an edited book. He is currently supervising 4 Ph.D. students at Bradford (3 as their principal supervisor) to investigate river hydrodynamics and sediment transport challenging applications. He is currently a Visiting Scientist at Tsinghua University and Nanyang Technological University. He has also served as a reviewer for several internationally well-reputed journals.

Preface to "Advances in Modelling and Prediction on the Impact of Human Activities and Extreme Events on Environments"

There is an emergent consensus that people living in cities, together with the existing infrastructure and ecology of these urban areas and global environments, are at risk from the impacts of climate change (e.g. heavy rainfall) and human activities (e.g. land use). There are multiple uncertainties linked with climate change scenarios: i) the first one refers to the quantification of future emissions of greenhouse gases, ii) the second one is related to scaling issues at local and global scales, and iii) the third one is due to the availability of datasets used as input to calibrate and validate predicting tools. Furthermore, territories across the world have been continuously modified and adapted by a variety of human activities for needs such as drinking water, as well as the research for renewable energies and other resources. The construction of large dams and reservoirs, or the excavation of canals in parallel to existing ones had a harmful impact on the environment. Across the world, human activities and engineering developments have also caused groundwater quality deterioration, waste disposal failures, erosion of coastal areas, and these actions have also endangered natural species and affected the habitats of ecological systems.

Flooding events, periods of droughts, debris flows induced by dam break, erosion of coastal areas, and river embankments are just some of the issues and disasters that continue to happen more frequently and science is gradually improving its predictive skills to enable a better understanding of what causes these phenomena and to establish early warning systems. The duration of each of these events varies from a few minutes to hours or even days, and, therefore, it is extremely challenging to identify early warning systems that require the possibility to generate targeted evacuation time for each scenario. This increasing pressure on environments across the world requires improvements in the understanding of cumulative impacts of human actions and climate change to enable more efficient management of these areas for local and national authorities.

The studies in this Special Issue provide timely inputs into growing needs across the world. Due to the continuously changing circumstances, future trends in the management of mountainous, river, and coastal environments will need to be dynamic processes based on adaptive management. It is a priority to keep assessing whether the existing management approaches are still effective in response to the increasing interactions between the environments, land use, and climate change on the global stage. Thus, studies in this Special Issue provide recent developments in effective tools that can better predict the causes of these natural hazards as well as the impacts that they can inflict on environments.

Songdong Shao, Min Luo, Matteo Rubinato, Xing Zheng, Jaan H. Pu

Editors

 *water*

Editorial

Advances in Modelling and Prediction on the Impact of Human Activities and Extreme Events on Environments

Matteo Rubinato [1,2,*], Min Luo [3], Xing Zheng [4,*], Jaan H. Pu [5] and Songdong Shao [6]

1   Faculty of Engineering, Environment & Computing, School of Energy, Construction and Environment, Coventry University, Coventry CV1 5FB, UK
2   Centre for Agroecology, Water and Resilience, Coventry University, Wolston Lane, Coventry CV8 3LG, UK
3   Zienkiewicz Centre for Computational Engineering, College of Engineering, Swansea University, Swansea SA1 8EN, UK; min.luo@swansea.ac.uk
4   College of Shipbuilding Engineering, Harbin Engineering University, Harbin 150001, China
5   Faculty of Engineering and Informatics, University of Bradford, Bradford BD7 1DP, UK; j.h.pu1@bradford.ac.uk
6   Department of Civil and Structural Engineering, University of Sheffield, Sheffield S1 3JD, UK; s.shao@sheffield.ac.uk
*   Correspondence: matteo.rubinato@coventry.ac.uk (M.R.); zhengxing@hrbeu.edu.cn (X.Z.); Tel.: +44-(0)24-7765-0887 (M.R.); +86-(0)451-8256-8147 (X.Z.)

Received: 8 June 2020; Accepted: 14 June 2020; Published: 22 June 2020

Abstract: Fast urbanization and industrialization have progressively caused severe impacts on mountainous, river, and coastal environments, and have increased the risks for people living in these areas. Human activities have changed ecosystems hence it is important to determine ways to predict these consequences to enable the preservation and restoration of these key areas. Furthermore, extreme events attributed to climate change are becoming more frequent, aggravating the entire scenario and introducing ulterior uncertainties on the accurate and efficient management of these areas to protect the environment as well as the health and safety of people. In actual fact, climate change is altering rain patterns and causing extreme heat, as well as inducing other weather mutations. All these lead to more frequent natural disasters such as flood events, erosions, and the contamination and spreading of pollutants. Therefore, efforts need to be devoted to investigate the underlying causes, and to identify feasible mitigation and adaptation strategies to reduce negative impacts on both the environment and citizens. To contribute towards this aim, the selected papers in this Special Issue covered a wide range of issues that are mainly relevant to: (i) the numerical and experimental characterization of complex flow conditions under specific circumstances induced by the natural hazards; (ii) the effect of climate change on the hydrological processes in mountainous, river, and coastal environments, (iii) the protection of ecosystems and the restoration of areas damaged by the effects of climate change and human activities.

Keywords: experimental modelling; numerical modelling; scouring; sediment transport; smoothed-particle hydrodynamics; flooding; dam-break; debris flows; climate change; urban evolution; natural hazard

---

1. Introduction

During the last few decades, urbanization, which refers to an increase in population and the amount of industrialization of a settlement, has become one of the dominant forms of landscape disturbance [1]. Over the years, more people are moving from rural to urban areas, increasing their density and inducing continuous land changes that indirectly inflict many unwanted consequences such as land insecurity [2,3], worsening water quality [4], poor air quality [5,6], and adverse hydrological processes [7–9] and local ecosystems [10,11]. The impacts of anthropogenic development have affected over 75% of Earth's land surface [12] and recent studies have confirmed that urbanization has also had an impact on precipitation rates, increasing their spatial variability and intensification [13,14].

Climate interacts to influence the quantity and quality of Earth's environment, aggravating the entire scenario [15–17]. Extreme events are becoming more frequent and their consequences are spreading worldwide. Due to a rise in sea levels, which are expected to increase by between 0.07 and 0.12 m during the 21st century [18], communities living near coastal areas are facing potential multi-hazard threats. Furthermore, due to the increased extreme rainfall events, both urban and rural catchments are subject to continuous risks of flooding [19], where properties can be inundated and, in combination with the critical events and geographical characteristics such as steep slopes and proximity of houses near mountainous rivers, human losses caused by the landslides and debris flows can be escalated [20–24]. About 3.8 million km^2 and 790 million people in the world are living in relatively high exposure to at least two hazards, while about 0.5 million km^2 and 105 million people to three or more hazards (e.g., floods, droughts, tropical storms, earthquakes, volcanoes, and landslides), according to a report of the World Bank on the main hotspots of natural hazards [25]. Considering the variety of effects and areas at risk, climate change is likely to further increase community exposure to multiple risks, affecting the magnitude, frequency, and spatial distribution of hazardous and disastrous events [25,26].

Dated back in 2012, governments had established a set of Sustainable Development Goals (SDGs) that were integrated into the follow-up of Millennium Development Goals (MDGs) after their 2015 deadline. SDG 13 considers both adaptation and mitigation to climate change and targets at strengthening resilience, improving monitoring, and integrating measures into local and national planning policies [27]. Additionally, SDG 11 aims at making cities and human settlements inclusive, safe, resilient, and sustainable [28].

To contribute to filling the needs identified within these SDGs, this Special Issue aimed at gathering the latest developments in advanced numerical and experimental modeling and other technologies, to provide a better understanding of the specific phenomena associated with natural disasters, and to predict and evaluate changes in river, coastal, and mountainous environments induced by extreme events and human activities. The articles published within this Special Issue aimed to aid local and national authorities on the design and implementation of mitigation strategies, thus providing tools that could secure a more accurate management of their environmental areas.

2. Summary of this Special Issue

Climate change has caused a transformation in the rainfall patterns, and the rainfall amount and intensity had a great impact on the hydrological dynamics across the world. To investigate this impact of climate change in karst regions, which are widely distributed in southwest China, Li et al. [29] have conducted experimental work to identify close relationship between subsurface and underground fissure flows with various rainfall intensities and bedrock degrees. Results obtained from this study have confirmed that under light rainfall conditions (30 mm h^{-1}), the hydrological processes observed were typical of Dunne overland flows; however, under moderate (60 mm h^{-1}) and high (90 mm h^{-1}) rainfall conditions, the hydrological processes were typical of Horton overland flows [29].

Climate change and catchment modification induced by human activities are the main drivers, and play a significant role worldwide in the dramatic variation of water levels in lakes. A case study on Qinghai Lake, the largest inland saline lake on the Tibetan Plateau, was presented by Fang et al. [30] and the meteorological and land use data collected between 1960 and 2016 have been analyzed to investigate the effect of climate change and human activities on this lake. Results obtained have demonstrated that the water level of Qinghai Lake declined between 1960 and 2004, and since then has risen continuously and gradually, due to changes in evaporation, precipitation, and consequent surface runoff associated with climate changes and catchment modifications. Moreover, the changes in rainfall patterns and magnitudes generated by climate change may have also had an effect on runoff and soil erosion processes. Ran et al. [31] have conducted a study aiming to understand how various rainfall patterns (constant, increasing, decreasing, rising–falling, and falling–rising) affect these processes, with a particular focus on the slopes with a wide range of gradients (5° to 40°) and length scales (25 m to 200 m). Results have identified a critical slope, 15°, which was independent of the rainfall pattern and slope length. However, it has been found that the critical slope of soil erosion amounts decreased from 35° to 25°, with an increasing projective slope length [31].

The increase of sediment erosions induced by higher rainfall intensities, together with the effects caused by human activities related to the deforestation and expansion of farmland, have recently been considered as a major issue for the normal functionality of dams and river embankments [32]. These engineering structures are important assets that need to be regularly checked because they could generate significant damages to the surrounding environment as well as to the people living in nearby areas once they fail. Multiple papers have been submitted on this topic within this Special Issue.

Check dams were numerically investigated by Tang et al. [33] To support the development of future planning and management strategies in terms of soil and water conservation, the hydrologic response changes and the sedimentary processes caused by a check dam system were simulated [33], and the results showed that a check dam can significantly alter water redistribution in the catchment and influence the groundwater table in different periods. Furthermore, Tang et al. [34] investigated the influences of filled check dams with six different deployment strategies in a Loess Plateau catchment and compared them with the no-dam and real scenarios. The results showed that the filled check dams were still able to effectively reduce the flood peak (Q_p) by 31% to 93% under different deployment strategies.

Dam as well as embankment failures can also be caused by the process of internal erosion. To date, various types of internal erosion have been identified: concentrated leak erosion, contact erosion, backward erosion, and suffusion. Suffusion constitutes a major threat to the foundation of a dam, and its likelihood is usually determined by the internal stability of soils and the interaction of soil particles with seepage flows. A study was presented by Feng et al. [35], where a numerical model was developed to simulate the suffusion process, thus assessing the internal stability of the dam. Results have confirmed that the suffusion effect is closely related to the specific grain size distribution of the soils [35].

Flooding due to dam break has potentially disastrous consequences and multiple studies have been conducted to replicate the hydrodynamics of this phenomenon [36–38]. Due to frequent field events during the last decade, it is becoming increasingly important to understand the key physics behind the dam break. Thanks to the latest advances in the use of high performance computing techniques to accelerate the computational fluid dynamics (CFD) codes, it is now possible to simulate the natural hazards associated with the rapidly varied flows of both water and dense granular mixtures, together with the sediment erosion and bed load transport [39,40]. Due to its Lagrangian nature, the Smoothed Particle Hydrodynamics (SPH) method has been used to solve a variety of fluid-dynamic processes with highly nonlinear features such as debris flow, wave breaking and impact, multi-phase mixing, jet impact, flooding and tsunami inundation, and fluid–structure interactions [41]. For example, Wu et al. [41] applied the SPH method to solve two-dimensional Shallow Water Equations (SWEs), and the solution proposed was validated against two open-source case studies of a dry-bed dam break and another dam break with a rectangular obstacle downstream. In addition to the improvement and optimization of the numerical algorithm, a CPU-OpenMP parallel computing technique was also implemented to enhance the model performance.

In this Special Issue, the SPH method was also vigorously applied to investigate another cause of the natural disaster, i.e., debris flows, which are characterized by high density, impact force and destructiveness, and complexity of the materials they are made of [42]. The numerical simulations involved three different soil configurations and the results obtained by applying the modifications included into the SPH model clearly demonstrated that the configuration where fine and coarse particles are fully mixed, with no specific layering, produced more fluctuations and instability of the debris flow. This SPH modeling work has provided a better understanding on the mechanism of intermittent debris flows [42].

Another study was also presented on the SPH method applied to simulate liquid sloshing in a 2D tank with water jet flows as presented by Jiang et al. [43] The study compared the liquid sloshing under different conditions to analyze the effects of excitation frequency and water jet on the impact pressures. Results obtained firstly confirmed the applicability of the SPH method to accurately replicate these features and, secondly, demonstrated that the water jet flows can significantly affect the impact pressures on the wall caused by the violent sloshing [43].

To protect residential areas from the river flooding, embankments are usually constructed as a defense. If well constructed, these riverside embankments can be relatively effective in stopping the water spilling onto the adjacent lands. These structures are equally a very good benefit to the wildlife habitat (e.g., species-rich grassland or riparian woodland). However, being similar to the dam-break, the failure of these systems can also lead to devastating and fatal consequences.

To address this knowledge gap, a study was published in this Special Issue where a numerical simulation was conducted to investigate the failure processes of a homogeneous embankment due to the flow overtopping [44]. The good agreement between experimental and numerical results confirmed the accuracy of the employed numerical approach based on a double-point two-phase material point method (MPM) considering the water–soil interactions and seepage effects [44]. Within the same river environment, sediment erosions can occur on the riverbed, on the side-bank of the river and even on some local areas, such as those around the piers that are built along the river to support the bridges. Pandey et al. [45] conducted experiments to analyze the maximum equilibrium between the scour depth and the scour process in an armored streambed. It was found that the variation of the maximum dimensionless scour depth with the dimensionless armor particle size depends on the densimetric particle Froude number (Fr_{d50}). Sediments are eroded and transported by the water along the river and their deposition, suspension and transport are regulated and continuously modified by the turbulence and other flow structures [45]. To provide better understanding on the coupled effects of sediment inertia and stratification on the pattern of secondary current in bend-flows, a study was published by Yang et al. [46] to evaluate a full 3D numerical model. The sediment inertia effect, as well as the stratification effect induced by the non-uniform distribution of suspended sediments, was accounted for by adopting the hydrodynamic equations without the Boussinesq approximation. The numerical results demonstrated that sediment stratification effects enhanced the intensity of secondary flow via reducing the eddy viscosity, while the sediment inertia effects suppressed it [46]. On the other hand, the sediments may be in contact with the emergent vegetations across the river, which also play an important role in affecting the turbulence, velocity pattern, resistance, and sediment transport, with consequent morphological changes. A well-conducted review on the hydrodynamics of free surface flows in an emergent vegetated channel was presented by Maji et al. [47] The authors have highlighted the progresses in a wide range of field, laboratory, and numerical investigations on the turbulent flow within different emergent vegetations, and focused on the vegetation-induced flow field, velocity distribution, and structure and drag effects [47]. This review is beneficial for the local and national authorities in charge of restoring the river environments and implementing the strategies for the attenuation of river flooding.

Rivers and lakes are the environments that provide a complex interaction among dissimilar water resources, which can be affected by both climate change and human activities, and they are rich of various ecosystems. A study provided by Ma et al. [48] focused on the ecosystem services in Dongting Lake area, investigating snail control and schistosomiasis prevention, water yield, soil conservation, and carbon storage. They evaluated plenty of data collected between 2005 and 2015 by using ArcGIS 10.2 and InVEST models [48]. This study has confirmed that the evapotranspiration, precipitation, soil erodibility, and rainfall erosivity significantly influenced some of the ecosystem services in the study area. Nevertheless, to implement satisfactory management strategies in these environments, it is necessary to fully understand the magnitude of the flows, and their duration and frequency, because these could have a huge impact on the ecological water demand. To further analyze these features, Zhou et al. [49] created a multi-scale coupled ecological dispatching model based on the decomposition-coordination principle, where they considered the multi-scale features of the ecological water demand. Results showed that the degree of hydrologic alteration of small-scale ecological flow regimes and the daily stream flows can be accurately predicted by their proposed model, demonstrating the impact of hydrologic alterations on the reliability of the water supply [49].

Hydrodynamic modelling is also one of the most relevant challenges in marine and ocean engineering for protecting the coastal areas from erosion and flooding [50], as well as identifying the sources of energy that could be generated by the specific wave conditions. In this Special Issue, the popular mesh-free SPH method was applied to investigate the flow behavior inside and outside a porous structure under continuous wave actions [51]. To construct an accurate and efficient model, a unified set of flow equations was solved for both the porous flow region and the outside free flow region, with the interface boundary condition being automatically satisfied. The SPH simulations have been used to analyze the flow structures near the porous obstacle, with a focus on the longitudinal and vertical velocity distributions in the complex vortex and eddy areas. The study provided an innovative insight into the mathematical modelling of fluid–structure interactions (FSI) in a practical coastal environment [51].

Oceans are subjected to a very challenging environment, because they are characterized by a variety of wave conditions, from freak to tsunami waves. Therefore, Xu et al. [52] conducted an experimental work aiming to better understand the generation and mechanism of those extreme waves, as well as their potential hydrodynamic loads on the floating or fixed ocean structures in extreme sea environments. In this study, a series of focusing waves based on the two newly proposed wave amplitude spectra (i.e., QCWA and QCWS spectra) were tested in a physical wave tank, demonstrating that different spectra can lead to different wave crest elevations and locations. The spectral analysis results showed that the wave nonlinearity also plays an important role in the focusing wave generation for one type of the spectrum, whereas the redistribution of wave energy in the input frequency range significantly affects the focusing wave generation for another type of the spectrum. In addition, Jin et al. [53] presented a 2D numerical model to investigate the vortex-induced vibrations (VIVs) for a submerged floating tunnel (SFT) with different Reynolds numbers (Re), by solving the incompressible viscous Reynolds-averaged Navier-Stokes (RANS) equations in the frame of the Abitrary Lagrangian Eulerian (ALE) approach. The computational results showed that the Re numbers have a great influence not only on the vibration amplitude and the lock-in region, but also on the force coefficient on the SFT. Then it was further concluded that when the size of SFT is small, or when the flow velocity action on the structure is slow, the force coefficient and the lock-in region are relatively large, while when the size is large or when the velocity is fast, these key parameter values are relatively small [53].

Last but not the least, aligned with the modern needs to identify renewable energy/energy efficiency and expand energy access, Yu et al. [54] focused on wind turbines and the effect of turbine blade deformations with relevant lifespan issues. The developed technique combined the actuator line model (ALM) with a beam solver for use in the wind turbine blade design. A popular open source code, OpenFOAM, was used to investigate the performance of the National Renewable Energy Laboratory 5 MW wind turbine in terms of its power, thrust, and blade tip displacement, leading to insights on the negative influences of tower shadow effect on the power production. The well-calibrated model has been found to be capable of obtaining acceptable predictions on a range of wind turbine parameters for practical purposes.

Besides, it is worth mentioning that other state-of-the-art numerical schemes are also being developed by the same research group aiming to improve the applicability of the mesh-free methods in coastal and ocean engineering, including the recent benchmark works on implementing the Taylor series consistency principle to treat the pressure gradient term in 3D Navier-Stokes equations [55] and developing the advanced macroscopic equations of mass and momentum for the interaction at an interface of flow with porous media [56].

3. Conclusions

This Special Issue has covered a wide range of contemporary issues on the impacts to the environments generated by global climate changes and human activities. The studies in this Special Issue provide timely inputs into growing needs across the world. Due to the continuously changing circumstances, future trends in the management of mountainous, river, and coastal environments will need to be dynamic processes based on adaptive management. Consequently, it is a priority to keep assessing whether the existing management approaches are still effective in response to the increasing interactions between the environments, the land use, and climate change on the global stage.

Further research should focus on understanding the responses of environments under the influence of climate change and human activities separately, as well as their combined effects, in order to develop more robust adaptation strategies and policies for environmental protection, planning, and management. Decision-makers need to have effective tools to better predict the causes of these natural hazards as well as the impacts that they can inflict onto environments. Policy-makers should then direct significant efforts to shape their guidelines towards a more sustainable societal progress including both the protection of the environment and the restoration of the areas that have been damaged by climate change and anthropogenic development; indeed, excessive deforestation or installation of dams with improper regulation could lead to negative impacts of a similar magnitude to those caused by climate change.

Targeting solutions to adapt to climate change and human impact could also lead to an improvement of the quality of life across the world. For example, to tackle urban, coastal, and river flooding, living communities can be more wisely managed by expanding green spaces that could help reduce flood risks and attenuate the peak flows. At the same time, these green areas could also serve as recreational spaces that contribute to human health and well-being, and could help improve local biodiversity. It is imperative to note that uncertainties should always be included so as not to underestimate the adverse impacts on the environments and there should always be frequent communications among stakeholders on a global scale to nurture the exchange of knowledge towards more joint collaborations.

Acknowledgments: The authors of this editorial and the guest editors of this Special Issue would like to thank all the authors for their notable contributions as well as all the reviewers for devoting their time and effort to reviewing the manuscripts. Finally, our special thanks go to the Water Editorial team, in particular Senior Assistant Editor Janelee Li, for their great support during the processing of the submitted manuscripts.

Funding: The authors want to acknowledge the following grants: (1) The National Natural Science Foundation of China (Grant No: 51879051); (2) The Belt and Road Special Foundation of the State Key Laboratory of Hydrology-Water Resources and Hydraulic Engineering (Grant No: 2019491711); (3) The Open Research Fund of the State Key Laboratory of Hydraulics and Mountain River Engineering in Sichuan University (Grant No: SKHL1710 and SKHL1712); and (4) Qinghai Science and Technology Projects (Grant No: 2017-ZJ-Y01).

Conflicts of Interest: The authors declare no conflict of interest.

References

1. Price, S.J.; Dorcas, M.E.; Gallant, A.L.; Klaver, R.W.; Wilson, J.D. Three decades of urbanization: Estimating the impact of land-cover change on stream salamander populations. *Biol. Conserv.* **2006**, *133*, 436–441. [CrossRef]

2. Abass, K.; Adanu, S.K.; Agyemang, S. Peri-urbanisation and loss of arable land in Kumasi Metropolis in three decades: Evidence from remote sensing image analysis. *Land Use Policy* **2018**, *72*, 470–479. [CrossRef]

3. Sussman, H.S.; Raghavendra, A.; Zhou, L. Impacts of increased urbanization on surface temperature, vegetation, and aerosols over Bengaluru, India. *Remote Sens. Appl. Soc. Environ.* **2019**, *16*, 100261. [CrossRef]

4. Singh, S.; Hassan, S.M.T.; Hassan, M.; Bharti, N. Urbanisation and water insecurity in the Hindy Kush Himalaya: Insights from Bangladesh, India, Nepal and Pakistan. *Water Policy* **2020**, *22*, 9–32. [CrossRef]

5. Yu, S.; Lu, H. Relationship between urbanization and pollutant emissions in transboundary river basins under the strategy of the Belt and Road Initiative. *Chemosphere* **2018**, *203*, 11–20. [CrossRef]

6. Xu, X.; Gonzalez, J.E.; Shen, S.; Miao, S.; Dou, J. Impacts of urbanization and air pollution on building energy demands—Beijing case study. *Appl. Energy* **2018**, *225*, 98–109. [CrossRef]
7. Locatelli, L.; Mark, O.; Mikkelsen, P.S.; Nielsen, K.A.; Deletic, A.; Roldin, M.; Binning, P.J. Hydrologic impact of urbanization with extensive stormwater infiltration. *J. Hydrol.* **2017**, *544*, 524–537. [CrossRef]
8. Oudin, L.; Salavati, B.; Furusho-Percot, C.; Ribstein, P.; Saadi, M. Hydrological impacts of urbanization at the catchment scale. *J. Hydrol.* **2018**, *559*, 774–786. [CrossRef]
9. Luo, J.; Zhou, X.; Rubinato, M.; Li, G.; Tian, Y.; Zhou, J. Impact of multiple vegetation covers on surface runoff and sediment yield in the small basin of Nverzhai, Hunan Province, China. *Forests* **2020**, *11*, 329. [CrossRef]
10. Garcia-Nieto, A.P.; Geizendorffer, I.R.; Baro, F.; Roche, P.K.; Bondeau, A.; Cramer, W. Impacts of urbanization around Mediterranean cities: Changes in ecosystem service supply. *Ecol. Indic.* **2018**, *91*, 589–606. [CrossRef]
11. Hodges, M.N.; McKinney, M.L. Urbanization impacts on land snail community composition. *Urban Ecosyst.* **2018**, *21*, 721–735. [CrossRef]
12. Ellis, E.C.; Ramankutty, N. Putting people in the map: Anthropogenic biomes of the World. *Front. Ecol. Environ.* **2008**, *6*, 439–447. [CrossRef]
13. Song, S.; Xu, X.P.; Wu, Z.F.; Deng, X.J.; Wang, Q. The relative impact of urbanization and precipitation on long-term water level variations in the Yangtze River Delta. *Sci. Total Environ.* **2019**, *648*, 460–471. [CrossRef] [PubMed]
14. Supantha, P.; Subimal, G.; Micky, M.; Anjana, D.; Subhankar, K.; Dev, N. Increased spatial variability and intensification of extreme monsoon rainfall due to urbanization. *Sci. Rep.* **2018**, *8*, 3918. [CrossRef]
15. Costa, A.C.; Santos, J.A.; Pinto, J.G. Climate change scenarios for precipitation extremes in Portugal. *Theor. Appl. Climatol.* **2011**, *108*, 217–234. [CrossRef]
16. Pyke, C.; Warren, M.P.; Johnson, T.; Lagro, J., Jr.; Scharfenberg, J.; Groth, P.; Freed, R.; Schroeer, W.; Main, E. Assessment of low impact development for managing stormwater with changing precipitation due to climate change. *Landsc. Urban Plan.* **2011**, *103*, 166–173. [CrossRef]
17. Xu, Y.-P.; Zhang, X.; Ran, Q.; Tian, Y. Impact of climate change on hydrology of upper reaches of Qiantang River Basin, East China. *J. Hydrol.* **2013**, *483*, 51–60. [CrossRef]
18. Mavromatidi, A.; Briche, E.; Claeys, C. Mapping and analyzing socio-environmental vulnerability to coastal hazards induced by climate change: An application to coastal Mediterranean cities in France. *Cities* **2018**, *72*, 189–200. [CrossRef]
19. Rubinato, M.; Nichols, A.; Peng, Y.; Zhang, J.; Lashford, C.; Cai, Y.; Lin, P.; Tait, S. Urban and river flooding: Comparison of flood risk management approaches in the UK and China and an assessment of future knowledge needs. *Water Sci. Eng.* **2019**, *12*, 274–283. [CrossRef]
20. Shu, A.; Tian, L.; Wang, S.; Rubinato, M.; Zhu, F.; Wang, M.; Sun, J. Hydrodynamic characteristics of the formation and movement processes for non-homogeneous debris-flow. *Water* **2018**, *10*, 452. [CrossRef]
21. Shu, A.; Duan, G.; Rubinato, M.; Wang, S.; Zhu, F. Collapsing mechanism of the typical cohesive riverbank along Ningxia-Inner Mongolia catchment along the Yellow River. *Water* **2018**, *10*, 1272. [CrossRef]
22. Kumar, A.; Asthana, A.K.L.; Priyanka, R.S.; Jayangondaperumal, R.; Gupta, A.; Bhakuni, S.S. Assessment of landslide hazards induced by extreme rainfall events in Jammu and Kashmir, Himalaya, northwest India. *Geomorphology* **2017**, *284*, 72–87. [CrossRef]
23. Alvioli, M.; Melillo, M.; Guzzetti, F.; Rossi, M.; Palazzi, E.; von Hardenberg, J.; Brunetti, M.T.; Peruccacci, S. Implications of climate change on landslide hazard in Central Italy. *Sci. Total Environ.* **2018**, *630*, 1528–1543. [CrossRef] [PubMed]
24. Gariano, S.L.; Guzzetti, F. Landslides in a changing climate. *Earth Sci. Rev.* **2016**, *162*, 227–252. [CrossRef]
25. Dilley, M.; Chen, U.; Deichmann, R.S.; Lerner-Lam, A.; Arnold, M. *Natural Disaster Hotspots: A Global Risk Analysis*; Disaster Risk Management Series, 5; The World Bank: Washington, DC, USA, 2005.
26. IPCC. *2014: Climate Change 2014: Synthesis Report. Contribution of Working Groups I, II and III to the Fifth Assessment Report of the Intergovernmental Panel on Climate Change*; Pachauri, R.K., Meyer, L.A., Eds.; IPCC: Geneva, Switzerland, 2014; 151p.
27. Campbell, B.M.; Hansen, J.; Rioux, J.; Stirling, C.M.; Twomlow, S.; Wollenberg, E.L. Urgent action to combat climate change and its impacts (SDG13): Transforming agriculture and food systems. *Curr. Opin. Environ. Sustain.* **2018**, *34*, 13–20. [CrossRef]
28. Arfvidsson, H.; Simon, D.; Oloko, M.; Moodley, N. Engaging with and measuring informality in the proposed Urban Sustainable Development Goal. *Afr. Geogr. Rev.* **2017**, *36*, 100–114. [CrossRef]

29. Li, G.; Rubinato, M.; Wan, L.; Wu, B.; Luo, J.; Fang, J.; Zhou, J. Preliminary characterization of underground hydrological processes under multiple rainfall conditions and rocky desertification degrees in Karst regions of Southwest China. *Water* **2020**, *12*, 594. [CrossRef]

30. Fang, J.; Li, G.; Rubinato, M.; Ma, G.; Zhou, J.; Jia, G.; Yu, X.; Wang, H. Analysis of long-term water level variations in Qinghai Lake in China. *Water* **2019**, *11*, 2136. [CrossRef]

31. Ran, Q.; Wang, F.; Gao, J. Modelling effects of rainfall patterns on runoff generation and soil erosion processes on slopes. *Water* **2019**, *11*, 2221. [CrossRef]

32. Shu, A.; Duan, G.; Rubinato, M.; Tian, L.; Wang, M.; Wang, S. An experimental study on mechanisms for sediment transformation due to riverbank collapse. *Water* **2019**, *11*, 529. [CrossRef]

33. Tang, H.; Ran, Q.; Gao, J. Physics-based simulation of hydrologic response and sediment transport in a hilly-gully catchment with a check dam system on the Loess Plateau, China. *Water* **2019**, *11*, 1161. [CrossRef]

34. Tang, H.; Pan, H.; Ran, Q. Impacts of filled check dams with different deployment strategies on the flood and sediment transport processes in a Loess Plateau catchment. *Water* **2020**, *12*, 1319. [CrossRef]

35. Feng, Q.; Ho, H.C.; Man, T.; Wen, J.; Jie, Y.; Fu, X. Internal stability evaluation of soils. *Water* **2019**, *11*, 1439. [CrossRef]

36. Liang, Q.; Borthwick, A.G.L.; Stelling, G. Simulation of dam and dyke break hydrodynamics on dynamically adaptive quadtree grids. *Int. J. Numer. Methods Fluids* **2004**, *46*, 127–162. [CrossRef]

37. Chang, T.J.; Kao, H.M.; Chang, K.H.; Hsu, M.H. Numerical simulation of shallow water dam break flows in open channels using smoothed particle hydrodynamics. *J. Hydrol.* **2011**, *408*, 78–90. [CrossRef]

38. Zhang, Y.; Rubinato, M.; Kazemi, E.; Pu, J.H.; Huang, Y.; Lin, P. Numerical and experimental analysis of shallow turbulent flow over complex roughness beds. *Int. J. Comput. Fluid Dyn.* **2019**, *33*, 202–221. [CrossRef]

39. Manenti, S.; Wang, D.; Domingues, J.M.; Li, S.; Amicarelli, A.; Albano, R. SPH modeling of water-related natural hazards. *Water* **2019**, *11*, 1875. [CrossRef]

40. Pu, J.H.; Hussain, K.; Shao, S.; Huang, Y. Shallow sediment transport flow computation using time-varying sediment adaptation length. *Int. J. Sediment. Res.* **2014**, *29*, 171–183. [CrossRef]

41. Wu, Y.; Tian, L.; Rubinato, M.; Gu, S.; Yu, T.; Xu, Z.; Cao, P.; Wang, X.; Zhao, Q. A new parallel framework of SPH-SWE for dam break simulation based on OpenMP. *Water* **2020**, *12*, 1395. [CrossRef]

42. Wang, S.; Shu, A.; Rubinato, M.; Wang, M.; Qin, J. Numerical simulation of non-homogeneous viscous debris-flows based on the Smoothed Particle Hydrodynamics (SPH) method. *Water* **2019**, *11*, 2314. [CrossRef]

43. Jiang, H.; You, Y.; Hu, Z.; Zheng, X.; Ma, Q. Comparative study on violent sloshing with water jet flows by using the ISPH method. *Water* **2019**, *11*, 2590. [CrossRef]

44. Yang, Y.S.; Yang, T.T.; Qiu, L.C.; Han, Y. Simulating the overtopping failure of homogeneous embankment by a Double-Point Two-Phase MPM. *Water* **2019**, *11*, 1636. [CrossRef]

45. Pandey, M.; Chen, S.C.; Sharma, P.K.; Ojha, C.S.P.; Kumar, V. Local scour of armor layer processes around the circular pier in non-uniform gravel bed. *Water* **2019**, *11*, 1421. [CrossRef]

46. Yang, F.; Shao, X.; Fu, X.; Kazemi, E. Simulated flow velocity structure in meandering channels: Stratification and inertia effects caused by suspended Sediment. *Water* **2019**, *11*, 1254. [CrossRef]

47. Maji, S.; Hanmaiahgari, P.R.; Balachandar, R.; Pu, J.H.; Ricardo, A.M.; Ferreira, R.M.L. A review on hydrodynamics of free surface flows in emergent vegetated channels. *Water* **2020**, *12*, 1218. [CrossRef]

48. Ma, L.; Sun, R.; Kazemi, E.; Pang, D.; Zhang, Y.; Sun, Q.; Zhou, J.; Zhang, K. Evaluation of ecosystem services in the Dongting Lake wetland. *Water* **2019**, *11*, 2564. [CrossRef]

49. Zhou, T.; Dong, Z.; Wang, W.; Shi, R.; Gao, X.; Huang, Z. Study on multi-scale coupled ecological dispatching model based on the decomposition-coordination principle. *Water* **2019**, *11*, 1443. [CrossRef]

50. Pu, J.H.; Shao, S. Smoothed Particle Hydrodynamics simulation of wave overtopping characteristics for different coastal structures. *Sci. World J.* **2012**, *2012*, 163613. [CrossRef]

51. Wu, S.; Rubinato, M.; Gui, Q. SPH simulation of interior and exterior flow field characteristics of porous media. *Water* **2020**, *12*, 918. [CrossRef]

52. Xu, G.; Hao, H.; Ma, Q.; Gui, Q. An experimental study of focusing wave generation with improved Wave Amplitude Spectra. *Water* **2019**, *11*, 2521. [CrossRef]

53. Jin, R.; Liu, M.; Geng, B.; Jin, X.; Zhang, H.; Liu, Y. Numerical investigation of vortex induced vibration for submerged floating tunnel under different Reynolds numbers. *Water* **2020**, *12*, 171. [CrossRef]

54. Yu, Z.; Hu, Z.; Zheng, X.; Ma, Q.; Hao, H. Aeroelastic performance analysis of wind turbine in the wake with a new Elastic Actuator Line model. *Water* **2020**, *12*, 1233. [CrossRef]

55. Zheng, X.; Shao, S.; Khayyer, A.; Duan, W.; Ma, Q.; Liao, K. Corrected first-order derivative ISPH in water wave simulations. *Coast. Eng. J.* **2017**, *59*, 1750010. [CrossRef]

56. Kazemi, E.; Tait, S.; Shao, S. SPH-based numerical treatment of the interfacial interaction of flow with porous media. *Int. J. Numer. Methods Fluids* **2020**, *92*, 219–245. [CrossRef]

Article

Preliminary Characterization of Underground Hydrological Processes under Multiple Rainfall Conditions and Rocky Desertification Degrees in Karst Regions of Southwest China

Guijing Li [1,2,†], Matteo Rubinato [3], Long Wan [1,2,†], Bin Wu [2], Jiufu Luo [1,2], Jianmei Fang [2] and Jinxing Zhou [1,2,*]

1   Jianshui Research Station, School of Soil and Water Conservation, Beijing Forestry University, Beijing 100083, China; lgj8023lhy@163.com (G.L.); wanlong255@sina.com (L.W.); yijiuyuan@yeah.net (J.L.)
2   Key Laboratory of State Forestry and Grassland Administration on Soil and Water Conservation, Beijing Forestry University, Beijing 100083, China; wubin@bjfu.edu.cn (B.W.); jmf46@163.com (J.F.)
3   School of Energy, Construction and Environment & Centre for Agroecology, Water and Resilience, Coventry University, CV1 5FB Coventry, UK; matteo.rubinato@coventry.ac.uk
*   Correspondence: zjx001@bjfu.edu.cn; Tel.: +86-10-62338561
†   These authors contributed equally to this work.

Received: 22 January 2020; Accepted: 18 February 2020; Published: 21 February 2020

Abstract: Karst regions are widely distributed in Southwest China and due to the complexity of their geologic structure, it is very challenging to collect data useful to provide a better understanding of surface, underground and fissure flows, needed to calibrate and validate numerical models. Without characterizing these features, it is very problematic to fully establish rainfall–runoff processes associated with soil loss in karst landscapes. Water infiltrated rapidly to the underground in rocky desertification areas. To fill this gap, this experimental work was completed to preliminarily determine the output characteristics of subsurface and underground fissure flows and their relationships with rainfall intensities (30 mm h^{-1}, 60 mm h^{-1} and 90 mm h^{-1}) and bedrock degrees (30%, 40% and 50%), as well as the role of underground fissure flow in the near-surface rainfall–runoff process. Results indicated that under light rainfall conditions (30 mm h^{-1}), the hydrological processes observed were typical of Dunne overland flows; however, under moderate (60 mm h^{-1}) and high rainfall conditions (90 mm h^{-1}), hydrological processes were typical of Horton overland flows. Furthermore, results confirmed that the generation of underground runoff for moderate rocky desertification (MRD) and severe rocky desertification (SRD) happened 18.18% and 45.45% later than the timing recorded for the light rocky desertification (LRD) scenario. Additionally, results established that the maximum rate of underground runoff increased with the increase of bedrock degrees and the amount of cumulative underground runoff measured under different rocky desertification was SRD > MRD > LRD. In terms of flow characterization, for the LRD configuration under light rainfall intensity the underground runoff was mainly associated with soil water, which was accounting for about 85%–95%. However, under moderate and high rainfall intensities, the underground flow was mainly generated from fissure flow.

Keywords: rock–soil contact area; fissure flow; karst rocky desertification; runoff; rainfall simulation

1. Introduction

Karst landscapes represent a crucial feature of the earth's geodiversity [1], and they account for 12% of the total land area in the world [2]. Southwest China is highly characterized by karst areas. In fact, the degraded vegetation and widely exposed limestone in Southwest China induced severe rocky desertification, generating a fragile ecosystem that represent a severe environmental and social issue [3–5]. Furthermore, human activities and forest clearance in western Ireland caused severe soil loss [6]. To date, several studies [7–17] have been conducted to provide a solution to the management of these areas, and water has been identified as the major drive of the hydrological processes investigated [18–21]. In actual fact, due to the nature of these environmental layers typical of karst areas, rainfall is largely lost through underground fissures, karst caves and pipelines [22,23], causing major concerns to local authorities on how to prevent droughts and better manage water distribution within the region.

Hydrological processes generated by surface runoff in karst areas have been identified to be linked to land degradation [24–26] or vegetation restoration [27], and in recent years, thanks to the improvement of existing technology and a wider understanding of local phenomena, most studies have focused on the characterization of groundwater in epikarst zones [28–30].

Multiple experimental tests and field investigations of rock and soil structures have been conducted as well as numerical simulations of underground water flow induced by rainfall events [31–33]. However, these studies did not carry out an accurate experimental campaign incorporating the soil–rock interaction, typical of the natural situation; hence, all results obtained are limited to artificial systems that do not completely reflect the real case scenario. Fu et al. [20] made interesting field observations on the dynamic change of the water balance and the underground water flow component by digging trenches at the subsurface boundary of karst developed over dolomite, characterized by a flat depression surrounded by overlapping hills and ridges [20]. Furthermore, karsts systems without soil are different and have a dissimilar hydrogeological behavior because the soil is the main source for CO_2 production, and the surface of the carbonate rocks directly contacting the overlying soil suffers the most intense chemical energies of karstification [34]. These surficial karst processes in turn enhance solutional enlargement of fissures in carbonate rocks and produce an irregular pitted and etched epikarst subsurface [34]. Hence these karsts areas were characterized by a specific rock–soil ratio and permeability of rock–soil fractures, which are important factors influencing underground runoff [35]. If compared with dolomite, carbonate rock and barren soil provide a better material basis for rock desertification, enabling a richer fissure development and causing more serious issues related to soil erosion [36]. Therefore, the karst landforms developed by limestone more typical in karst areas in Southwest China have been the rock chosen to be investigated in this paper in order to provide a better understanding of underground hydrological processes in relation with rainfall events.

Rainfall is in fact another very active dynamic factor in the hydrological cycle [37]. The rainfall amount and the rainfall intensity have a great impact on hydrological dynamics, especially in karst areas [38,39]. Regional climate and hydrological conditions have been demonstrated to affect the soil infiltration and water storage in epikarst zones [40], and the response of infiltration and runoff to rainfall intensity is different for each type of rocky desertification. Hence, it is crucial to further study these underground water behaviors associated with each rainfall intensity to determine how water could be regulated or controlled in karst areas.

To fulfil this gap, this work included the simulation of a unique "dual structure" karst area focusing on the soil–rock integration and the experimental representation of underground runoff with the purposes of (1) providing a better understanding of underground runoff under different rainfall intensities and (2) discussing the influence of rocky desertification on underground runoff. Results obtained by this study have vital theoretical and practical significance to reveal hidden hydrological mechanisms, and can help to reduce water loss supporting healthy and sustainable growth in karst regions.

The paper is organized as follows: Section 2 introduces the description of the experimental setup describing the methods applied and the analysis conducted. Section 3 presents the effects of rainfall intensity and rocky desertification on the dynamics of underground runoff. Section 4 provides a discussion of the results obtained and Section 5 produces a brief summary and concluding remarks of the whole study.

2. Materials and Methods

2.1. Experimental Setup

The experimental apparatus used to conduct this research is composed of a pond, previously designed by [40], and a rainfall generator. The pond is made by a reinforced concrete structure with a length of 100 cm, a width of 100 cm and a height of 120 cm. The bottom of the pond is a 20-cm-thick permeable layer, and the permeable layer is filled with gravels (diameter ranging between 5 cm and 10 cm). The remaining 100 cm depth of the pond is used to fill soil and limestone blocks.

The rainfall generator is composed of a water tank which is 110 cm long, 110 cm wide and 25 cm high. A total of 441 holes with a diameter of 1 cm are located at the bottom of the water tank.

The test soil was collected from a 0–100 cm depression located in the typical karst area—Jianshui County, Yunnan province, China (Figure 1). This area is affected by a subtropical monsoon climate, hence characterized by wet and dry seasons. The annual average temperature of this area is 19.7 °C, and the annual average precipitation is 828.3 mm. Records states that the rainfall was unevenly distributed throughout the year, happening mainly during the rainy season (from May to October).

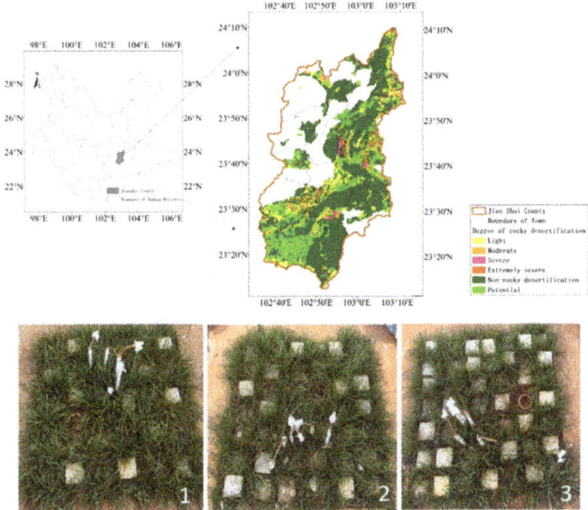

Figure 1. Geographical location of the study site and bed rock configurations tested (1, 2 and 3), whose characteristics are displayed in Table 1.

The distance between each hole is 5 cm, and the holes are covered with rubber plugs to facilitate the insertion of the needles (inner diameter 0.57 mm) to release the water. The rainfall intensity can be adjusted according to the pressure generated by the different water levels in the water tank. To prevent blockage of the needles, water used to simulate the rainfall is filtered in advance. Furthermore, 9 measuring glasses are located under the sprinkler with a distance between each other of 30 cm, and the volume of the water in the glass can be measured every 10 min to calibrate the water level in the tank and the corresponding rainfall intensity.

Due to the complexity of this phenomenon, we may have simplified the real phenomenon in terms of dimensions or configurations, but we have made reasonable assumptions to clearly distinguish the phenomena observed and how each parameter can affect the flows generated. If considering the main site, there may be issues in terms of operability of tests and too many variables could interfere with the results, making it very challenging to be accurately quantified. In order to accurately simulate the complex karst fissure environment, reducing artificial aspects, standard cuts were made on the limestone rock with the size 10 cm × 10 cm × 10 cm (length × width × height). Soil and rocks were filled in the pond according to the measured soil bulk density in the field from bottom to top. The soil was not sieved before filling and only large soil clumps were dispersed.

In order to reduce and limit the abnormal effects on the edges, the boundaries were artificially compacted manually. Regular sweeping was conducted with a wooden board designed on purpose for this study after each layer was filled. The soil volumetric water content (VWC) at 5 cm, 15 cm, 30 cm, 50 cm and 70 cm of the vertical profile was monitored using an automatic monitoring system because it can be used to calculate the time required for water to penetrate to a certain depth. The probe was buried horizontally perpendicular to the corresponding layer of the soil profile. After completing each layer of soil, probes were buried and calibrated, and the process was repeated for the next layer. Campbell CS616 soil moisture probes (Campbell Scientific, Inc.–Logan, UT, USA) (with an error of ±2.5%) were used for data acquisition, and the acquisition frequency was once every 5 min. The calibration method for the probes can be summarized as follows:

- An aluminum box was used to sample near the 4 vertexes and the center point of the soil layer where the probe was placed, and the sampling time was recorded in parallel.
- The samples were taken every 10 min for a total of 5 times. The VWC was obtained by multiplying the soil bulk density by the soil mass moisture content (M_w) of the layer. The soil mass water content M_w was obtained using the following equation:

$$M_w = \frac{M_a - M_b}{M_b - M_c} \times 100\% \tag{1}$$

  where M_a is the total weight of the aluminum box and the wet soil; M_b is the total weight of the dried aluminum box and the soil; and M_c is the weight of the aluminum box.

- A correlation analysis was then performed between the average value of VWC and the value measured by the sensor at each time. The results obtained showed that there was a linear relationship between these two parameters, and the correlation coefficient was above 0.90.

2.2. Experimental Testing Conditions

Based on the field investigation of 15 different rocky desertification profiles, and considering existing literature published to date, three configurations were set up with varying parameters: bedrock exposure, vegetation coverage, average soil thickness and rock–soil contact area (Table 1). Each configuration was repeated three times to confirm the repeatability of the tests and reduce experimental errors. According to the grading evaluation criteria for the degree of rocky desertification in the "Technical Regulations for Monitoring Rocky Desertification in the Southwest Karst Area" issued by the National Forestry and Grass Bureau in January 2005, the three configurations tested within this study represented light rocky desertification (LRD), moderate rocky desertification (MRD) and severe rocky desertification (SRD).

The bedrock exposure represented (Figure 1) the vertical projected area of exposed rocks per unit area. The parameter values were 30%, 40% and 50%, which were equivalent to 30, 40 and 50 limestone rocks exposed on the soil surface.

Ryegrass (*Lolium perenne* L.), a common herbaceous plant in the study area, was selected for this study and the vegetation coverage reproduced within the experimental model was referred to the proportion of the vertical projection area of the vegetation to the land surface area. The average thickness of the soil layer represented the ratio of the total volume of the filled soil to the base area of the pond. The rock–soil contact area represented the surface area per unit volume of rock in contact with soil. The larger the rock-soil contact area is, the greater is the degree of rock fragmentation and the karst fissures are more fully developed.

The rainfall intensity was designed as three gradients typical of light rain intensity (30 mm h^{-1}), moderate rain intensity (60 mm h^{-1}) and heavy rain intensity (90 mm h^{-1}). The simulated rainfall lasted 1h for each condition. Each rainfall event was repeated 3 times.

The underground runoff was collected every 15 min with the use of a plastic bucket. This procedure was conducted every 1 h after the underground runoff rate was stabilized. To calculate the underground runoff rate $U_r = (L \cdot h^{-1})$, the following equation was used:

$$Ur = \frac{V}{AT} \tag{2}$$

where V is the volume of underground runoff (L); A is the base area of the pond (m^2); and T is the duration of underground runoff (h).

Table 1. Experimental configurations.

Configuration	Bedrock Exposure (%)	Average Soil Thickness (m)	Rock–Soil Contact Area (m^2)	Vegetation Coverage (%)
1	40	0.55	22.06	65
2	50	0.40	26.85	55
3	60	0.25	32.14	25

3. Results

3.1. Effects of Rainfall Intensity and Rocky Desertification on the Dynamics of Underground Runoff

Figure 2 shows the dynamics of underground runoff for each of the three configurations tested (LRD (Figure 2a,d,g), MRD (Figure 2b,e,h) and SRD (Figure 2c,f,i)) under different rainfall intensities. It is possible to notice that despite under dissimilar rainfall intensity, underground runoff can be divided into four stages: (i) runoff rising stage, (ii) rapid declining stage, (iii) slow declining stage and (iv) recession stage.

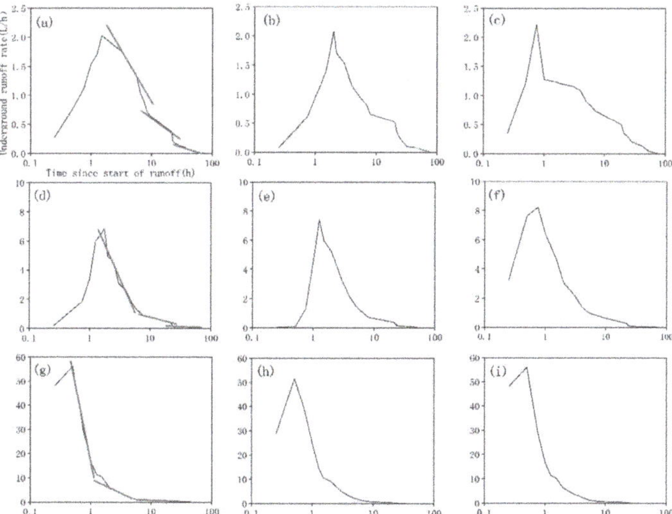

Figure 2. Dynamics of underground runoff for light rocky desertification (LRD) (**a,d,g**), moderate rocky desertification (MRD) (**b,e,h**) and severe rocky desertification (SRD) (**c,f,i**) under 30 mm/h (**a–c**), 60 mm/h (**d–f**) and 90 mm/h (**g–i**).

Additionally, different degrees of rocky desertification have an influence on the dynamic process of the groundwater flow. Considering, for example, a moderate rainfall intensity (60 mm/h), the response time to rainfall of underground runoff can be simplified as LRD < MRD < SRD (shown in Figure 3).

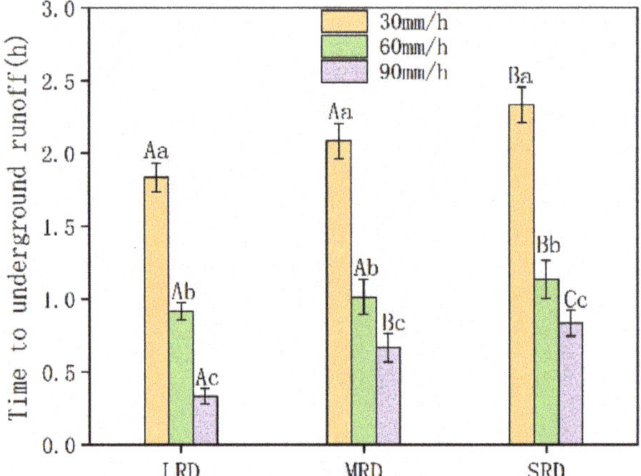

Figure 3. Time to underground runoff under different rainfall intensities. Note: In the figure, the capital letter represented the significant difference between different rainfall intensity under a certain degree of rocky desertification ($p < 0.05$), and the lowercase letter represented the significant difference between different degrees of rocky desertification under a certain degree of rainfall intensity ($p < 0.05$).

The initial time for the generation of underground runoff of LRD was recorded to be the shortest, 51 min. For MRD and SRD, this time was 18.18% and 45.45% later than the one recorded for LRD. The time to reach the runoff "flood" peak was 105 min, 75 min and 45 min, respectively, indicating that the time to reach the flood peak gradually shortened with the increase of the severity of rocky desertification. Furthermore, as shown in Figure 4, the maximum rate of underground runoff was achieved with the highest rocky desertification. The maximum runoff rate (8.24 L/h) of SRD was 1.20 times and 1.11 times higher than the ones recorded for the LRD and MRD scenarios, respectively.

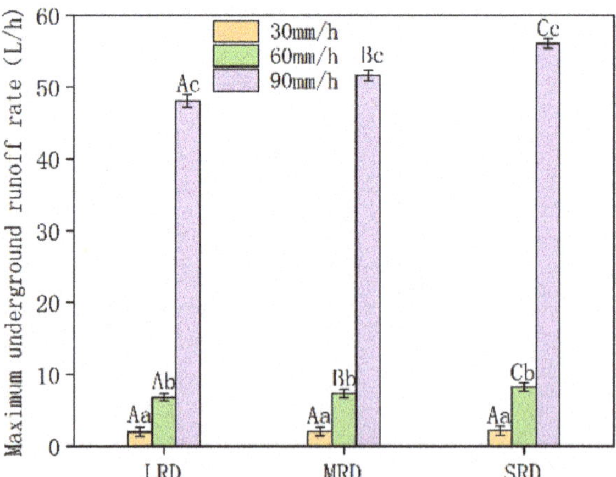

Figure 4. Maximun underground runoff rate under different rainfall intensities.

As initially stated in Section 1, rainfall was considered the main dynamic factor to drive the underground runoff [20], and the rainfall intensity confirmed to have an important impact on the characteristics of the underground runoff.

It can be seen from Figure 2 that with the increase in rainfall intensity, the duration of the rising stage to reach the maximum peak of flow decreased. In parallel, the underground runoff velocity reached the peak value in a shorter time, and the maximum rate of underground runoff significantly increased with the increase in rainfall intensity. Furthermore, the maximum rate of underground runoff in severe rocky desertification was the most affected by rainfall intensity (Figure 4).

For SRD, the maximum rate of underground runoff recorded under the condition of light rain increased 2.71 times under the condition of moderate rain and 24.26 times under the condition of heavy rain. The initial runoff generation duration of the underground runoff decreased significantly with the increase in rainfall intensity (Figure 3).

The underground runoff of LRD was mostly affected by rainfall intensity. Compared with the initial runoff duration under the condition of light rain, the initial runoff duration under the condition of medium rain and heavy rain was reduced by 50.00% and 81.82%, respectively.

3.2. Volume and Percentage of Underground Runoff Based on Different Degrees of Rocky Desertification

The amount of cumulative underground runoff for different rocky desertification was SRD > MRD > LRD (Figure 5 below). Under light rain conditions, the cumulative underground runoff of SRD was 2.11% and 0.77% larger than that of LRD and MRD. Under moderate rain condition, the cumulative underground runoff of SRD was 17.56% and 9.53% larger than that of LRD and MRD. Under heavy rain condition, the cumulative underground runoff of SRD was 5.47% and 2.67% larger than that of LRD and MRD. The cumulative underground runoff significantly increased with the increase in rainfall intensity, and the cumulative underground runoff of LRD was the most affected by the rainfall intensity (Figure 5). The cumulative underground runoff of LRD increased 1.75 times under heavy rain conditions than that of under light rain conditions.

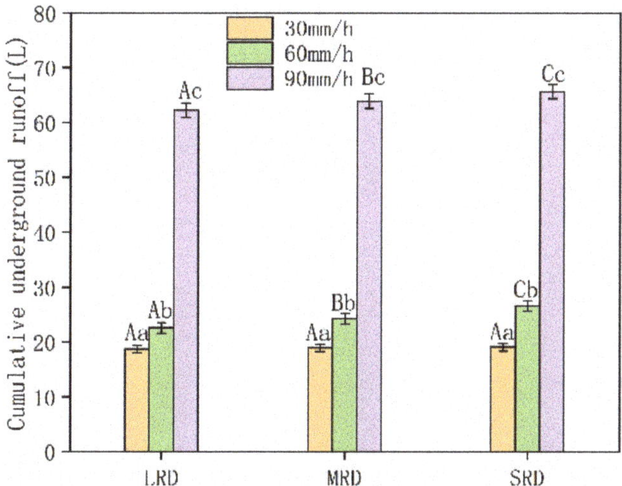

Figure 5. Cumulative underground runoff under different rainfall intensities.

The relationships identified between underground runoff duration and cumulative underground runoff indicated that there was a logarithmic function between the parameters considered (rainfall and rock desertification, as shown in Table 2) and the correlation coefficients were all above 0.85 (Table 2), confirming the trustworthiness of these results.

Table 2. Regression analysis of cumulative underground runoff and runoff duration.

Rainfall Intensity	Degree of Karst Rocky Desertification	Regression Equation	Correlation Coefficient
30 mm/h	LRD	V = 4.27lnT + 1.91	0.967
	MRD	V = 4.37lnT + 0.70	0.953
	SRD	V = 4.08lnT + 1.38	0.954
60 mm/h	LRD	V = 5.85lnT + 3.90	0.957
	MRD	V = 5.20lnT + 4.32	0.965
	SRD	V = 4.06lnT + 7.23	0.972
90 mm/h	LRD	V = 9.72lnT + 28.20	0.861
	MRD	V = 8.72lnT + 30.77	0.876
	SRD	V = 6.59lnT + 37.49	0.854

3.3. Characteristics of VWC and Runoff Sources for Different Degrees of Karst Rocky Desertification

The saturated VWC of each soil layer was between 0.45 and 0.52, as shown in Figure 6. Under the conditions of light, medium and heavy rain, the timings for soil on top (5 cm) to reach saturation were 20–30 min, 15–20 min and 10–15 min, respectively. Meanwhile, the timings for the bottom soil (70 cm) to reach saturation were about 100–150 min, 80–100 min and 45–55 min, respectively. Under light rain conditions, the time to generate underground runoff was 60 min after the soil reached the saturation, typical of Dunne overland flows [41,42]. However, under moderate and heavy rain conditions, the initiation of underground runoff happened before the soil reached the saturation, which is typical of Horton overland flows. The VWC began to decrease after the rainfall stopped for 5–10 min. For example, the VWC at 5 cm and 15 cm of LRD significantly decreased after reaching saturation, decreased by 0.05 in 20 min and finally continued to steadily decline. However, the VWC at 30 cm, 50 cm and 70 cm decreased slowly after saturation.

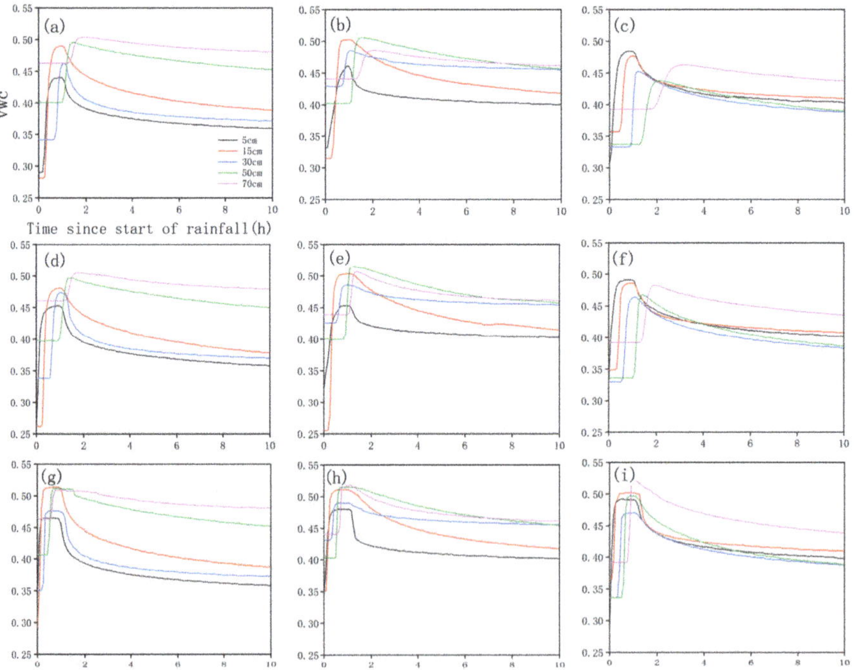

Figure 6. Dynamics of volumetric water content (VWC) for LRD (**a,d,g**), MRD (**b,e,h**) and SRD (**c,f,i**) under 30 mm/h (**a–c**), 60 mm/h (**d–f**), 90 mm/h (**g–i**).

The underground runoff (V) was mainly produced by the content of water within the soil and the infiltration rate through the fissures (F). The volume of fissure flow VF (in liters) can be obtained by applying a water balance, neglecting the effects of evaporation as follows:

$$VF = P - \Delta Vsoil - V \tag{3}$$

where P is the precipitation (mm); ΔVsoil is the change of soil water content (l); and V is the underground runoff (l).

The proportion of fissure flow increased significantly with the increase in rainfall intensity (Figure 7). Under the conditions of light and moderate rain intensities, the rate of water within the soil water was higher than the fissure flow, which only accounted for 10%–13% and 25%–35% of the total precipitation. However, under the condition of heavy rain, the proportion of fissure flow was relatively high, accounting for 47%–50% of the total precipitation. Moreover, the proportion of fissure flow increased with the increase of rocky desertification degree. The proportion of fissure flow of SRD was 3%–10% larger than that of LRD.

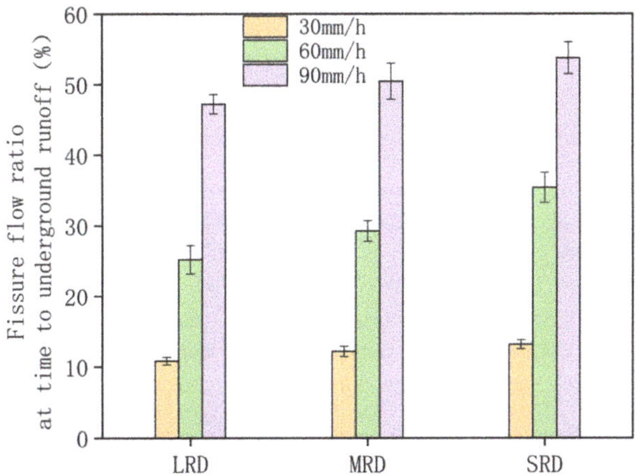

Figure 7. Fissure flow ratio under different rainfall intensities.

From the start of the underground runoff to its peak, the water balance analysis was continuously performed to obtain the ratio of soil water and fissure flow vs. the underground runoff. Under light rain conditions, underground runoff was mainly generated by soil water, (85%–95%), while fissure flow only accounted for 5%–15%. Under moderate rain conditions, runoff was mainly produced by fissure flow (72%–87%), while soil water accounted only for 13%–27%. Under heavy rain conditions, the soil was not saturated due to the short time (20 min) in which the underground runoff was formed, hence fissure flow was the main source of the underground runoff.

4. Discussion

Experimental studies conducted confirmed that underground fissures, which are unique to the karst areas around the world, create migration channels for groundwater and soil losses [32]. Results have established that soil and water losses were caused by factors such as the exposed rate of the bedrock, the degree of karst fissures and the rainfall intensity [43,44]. However, in order of impact, rainfall is the most important driving factor, followed by the degree of karst fissures [45,46].

There were significant differences in underground runoff characteristics between different degrees of rocky desertification; results confirmed that the response time of the underground runoff to the rainfall increased with the increase in degree of rocky desertification.

As shown in Table 1, with the increase in the degree of rocky desertification, the rock–soil contact area and the rock–soil ratio gradually increased, which indicated that the main channel of runoff migration, the soft–hard interface between rock and soil [47], gradually increased; meanwhile, the width of the cracks between the rocks gradually decreased. The size of the cracks between the rocks has demonstrated to have significant influence on the velocity of water moving through the rocks [48], and this velocity in the fractures was much slower than the pipelines [49].

As shown in Figure 6, the water infiltration rate of LRD was faster than the one measured for MRD and SRD. For example, under light rain conditions, at 50 cm, soil VWC for LRD, MRD and SRD started to increase after 65 min, 70 min and 85 min, respectively, while at 70 cm, soil VWC for LRD, MRD and SRD started to increase after 85 min, 90 min and 115 min, respectively. Furthermore, the rock–soil contact area of SRD was 45.69% higher than LRD, but the time needed for the water to move from soil at 50 cm to soil at 70 cm was 10 min longer. This behavior also indicated that although the rock–soil contact area and migration channels for groundwater were the largest, these limited the velocity of the water infiltrating. Therefore, the authors believe that this was a critical point for quantifying the effect of rock–soil contact area on the water infiltration rate. When the rock–soil contact area was greater than the critical value, the water infiltration rate decreased with the increase in water transport channels and the decrease in crack width, which consequently affected the initiation of underground runoff. However, once the underground runoff was produced, a preferred flow path had been established connecting the soil and rock interface. As the results showed, SRD firstly reached the maximum underground runoff rate due to the largest rock–soil contact area. The maximum underground runoff rate and the total amount of underground runoff were also higher than those recorded for LRD and MRD, confirming that the rock–soil contact area had an important influence on the characteristics of underground runoff. The increase in rock–soil contact area will worsen the soil and water losses in the karst area, reducing the time to generate underground runoff and increasing the water leakages.

5. Conclusions

Due to heavy rainfall events becoming more frequent due to the climate change across the world, it is necessary to carefully consider strategies to guarantee the safety of water resources and ecological systems, especially in karst landform areas, typical in Southwest China, which have not been addressed completely to date. This study included the simulation of a unique "dual structure" karst area focusing on the soil–rock integration, and the experimental representation of underground runoff with the purposes of (1) providing a better understanding of underground runoff under different rainfall intensities and (2) discussing the influence of rocky desertification on underground runoff. Results can be summarized as follows:

- The higher the rainfall intensity is, the shorter is the time needed for the formation of underground runoff. The time observed for the formation of underground runoff for MRD and SRD was 18.18% and 45.45% later than the one recorded for LRD.
- The maximum rate of underground runoff and cumulative underground runoff were characterized by the following trend: SRD > MRD > LRD.
- The rock–soil contact area is an important factor influencing the characteristics of underground runoff, as shown in Sections 3 and 4.

The severe loss of groundwater in Southwest China has severely restricted local ecological restoration and economic development, but which can, however, be controlled through sustainable engineering measures. Resources should be directed to reduce the amount of runoff leakage and preserving underground runoff, facilitating water storage. The authors suggest that techniques such as filling cracks with gravels [50] or filling cracks with mud [34] should be implemented to expedite this task. Although engineering measures can be effective, costs are high, and operations are difficult. Planting vegetation could definitely be another solution to improve soil characteristics and enable rainfall interception. At the same time, the distribution of plant roots in underground fissures could reduce the channels for the underground runoff to a certain extent [51].

Due to the nature of the stones used for this study, which are not irregular, the authors suggest that this limitation can be solved by considering this scenario within a new bespoke experimental setup, incorporating also larger field areas. Furthermore, future research should focus on (i) the effects and mechanisms of plant measures on groundwater loss control; (ii) the specific impact of the rock–soil interface; (iii) the definition of fissure characteristics, such as the density values or widths; and (iv) further strengthening the model research of the karst dual hydrological unit to replicate field observations and experimental simulations, crucial to provide a more comprehensive understanding of the complex underground runoff processes in the karst areas.

Author Contributions: Conceptualization, G.L., L.W., J.F. and J.Z.; data curation, G.L., L.W., J.Z. and M.R.; formal analysis, G.L., L.W., J.L., B.W., J.F. and J.Z.; funding acquisition, L.W. and J.Z.; investigation, G.L., L.W., J.L., J.F. and M.R.; supervision, G.L., L.W., J.L., B.W. and J.Z.; validation, G.L., L.W., J.L., B.W., M.R. and J.Z.; writing—original draft, G.L. and L.W.; writing—review and editing, G.L., L.W and M.R. All authors have read and agreed to the published version of the manuscript.

Funding: The research was supported by the Fundamental Research Funds for the Central Universities of China (BLX201709), the National Natural Science Foundation of China (31870707; 31700640), and the National Basic Research Program of China (2016YFC0502502; 2016YFC0502504).

Acknowledgments: We would like to thank Lechuan Qiu, Jianghua Liao, Jiatong Zhang and Weixin Zhang for their invaluable help with measurements taken in the field and in the laboratory.

Conflicts of Interest: The authors declare no conflict of interest.

References

1. Dai, Q.; Peng, X.; Yang, Z.; Zhao, L. Runoff and erosion processes on bare slopes in the Karts Rocky Desertification Area. *CATENA* **2017**, *152*, 218–226. [CrossRef]

2. Liu, C.Q.; Lang, Y.C.; Li, S.L.; Piao, H.C.; Tu, C.L.; Liu, T.Z.; Zhang, W.; Zhu, S.F. Researches on biogeochemical processes and nutrient cycling in karstic ecological systems, Southwest China: A review. Front. *Earth. Sci.* **2009**, *16*, 1–12.

3. Parise, M.; de Waele, J.; Gutierrez, F. Current perspectives on the environmental impacts and hazards in karst. *Environ. Geol.* **2009**, *58*, 235–237. [CrossRef]

4. Zhang, X.B.; Wang, S.J.; Cao, J.H.; Wang, K.L.; Meng, T.Y.; Bai, X.Y. Characteristics of water loss and soil erosion and some scientific problems on karst rocky desertification in Southwest China karst area. *Carsologica. Sin.* **2010**, *29*, 274–279.

5. Wang, S.J.; Li, Y.B.; Li, R.L. Karst rocky desertification: Formation background, evolution and comprehensive taming. *Quat. Sci.* **2003**, *23*, 657–666.

6. Drew, D.P. Accelerated soil erosion in a karst area: The Burren, Western Irelan. *J. Hydrol.* **1983**, *61*, 113–124. [CrossRef]

7. Calvo-Cases, A.; Boix-Fayos, C.; Imeson, A.C. Runoff generation, sediment movement and soil water behaviour on calcareous (limestone) slopes of some Mediterranean environments in southeast Spain. *Geomorphology* **2003**, *50*, 269–291. [CrossRef]

8. Imeson, A.C.; Lavee, H.; Calvo, A.; Cerdà, A. The erosional response of calcareous soils along a climatological gradient in Southeast Spain. *Geomorphology* **1998**, *24*, 3–16. [CrossRef]

9. Kheir, R.B.; Abdallah, C.; Khawlie, A. Assessing soil erosion in Mediterranean karst landscapes of Lebanon using remote sensing and GIS. *Eng. Geol.* **2008**, *99*, 239–254. [CrossRef]

10. Kosmas Danalatos, N.; Cammeraat, L.H.; Chabart, M.; Diamantopoulos, J.; Farand, R.; Gutierrez, L.; Jacob, A.; Marques, H.; Martinez Fernandez, J.; Mizara, A.; et al. The effect of land use on runoff and soil erosion rates under Mediterranean conditions. *CATENA* **1997**, *29*, 45–59. [CrossRef]

11. Cerdà, A. Effect of climate on surface flow along climatological gradient in Israel: A field rainfall simulation approach. *J. Arid Environ.* **1998**, *38*, 145–159. [CrossRef]

12. Miller, T.E. Geologic and Hydrologic Controls on Karst and Cave Development in Belize. *J. Cave Karst Stud.* **1996**, *58*, 100–120.

13. Brook, G.A.; Ford, D.C. Hydrology of the Nahanni Kars, Northern Canada, and the importance of extreme summer storms. *J. Hydrol.* **1980**, *46*, 103–121. [CrossRef]

14. Andreichuk, V. Gypsum karts of the pre-rural region, Russia. *Int. J. Speleol.* **1996**, *25*, 285–292. [CrossRef]

15. Jurkovšek, B.; Biolchi, S.; Furlani, S.; Kolar-Jurkovšek, T.; Zini, L.; Jež, J.; Tunis, G.; Bavec, M.; Cucchi, F. Geology of the Classical Karst Region (SW Slovenia–NE Italy). *J. Maps* **2016**, *12*, 352–362. [CrossRef]

16. Xanke, J.; Liesch, T.; Goeppert, N.; Klinger, J.; Gassen, N.; Goldscheider, N. Contamination risk and drinking water protection for a large-scale managed aquifer recharge site in a semi-arid karst region, Jordan. *Hydrogeol. J.* **2017**, *25*, 1795–1809. [CrossRef]

17. Artugyan, L. Geomorphosites Assessment in Karst Terrains: Anina Karst Region (Banat Mountains, Romania). *Geoheritage* **2017**, *9*, 153–162. [CrossRef]

18. Legrand, H.E. Hydrological and Ecological Problems of Karst Regions. *Science* **1973**, *179*, 859–864. [CrossRef]

19. Bakalowicz, M. Karst groundwater: A challenge for new resources. *Hydrol. J.* **2005**, *13*, 148–160. [CrossRef]

20. Fu, Z.Y.; Chen, H.S.; Zhang, W.; Xu, Q.X.; Wang, S.; Wang, K.L. Subsurface flow in a soil-mantled subtropical dolomite karst slope: A field rainfall simulation study. *Geomorphology.* **2015**, *250*, 1–14. [CrossRef]

21. Wang, S.J.; Liu, Q.M.; Zhang, D.F. Karst rocky desertification in south western China: Geomorphology, land use, impact and rehabilitation. *Land Degrad. Dev.* **2004**, *15*, 115–121. [CrossRef]

22. Williams, P.W. Geomorphic inheritance and the development of tower karst. Earth Surf. Process. *Earth Surf. Process. Landf.* **1987**, *12*, 453–465. [CrossRef]

23. Dai, Q.H.; Liu, Z.T.; Shao, H.B. Karst bare slope soil erosion and soil quality: A simulation case study. *Solid Earth* **2015**, *6*, 985–995. [CrossRef]

24. Bai, X.Y.; Wang, S.J.; Xiong, K.N. Assessing spatial–temporal evolution processes of karst rocky desertification land: Indications for restoration strategies. *Land Degrad. Dev.* **2013**, *24*, 47–56. [CrossRef]

25. Xu, E.Q.; Zhang, H.Q. Characterization and interaction of driving factors in karst rocky desertification: A case study from Changshun, China. *Solid Earth* **2014**, *5*, 1329–1340. [CrossRef]

26. Yan, X.; Cai, Y.L. Multi-scale anthropogenic driving forces of karst rocky desertification in Southwest China. *Land Degrad. Dev.* **2015**, *26*, 193–200. [CrossRef]

27. Nie, Y.P.; Chen, H.S.; Wang, K.L.; Ding, Y.L. Rooting characteristics of two widely distributed woody plant species established in rocky karst habitats of Southwest China. *Plant. Ecol.* **2014**, *215*, 1099–1109. [CrossRef]

28. Zhang, H.; Cai, Z.; Hao, F.H.; Qi, L.; Yun, L.; Jiang, L. Hydrogeomorphologic architecture of epikarst reservoirs in the Middle-Lower Ordovician, Tazhong Uplift, Tarim Basin, China. *Mar. Pet. Geol.* **2018**, *98*, 146–161. [CrossRef]

29. Peng, X.; Dai, Q.; Li, C.; Zhao, L. Role of underground fissure flow in near-surface rainfall-runoff process on a rock mantled slope in the karst rocky desertification area. *Eng. Geol.* **2018**, *243*, 10–17. [CrossRef]

30. Wang, S.; Fu, Z.; Chen, H.; Nie, Y.; Xu, Q. Mechanisms of surface and subsurface runoff generation in subtropical soil-epikarst systems: Implications of rainfall simulation experiments on karst slope. *J. Hydrol.* **2020**, *580*, 124370. [CrossRef]

31. Shen, Z.Z.; Chen, F.; Zhao, J. Experimental study on seepage characteristics of the intersection of tubular karst passage and fissure. *J. Hydraul. Eng.* **2008**, *39*, 137–145.

32. Peng, T.; Wang, S.J. Effects of land use, land cover and rainfall regimes on the surface runoff and soil loss on karst slopes in southwest China. *CATENA* **2012**, *90*, 53–62. [CrossRef]

33. Zhang, Z.C.; Chen, X.; Cheng, Q.B.; Peng, T.; Zhang, Y.F.; Ji, Z.H. Hydrogeology of epikarst in karst mountains-A case study of the Chenqi catchment. *Earth. Env.* **2011**, *39*, 19–25.

34. Zhou, W.; Beck, B.F. Engineering issues on karst. In *Karst Management*, 1st ed.; Van Beynen, P., Ed.; Springer: Berlin/Heidelberg, Germany, 2011.

35. Graham, C.B.; Woods, R.A.; McDonnell, J.J. Hillslope threshold response to rainfall: A field based forensic approach. *J. Hydrol.* **2010**, *393*, 65–76. [CrossRef]

36. Xiong, P.S.; Yuan, D.X.; Xie, S.Y. Progress of research on rocky desertification in south China karst mountain. *Carsologica Sin.* **2010**, *29*, 355–362.

37. Cuomo, S.; Della, S.M. Rainfall-induced infiltration, runoff and failure in steep unsaturated shallow soil deposits. *Eng. Geol.* **2013**, *162*, 118–127. [CrossRef]

38. De Lima, J.L.M.P.; Singh, V.P. The influence of the pattern of moving rainstorms on overland flow. *Adv. Water. Resour.* **2002**, *25*, 817–828. [CrossRef]

39. Wei, H.; Nearing, M.A.; Stone, J.J.; Guertin, D.P.; Spaeth, K.E.; Pierson, F.B.; Nichols, M.H.; Moffett, C.A. A new splash and sheet erosion equation for rangelands. *Soil. Sci. Soc. Am. J.* **2009**, *73*, 1386–1392. [CrossRef]

40. Klimchouk, A. Toward defining, delimiting and classifying epikarst: Its origin, processes and variants of geomorphic evolution Speleogenesis. Evol. *Karst Aquifers* **2003**, *2*, 1–13.

41. Stewart, R.D.; Bhaskar, A.; Parolari, A.J.; Hermann, D.L.; Jian, J.; Schifman, L.A.; Shuster, W.D. An analytical approach to ascertain saturation-excess versus infiltration-excess overland flow in urban and reference landscapes. *Hydrol. Process.* **2019**, *33*, 3349–3363. [CrossRef]

42. Suseno, D.P.Y.; Yamada, T.J. Simulating Flash Floods Using a Geostationary Satellite-Based Rainfall Estimation Coupled with a Land Surface Model. *Preprints* **2019**, *7*, 9. [CrossRef]

43. Dasgupta, S.; Mohanty, B.P.; Kohne, J.M. Impacts of Juniper vegetation and karst geology on subsurface flow processes in the Edwards Plateau,Texas. *Vadose. Zone. J.* **2006**, *5*, 1076–1085. [CrossRef]

44. Sohrt, J.; Ries, F.; Sauter, M.; Lange, J. Significance of preferential flow at the rock soil interface in a semi-arid karst environment. *CATENA* **2014**, *123*, 1–10. [CrossRef]

45. Bögli, A. *Karst Hydrology and Physical Speleology*, 1st ed.; Springer: Berlin/Heidelberg, Germany, 1980.

46. Yan, Y.J.; Dai, Q.H.; Yu, Y.F.; Peng, X.D.; Zhao, L.S.; Yang, J. Effects of rainfall intensity on runoff and sediment yields on bare slopes in a karst area, SW China. *Geoderma* **2018**, *330*, 30–40. [CrossRef]

47. Feng, T.; Chen, H.; Polyakov, V.O.; Wang, K.L.; Zhang, X.B.; Zhang, W. Soil erosion rates in two karst peak-cluster depression basins of north west Guangxi, China: Comparison of the RUSLE model with 137Cs measurements. *Geomorphology* **2016**, *253*, 217–224. [CrossRef]

48. Goldscheider, N.; Drew, D. *Methods in Karst Hydrogeology*, 1st ed.; Taylor and Fracis: London, UK, 2007.

49. Matthai, S.K.; Belayneh, M. Fluid flow partitioning between fractures and a permeable rock matrix. *Geophys. Res. Lett.* **2004**, *31*, L07602. [CrossRef]

50. Ghasemizadeh, R.; Hellweger, F.; Butscher, C.; Padilla, I.; Vesper, D.; Field, M. Review: Groundwater flow and transport modeling of karstaquifers, with particular reference to the North Coast Limestone aquifer system of Puerto Rico. *Hydrol. J.* **2012**, *20*, 1441–1461. [CrossRef]

51. Demenois, J.; Carriconde, F.; Bonaventure, P.; Maeght, J.L.; Stokes, A.; Rey, F. Impact of plant root functional traits and associated mycorrhizas on the aggregate stability of a tropical Ferralsol. *Geoderma* **2018**, *312*, 6–16. [CrossRef]

Article

Analysis of Long-Term Water Level Variations in Qinghai Lake in China

Jianmei Fang [1], Guijing Li [1], Matteo Rubinato [2], Guoqing Ma [3], Jinxing Zhou [1], Guodong Jia [1], Xinxiao Yu [1,*] and Henian Wang [4]

[1] College of Soil and Water Conservation, Beijing Forestry University, Beijing 100083, China; jmf46@163.com (J.F.); lgj8023lhy@163.com (G.L.); zjx9277@126.com (J.Z.); jgd3@163.com (G.J.)

[2] School of Energy, Construction and Environment & Centre for Agroecology, Water and Resilience, Coventry University, Coventry CV1 5FB, UK; matteo.rubinato@coventry.ac.uk

[3] World Bank Loan Project Management Center of State Forestry and Grassland Administration, Beijing 100714, China; maguoqing8042@sina.com

[4] Institute of Wetland Research, Chinese Academy of Forestry, Beijing 100091, China; wanghenian2006@126.com

* Correspondence: yuxinxiao111@126.com

Received: 17 September 2019; Accepted: 9 October 2019; Published: 14 October 2019

Abstract: Qinghai Lake is the largest inland saline lake on the Tibetan Plateau. Climate change and catchment modifications induced by human activities are the main drivers playing a significant role in the dramatic variation of water levels in the lake (Δh); hence, it is crucial to provide a better understanding of the impacts caused by these phenomena. However, their respective contribution to and influence on water level variations in Qinghai Lake are still unclear and without characterizing them, targeted measures for a more efficient conservation and management of the lake cannot be implemented. In this paper, data monitored during the period 1960–2016 (e.g., meteorological and land use data) have been analyzed by applying multiple techniques to fill this gap and estimate the contribution of each parameter recorded to water level variations (Δh). Results obtained have demonstrated that the water level of Qinghai Lake declined between 1960 and 2004, and since then has risen continuously and gradually, due to the changes in evaporation rates, precipitation and consequently surface runoff associated with climate change effects and catchment modifications. The authors have also pinpointed that climate change is the main leading cause impacting the water level in Qinghai Lake because results demonstrated that 93.13% of water level variations can be attributable to it, while the catchment modifications are responsible for 6.87%. This is a very important outcome in the view of the fact that global warming clearly had a profound impact in this sensitive and responsive region, affecting hydrological processes in the largest inland lake of the Tibetan Plateau.

Keywords: climate change; water levels; causes and implications; Qinghai Lake, Tibetan Plateau

---

1. Introduction

The observed climate changes [1] had a significant impact on physical and natural processes on Earth during the past decade. The IPCC's (Intergovernmental Panel on Climate Change) latest report pointed out that if global warming will continue at its current rate, it could reach an increase in temperature up to 1.5 °C between 2030 and 2052 [2], causing the rising of sea levels as well as warming of water surfaces in oceans and lakes. Furthermore, human activities also had a strong impact on hydrological processes considering increased water consumption and situations of water shortage recorded around the world during the last decade. Hence, recent research focused on the impact of climate change and human activities on hydrological processes because this topic was identified by many researchers as a priority [3,4] when planning for the future and making new developments more sustainable. Lake ecosystems usually provide indicators (i.e., water temperature, water levels, dissolved organic carbon (DOC)) of climate change, either directly or indirectly [5]. Recently, many studies have also been completed to investigate what has caused water level variations of lakes, such as the ones conducted on (i) the North American Great Lakes [6–8]; (ii) Lake Chad [9]; (iii) the Salton Sea [10] in the United States; (iv) Lake Lisan, Dead Sea rift [11]; and (v) Poyang Lake in China [12–14]. Summarizing the results achieved to date, hydrological conditions of each lake are affected by the lake's location, upstream boundaries, geographical climate and specific human activities undertaken on it such as residential developments, industry and irrigation tasks; hence, it is necessary to figure out the key factors that affect water levels to develop methods and procedures that can regulate those that alter or alleviate hydrological extreme processes in lakes.

The Tibetan Plateau (TP), known as "the roof of the world", "the third pole" and "the water tower of Asia", is the largest plateau in China and the highest plateau in the world [5,15], and it is considered the perfect location to identify the effects of global climate change [16–18]. Qinghai Lake is the largest inland saline lake on TP, and it has attracted extensive attention due to its special geographical location and its wide area characterized by fragile ecosystems. Over the last years, researchers have reported a drastic change in water levels in Qinghai Lake, indicating that the ecological environment around it is undergoing a rapid evolution [19–22]. Typically, inland closed lakes with no outlet streams are ideal to distinguish hydrologic processes and phenomena affecting the water balance because changes in water levels can result from limited factors such as precipitation, evaporation, groundwater infiltration [23–25] and the presence of specific vegetation [19].

The most effective way to estimate water levels in lakes is by applying the water-balance equation model, where the gain or loss of water directly reflects the changes in water levels [26,27]. Multiple inland lakes, due to the lack of funding and therefore resilient and accurate equipment to monitor basic data, are considered not appropriate to identify and quantify objectively the factors affecting the water balance. Nevertheless, Qinghai Lake, being a closed one with inlet river streams and without outlet river streams, is an ideal place to study, especially having the availability of meteorological and hydrological monitored data. To date, the study conducted by Li et al. [28] calculated the main water balance estimation of Qinghai Lake, while Cui et al. [20] preliminarily analyzed the climatic factors that affect the water level variations of Qinghai Lake. Despite this, there is still a need for new studies to fully distinguish and assess the relative contribution of anthropogenic activities and climate variability to water level variations in Qinghai Lake and their impacts on the corresponding water balance.

This paper present the analysis of water level variations recorded in Qinghai Lake during the last 57 years, examining the evolution and interpreting the impacts of driving factors to better understand the hydrological process of this inland lake basin in the northeast of TP, enhancing the present understanding of climatic variations on surface changes to provide a reference for local and regional water management.

The paper is organized as follows: Section 2 presents the description of the study area, introducing the data monitored and describing the statistical analysis used. Section 3 includes the estimation of long-term variations in Qinghai Lake as well as the impacts of climate change and catchment modifications for inflow runoff to Qinghai Lake. Section 4 provides a discussion of the results obtained, and Section 5 produces a brief summary and concluding remarks of the whole study.

2. Materials and Methods

2.1. Study Area and Data Availability

Qinghai Lake basin (97°50′~101°20′ E, 36°15′~38°20′ N) is located in the northeast of TP, covering an area of 29,664 km². The average annual air temperature ranges from −0.8 °C to 1.1 °C, and the average annual precipitation ranges from 327 to 423 mm. However, the annual precipitation is unevenly distributed, decreasing from the east and south to the west and north. The total surface runoff of local main rivers including Buha River, Shaliu River, Haergai River, Heima River and Daotang River accounts for 83% of the total surface runoff into Qinghai Lake [28,29], with the first two (Buha & Shaliu) constituting the 64% of the surface runoff for the entire basin [30]. The main vegetation types in the basin are alpine meadows and alpine grasslands.

Qinghai Lake is the largest inland saline lake in this basin with an area of 4400 km^2 (in 2016), and it is located at an altitude of 3193 m above sea level. As previously mentioned in Section 1, this lake is a closed one with no surface water outflow. It is about 106 km in length from east to west, and 63 km in width from north to south, and 360 km in circumference [30]. The average annual air temperature above the lake is about 1.2 °C, and the average annual precipitation near its proximity is about 357 mm.

The datasets available used for this study can be summarized as follows:

- Daily water levels in Qinghai Lake at Xiashe station (36°35′ N, 100°29′ E) from 1959 to 2016, obtained by the Information Center of Qinghai Hydrographic Bureau, China (ICQHB).
- Daily surface runoff of Buha River and Shaliu River, observed at the estuary of Buha River station (37°18′ N, 99°44′ E, from 1960 to 2016), at Gangcha station (37°17′ N, 100°19′ E, from 1960 to 1975) and at Gangcha II station (36°19′ N, 100°18′ E, from 1976 to 2016, obtained as well by ICQHB.
- Meteorological data:

  i. Daily meteorological data of 14 national meteorological stations from 1960 to 2016, obtained by the China Meteorological Information Center.
  ii. Monthly meteorological data from 1960 to 2010 at three meteorological stations, obtained by Qinghai Meteorological Bureau in China.
  iii. Daily precipitation data of Buha River rain station from 1962 to 2016 obtained by ICQHB.
  iv. Daily evaporation data from 1984 to 2016 at Xiashe station obtained from ICQHB.
  v. Yearly evaporation data from 1960 to 1988 obtained from the literature [29]. (Figure 1 and Table 1).

- Environmental and physical details of Qinghai Lake, and these datasets were obtained from ICQHB and the literature [30].
- Land use data from 1980 to 2015, obtained by the Data Center of Resources and Environmental Sciences, Chinese Academy of Sciences.

Table 1. Detailed information of the meteorological stations in and around Qinghai Lake basin used to collect the datasets previously described.

No.	Station Number	Station Name	Latitude (°N)	Longitude (°E)	ASL (m)	Data Collection Frame
1	52,645	Yeniugou	38.43	99.60	3315	1960–2016
2	52,842	Chaka	36.78	99.08	3088	1960–2016
3	52,633	Tuole	38.82	98.42	3368	1960–2016
4	52,833	Wulan	36.93	98.48	2951	1960–2016
5	52,836	Dulan	36.30	98.10	3190	1960–2016
6	52,737	Delingha	37.37	97.38	2982	1960–2016
7	52,868	Guide	36.02	101.37	2274	1960–2016
8	52,657	Qilian	38.18	100.25	2788	1960–2016
9	52,754	Gangcha	37.33	100.13	3302	1960–2016
10	52,856	Gonghe	36.27	100.62	2836	1960–2016
11	52,943	Xinghai	35.58	99.98	3324	1960–2016
12	52,765	Menyuan	37.38	101.62	2851	1960–2016
13	52,866	Xining	36.73	101.75	2296	1960–2016
14	52,955	Guinan	35.58	100.73	3121	1960–2016
15	52,745	Tianjun	37.30	99.02	3417	1961–2010
16	52,855	Huangyuan	36.68	101.25	2675	1961–2010
17	52,853	Haiyan	36.90	100.98	3010	1961–2010
18	1,329,500	The estuary of Buha River	37.03	99.73	3191	1962–2016

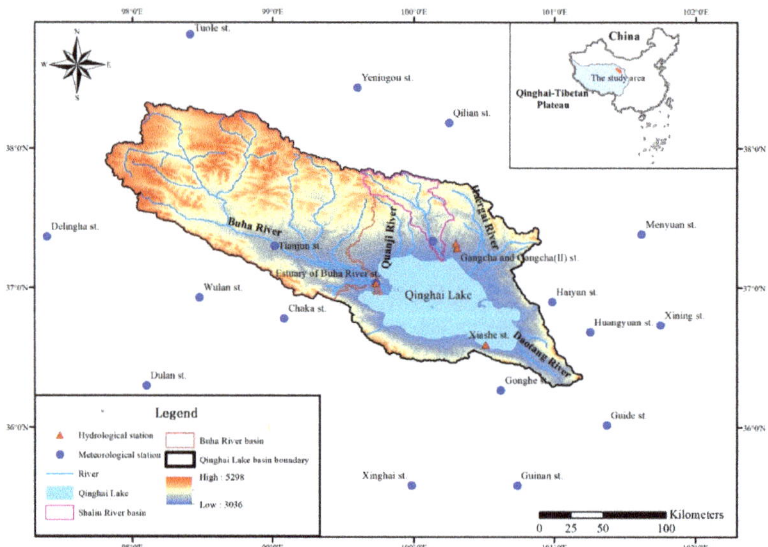

Figure 1. The location of Qinghai Lake basin in China (top right corner), physical characteristics of Qinghai Lake in Qinghai Lake basin (centre) and the spatial distribution of the hydrological stations (red triangles) and meteorological stations (blue dots).

2.2. Governing Equations

2.2.1. Lake Water Balance Model

As Qinghai Lake is a closed-catchment with no surface water outflow, the annual hydrological water balance equation can be expressed as follows:

$$\Delta h = P_l - E_l + R_{ls} + R_{lg} \pm \varepsilon \tag{1}$$

where Δh is the yearly water level variation, (mm); P_l is the yearly precipitation on the lake surface, (mm); E_l is the yearly evaporation from the lake surface, (mm); R_{ls} is the yearly surface runoff into the lake, (mm); R_{lg} is yearly underground runoff on the lake bottom, (mm); ε is the error, (mm). For this watershed scenario, the surface runoff is almost equivalent to the river runoff and the slope surface runoff can be considered negligible. Δh can be quantified as well as follows:

$$\Delta h = h_i - h_{i-1} \tag{2}$$

where h_i and h_{i-1} are the lake level at i year and at $i - 1$ year.

The yearly average water level of the lake was obtained from the daily water level data at Xiasha station. P_l was calculated by applying the Thiessen Polygon method focusing in the area between Buha River station, Gangcha station, Haiyan station and Gonghe station, which are the nearest stations to the lake. E_l was obtained from evaporation pan (type of E20) data at Xiashe station [29]. The yearly total surface runoff (R_{ls}, mm) into the lake was obtained from the surface runoff (Q_{ls}, m^3) of Buha River and Shaliu River by using the proportion amplification method [31].

In this paper, Δh is subject to a combination of climate and human activities effects (such as farmland reclamation, grazing, afforestation which indirectly influence runoff and catchment characteristics). P_l and E_l represents the climate variability, R_{ls} and R_{lg} are the results of the combination of climate and catchment modifications. Hence, to correctly quantify the contribution rate of climate and catchment characteristics (human induced) to Δh, it is necessary to calculate accurately the contribution rate of climate and catchment characteristics (human induced) to R_{ls} and R_{lg}. Therefore, the calculation method can be applied as follows:

$$R(\Delta h)_c = R(\Delta h_{Pl})_c + R(\Delta h_{El})_c + R(\Delta h_{Rls})_c + R(\Delta h_{Rlg\pm\varepsilon})_c \tag{3}$$

$$R(\Delta h)_u = R(\Delta h_{Rls})_u + R(\Delta h_{Rlg\pm\varepsilon})_u \tag{4}$$

where $R(\Delta h)_c$, $R(\Delta h)_u$ respectively represent the contribution rate of climate and catchment modifications to Δh. $R(\Delta h_{Pl})_c$, $R(\Delta h_{El})_c$ respectively represent the climate change contribution of P_l and E_l to Δh. $R(\Delta h_{Rls})_c$, $R(\Delta h_{Rls})_u$ respectively represent the R_{ls} contribution rate of climate change and catchment change to Δh. $R(\Delta h_{Rlg\pm\varepsilon})_c$, $R(\Delta h_{Rlg\pm\varepsilon})_u$ respectively represent the $R_{lg} \pm \varepsilon$ contribution rate of climate change and catchment change to Δh.

2.2.2. Land Use Dynamic Index

Land use change can reflect the effect as well as the intensity of human activities. The Land Use Dynamic Index was proposed by Chen et al. [32] and was adopted in this study to describe the change of land use types in the research area for a certain period (1980–2015). The calculation method was completed as follows:

$$LC = \frac{\sum\limits_{i=1}^{n} \Delta U_{iin}}{2\sum\limits_{i=1}^{n} U_{i0}} \times \frac{1}{T} \times 100\% \tag{5}$$

where LC is the Land Use Dynamic Index in a certain period of time in the research area (%), ΔU_{iin} refers to the area of type i land use converted into the non-i type land use within a certain period of time in the research area (km^2), U_{i0} is the area of type i land use at the beginning of the study period (km^2), T is the research period (years).

2.2.3. Statistical Analysis

(1) The Non-parametric Mann-Kendall Test

The non-parametric Mann-Kendall test [33,34] (M-K test) and the cumulative anomaly method were used to detect any point of abrupt changes in the variables considered. The M-K test has been widely applied to identify the point at which the hydrological processes change significantly due to the climate [35,36]. The details about this statistical method can be obtained in the relevant literature [37].

First, the partial M-K test statistics are calculated as:

$$S_k = \sum_{i=1}^{k} \sum_{j=1}^{i-1} \alpha_{ij} \quad (k = 2,3,4,\ldots,n) \tag{6}$$

$$\alpha_{ij} = \begin{cases} 1 & x_i > x_j \\ 0 & x_i \leq x_j \end{cases} \quad 1 \leq j \leq i \tag{7}$$

Statistical variable UF is adopted and defined as:

$$UF = \frac{S_k - E(S_k)}{\sqrt{Var(S_k)}} \quad (k = 1,2,3,\ldots,n) \tag{8}$$

$$E(S_k) = \frac{k(k-1)}{4} \tag{9}$$

$$Var(S_k) = \frac{k(k-1)(2k+5)}{72} \tag{10}$$

Proceed to Equation (11) putting the data sequence x in reverse order:

$$\begin{cases} UB_k = -UF_{k'} \\ k' = n+1-k \end{cases} \quad (k = 1,2,3,\ldots,n\ ) \tag{11}$$

In the M-K curve, if the value of the intersection of the curve forward (UF) or the curve backward (UB) is greater than 0, this suggests that the record sequence shows an upward trend; less than 0 suggests a downward trend. When the record exceeds the critical line (Given the significance level $\alpha = 0.05$, the critical lines $U_{0.05} = \pm 1.96$), this suggests that an increase or decrease in the trend may be significant. The range of exceeding the critical line is the time zone in which the abrupt change occurs. If there is an intersection between the curves of UF and UB in the range of the critical lines, the time of the intersection is the time of the abrupt change started [38].

(2) The Cumulative Anomaly Method

The cumulative anomaly method is widely used to indicate the runoff [39], precipitation and other factors that deviate from the normal situations, focusing on the difference between a certain value and the average value of a series [38].

(3) The Principal Component Regression Analysis

The principal component regression (PCR) analysis [40] is a combination of principal component analysis and regression analysis. Typically, this method considers regressing the outcome on a set of covariates based on a standard linear regression model, using PCA (principal component analysis) for estimating the unknown regression coefficients. Generally, only a subset of all the principal components for regression is used; hence, PCR tends to act as a regularized procedure.

(4) The Grey Relational Analysis

The grey relational analysis [41] is adopted in this study to solve uncertain problems such as limited data and incomplete information by calculating the grey correlation degree γ_i, quantifying the correlation degree among the influential factors of underground runoff.

(5) The Least Square Method

The least square method [42] is applied to procure unknown data and minimize the sum of squared errors between the obtained data and the actual data. The least square method can also be used for curve fitting.

(6) The Partial Least Squares Regression Method

The partial least squares regression method [43] is a combination of multiple linear regression analysis, canonical correlation analysis and principal component analysis, reflecting the influence of the sample population on the predicted values and fully considering the influence of the comprehensive effect between individual factors on the predicted ones.

2.2.4. Sensitivity Analysis Based on the Budyko Framework

Climate change and human activities are the most important drivers to determine the river hydrological process of the catchment [44]. In this study, the sensitivity coefficient method [45] based on Budyko Theory [46] was used to quantitatively separate the impacts of climate change and human activities on the variations of streamflow into Qinghai Lake. The theoretical equation of Budyko curve [47] can be applied as follows:

$$\frac{ET}{P} = 1 + \frac{ET_0}{P} - \left[1 + \left(\frac{ET_0}{P} \right)^\omega \right]^{1/\omega} \tag{12}$$

where ET is the evapotranspiration of the upper catchment area, (mm); P is the precipitation of the catchment area, (mm); ET_0 is the potential evapotranspiration of the catchment area, (mm); the empirical parameter ω represent catchment characteristics, such as human activities, land use, vegetation, topography, and properties of soil [48,49], [$\omega \in (1, \infty)$].

The change of surface runoff in a given basin can be characterized by climate and human activities changes as follows:

$$\Delta Q = \Delta Qc + \Delta Qu \tag{13}$$

where ΔQc and ΔQu represent the surface runoff variation caused by climate change and human activities changes, respectively. The surface runoff variation caused by climate change can be expressed by the following formula [45]:

$$\Delta Qc = \frac{\partial Q}{\partial P} \times \Delta P + \frac{\partial Q}{\partial ET_0} \times \Delta ET_0 \tag{14}$$

The surface runoff variation caused by human activities can be expressed by the following formula:

$$\Delta Qu = \frac{\partial Q}{\partial \omega} \times \Delta \omega \tag{15}$$

where ΔP is the variation of precipitation, $\Delta \omega$ is the variation of the empirical parameter ω of a given catchment; ΔET_0 is the potential evapotranspiration variation; $\frac{\partial Q}{\partial P}$, $\frac{\partial Q}{\partial ET_0}$, $\frac{\partial Q}{\partial \omega}$ respectively represent the sensitivity coefficient of runoff to precipitation, runoff to potential evapotranspiration, runoff to precipitation, runoff to the empirical parameter represent catchment characteristics. All of the sensitivity coefficients can be calculated as follows:

$$\frac{\partial Q}{\partial P} = \left[1 + \left(\frac{ET_0}{P}\right)^\omega\right]^{(1/\omega - 1)} \tag{16}$$

$$\frac{\partial Q}{\partial ET_0} = \left[1 + \left(\frac{P}{ET_0}\right)^\omega\right]^{(1/\omega - 1)} - 1 \tag{17}$$

$$\frac{\partial Q}{\partial \omega} = [P^\omega + ET_0{}^\omega]^{1/\omega} \cdot \left[(-\frac{1}{\omega^2}) \cdot \ln(P^\omega + ET_0{}^\omega) + \frac{1}{\omega} \frac{1}{P^\omega + ET_0{}^\omega} \cdot (\ln P \cdot P^\omega + \ln ET_0 \cdot ET_0{}^\omega) \right] \tag{18}$$

In this paper, the potential evapotranspiration at the meteorological stations was calculated by applying the FAO56 method, Penman-Monteith model [50], because previous literature [51,52] has demonstrated how these methods are reliable to estimate potential effects of climate change on the calculation of the evaporation as well as the influence of climate change on water cycles. P and ET_0 of the entire basin were obtained by applying the area-weight method of Tyson Polygon.

3. Results and Analysis

3.1. Long-Term Variations in Water Levels and the Hydro-Climatic Factors

3.1.1. Long-Term Variations in Water Levels

Figure 2 shows the annual water levels of Qinghai Lake recorded during the period 1960–2016. Overall, comparing datasets within other locations in the semi-arid areas of Western China, the water level varied significantly ($\approx$3.5 m) and it is possible to notice a clear inflection point recorded in 2004. Herein, the analysis of the graph was divided into two periods to simplify the procedure: period I was selected to be between 1960 and 2004, while period II was selected to be between 2005 and 2016. The annual water level of the lake declined at the rate of 7.84 cm/year ($P < 0.001$), with a total decrease of 3.46 m in period I, while the annual water level of the lake has risen at the rate of 13.80 cm/year ($P < 0.001$), with a total increase of 1.49 m in period II.

The variation of the water levels of the lake (Δh) reflected the acquisition and loss of water volume over the years due to multiple factors, and the trend is shown in Figure 3. According to the results obtained, Δh tended to increase during period I ($R^2 = 0.0105$, $P < 0.01$) as well as during period II ($R^2 = 0.0291$, $P < 0.01$). The increasing rate of the water level in period II was notably faster than the one in period I, and the water level of the lake in 1960 could be reached again by 2030 if the present increasing rate continues constantly.

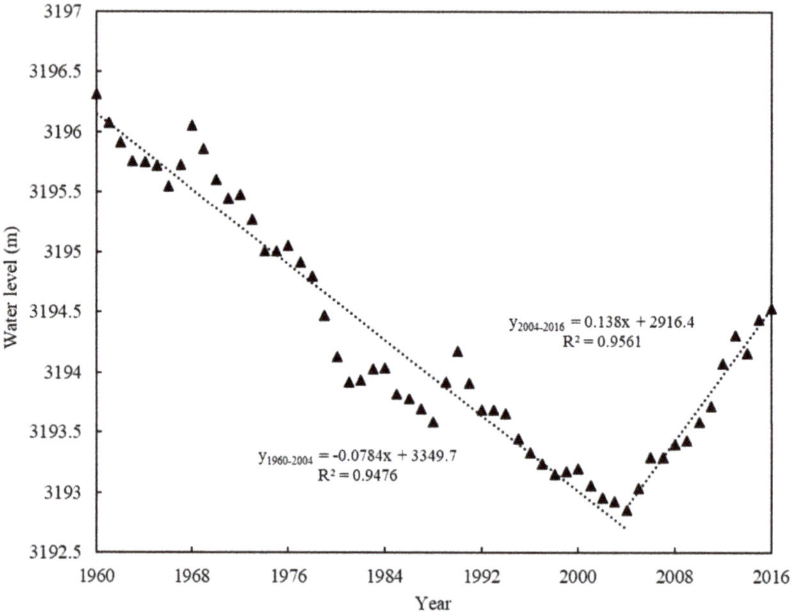

Figure 2. The annual water levels of Qinghai Lake recorded during the period 1960–2016.

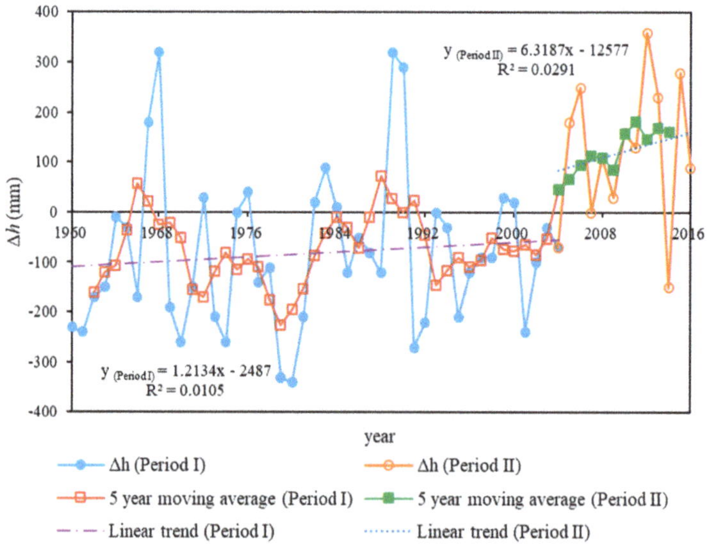

Figure 3. The trend of water level variation (Δh) during the period of 1960–2016.

3.1.2. Analysis of Hydro-Climatic Factors Influencing Water Levels

Δh was dependent on the lake hydro-climatic conditions P_l, E_l, R_{ls} and $R_{lg} \pm \varepsilon$. All together, these variables affected the rise or fall of the water level in the lake, and the relationship between them and the corresponding variations in the water levels of the lake are shown in Figure 4 and Table 2.

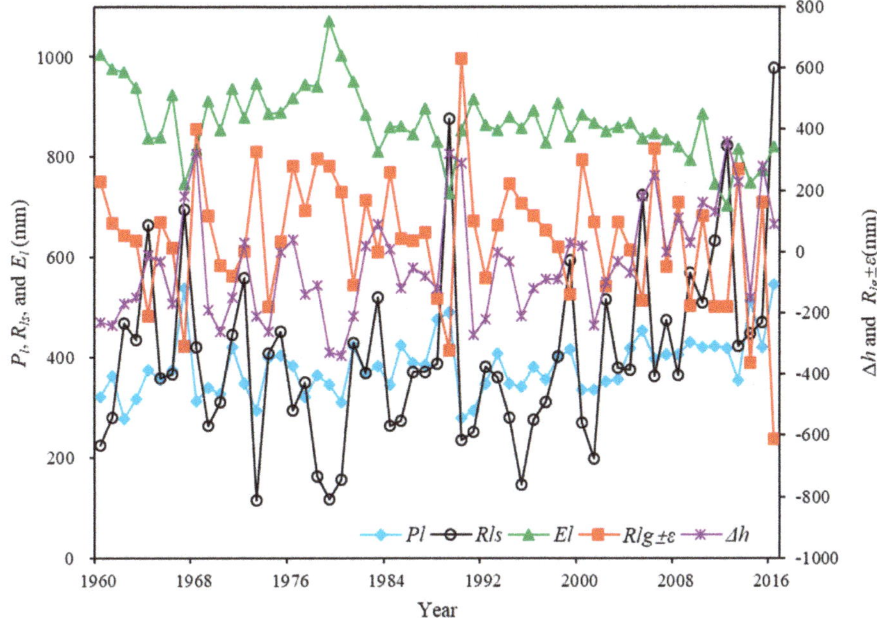

Figure 4. Variations of the hydro-climatic factors of Qinghai Lake during the period of 1960–2016. P_l is the precipitation over the lake (mm), E_l is the evaporation from the lake surface (mm), R_{ls} is the surface runoff into the lake (mm), $R_{lg} \pm \varepsilon$ is the underground runoff and study error (mm), Δh is lake level variation (mm).

Table 2. The hydro-climatic factors calculated in different periods (mm) *.

Periods	P_l	R_{ls}	$R_{lg} \pm \varepsilon$	E_l	Δh
I (1960–2004)	367.94 (+45.67%)	364.02 (+45.18%)	73.68 (+9.15%)	887.64 (−100%)	−82.00
II (2005–2016)	432.77 (+43.38%)	564.93 (+56.62%)	−55.81 (−6.50%)	802.73 (−93.50%)	+139.17
1960–2016	381.59 (+45.74%)	406.32 (+48.70%)	46.42 (+5.56%)	869.77 (−100.00%)	−35.44

* The bracketed values refer to the percentage of total input or output represented by average yearly volumetric flux (mm) changes at different periods. Sign + represents water in, while sign - represents water out.

During the period under investigation, surface runoff into the lake (R_{ls}) was mainly due to the Buha River and Shaliu River, which contributed about 48.70% of the total water in the lake. The underground runoff on the lake bed and the associated error ($R_{lg} \pm \varepsilon$) accounted for 5.56% of the total water intake of the lake, and this value fluctuated significantly. The precipitation on the lake surface (P_l) contributed about 45.74% of the total water in the lake while the evaporation from the lake surface (E_l) contributed about 100% of the total water removed from the lake. As possible to notice from Figure 4, P_l, and R_{ls} had similar trends, while this could not be confirmed for E_l and R_{lg}. The peak of Δh often corresponded to the peak of P_l, R_{ls}.

The annual average values for each hydro-climatic variable P_l, R_{ls}, $R_{lg} \pm \varepsilon$, and E_l were 381.59 (mm), 406.32 (mm), 46.42 (mm), and 869.77 (mm), respectively. In period I, approximately 45.67% of the total water input into the lake came via P_l, with 45.18% of water input coming from R_{ls}, and a small fraction of water input was due to $R_{lg} \pm \varepsilon$. The whole outflow was estimated to be associated with E_l. In period II, approximately 43.38% and 56.62% of the total water input was associated with P_l and R_{ls}, respectively, while E_l contributed 93.5% to the outflow with a small fraction of water escaping the lake attributed to $R_{lg} \pm \varepsilon$ (6.5%). This indicated that the water balance of Qinghai Lake was mainly determined by P_l, R_{ls} and E_l. Therefore, the authors can confirm that $R_{lg} \pm \varepsilon$, always being a small percentage, accounted for a small proportion of the water balance of Qinghai Lake.

3.2. Causes of Changes in Water Levels of the Lake

3.2.1. Impact of Climate Change on Water Levels

Figure 4 shows the relationship between P_l, E_l and Δh. The correlation coefficient between P_l and Δh calculated was 0.356 ($P < 0.01$). The fluctuation range for P_l was estimated between 277.2 and 546.4 (mm), where the mean value was obtained equal to 381.6 (mm), showing an upward trend at the rate of 1.4347 mm/year ($R^2 = 0.164$, $P < 0.01$). The correlation coefficient between E_l and Δh was -0.705 ($P < 0.01$) and the fluctuation range for E_l was between 702.6 and 1070.5 (mm), where the mean value was 869.8 (mm), showing a downward trend at the rate of 2.2823 mm/year ($R^2 = 0.290$, $P < 0.01$). Changes in behavior for P_l were consistent with the fluctuations of Δh, and the peaks of Δh were noticed to be delayed by 1 year when comparing them with those associated with P_l. However, E_l was generally contrary to the fluctuations of Δh.

Based on the results achieved, E_l and P_l were identified as the main important climate factors affecting the water level changes, and the correlation relationships between E_l and Δh estimated were of higher quality than those obtained between P_l and Δh.

It was clear that at peaks in the rising periods for Δh corresponded to higher values of P_l but lower values of E_l (e.g., 1968, 1989, and 2012). Furthermore, decreasing peaks of Δh corresponded to lower values of P_l but higher values of E_l (e.g., 1980 and 1991). Precipitation rates were quantified to affect both the runoff of the inflow rivers and underground runoff acting on the water level changes. Finally, evaporation was selected as the only factor of climate influencing water "exiting" the lake, playing a significant role in the fluctuation of the water level.

3.2.2. Impact of Human Activities on Catchment Modifications and Consequently on Water Levels

Figure 5 illustrates 7 different years spanning 1980 and 2015, to highlight the land use changes from 1980 to 2015 obtained by the superposition function fitted within ArcGIS10.2. Despite being present and noticeable, land use changes observed were not particularly significant as possible to notice in Figure 5 (Tables 3 and 4).

Table 3. The land use dynamic attitude (LC) from 1980 to 2015.

	Period					
Period	1980–1990	1190–1995	1995–2000	2000–2005	2005–2010	2010–2015
LC (%)	0.06	0.04	0.05	0.03	0.01	0.03

Table 4. Land use change parameters from 1980 to 2015 (km^2).

Type	Farmland	Forestland	Grassland	Water Area	Constructive Land	Unused Land	Total
Farmland	493	/	1	1	4	1	500
Forestland	/	1371	10	2	1	1	1385
Grassland	64	5	17,475	30	7	16	17,596
Water area	/	/	142	4842	/	157	5141
Constructive land	/	/	/	/	26	/	26
Unused land	/	/	15	27	/	4975	5017
Total	557	1376	17,643	4901	37	5149	29,664

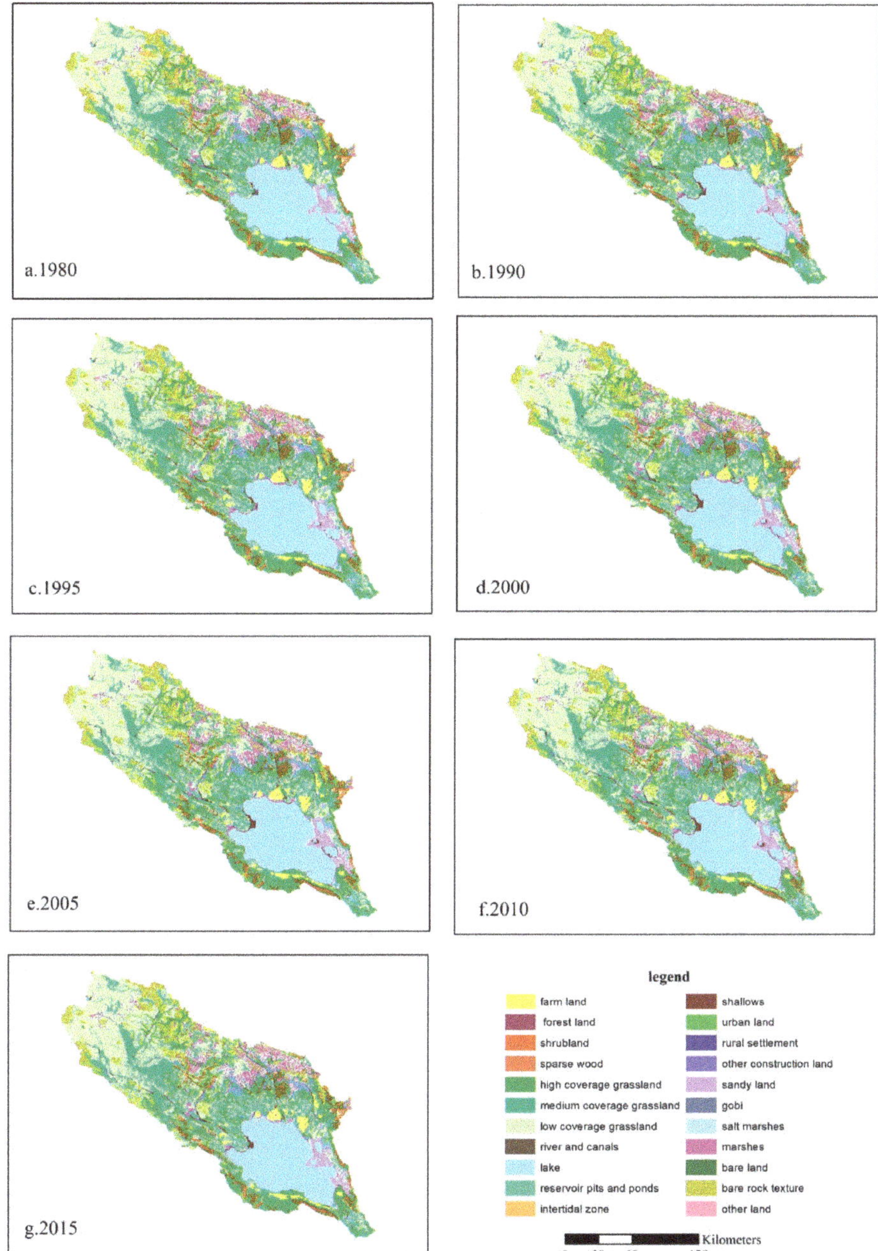

Figure 5. The changes in land use in Qinghai Lake basin from 1980 to 2015.

As the overall topography of Qinghai Lake basin is relatively gentle, the production and confluence of the runoff typically have longer durations, highly dependent on the catchment. The empirical parameter representing land surface characteristics of the basin (ω) was calculated by applying the least square method [45] according to Equation (6) and results are displayed in Figure 6. The correlation coefficient between ω and Δh was -0.262 ($P < 0.05$). Figure 6 and Table 4 confirmed as previously stated that there were little changes in the land use, and the main reasons causing ω changes were not associated with the land change use, but were probably due to changes in local vegetation and soil conditions.

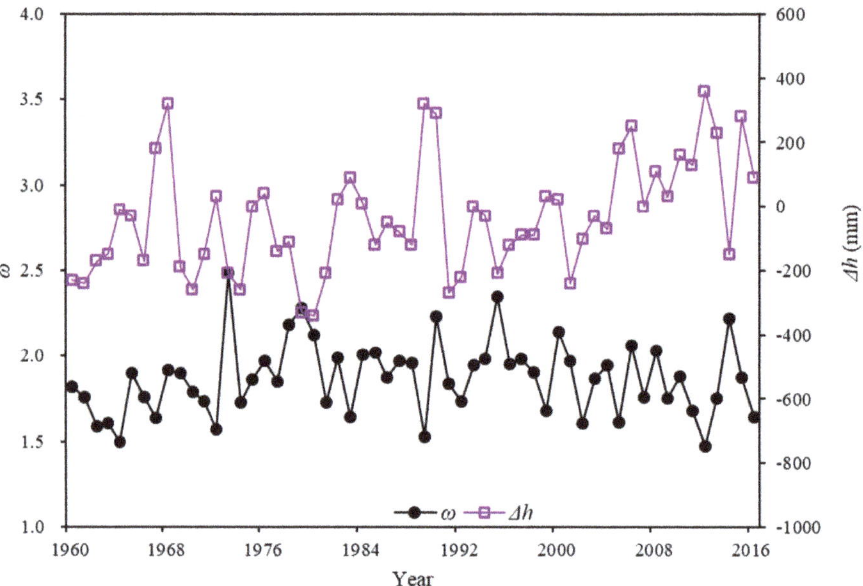

Figure 6. Variation the empirical parameter representing catchment characteristics (ω) and Δh in Qinghai Lake basin during the period of 1960–2016.

By summarizing all the contributions, results have also confirmed that affecting factors Δh, P_l and E_l were entirely attributable to climate change, while R_{ls} and $R_{lg} \pm \varepsilon$ were the results of the joint action between climate and catchment modifications. Therefore, it became necessary to focus on R_{ls} and $R_{lg} \pm \varepsilon$ and to quantify their impact due to climate change and catchment change.

3.2.3. Impact of Climate and Catchment Modifications on the Surface Runoff (R_{ls})

Surface runoff is a crucial variable to consider when completing any lake water balance [20,53] and in this study it accounted for 45.18~56.62% of the total lake inflow during the study period, which demonstrates how this parameter was a key factor affecting the water level variations in the lake. The correlation coefficient between R_{ls}, and Δh was calculated to be 0.590 ($P < 0.01$).

Surface runoff was generated from the surrounding catchment area of about 25,000 km^2 (obtained by subtracting the lake area from the total basin area). Surface runoff (mm) was obtained by dividing the annual total runoff of the basin (m^3) by the annual catchment area. The catchment area of annual maximum, annual minimum and mean annual values was 25,439.71 km^2 (in 2004), 25,136.70 km^2 (in 1960), and 25,308.56 km^2, respectively. In this paper, the Mann-Kendall test method (M-K test) and the cumulative anomaly method were used to identify remarkable changes in the variables' behaviors (i.e., surface runoff, precipitation and potential evapotranspiration). As shown in Figure 7a, the precipitation in the basin is always rising ($UF > 0$), and the precipitation trend shows a significant change from 2005 ($UF > 1.96$). The intersection of UF and UB curves indicates an abrupt change point in 2003. Furthermore, when focusing on the evapotranspiration trends in Figure 7b, and intersection point was noticed between the two curves in 1998. The M-K test failed to identify any abrupt change point in the trend of the surface runoff recorded, while the cumulative anomaly analysis method correctly estimated it as observed in 2004 (Figure 7c). The results indicated that surface runoff, precipitation and lake water levels were closely related. Additionally, the year corresponding to the anomalies noted in the variation of the lake water levels and the surface runoff slightly lagged behind the year corresponding to the abrupt change point related to the precipitation. Based on the cumulative anomaly analysis method, the runoff series were divided into two periods like the variation of water levels: period I (1960–2004) and period II (2005–2016), which enabled the authors to calculate the basin characteristic parameters and sensitivity coefficients for period I and period II that are presented in Table 5.

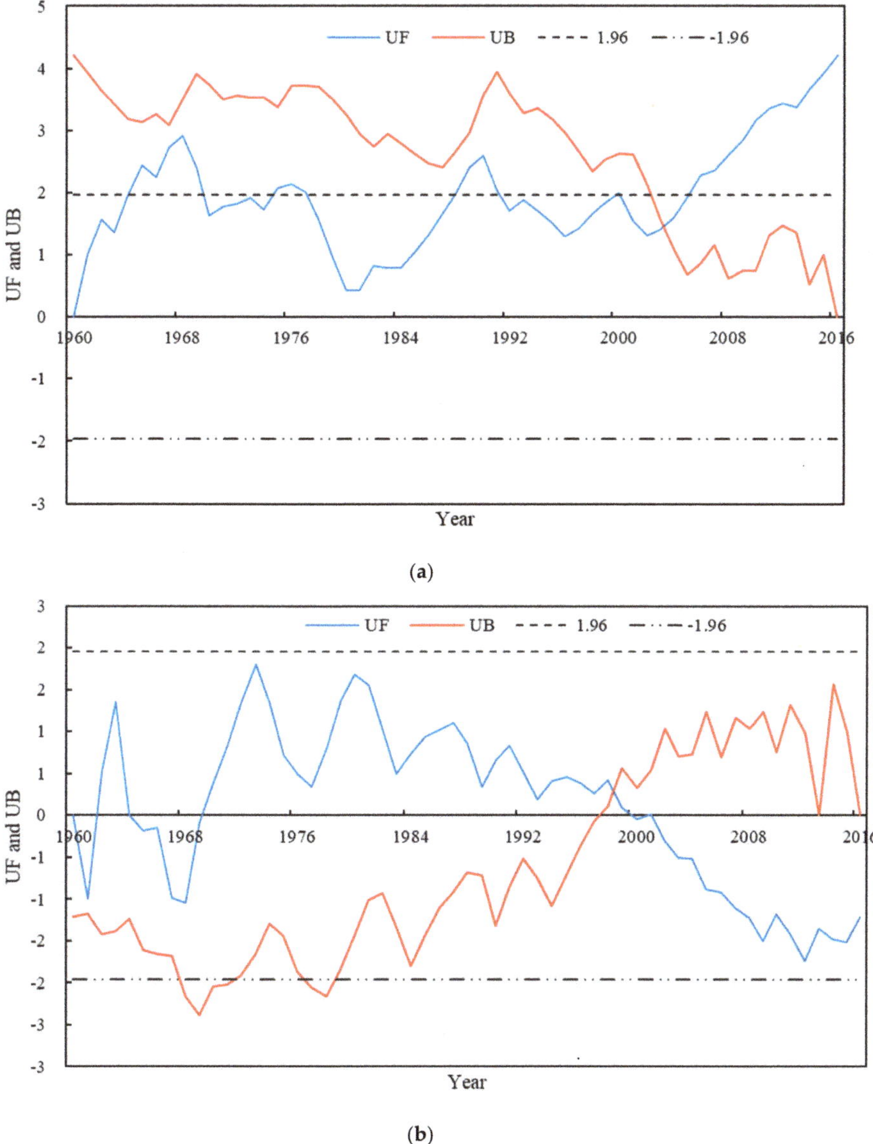

Figure 7. *Cont.*

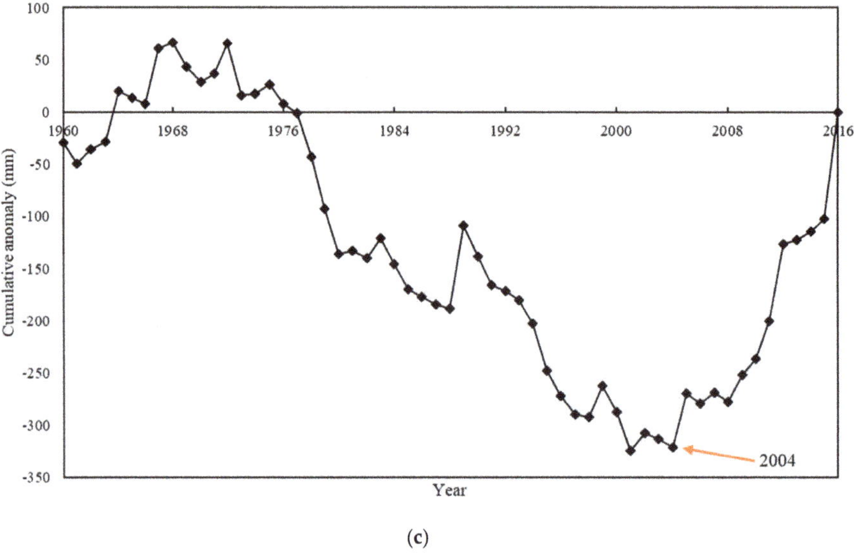

Figure 7. Mann-Kendall test of annual precipitation from 1960 to 2016 (**a**), Mann-Kendall test of annual potential evapotranspiration (**b**), and Cumulative anomaly of the annual surface runoff (**c**) in Qinghai Lake basin.

Table 5. Basin characteristic parameters and sensitivity coefficients in each study period.

Variable	Study Period		
	1960 to 2016	I	II
Q (mm)	69.96	62.84	96.69
P (mm)	349.13	334.65	403.42
ET_0 (mm)	1078.55	1081.23	1068.49
ω	1.85	1.87	1.82
$\partial Q/\partial P$	0.36	0.34	0.42
$\partial Q/\partial ET_0$	−0.05	−0.05	−0.07
$\partial Q/\partial \omega$	−115.75	−106.43	−146.57

According to the results, the total surface runoff variation was measured as +33.9 mm. Contribution rates of climate change and catchment modifications to this variation were calculated using Equations (14) and (15) and the results show that the main cause of runoff change from 1960 to 2016 was climate change (producing an increased surface runoff by 25.54 mm for a corresponding contribution rate of 80.19%). On the other hand, the effect caused by catchment modifications was not to be considered a negligible factor, considering that it generated an increase in surface runoff of 6.31 mm and hence its contribution rate was 19.81%. Error was estimated to be 2 mm, corresponding to 5.91% of the total variation (Table 6).

Table 6. The results of attribution analysis of runoff change.

	ΔQ	ΔQc	ΔQu	Error
Contribution Amount (mm)	33.85	25.54	6.31	2.00
Contribution Rate (%)	100	80.19	19.81	5.91

3.2.4. Impact of Climate and Catchment Modifications on the Underground Runoff ($R_{lg} \pm \varepsilon$)

$R_{lg} \pm \varepsilon$ was included in the water balance equation model but due to limitations in data availability, it was more challenging to accurately estimate the effects on its variations due to climate and catchment modifications. However, by applying the equations described in Section 2, the correlation between $R_{lg} \pm \varepsilon$ and Δh was calculated to be 0.143.

Underground runoff could be affected directly and indirectly by multiple factors; therefore, the authors believed it was not appropriate to complete a comprehensive and systematic analysis of its dynamic changes by using a single factor analysis. Hence, the principal component regression method [40] was used. The factors x_i identified to influence the underground runoff (y) can be summarized as follows:

- water level variation in the lake (x_1);
- precipitation on the lake surface area (x_2);
- surface runoff of the basin (x_3);
- evaporation from the lake surface (x_4);
- precipitation across the entire basin area (x_5);
- empirical parameter representing land surface characteristics of the basin (x_6).

The correlation analysis is shown in Table 7; the multi-collinearity among the influential factors, and the grey relational degree analysis (Table 8) show that each factor had a closed relationship with y, and the grey relational degree of each factor x_i is displayed as γ_i.

Table 7. Correlation coefficient matrix of influential factors.

Correlation Coefficient	x_1	x_2	x_3	x_4	x_5	x_6
x_1	1	0.36 **	0.51 **	−0.71 **	0.39 **	−0.26 *
x_2	0.36 **	1	0.61 **	−0.56 **	0.97 **	−0.28 *
x_3	0.51 **	0.61 **	1	−0.56 **	0.64 **	−0.59 **
x_4	−0.71 **	−0.56 **	−0.56 **	1	−0.55 **	0.31 *
x_5	0.39 **	0.97 **	0.64 **	−0.55 **	1	−0.35 **
x_6	−0.26 *	−0.28 *	−0.59 **	0.31 *	−0.35 **	1

$* P < 0.05, ** P < 0.01.$

Table 8. Grey relational degree matrix of influential factors on subsurface runoff.

Incidence Matrix	γ_1	γ_2	γ_3	γ_4	γ_5	γ_6
y	0.8205	0.7683	0.7739	0.8295	0.7609	0.8441

It can be seen from Table 9 that the cumulative contribution rates of the first, second and third principal components (F_1, F_2, F_3) were more than 80%, indicating that they basically contained all the information of the original impact factors. Subsequently, the three principal components were used to analyse y. According to Table 10, the linear equations were obtained as follows:

$$F_1 = 0.3589x_1 - 0.5745x_2 + 0.3797x_3 + 0.2533x_4 + 0.5738x_5 + 0.0598x_6 \tag{19}$$

$$F_2 = 0.4464x_1 + 0.5176x_2 + 0.1273x_3 - 0.0852x_4 + 0.119x_5 + 0.7037x_6 \tag{20}$$

$$F_3 = 0.4448x_1 - 0.0788x_2 - 0.2944x_3 + 0.6988x_4 - 0.4700x_5 - 0.0069x_6 \tag{21}$$

Principal component F_1 could be almost interpreted as precipitation on the lake surface area (x_2) and precipitation across the entire basin area (x_5), principal component F_2 as the empirical parameter representing catchment characteristics (x_6), and principal component F_3 as lake surface evaporation (x_4). The *catchment factor* represents the catchment change; the precipitation and lake evaporation factor represents the climate change. Taking F_1 (precipitation factor), F_2 (catchment factor), and F_3 (evaporation factor) as independent variables and y as a dependent variable, multiple linear regression was carried out, and the partial least squares [43] was adopted to obtain the following equation:

$$y = 384.0088 - 0.3556F_1 - 0.3061F_2 + 1.9858F_3 \tag{22}$$

Table 9. Eigenvalue of the correlation coefficient matrix and variance contribution rate.

Principal Components	The Eigenvalue	Contribution Rate (%)	Cumulative Contribution Rate (%)
F_1	3.6055	60.0919	60.0919
F_2	0.9272	15.4527	75.5446
F_3	0.8867	14.7779	90.3225
F_4	0.3052	5.0872	95.4097
F_5	0.249	4.1507	99.5604
F_6	0.0264	0.4396	100

Table 10. Eigenvectors of the correlation coefficient matrix.

Principal Components	x_1	x_2	x_3	x_4	x_5	x_6
F_1	0.3589	−0.5745	0.3797	0.2533	0.5738	0.0598
F_2	0.4464	0.5176	0.1273	−0.0852	0.119	0.7037
F_3	0.4448	−0.0788	−0.2944	0.6988	−0.47	−0.0069

By converting the influence of principal component factors on underground runoff into percentages, it could be known that evaporation had the largest influence on underground runoff (75%), while precipitation and catchment modifications had a significant influence on the underground runoff (−13.43% and −11.57%, respectively) with an inverse relationship.

Therefore, the contribution rate of climate change to underground runoff can be estimated to be 83.43%, while catchment modifications correspond to −11.57%.

3.2.5. Summary

The partial least squares method was used to analyze the contribution rate of P_l, R_{ls}, E_l and $R_{lg} \pm \varepsilon$ to Δh according to Equation (1), and the relative contribution rate obtained was 20.86%, 29.58%, −41.28% and 9.36%, respectively.

The contribution rates of climate change and catchment variability to (Δh) were obtained by the Equations (3) and (4). It was concluded that the contribution rate to the lake water level variations caused by climate and catchment factors was 93.13%, and 6.87%, respectively (Table 11).

$$R(\Delta h)_c = 20.64\% + 40.84\% + 29.26\% \times 80.19\% + 9.26\% \times 88.44\% = 93.13\% \tag{23}$$

$$R(\Delta h)_u = 29.26\% \times 19.81\% + 9.26\% \times 11.56\% = 6.87\% \tag{24}$$

Table 11. Contribution rate of the hydro-climatic factors.

Contribution Rate	P_l	R_{ls}	E_l	$R_{lg} \pm \varepsilon$	Δh
Climate Changes	100	80.19	100	88.44	93.13
Catchment Modifications	0	19.81	0	11.56	6.87

4. Discussion

4.1. Relationship between the Hydro-Climatic Factors and Lake Water Level Variations

In line with global warming consequences, high temperatures enhanced water vapor transport and redistribution across the entire catchment area, increasing the precipitation rates on the TP (the precipitation recorded in the basin under investigation increased by 1.4347 mm/year). Furthermore, datasets demonstrated that when temperatures increased, the potential evapotranspiration showed a decreasing trend, which confirmed the theory of the "Evaporation Paradox" [54]. Furthermore, according to datasets recorded, the annual maximum temperature and annual minimum temperatures in the basin had risen during the period of study, while the solar duration and wind speed had significantly decreased [55,56]; therefore, these factors may have had a crucial impact on the reduction of potential evapotranspiration [57]. The evaporation rate in the lake decreased at the beginning (1960–1967), then increased (1968–1979), then declined again towards the end of the period under investigation (1980–2016). During the period from 1960 to 2004, the main reasons for the decline of water levels were the overall strong evaporation, the lack of rain and runoff, and the decrease of evaporation since 1980 could not reverse this negative trend. On the other hand, since 2005, the water level of the lake increased and this could have been due to increased precipitation recorded, combined with more runoff and lower evaporation rates.

Song et al. pointed out that runoff in most parts of the world has been decreasing significantly [58], such as in southern Australia, southern Europe, the southern region of South America and the western region of North America [59], as well as in most areas in the North of China (such as the Huaihe River [60]). While previous studies completed by Zhao et al. [61] showed that the annual runoff reduction at four main hydrologic stations in the Yellow River basin (a neighborhood area adjacent to the one investigated by this study) ranged from 17.93% to 40.79%, the results of this study showed that runoff in Qinghai Lake basin presented an upward trend, which was similar to the research of Wang et al. [62]. The results of runoff evolution attribution analysis showed that the increase in precipitation and the decrease of evaporation are the main factors leading to the increase in runoff. The trends of surface runoff and water level variations of lake were strongly consistent, and water level variations were largely affected by the effects of the climate factors. The change in precipitation had a more obvious influence on the runoff in the basins of TP, which are relatively arid, than in the humid area.

It was found that fluctuation of the annual underground runoff was not only affected by precipitation, evaporation and infiltration of surface runoff in the lake area and surrounding areas, but was also related to the fluctuation of water levels of the lake [29], and there was a noticeable connection between surface water and groundwater [63,64], showing that the runoff into the lake had positive and negative values. Since 2005, the decrease of evaporation and the increase in precipitation changed the conversion process of surface runoff and underground runoff, and the negative values increased significantly, indicating that more and more water in the lake was replenishing groundwater.

4.2. Relationship between the Catchment Modifications and Water Level Variations

In general, water level variations in the lake were the result of combined effects due to climate change and human activities. Among them, direct water intake (e.g., agricultural irrigation and drinking water for livestock) only affected the inflow rates into the lake for 4.8% of the total river discharge [28]; hence, it can be considered a negligible factor. By also developing farming areas and reducing forests (especially with local projects started in 2000), direct water intake dropped even more. Therefore, this paper did not consider the influence of direct water intake on water level changes but mainly focused on the influence of climate change and catchment change on water levels.

In those areas potentially affected by major human activities, the changes in ω were mainly manifested in land use changes and vegetation changes. It was found that 72% of the total grassland showed significant improvement in Qinghai Lake basin [65]; however, Qinghai lake basin is located at a high altitude and it is affected by a cold climate, and has low population density (4.08 people/km^2), so there was little impact due to the land use changes. With the implementation of the returning pasture (farmland) to grass project since 1999 and the comprehensive management project in Qinghai Lake basin since 2008, the vegetation condition had been improved, and changes are reflected in ω trends.

According to the research conducted by Yuan [66], the annual average ground temperature in Qinghai Lake area increased by a rate of 0.74 °C/10 years. The depth of the annual average maximum permafrost region was then reduced by the rate of 11.7 cm/10 years and the change of permafrost layers [25,67] could definitely change the hydrologic processes under investigation. By becoming smaller, the permafrost area could not contribute consistently as previously to regulate the runoff of the catchment.

4.3. Uncertainty

The range of hydrological processes typical of great lakes is inherently uncertain, plus data scarcity adds uncertainty and methodological limitations. Firstly, this study assumed that climate and catchment were independent of each other, but the two factors were interacting in nature [68], and the effects could have also cancelled each other out. Secondly, the mathematical statistics method was used to obtain the contribution rate of climate and catchment modifications to underground runoff, but these methods had some limitations within the assumptions. Finally, the presence of permafrost complicated the investigation of hydrological processes and the characterization of their anomaly behaviors associated with climate warming.

Despite these limitations, the main purpose of this study was to use existing monitoring data to analyze the evolution law of Qinghai Lake level, separating the contribution rate of climate and catchment change to the water level variation, and better guide the future water resource management and rational utilization. From 1960 to 2016, the maximum lake area of Qinghai Lake was 4527.3 km^2 (in 1960), while the minimum was 4224.3 km^2 (in 2004), with a difference of 303 km^2, which is equivalent to the size of Co Nag Lake in China, the highest fresh water lake in the world. Therefore, this study can be very useful as a pilot case to associate with other behaviors recorded in lakes with similar and different conditions.

5. Conclusions

This study analyzed the trend of water level variation and hydro-climatic factors in Qinghai Lake Basin from 1960 to 2016 and revealed the main causes affecting the lake water levels. The paper provided a reference base for the development and management of water management in this region and provided important insights that could be applied to other basins.

Conclusions can be summarized as follows:

(1) Qinghai lake experienced severe water level fluctuations in the past 57 years. In period I (1960–2004), the annual water level of the lake declined by 3.46 m at the rate of 7.84 cm/year ($P < 0.001$), while it rose by 1.49 m at the rate of 13.80 cm/year ($P < 0.001$) in period II (2005 to 2016). The variation in water level Δh mainly tended to increase during the study period, and the water quantity of the lake increased, passing temporarily from a deficit rate to a surplus one.

(2) The correlation relationships between E_1, P_1, R_{ls}, $R_{lg} \pm \varepsilon$, ω and Δh followed this order: E_1 (-0.705) $> R_{ls}$ (0.590) $> P_1$ (0.356) $> \omega$ (-0.262) $> R_{lg} \pm \varepsilon$ (0.143). Overall, the major cause of water level change in Qinghai Lake was the combined effect of evaporation (causing a reduction in water quantities), and precipitation (causing a surface runoff increase).

(3) The contribution rate of multiple factors to the water balance of Qinghai Lake Basin to Δh was quantified and it can be classified as follows: E_1 (-49.34%) $> P_1$ (29.82%) $> R_{ls}$ (16.76%) $> R_{lg} \pm \varepsilon$ (4.08%). Among all the factors investigated, E_1 and P_1 belong to climate change factors; hence, by combining the contribution rates of climate change and catchment change induced by human activities to R_{ls}, the results obtained were 80.19%, 19.81%, respectively, and those related to $R_{lg} \pm \varepsilon$ were 8.44%, -11.56%, respectively. Therefore, the contribution rate for both groups of parameters to Δh was in total 93.13%, 6.87%, respectively. The results showed that climate change was the leading cause of significant changes in water levels in the lake.

(4) The impact of global climate change on the hydrology and environment of the Tibetan Plateau was clear, strongly confirming the high sensitivity of great lakes on the Tibetan Plateau to climate change, and solutions need to be adopted to enable strategies to deal and cope with future climate change scenarios.

Author Contributions: J.F. and G.L. and J.Z. designed the study. J.F. performed the analysis and wrote the first draft of the manuscript. M.R., G.M., X.Y. and H.W. contributed to reviewing and editing the final version of the manuscript, software: G.J.

Water **2019**, *11*, 2136

Funding: This research was supported by the National Key Research and Development Program of China (2016YFC0500802), and the Beijing Municipal Education Commission (CEFF-PXM2019_014207_000099), and Special Funds for Scientific Research of Forestry Public Welfare Industry (201404308), and Driving Analysis of Extreme Climate Events on Variation of Runoff and Sediment Discharge in Jinsha River Basin, the National Natural Science Foundation of China (Grant No. 41601279).

Acknowledgments: The authors would like to give special thanks to the Institute of Information Center of Qinghai Hydrographic Bureau, China (ICQHB) for providing data on water levels, evaporation and surface runoff.

Conflicts of Interest: The authors declare no conflict of interest.

References

1. IPCC. *Climate Change 2014: Synthesis Report. Contribution of Working Groups I, II and III to the Fifth Assessment Report of the Intergovernmental Panel on Climate Change*; Core Writing Team; Pachauri, R.K., Meyer, L.A., Eds.; IPCC: Geneva, Switzerland, 2014; pp. 39–55.

2. IPCC. *Special Report on Global Warming of 1.5 °C (SR15)*; Cambridge University Press: Cambridge, UK, 2018; pp. 53–68.

3. Vorosmarty, C.J.; Green, P.; Salisbury, J.; Lammers, R.B. Global water resources: Vulnerability from climate change and population growth. *Science* **2000**, *289*, 284–288. [CrossRef] [PubMed]

4. Scanlon, B.R.; Jolly, I.; Sophocleous, M.; Zhang, L. Global impacts of conversion from natural to agricultural ecosystem on water resources: Quantity versus quality. *Water Resour. Res.* **2007**, *43*, W03437. [CrossRef]

5. IJmker, J.; Stauch, G.; Pötsch, S.; Diekmann, B.; Wünnemann, B.; Lehmkuhl, F. Dry periods on the NE Tibetan Plateau during the late Quaternary. *Palaeogeogr. Palaeoclimatol. Palaeoecol.* **2012**, *346–347*, 108–119. [CrossRef]

6. Clites, A.H.; Smith, J.P.; Hunter, T.S.; Gronewold, A.D. Visualizing relationships between hydrology, climate, and water level fluctuations on Earth's largest system of lakes. *J. Gt. Lakes Res.* **2014**, *40*, 807–811. [CrossRef]

7. Assani, A.A.; Landry, R.; Azouaoui, O.; Massicotte, P.; Gratton, D. Comparison of the Characteristics (Frequency and Timing) of Drought and Wetness Indices of Annual Mean Water Levels in the Five North American Great Lakes. *Water Resour Manag.* **2016**, *30*, 359–373. [CrossRef]

8. Gronewold, A.D.; Fortin, V.; Lofgren, B.; Clites, A.; Stow, C.A.; Quinn, F. Coasts, water levels, and climate change: A Great Lakes perspective. *Clim. Chang.* **2013**, *120*, 697–711. [CrossRef]

9. Zhu, W.B.; Jia, S.F.; Lall, U.; Cao, Q.; Mahmood, R. Relative contribution of climate variability and human activities on the water loss of the Chari/Logone River discharge into Lake Chad: A conceptual and statistical approach. *J. Hydrol.* **2019**, *569*, 519–531. [CrossRef]

10. Paillisson, J.; Marion, L. Water level fluctuations for managing excessive plant biomass in shallow lakes. *Ecol. Eng.* **2011**, *37*, 241–247. [CrossRef]

11. Rohling, E.J. Quantitative assessment of glacial fluctuations in the level of Lake Lisan, Dead Sea rift. *Quat. Sci. Rev.* **2013**, *70*, 63–72. [CrossRef]

12. Ye, X.C.; Li, Y.L.; Li, X.H.; Zhang, Q. Factors influencing water level changes in China's largest freshwater lake, Poyang Lake, in the past 50 years. *Water Int.* **2014**, *39*, 983–999. [CrossRef]

13. Dai, X.; Wan, R.R.; Yang, G.S. Non-stationary water-level fluctuation in China's Poyang Lake and its interactions with Yangtze River. *J. Geogr. Sci.* **2015**, *25*, 274–288. [CrossRef]

14. Dai, X.; Wan, R.R.; Yang, G.S.; Wang, X.L.; Xu, L.G. Responses of wetland vegetation in Poyang Lake, China to water-level fluctuations. *Hydrobiologia* **2016**, *773*, 35–47. [CrossRef]

15. You, Q.L.; Kang, S.C.; Aguilar, E.; Yan, Y.P. Changes in daily climate extremes in the eastern and central Tibetan Plateau during 1961–2005. *J. Geophys. Res.* **2008**, *113*, D07101. [CrossRef]

16. Chen, S.Y.; Liang, T.G.; Xie, H.J.; Feng, Q.S.; Huang, X.D.; Yu, H. Interrelation among climate factors, snow cover, grassland vegetation, and lake in the Nam co basin of the Tibetan plateau. *J. Appl. Remote. Sens.* **2014**, *8*, 084694. [CrossRef]

17. Ke, L.H.; Song, C.Q. Remotely sensed surface temperature variation of an inland saline lake over the central Qinghai–Tibet Plateau. *ISPRS J. Photogramm. Remote Sens.* **2014**, *98*, 157–167. [CrossRef]

18. Zhang, G.Q.; Yao, T.; Xie, H.J.; Zhang, K.; Zhu, F.J. Lakes' state and abundance across the Tibetan Plateau. *Chin. Sci. Bull.* **2014**, *24*, 3010–3021. [CrossRef]

19. Zhang, Z.X.; Chang, J.; Xu, C.Y.; Zhou, Y.; Wu, Y.H.; Chen, X.; Jiang, S.S.; Duan, Z. The response of lake area and vegetation cover variations to climate change over the Qinghai-Tibetan Plateau during the past 30 years. *Sci. Total Environ.* **2018**, *635*, 443–451. [CrossRef]

20. Cui, B.L.; Li, X.Y. The impact of climate changes on water level of Qinghai Lake in China over the past 50 years. *Hydrol. Res.* **2016**, *47*, 532–542. [CrossRef]

21. Lu, F.; Li, X. Climate change and tectonic activity during the early Pliocene Warm Period from the ostracod record at Lake Qinghai, northestern Tibetan Plateau. *J. Asian Earth Sci.* **2017**, *138*, 446–476. [CrossRef]

22. Tang, L.; Duan, X.; Kong, F.; Zhang, F.; Zheng, Y.; Li, Z.; Mei, Y.; Zhao, Y.; Hu, S. Influences of climate change on area variation of Qinghai Lake on Qinghai-Tibetan Plateau since 1980s. *Sci. Rep.* **2018**, *8*, 7331. [CrossRef]

23. Karthe, D.; Malsy, M.; Kopp, B.J.; Minderlein, S.; Hülsmann, L. Assessing water availability and drivers in the context of an integrated water resources management (IWRM): A case study from the Kharaa river basin. *Mong. Open-File Rep.* **2013**, *34*, 5–26.

24. Hampton, S.E.; Izmest'Eva, L.R.; Moore, M.V.; Katz, S.L.; Dennis, B.; Silow, E.A. Sixty years of environmental change in the world's largest freshwater lake-Lake Baikal, Siberia. *Glob. Chang. Biol.* **2008**, *14*, 1947–1958. [CrossRef]

25. Magnuson, J.J.; Robertson, D.M.; Benson, B.J.; Wynne, R.H.; Livingstone, D.M.; Arai, T.; Assel, R.A.; Barry, R.G.; Card, V.V.; Kuusisto, E.; et al. Historical trends in lake and river ice cover in the Northern Hemisphere. *Science* **2000**, *289*, 1743–1746. [CrossRef]

26. Crapper, P.F.; Fleming, P.M.; Kalma, J.D. Prediction of lake levels using water balance models. *Environ. Softw.* **1996**, *11*, 251–258. [CrossRef]

27. Soja, G.; Züger, J.; Knoflacher, M.; Kinner, P.; Soja, A. Climate impacts on water balance of a shallow steppe lake in Eastern Austria (Lake Neusiedl). *J. Hydrol.* **2013**, *480*, 115–124. [CrossRef]

28. Li, X.Y.; Xu, H.Y.; Sun, Y.L.; Zhang, D.S. Lake-level change and water balance analysis at lake Qinghai, west China during recent decades. *Water Resour. Manag.* **2007**, *21*, 1505–1516. [CrossRef]

29. Qu, Y.G. Water balance and forecasting of water level change in Qinghai Lake. *J. Lake Sci.* **1994**, *6*, 298–307.

30. Ma, F.Y. *Reason Analyses and Strategies for the Water Label Drop-Off of the Qinghai Lake*; Xi'an University of Technology: Xi'an, China, 2002; pp. 22–27.

31. Ding, Y.J.; Liu, F.J. Estimating water balance elements in the drainage basin of Qinghai Lake. *Arid Land Geogr. J.* **1993**, *16*, 25–30.

32. Chen, S.P.; Tong, Q.X.; Guo, H.D. *Research on the Mechanism of Remote Sensing Information*; Science Press: Beijing, China, 1998; pp. 84–113.

33. Mann, H.B. Nonparametric tests against trend. *Econom. J. Econom. Soc.* **1945**, *13*, 245–259. [CrossRef]

34. Kendall, M.G. *Rank Correlation Methods*, 3rd ed.; Hafner Publishing Company: New York, NY, USA, 1962; pp. 92–126.

35. Mitchell, J.M.; Dzerdzeevskii, B.; Flohn, H.; Hofmeyr, W.L.; Lamb, H.H.; Rao, K.N.; Wallen, C.C. *Climatic Change, WMO Technical Note 79*; World Meteorological Organization: Geneva, Switzerland, 1966; pp. 1–79.

36. Liang, L.Q.; Li, L.J.; Liu, Q. Temporal variation of reference evapotranspiration during 1961–2005 in the Taoer River basin of Northeast China. *Agric. For. Meteorol.* **2010**, *150*, 298–306. [CrossRef]

37. Wei, F.Y. *Statistical Techniques of Modern Climatic Diagnosis and Forecasting*; China Meteorological Press: Beijing, China, 1999; pp. 63–65.

38. Shang, X.X.; Jiang, X.H.; Jia, R.N.; Chen, W. Land Use and Climate Change Effects on Surface Runoff Variations in the Upper Heihe River Basin. *Water* **2019**, *11*, 344. [CrossRef]

39. Yang, L.S.; Feng, Q.; Yin, Z.L.; Wen, X.H.; Si, J.H.; Deo, R.C. Identifying separate impacts of climate and land use/cover change on hydrological processes in upper stream of Heihe River, Northwest China. *Hydrol. Process.* **2017**, *31*, 1100–1112. [CrossRef]

40. Jolliffe, I.T. *Principal Component Analysis*; Springer: New York, NY, USA, 1986; pp. 111–137.

41. Deng, J.L. Control problems of grey systems. *Syst. Control Lett.* **1982**, *1*, 288–294.

42. Plackett, R.L. The discovery of the method of Least Squares. *Biometrika* **1972**, *59*, 239–251.

43. Wold, S.; Ruhe, A.; Wold, H.; Dunn, I.W.J. The collinearity problem in linear regression, the partial least squares (PLS) approach to generalized inverses. *Siam J. Sci. Stat. Comput.* **1984**, *5*, 735–743. [CrossRef]

44. Liu, X.M.; Zhang, D.; Luo, Y.Z.; Liu, C.M. Spatial and temporal changes in aridity index in northwest China: 1960 to 2010. *Theor. Appl. Climatol.* **2013**, *112*, 307–316. [CrossRef]

45. Li, B.; Li, L.J.; Qin, Y.C.; Liang, L.Q.; Li, J.Y.; Liu, Y.M. Impact of climate variability on streamflow in the upper and middle reaches of the Taoer River based on the Budyko hypothesis. *Resour. Sci.* **2011**, *33*, 70–76.

46. Budyko, M.I. *Climate and Life*; Academic Press: New York, NY, USA, 1974; pp. 317–508.

47. Fu, B.P. On the calculation of the evaporation from land surface. *Sci. Atmos. Sin.* **1981**, *5*, 23–31.

48. Yang, D.W.; Shao, W.W.; Yeh, P.J.F.; Yang, H.B.; Kanae, S.; Oki, T.K. Impact of vegetation coverage on regional water balance in the nonhumid regions of China. *Water Resour. Res.* **2009**, *45*, W00A14. [CrossRef]

49. Donohue, R.J.; Roderick, M.L.; McVicar, T.R. Roots, storms and soil pores: Incorporating key ecohydrological processes into Budyko's hydrological model. *J. Hydrol.* **2012**, *436–437*, 35–50. [CrossRef]

50. Allen, R.G.; Pereira, L.S.; Raes, D.; Smith, M. *Crop Evapotranspiration: Guidelines for Computing Crop Water Requirements-FAO Irrigation and Drainage Paper 56*; FAO: Rome, Italy, 1998; Volume 300, p. D05109.

51. Liu, C.M.; Zhang, D. Temporal and spatial change analysis of the sensitivity of potential evapotranspiration to meteorological influencing factors in China. *Acta Geogr. Sin.* **2011**, *66*, 579–588.

52. Zhu, G.F.; He, Y.Q.; Pu, T.; Wang, X.F.; Jia, W.X.; Li, Z.S.; Xin, H.J. Spatial distribution and temporal trends in potential evaporation over Hengduan Mountains Region from 1960 to 2009. *Acta Geogr. Sin.* **2011**, *66*, 905–916.

53. El-Zehairy, A.A.; Lubczynski, M.W.; Gurwin, J. Interactions of artificial lakes with groundwater applying an integrated MODFLOW solution. *Hydrogeol. J.* **2018**, *26*, 109–132. [CrossRef]

54. Michael, L.R.; Graham, D.F. The cause of decreased pan evaporation over the past 50 years. *Science* **2002**, *298*, 1410–1411.

55. Hao, X.N.; Li, Y.T.; Li, B.Y. Variation Characteristics of Pan Evaporation and Its Influence Factor. *J. Anhui Agric. Sci.* **2011**, *39*, 19405–19409.

56. Liu, B.K. *Spatial and Temporal Variation Characteristics of Grassland and Lake in Qinghai Lake Basin under Climate Change*; Lanzhou University: Lanzhou, China, 2016; pp. 16–19.

57. Yin, Y.H.; Wu, S.H.; Dai, E.F. Determining factors in potential evapotranspiration changes over China in the period 1971–2008. *Chin. Sci Bull.* **2010**, *55*, 2226–2234. [CrossRef]

58. Song, X.M.; Zhang, J.Y.; Zhan, C.S.; Liu, C.Z. Review for impacts of climate change and human activities on water cycle. *J. Hydraul. Eng.* **2013**, *44*, 779–790.

59. Milly, P.C.D.; Dunne, K.A.; Vecchia, A.V. Global pattern of trends in streamflow and water availability in a changing climate. *Nature* **2005**, *438*, 347–350. [CrossRef]

60. Committee of China's National Assessment Report on Climate Change. *China's National Assessment Report on Climate Change*; Science Press: Beijing, China, 2007; pp. 202–211.

61. Zhao, Y.; Hu, C.H.; Zhang, X.M.; Wang, Y.S.; Cheng, C.; Yin, X.L.; Xie, M. Analysis on runoff and sediment regimes and its causes of the Yellow River in recent 70 years. *Trans. Chin. Soc. Agric. Eng.* **2018**, *34*, 112–119.

62. Wang, H.; Liu, J.F.; Xie, Z.Y.; Ma, L.J. Trend and attribution analysis of runoff in Qinghai Lake basin. *Water Resour. Power* **2018**, *36*, 19–21, 32.

63. Carolina, G.A.; Jackson, C.R. Potential Impacts of Climate Change on Groundwater Supplies to the Doñana Wetland, Spain. *Wetlands* **2011**, *31*, 907–920.

64. Kirshen, P.H. Potential Impacts of Global Warming on Groundwater in Eastern Massachusetts. *J. Water Resour. Plan. Manag.* **2002**, *6*, 216–219. [CrossRef]

65. Wang, X.L.; Liang, T.G.; Xie, H.J.; Huang, X.D.; Lin, H.L. Climate-driven changes in grassland vegetation, snow cover, and lake water of the Qinghai Lake basin. *J. Appl. Remote Sens.* **2016**, *10*, 036017. [CrossRef]

66. Yuan, Y.M. Analysis of temperature, ground temperature and permafrost variation in Qinghai Lake area in recent 30 years. *Qinghai Meteorol.* **2016**, *2*, 20–22.

67. Törnqvist, R.; Jarsjö, J.; Pietroń, J.; Bring, A.; Rogberg, P.; Asokan, S.M.; Destouni, G. Evolution of the hydro-climate system in the Lake Baikal basin. *J. Hydrol.* **2014**, *519*, 1953–1962. [CrossRef]

68. Wang, F.Y.; Duan, K.Q.; Fu, S.Y.; Gou, F.; Liang, W.; Yan, J.W.; Zhang, W.B. Partitioning climate and human contributions to changes in mean annual streamflow based on the Budyko complementary relationship in the Loess Plateau, China. *Sci. Total Environ.* **2019**, *665*, 579–590. [CrossRef]

Article

Modelling Effects of Rainfall Patterns on Runoff Generation and Soil Erosion Processes on Slopes

Qihua Ran [1], Feng Wang [1] and Jihui Gao [1,2,3,*]

[1] Institute of Hydrology and Water Resources, Zhejiang University, Hangzhou 310058, China; ranqihua@zju.edu.cn (Q.R.); wfno1@163.com (F.W.)
[2] State Key Laboratory of Hydraulics and Mountain River Engineering, Sichuan University, Chengdu 610065, China
[3] School of Water Resource & Hydropower, Sichuan University, Chengdu 610065, China
* Correspondence: jgao@scu.edu.cn

Received: 19 September 2019; Accepted: 22 October 2019; Published: 25 October 2019

Abstract: Rainfall patterns and landform characteristics are controlling factors in runoff and soil erosion processes. At a hillslope scale, there is still a lack of understanding of how rainfall temporal patterns affect these processes, especially on slopes with a wide range of gradients and length scales. Using a physically-based distributed hydrological model (InHM), these processes under different rainfall temporal patterns were simulated to illustrate this issue. Five rainfall patterns (constant, increasing, decreasing, rising-falling and falling-rising) were applied to slopes, whose gradients range from 5° to 40° and projective slope lengths range from 25 m to 200 m. The rising-falling rainfall generally had the largest total runoff and soil erosion amount; while the constant rainfall had the lowest ones when the projective slope length was less than 100 m. The critical slope of total runoff was 15°, which was independent of rainfall pattern and slope length. However, the critical slope of soil erosion amount decreased from 35° to 25° with increasing projective slope length. The increasing rainfall had the highest peak discharge and erosion rate just at the end of the peak rainfall intensity. The peak value discharges and erosion rates of decreasing and rising-falling rainfalls were several minutes later than the peak rainfall intensity.

Keywords: rainfall patterns; rainfall-runoff; soil erosion; slope length; slope gradient; InHM

---

1. Introduction

Rainfall patterns and landform characteristics are controlling factors of the runoff and soil erosion processes in natural catchments [1,2]. Due to climatic change and climatic variability, rainfall events commonly show great temporal variation in intensity, especially in hilly areas [3,4] and the peak rainfall rates within an event may dozens of times higher than the mean event rate [1,5]. Although the temporal distribution of an individual rainfall event is diverse, some patterns of such distribution in a region can be derived based on historical data (e.g., [6,7]).

Previous studies have recognized that the rainfall patterns greatly affect the runoff generation and soil erosion processes (e.g., [8,9]). Parsons and Stone [10] adopted five rainfalls with different patterns but the same total kinetic energy to the soil surface. They found that the soil erosion amount under a constant-intensity storm are reduced by about 25% compared to varied-intensity storms, and that the eroded sediments are coarser under the constant-intensity pattern. An et al. [8] used the similar rainfall patterns and indicated that, although the total runoff was nearly not affected by the rainfall pattern, the varied intensity patterns yield 1–5 times more soil losses than even-intensity patterns and the rising pattern resulted in a consistently higher soil loss relative to the other four rainfall patterns. Conversely, Dunkerley [3] performed rainfall simulations of varying intensity profile in a dryland intergrove (runoff source area) and discovered that the late peak events showed runoff ratios that were more than double those of the early peak events and the constant rainfall yielded the lowest total runoff, the lowest peak runoff rate. The reason was inferred to be the reductions in soil infiltration capacity during late rainfall. Zhai et al. [11] applied a distributed hydrological model at the basin scale, and found that the rainfall patterns have significant impact on the rainfall threshold of flood warning, which the flood rainfall threshold of advanced rainfall is the highest.

However, in most studies on rainfall pattern at plot scale, spatially distributed results of infiltration and soil erosion processes were not carefully considered. The temporal variation of precipitation can lead to corresponding spatial and temporal variations of infiltration, overland flow generation, and further soil erosion. Only considering runoff and soil erosion data at plot outlet, like many previous study did, will miss some important information (e.g., distributed cumulative infiltration or erosion depth) within the study area for comprehensive interpreting the influence of rainfall pattern on runoff and soil erosion processes.

In recent years, many studies on slopes have reported that observed runoff coefficient in Hortonian runoff processes decreases with increasing slope length, and variance of runoff reduces as slope scale increases (e.g., [12,13]). A reason was that the runoff generated upslope can infiltrate in downslope areas, which was called the run-on infiltration [14] or the re-infiltration [15]. Although rainfall characteristics such as duration were one of the major factors affecting runoff generation at different slope scales (e.g., [16]), it is still unknown how slope length influences the effect of temporal rainfall pattern on rainfall-runoff and soil erosion processes.

Slope steepness was an important topographic factor of hillslope rainfall-runoff and soil erosion processes. At plot scale, contradictory results were derived regarding slope effects on infiltration (e.g., [17,18]) and soil erosion (e.g., [19,20]). Besides, some researchers observed that runoff volume and soil loss on slopes increases with increasing slope angle till a critical slope angle of 20°–30° (e.g., [21]), while others reported that soil erosion is not correlated with slope gradient in tilled fields (e.g., [22]). However, a majority of the studies focusing on slope steepness neglected the influence of rainfall temporal variation. There is a lack of systematically studies on the effect of slope gradient under different rainfall patterns.

Numerical modelling is an effective approach to reveal spatial and temporal impacts of rainfall patterns on infiltration, overland flow and soil erosion processes at slopes with wide ranges of steepness and length, which can broaden the limitation of the artificial rainfall experiment (e.g., a plot with a few meters long [8]). Further, strictly controlling factors such as initial condition and soil property, the effect of rainfall patterns can be specifically focused. As a mature hydraulic model, Integrated Hydrology Model (InHM) can quantitatively simulate surface (2D) and subsurface (3D) hydrologic responses to rainfall in a fully coupled approach [23,24]. Previously, InHM has been successfully applied in the simulations of hillslope hydrology and slope failure (e.g., [23,25]). As this physics-based hydrological model employs fundamental physics laws to describe natural processes [26], its output results have clear physical meanings and can be used to generalize our understanding of rainfall pattern effects on runoff and soil erosion processes.

The main purpose of this study is to investigate the impact of rainfall temporal patterns on infiltration, runoff generation and soil erosion on slopes with a range of slope lengths and gradients, using a physically based modelling approach. These modelling results are expected to improve the theoretical basis for hillslope runoff and soil erosion prediction, which will be further helpful in soil conservation planning and land management.

2. Methodology

2.1. InHM Model

The Integrated Hydrology Model (InHM) was originally developed by VanderKwaak [27], which exceeds the specifications of the hydrologic-response model proposed by Freeze and Harlan (1969) that the model based upon the numerical solution to an almost-complete set of coupled partial differential equations which describe water movement processes at surface and in unsaturated and saturated subsurface [26]. With the advantage of the model that it doesn't need a priori assumption of a dominant runoff-generation mechanism [28,29], InHM is capable of accurately simulating dynamic infiltration, runoff and sediment processes under temporal varying rainfall. Previous studies have shown that the calibrated model reproduced accurately measured runoff and soil erosion results on semiarid hillslopes during constant-intensity rainfall-simulation events [28,30]. The equations and a detailed description of InHM can be found in VanderKwaak [27], VanderKwaak and Loague [26] and Heppner et al. [30].

2.2. Model Setup

Runoff and soil erosion processes were simulated and analysed on slopes with four horizontal projective slope lengths (25 m, 50 m, 100 m and 200 m), which were all 40 m wide and 3 m deep. For each horizontal projective length, nine slope gradients from 5° to 40° in 5° increments were considered, and identical rainfall amount reviewed on slope surfaces was ensured for different slope gradients due to the constant horizontal projective length. In total, for each rainfall scenario the runoff and erosion processes were simulated for 36 slopes. The schematic representation of the 200 m slope used in the simulation was shown in Figure 1 as an example. To avoid the influence of the downstream outlet boundary, the overall projective slope length of the 3D finite element meshes were 220 m. The vertical nodal spacing (Dz) in the mesh varies from 0.01 to 0.1 m; the horizontal nodal spacing (Dx, Dy) is 0.5 m. The 3-m mesh depth was sufficient for short-time simulation of rainfall-runoff events and deep groundwater movement was not considered in this study. The boundary conditions contain impermeable boundaries (A-E-H-D, B-F-G-C, E-F-G-H), flux boundary (A-E-F-B) and permeable boundary (A-B-C-D, C-G-H-D). The total numbers of the nodes and the elements of the 200 m slope 3D meshes are 35,721 and 70,400, respectively.

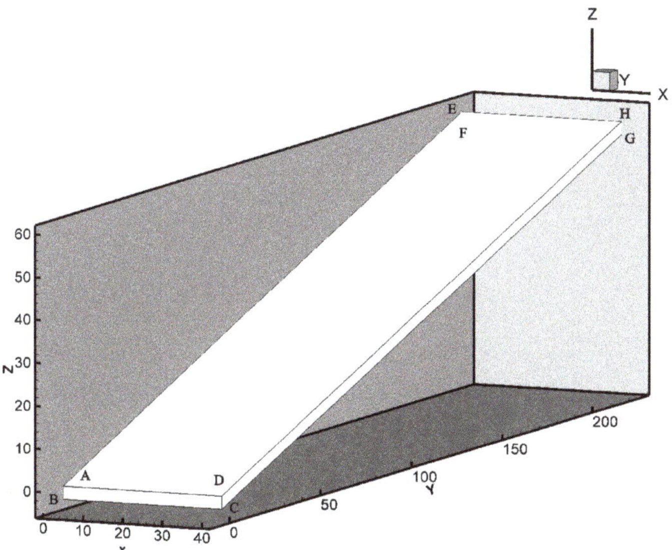

Figure 1. The schematic representation of the 200 m slope used the InHM modeling, and A-H represents each of the boundary nodes.

In this modelling study, the parameters of the plots (Table 1) were obtained from Ran et al. [28] who calibrated and validated the InHM parameters via the plot-scale experiments of Horton overland flow and surface erosion on the silty clay loam slopes within the Los Alamos National Laboratory [31]. The slope gradient of their experimental plot was 25.8° and its vegetation coverage was 61%, which was an ideal condition for hillslope runoff and soil erosion study. A 1-h 40 mm h^{-1} rainfall event is approximately equivalent to a 5-year return period event in that area.

Table 1. Parameters for the simulation.

Parameters	Value
Porosity	0.46
Species average grain diameter	2.0×10^{-5} m
Manning coefficient	0.275
Initial water table	−5 m
Mobile water depth (i.e., depression storage)	5×10^{-4} m
Height of microtopography	0.01 m
Soil-water retention function (van Genuchten approach [32])	α: 1.0 n: 1.23 θ_r: 0.088
Saturated hydraulic conductivity (Ks)	10.8 mm h^{-1}
Rainsplash coefficient	2.93

2.3. Rainfall Scenarios

In this study, five temporal patterns of rainfall intensity were designed: constant rainfall intensity, increasing rainfall intensity, decreasing rainfall intensity, rising-falling rainfall intensity and falling-rising rainfall intensity (Figure 2).

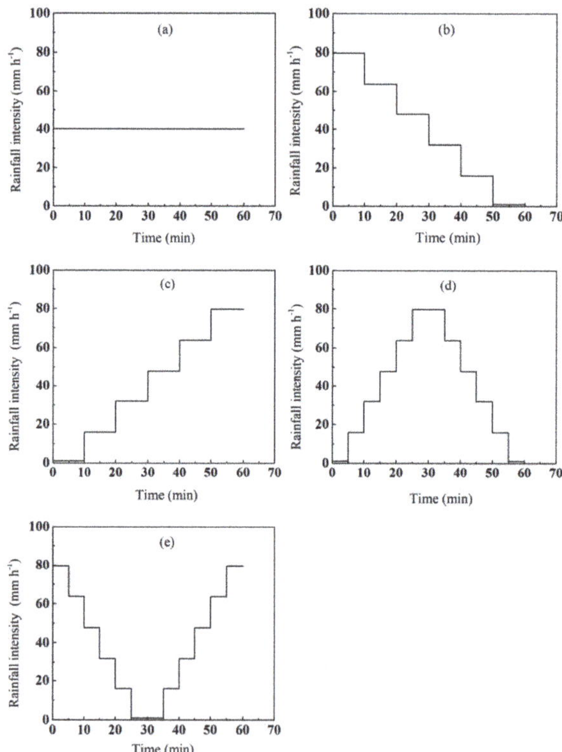

Figure 2. Five different rainfall patterns. (**a**) constant, (**b**) decreasing, (**c**) increasing, (**d**) rising-falling and (**e**) falling-rising.

Similar rainfall scenarios were also adopted by other researchers foucusing on temporal patterns of rainfall (e.g., [3,33]). All the five rainfall patterns had a 1-h duration and 40 mm rainfall depth. Previously, many studies focusing on rainfall patterns adopted extremely high rainfall intensities in their rainfall simulations [1]. For instance, Flanagan et al. [34] used rainfall rates that peaked at 250 mm h^{-1} and Parsons and Stone [10] adopted rainfall rates in the range 46.4–170.8 mm h^{-1}. The rainfall intensities of this study were within the range of 1–80 mm h^{-1}, which reprensents a more general condition.

3. Results

3.1. Hydrological Responses

The modelling results of runoff in the rainfall scenarios at the different projective slope lengths were summarized in Table 2. Generally, the increasing and rising-falling rainfalls had the largest total runoff, and the constant and falling-rising rainfalls had the least total runoff. The constant rainfall had the lowest total runoff when the slope length was shorter than 100 m. At a same slope gradient, the relative difference between the total runoff of different rainfall patterns was up to 111% (Table 2). The runoff coefficient at different projective slope lengths were shown in Figure 3. The runoff coefficient increased with increasing slope gradient until 15° and then decreased to the lowest value at 40° slope. The runoff coefficients of the rising-falling rainfalls were close to those of the increasing rainfalls at the 25 m slope, then they gradually became higher as slope length increased. As the projective slope length increased, the runoff coefficient of all the five rainfall patterns decreased, and that of the falling-rising rainfall decreased most greatly.

The peak discharge of the constant rainfall was the lowest (e.g., Figure 4). For increasing rainfall, the discharge rate kept increasing with time and reached the highest peak value among the five rainfall patterns at the end of the event. For the decreasing and rising-falling rainfalls, the peak discharges (Figure 4) were several minutes after the rainfall intensity decreased (Figure 2). The peak discharge of the falling-rising rainfall dropped a lot compared with other rainfalls as the projective slope length increased from 25 m to 200 m.

Table 2. Simulated results of total runoff (m^3) at different projective slope lengths.

Projective Slope Length (m)	Rainfall Pattern	5°	10°	15°	20°	25°	30°	35°	40°
	Constant	13.3	14.1	14.3	14.3	14.0	13.5	12.7	11.8
	Decreasing	15.9	16.5	16.6	16.5	16.1	15.5	14.6	13.5
25	Increasing	17.3	18.1	18.3	18.3	18.0	17.4	16.7	15.7
	Rising-falling	17.6	18.2	18.3	18.1	17.8	17.2	16.4	15.3
	Falling-rising	14.3	15.4	15.8	15.8	15.5	15.0	14.3	13.3
	Constant	21.7	23.4	23.7	23.4	22.5	21.1	19.2	16.9
	Decreasing	27.8	29.2	29.3	28.7	27.7	26.1	24.0	21.4
50	Increasing	29.4	31.3	31.9	31.7	31.0	29.7	28.0	25.7
	Rising-falling	30.5	32.3	32.6	32.3	31.4	30.1	28.2	25.7
	Falling-rising	22.0	24.6	25.5	25.4	24.7	23.5	21.7	19.6
	Constant	32.8	36.9	37.9	37.2	35.3	32.3	28.3	23.5
	Decreasing	45.0	49.3	50.3	49.5	47.4	44.1	39.6	33.9
100	Increasing	47.3	51.8	53.2	52.9	51.3	48.5	44.6	39.5
	Rising-falling	49.6	54.3	55.6	55.3	53.6	50.8	46.8	41.5
	Falling-rising	30.0	35.6	37.6	37.7	36.4	33.9	30.4	26.2
	Constant	43.4	51.5	53.9	52.9	49.2	43.3	35.7	27.4
	Decreasing	65.9	74.9	77.8	76.8	73.0	66.3	57.2	45.7
200	Increasing	69.7	79.9	82.8	82.0	78.3	72.1	63.4	52.2
	Rising-falling	75.4	84.6	87.3	86.6	83.0	76.9	68.0	56.6
	Falling-rising	35.8	44.9	48.4	48.4	46.0	41.4	35.5	29.1

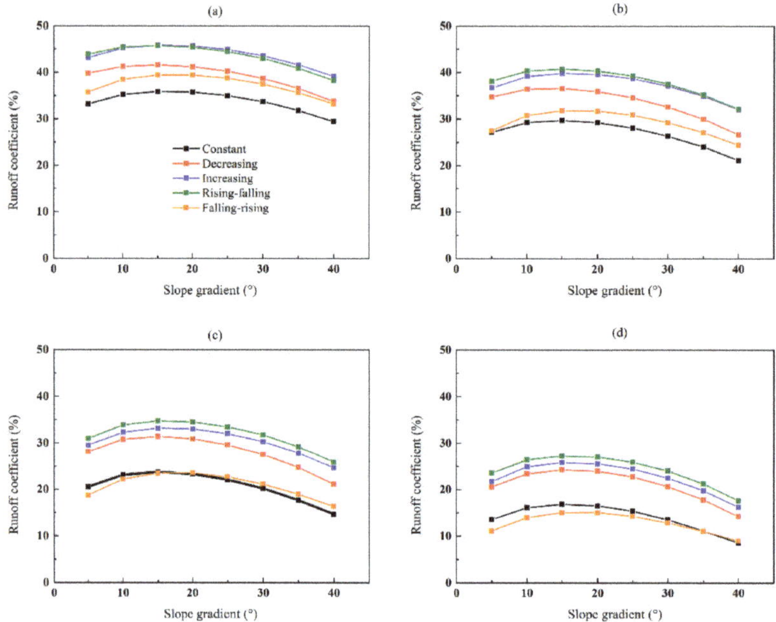

Figure 3. The runoff coefficient at different slopes with projective slope lengths of (**a**) 25 m, (**b**) 50 m, (**c**) 100 m and (**d**) 200 m.

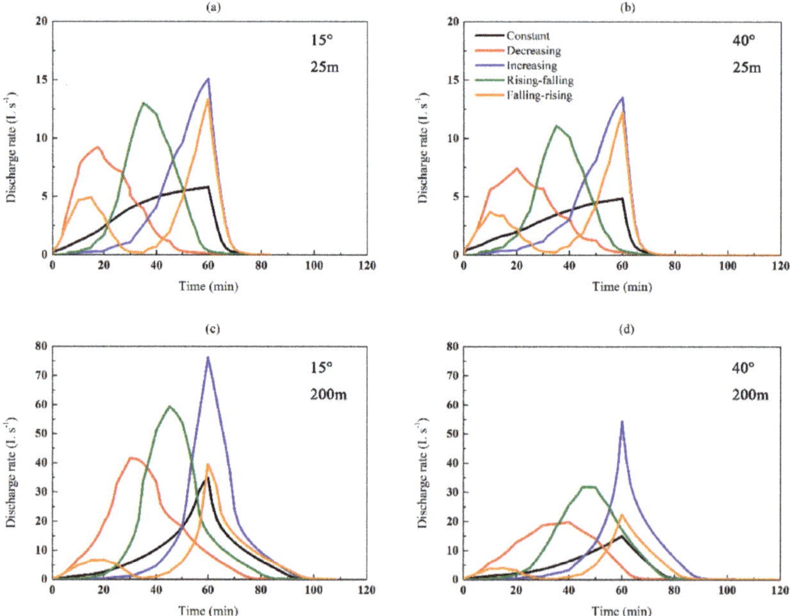

Figure 4. The hydrographs of five different rainfall patterns with 25 m and 200 projective slope lengths at 15° slope (**a**,**c**) and 40° slope (**b**,**d**), respectively.

3.2. Soil Erosion

The soil erosion results of the rainfall scenarios at different projective slope lengths were summarized in Table 3. Similar to the hydrologic-response results, the increasing and rising-falling rainfalls had the largest total soil erosion, and the constant and falling-rising rainfalls had the least total soil erosion. This is due to the fact that rainfall-runoff was the controlling factor for soil erosion at the plot scale. Besides, at a same slope gradient, the relative difference between the soil erosion amounts of different rainfall patterns was up to 381% (Table 3). In general, relative differences in soil erosion among the five rainfall patterns were higher than relative differences in runoff under the same condition. The soil erosion amount at the different projective slope lengths were shown in Figure 5. Different from the runoff coefficient, the slope gradients that had the peak values of soil erosion amount were around 25°–40°, and this critical slope gradient decreased as projective slope length increased. Soil erosion amount of the rising-falling rainfalls became higher than that of the increasing rainfalls when slope length over 50 m. When the slope lengths were 25–100 m, soil erosion amount of the constant rainfalls were the lowest among the five rainfall patterns; while it became higher than that of the falling-rising rainfall when the slope length was 200 m. The sedigraphs were similar with the corresponding hydrographs. However, the erosion rate at the 40° slope with 25 m slope length was much higher than that at the 15° slope (Figure 6a,b), even the hydrographs at the two slopes only had small differences (Figure 4a,b). An example under rising-falling rainfall with 25 m slope length was shown in Figure 7.

Table 3. Simulated results of total soil erosion (kg) at different projective slope lengths.

Projective Slope Length (m)	Rainfall Pattern	5°	10°	15°	20°	25°	30°	35°	40°
	Constant	66	201	385	610	867	1149	1426	1666
	Decreasing	116	364	703	1116	1586	2087	2546	2574
25	Increasing	179	599	1198	1961	2888	3715	3965	3888
	Rising-falling	175	567	1114	1797	2607	3446	3717	3620
	Falling-rising	102	352	722	1191	1757	2329	2592	2618
	Constant	227	794	1590	2530	3527	4465	4800	4248
	Decreasing	411	1452	2935	4729	6704	7837	7534	6590
50	Increasing	586	2116	4442	7445	9636	10,154	9842	9102
	Rising-falling	581	2127	4426	7338	10,017	10,480	9993	9123
	Falling-rising	261	1058	2314	3931	5598	6258	6210	5762
	Constant	547	2199	4662	7566	10,401	10,845	9622	7629
	Decreasing	1060	4111	8795	14,523	18,102	17,590	15,878	13,242
100	Increasing	1571	6053	12,787	17,745	19,753	19,720	18,447	16,318
	Rising-falling	1559	5962	12,579	19,620	21,465	21,173	19,703	17,396
	Falling-rising	487	2186	4993	8389	10,827	11,320	10,589	9150
	Constant	955	4250	9193	14,618	17,358	16,384	13,373	9548
	Decreasing	2205	9037	19,436	28,850	30,831	28,977	24,982	19,345
200	Increasing	2895	13,002	27,699	32,262	33,449	31,753	28,030	22,777
	Rising-falling	3179	13,562	29,359	34,865	36,063	34,389	30,605	25,198
	Falling-rising	660	3370	7866	12,564	15,046	14,901	13,047	10,428

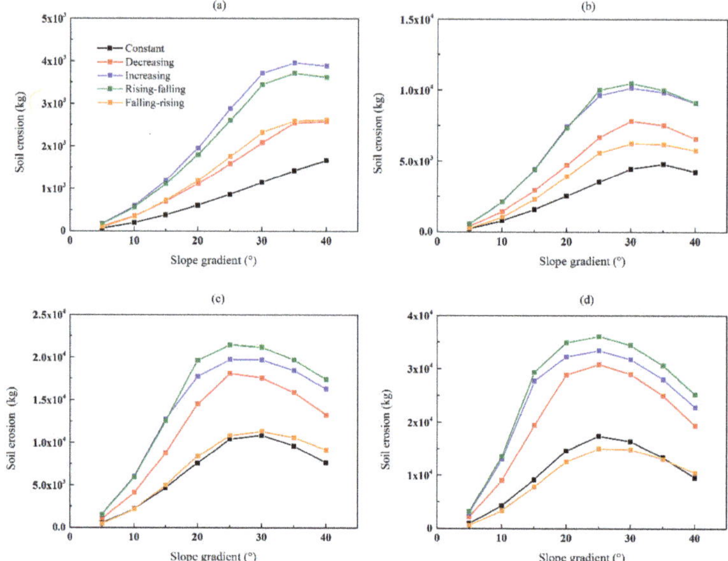

Figure 5. The soil erosion amount at different slopes with projective slope lengths of (**a**) 25 m, (**b**) 50 m, (**c**) 100 m and (**d**) 200 m.

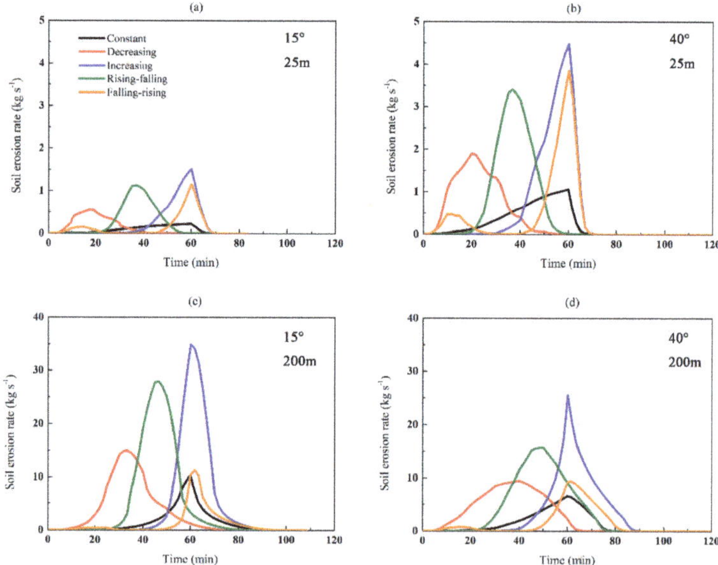

Figure 6. The sedigraphs of five different rainfall patterns with 25 m and 200 projective slope lengths at 15° slope (**a**,**c**) and 40° slope (**b**,**d**), respectively.

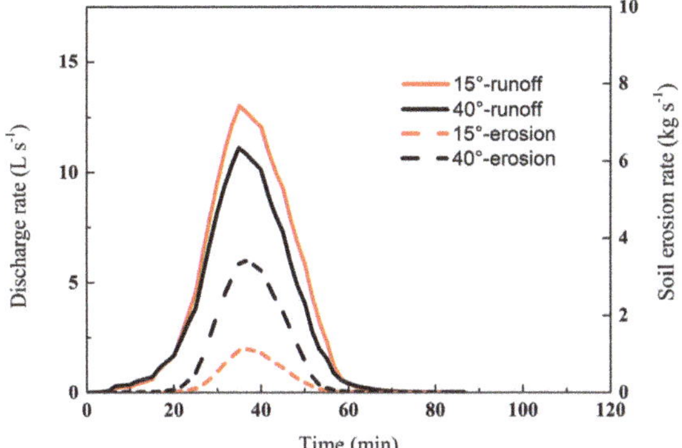

Figure 7. The comparison between the hydrographs and the sedigraphs under rising-falling rainfall at 25 m slope.

4. Discussion

4.1. Effect of Rainfall Patterns on Total Runoff and Soil Erosion at Different Slope Lengths

For constant rainfalls, their total runoff and soil erosion were lower than those of increasing rainfalls, decreasing rainfalls and rising-falling rainfalls. It was consistent with previous studies. Dunkerley [3] observed that the runoff ratios of varying intensity rainfalls were 85–570% larger than that of constant rainfall and Wang et al. [35] found that the constant rainfall produced the lowest sediment yield at around 61.8% of the average soil loss for the increasing rainfall. Comparing the cumulative infiltration of increasing rainfalls, decreasing rainfalls and rising-falling rainfall with that of constant rainfalls (Figure 8), their gaps almost reached the highest value when slope was around 12 m, and then gradually stabilized as projective slope length increased. Thus, the differences in total runoff and soil erosion between these inconstant and constant rainfalls increased with increasing slope length. However, the total runoff and soil erosion amount of the constant rainfall become larger than those of falling-rising rainfall when the projective slope length was over 100 m (Figures 3 and 5). Previous experimental studies did not find this as their plots were much shorter than 100 m. For instance, Wang et al. [35] adopted 2 m-long flume and Parsons and Stone [10] used 2.45 m-long flume in their rainfall exerpiments.

The cumulative infiltration of the constant rainfall along the slope axis was the highest, until the projective slope length was around 50 m as the cumulative infiltration of falling-rising rainfall became larger (Figure 8). When the slope was short, compared with the constant rainfall, there was not much water infiltrated downstream the slope during the low rainfall intensity period (i.e., rainfall intensity = 1 mm h⁻¹) under the falling-rising rainfall, because the runoff generated during the first rainfall peak quickly flowed out of the slope. Thus, the cumulative infiltration of falling-rising rainfall was lower than the constant rainfall when projective slope length was shorter than 50 m (Figure 8), leading to larger total runoff, peak discharge (Figures 3a and 4a) and erosion depth (Figure 9) than those of the constant rainfall. As slope length increased, for the falling-rising rainfall, the runoff generated during the first half of the event lasted longer before flowed out so that more water infiltrated downstream the slope during the low rainfall intensity period. Meanwhile, the recession period of the falling-rising rainfall became much longer than that of the constant rainfall (e.g., Figure 4c,d), which dramatically increased the cumulative infiltration (Figure 8). Due to these reasons, when slope length was over 100 m, the runoff coefficient of the constant rainfall became higher than that of the falling-rising rainfall (Figure 3c,d). Because of less infiltration downstream, less sediment deposited downstream the slope under the constant rainfall than the falling-rising rainfall (Figure 9), resulting in higher soil erosion amount when the slope length was over 100 m (Figure 5c,d). The results of the falling-rising rainfall indicated that, the runoff and soil erosion amount of such multi-peak rainfall may even lower than those of the uniform rainfall, especially when the slope length was long.

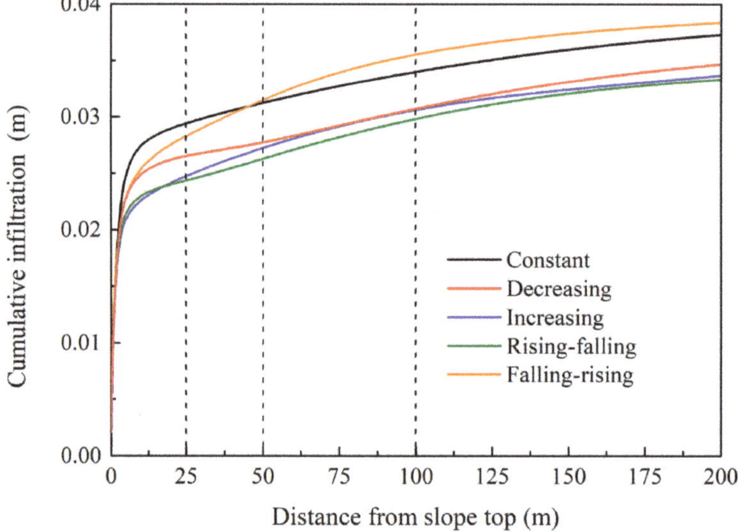

Figure 8. The cumulative infiltration distribution along the slope axis at the 15° slope.

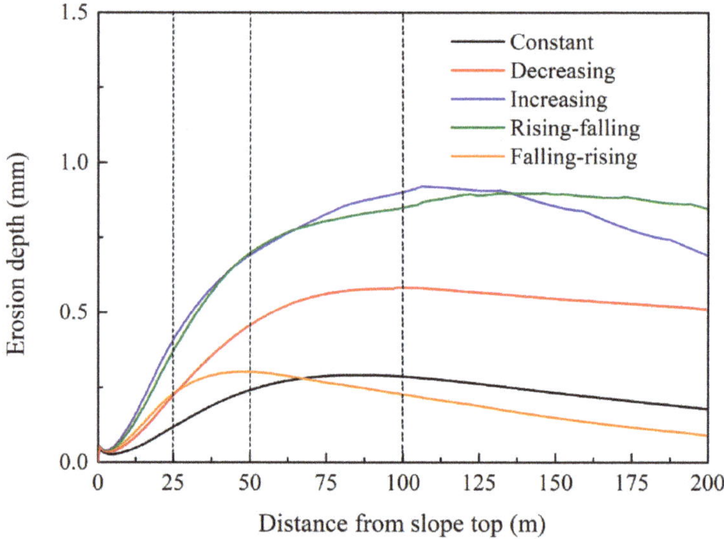

Figure 9. The erosion depth distribution along the slope axis at the 15° slope.

The total runoff and soil erosion of decreasing rainfall were much lower than those of increasing rainfall and rising-falling rainfall. It was in agreement with the experimental research on small plots by Dunkerley [3], which found the runoff ratio of late peak rainfall was double that of early peak rainfall, and numerical simulation research by Zhai et al. [11], which reported that the delayed rainfall pattern yield higher flood volume and peak than the early peak pattern. The reason was that the soil infiltrability remained high in the early part of the event under decreasing rainfall [36].

Higher cumulative infiltration of decreasing rainfall compared with those of increasing rainfall and rising-falling rainfall was obvious, especially when the projective slope length was shorter than 50 m (e.g., Figure 8). Moreover, the smaller total runoff and peak discharge under decreasing rainfall (e.g., Figure 4a,c) led to shallower erosion depth (Figure 8), due to much lower stream power and sediment transport capacity.

The simulation results also indicated that the rising-falling rainfall had the highest runoff and soil erosion amount than other rainfall patterns when projective slope length was over 50 m (Tables 2 and 3), which was not consistent with previous studies. Dunkerley [3] indicated that the late peak rainfall had the highest peak runoff rate and runoff ratio. An et al. [8] reported that in their rainfall experiments the soil loss under increasing rainfall were the highest. The main reason was that, compared with rising-falling rainfall, the increasing rainfall had much longer recession period when slope was long (e.g., Figure 4c,d), leading to larger amount of infiltration and sediment deposition.

4.2. The Impact of Slope Gradient on Total Runoff and Erosion under Five Rainfall Patterns

From Figures 3 and 5 it can be seen that, for all the five different rainfall patterns, total runoff or total soil erosion showed a trend that it increased with increasing slope gradient and then gradually decreased after a critical slope. The critical slope of the total runoff was 15°, which was independent of rainfall pattern and projective slope length. Taking the rising-falling rainfall as an example, Figure 10 shows the cumulative infiltration distribution along the slope axis at 10°–20° slopes. For slopes lower than 15° (e.g., 10°), overland flow velocity was slower than that on the 15° slope, thus leading to more infiltration, especially when the projective slope length was over 30 m (Figure 10). For slope steeper than 15° (e.g., 20°), because the slope length was longer than the 15° slope, overland flow had to travelled longer path to reach the outlet and caused more infiltration. It can be seen in Figure 10 that cumulative infiltration difference between 15° and 20° slopes was mainly lay in area 10–50 m and 150–200 m from the slope top. Wu et al. [37] also found the critical slope for runoff rate was around 11° regardless of rainfall duration and slope length through a modified Green-Ampt model. The critical slope of total runoff may be affected by the surface condition (e.g., vegetation coverage, surface roughness) and the soil property (e.g., permeability, soil surface sealing), which worth further investigation.

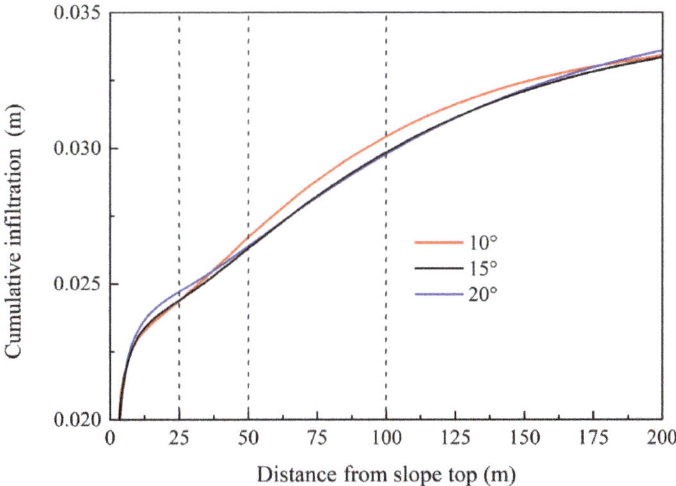

Figure 10. The cumulative infiltration distribution along the slope axis under rising-falling rainfall at 10°–20° slopes.

The critical slope of soil erosion amount decreased from 35° to 25° when projective slope length increased under five different rainfall patterns (Figure 5), except for constant rainfall at 25 m slope length. Such simulation result was close to the range of critical slope of soil loss often observed in the field, which was 20°–30° (e.g., [21,38]). The smaller critical slope maybe because their rainfall experiments adopted slopes with equal length. Generally, the critical slope for the constant rainfall was 5° larger than those of other rainfalls. Taking the rising-falling rainfall as an example, Figure 11 shows the erosion depth distribution along the slope axis at 25°–35° slopes. At 35° slope, the erosion depth curve rose more quickly than other slopes as projective slope length increased to 25 m, due to higher flow velocity and shear stress. Thus, the critical slope of soil erosion amount was 35° for the slope shorter than 25 m. As mentioned above, for slope steeper than 15°, the increase of slope gradient resulted in longer slope length and more infiltration. As the projective slope length increased, for each rainfall pattern, the reduction of runoff from 25° slope to 35° slope became larger (Figure 3) so that more sediment deposited on the slope. In consequence, the erosion depth curve at 25° slope finally reached the highest peak and decreased much slower than the other slopes when the projective slope length over 50 m (Figure 11). The critical slope of soil erosion amount was 25° for the slope longer than 100 m.

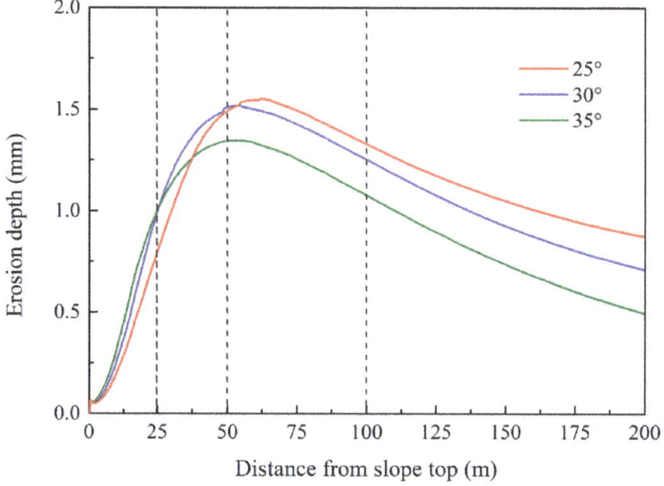

Figure 11. The erosion depth distribution along the slope axis under rising-falling rainfall at 25°–35° slopes.

4.3. Effect of Rainfall Patterns on Runoff and Soil Erosion Peaks

The time and value of the peak discharge as well as peak erosion rate were greatly influenced by the rainfall pattern. For the non-constant rainfall patterns, the increasing rainfall had the highest peak discharges and peak erosion rates, which was also mentioned in previous studies (e.g., [33]). Under increasing rainfall, as the rainfall intensity gradually increased and the surface gradually became saturated, the discharge rate and soil erosion rate kept increasing and reached the highest peak discharge and erosion rate (e.g., Figures 4 and 6) [39].

Because the infiltrability of the surface soil was high in the early part of the event, the decreasing and rising-falling rainfalls generally had lower peak discharges and peak erosion rates than the increasing rainfall (Figures 4 and 6). The peak discharge and erosion rate of increasing rainfall were reached just at the end of the peak rainfall intensity, while those of decreasing rainfall were several minutes later than the end of the peak rainfall intensity. High infiltrability of the surface soil in the early part of the event may be also the reason for the delay of the peak discharge and erosion rate, which dramatically slowed down the runoff generation process. Under rising-falling rainfall, the time of the peak discharge and erosion rate was also later than the end of the peak rainfall intensity, but the time was shorter than that under decreasing rainfall as its peak time was during the middle of the event.

For the falling-rising rainfall, as the two high rainfall intensity periods were separated by the low-rainfall-intensity period (Figure 2), the rainfall amount for peak discharge was much less than other non-constant rainfalls. Thus, the peak discharges and erosion rates were lower than those the increasing rainfall (Figures 4 and 6). As projective slope length increased, the effect of rainfall amount was more important so that the peak discharge and erosion rate under falling-rising rainfall was even lower than those of the decreasing rainfall on the 200 m slope.

4.4. Benefits and Future Work

This research work provided comprehensive theoretical studies on effects of rainfall patterns at slope scale. Even though it lacked field measurements as validation, the parameters of the slope that used in this study were well validated previously so the simulation results were rational and realistic for runoff and sediment research. The lumped and distributed simulation results showed how rainfall patterns affected runoff generation and soil erosion processes on the wide ranges of slope gradient (5° to 40°) and length (25–200 m), which can improve the accuracy of hillslope runoff and soil erosion prediction and be helpful for catchment flood management.

Table 4 illustrates the comparison between this study and previous studies, aiming at identifying the differences and emphasizing the findings of this study. In the future, the effect of rainfall patterns on hydrological responses at catchment scale will be explored. This study indicated that slope length and steepness may have great influence on the impact of rainfall patterns, and different features of hillslope length and steepness in a natural catchment will be carefully considered. Besides, rainfall patterns with multi-peak (e.g., falling-rising rainfall) showed great variety in runoff and soil erosion amount at different slope length compared with other rainfall patterns so more research effort can be put into it.

Table 4. Comparison of the findings between the current and previous studies.

Source	Experiment Setup	Scenario Arrangement	Finding	Comparison with This Study
Parsons and Stone [10]	2.43 m-long, 0.9 m-width, 0.2 m-depth flume, 10° slope, three soil types.	Intensity from 46.4 to 170.8 mm h^{-1}, 93.9 mm h^{-1} on average.	A constant-intensity storm are reduced by about 25% compared to varied-intensity storms.	
An et al. [8]	8 m-long, 1.6 m-width, 0.2 m-depth flume, 5° and 10° slope, pre wetted silt loam soil.	Intensity from 50 to 100 mm h^{-1}, 75 mm h^{-1} on average.	Soil loss from varying-intensity rainfalls was 1.13 to 5.17 times greater than that from even-intensity rainfall. Soil loss under increasing rainfall were the highest.	1. Conform part: Total runoff and soil erosion were lower than those of increasing, decreasing and rising-falling rainfalls. 2. Non conform part: The rising-falling rainfall generally had the largest runoff and soil erosion amount, while the constant rainfall did not have the lowest ones when the projective slope length was over 100 m.
Dunkerley [3]	0.5 m × 0.5 m plot, 0.2° slope, loam soil.	Intensity peaked at 30 mm h^{-1}, 10 mm h^{-1} on average.	Late peak events had the highest peak runoff rate and runoff ratio, which were more than double those of the early peak events. The constant rainfall yielded the lowest total runoff and runoff rate.	
Wang et al. [35]	2 m-long, 1 m-width, 0.5 m-depth flume, 10° slope, pre wetted clay loam soil.	Intensity from 50 to 100 mm h^{-1}, 75 mm h^{-1} on average.	The constant rainfall produced the lowest sediment yield at around 61.8% of the average soil loss for the increasing rainfall, which had the highest soil loss.	

Table 4. *Cont.*

Source	Experiment Setup	Scenario Arrangement	Finding	Comparison with This Study
Zhai et al. [11]	Hydrological simulation study at catchment around 100 km^2.	Intensity from 7 to 69 mm h^{-1}, about 20 mm h^{-1} on average	The delayed rainfall pattern yield higher flood volume and peak than the early peak pattern.	Consensus
Wu et al. [37]	Theoretical framework work at 22.1 m long slopes from 0.5° to 60°.	No description.	Critical slope for runoff rate was around 11° regardless of rainfall duration and slope length	The critical slope of the runoff was close, and was also independent of slope length.
Cheng et al. [38]	2 m × 5 m plot, 5° to 25° slope, sandy loam soil.	72 mm h^{-1} rainfall for 30 min.	Soil loss increased with increasing slope angle till the critical slope angle of 20°–30°.	The range of critical slope of the soil erosion was close.

Additionally, in the modelling study a stable slope surface was assumed, which meant that evolution of rill was not considered on the surface. On steep and long hillslopes, rill may generate under heavy rainfall (e.g., [40]). As the surface flow and related soil erosion characteristics in rills are different from those in an interrill area [41], it may influence the runoff generation and erosion dynamics at various scales. Although addressing the influence of rill was beyond the scope of this study, it is worth further investigations.

5. Conclusions

In this study, the effect of rainfall pattern on runoff generation and soil erosion processes on slopes were analysed through numerical modelling. The modelling work provides infiltration, runoff and soil erosion differences among five rainfall patterns on wide ranges of slope gradient (5° to 40°) and slope length (25–200 m). The simulation result indicated that the rising-falling rainfall generally had the largest total runoff and soil erosion amount. The constant rainfall did not have the lowest total runoff and soil erosion amount when the projective slope length was over 100 m, which was higher than the falling-rising rainfall. The critical slope of the total runoff was 15°, which was independent of rainfall pattern and slope length. However, the critical slope of the soil erosion amount varied, which decreased with increasing projective slope length from 35° to 25°. And the critical slope for the soil erosion of the constant rainfall was generally 5° larger than that of other rainfalls. The increasing rainfall had the highest peak discharge and erosion rate just at the end of the peak rainfall intensity, while those of the decreasing and rising-falling rainfalls were lower and were several minutes later than the end of peak rainfall intensity.

These findings are helpful to improve the knowledge of the characteristics in runoff generation and soil erosion processes under various rainfall patterns at slopes, and they may be also beneficial for further understanding of hillslope morphology and ecology. Further work will be required for adequate meteorological and hydrological data to gain a more comprehensive understanding of rainfall pattern effects on hydrological processes at larger scale.

Author Contributions: Conceptualization, Q.R. and F.W.; Investigation, F.W. and J.G.; Writing—Original Draft Preparation, F.W. and J.G.; Writing—Review & Editing, Q.R. and J.G.

Funding: This research was funded by National Key Research and Development Program of China (2016YFC0402404) and National Natural Scientific Foundation of China (51679209).

Conflicts of Interest: The authors declare no conflict of interest.

References

1. Dunkerley, D. Rain event properties in nature and in rainfall simulation experiments: A comparative review with recommendations for increasingly systematic study and reporting. *Hydrol. Process.* **2008**, *22*, 4415–4435. [CrossRef]
2. Farrick, K.K.; Branfireun, B.A. Soil water storage, rainfall and runoff relationships in a tropical dry forest catchment. *Water Resour. Res.* **2014**, *50*, 9236–9250. [CrossRef]
3. Dunkerley, D. Effects of rainfall intensity fluctuations on infiltration and runoff: Rainfall simulation on dryland soils, fowlers gap, australia. *Hydrol. Process.* **2012**, *26*, 2211–2224. [CrossRef]
4. Singh, V.P. Effect of spatial and temporal variability in rainfall and watershed characteristics on stream flow hydrograph. *Hydrol. Process.* **1997**, *11*, 1649–1669. [CrossRef]
5. Cerdan, O.; Bissonnais, Y.L.; Couturier, A.; Bourennane, H.; Souchère, V. Rill erosion on cultivated hillslopes during two extreme rainfall events in normandy, france. *Soil Tillage Res.* **2002**, *67*, 99–108. [CrossRef]
6. Dolšak, D.; Bezak, N.; Šraj, M. Temporal characteristics of rainfall events under three climate types in slovenia. *J. Hydrol.* **2016**, *541*, 1395–1405. [CrossRef]
7. Huff, F.A. Time distribution of rainfall in heavy storms. *Water Resour. Res.* **1967**, *3*, 1007–1019. [CrossRef]
8. An, J.; Zheng, F.L.; Han, Y. Effects of rainstorm patterns on runoff and sediment yield processes. *Soil Sci.* **2014**, *179*, 293–303. [CrossRef]
9. Dunkerley, D. An approach to analysing plot scale infiltration and runoff responses to rainfall of fluctuating intensity. *Hydrol. Process.* **2017**, *31*, 191–206. [CrossRef]
10. Parsons, A.J.; Stone, P.M. Effects of intra-storm variations in rainfall intensity on interrill runoff and erosion. *Catena* **2006**, *67*, 68–78. [CrossRef]
11. Zhai, X.Y.; Guo, L.; Liu, R.H.; Zhang, Y.Y. Rainfall threshold determination for flash flood warning in mountainous catchments with consideration of antecedent soil moisture and rainfall pattern. *Nat. Hazards* **2018**, *94*, 605–625. [CrossRef]
12. Chen, L.; Sela, S.; Svoray, T.; Assouline, S. Scale dependence of hortonian rainfall-runoff processes in a semiarid environment. *Water Resour. Res.* **2016**, *52*, 5149–5166. [CrossRef]
13. Parsons, A.J.; Brazier, R.E.; Wainwright, J.; Powell, D.M. Scale relationships in hillslope runoff and erosion. *Earth Surf. Process. Landf.* **2006**, *31*, 1384–1393. [CrossRef]
14. Langhans, C.; Govers, G.; Diels, J.; Stone, J.J.; Nearing, M.A. Modeling scale-dependent runoff generation in a small semi-arid watershed accounting for rainfall intensity and water depth. *Adv. Water Resour.* **2014**, *69*, 65–78. [CrossRef]
15. Guntner, A.; Bronstert, A. Representation of landscape variability and lateral redistribution processes for large-scale hydrological modelling in semi-arid areas. *J. Hydrol.* **2004**, *297*, 136–161. [CrossRef]
16. Stomph, T.J.; Ridder, N.D.; Steenhuis, T.S.; Giesen, N.C.V.D. Scale effects of hortonian overland flow and rainfall-runoff dynamics: Laboratory validation of a process-based model. *Earth Surf. Process. Landf.* **2002**, *27*, 847–855. [CrossRef]
17. Assouline, S.; Ben-Hur, M. Effects of rainfall intensity and slope gradient on the dynamics of interrill erosion during soil surface sealing. *Catena* **2006**, *66*, 211–220. [CrossRef]
18. Essig, E.T.; Corradini, C.; Morbidelli, R.; Govindaraju, R.S. Infiltration and deep flow over sloping surfaces: Comparison of numerical and experimental results. *J. Hydrol.* **2009**, *374*, 30–42. [CrossRef]
19. Defersha, M.B.; Quraishi, S.; Melesse, A. The effect of slope steepness and antecedent moisture content on interrill erosion, runoff and sediment size distribution in the highlands of ethiopia. *Hydrol. Earth Syst. Sci.* **2011**, *15*, 2367–2375. [CrossRef]
20. Shi, Z.H.; Fang, N.F.; Wu, F.Z.; Wang, L.; Yue, B.J.; Wu, G.L. Soil erosion processes and sediment sorting associated with transport mechanisms on steep slopes. *J. Hydrol.* **2012**, *454*, 123–130. [CrossRef]
21. Cheng, Q.; Ma, W.; Cai, Q. The relative importance of soil crust and slope angle in runoff and soil loss: A case study in the hilly areas of the loess plateau, north china. *GeoJournal* **2008**, *71*, 117–125. [CrossRef]
22. Chaplot, V.A.M.; Le Bissonnais, Y. Runoff features for interrill erosion at different rainfall intensities, slope lengths, and gradients in an agricultural loessial hillslope. *Soil Sci. Soc. Am. J.* **2003**, *67*, 844–851. [CrossRef]

23. Ebel, B.A.; Loague, K.; Montgomery, D.R.; Dietrich, W.E. Physics-based continuous simulation of long-term near-surface hydrologic response for the coos bay experimental catchment. *Water Resour. Res.* **2008**, *44*, 137–149. [CrossRef]

24. Ran, Q.; Keith, L.; Verkwaak, J.E. Hydrologic-response-driven sediment transport at a regional scale, process-based simulation. *Hydrol. Process.* **2012**, *26*, 159–167. [CrossRef]

25. Ebel, B.A.; Loague, K.; Borja, R.I. The impacts of hysteresis on variably saturated hydrologic response and slope failure. *Environ. Earth Sci.* **2010**, *61*, 1215–1225. [CrossRef]

26. VanderKwaak, J.E.; Loague, K. Hydrologic-response simulations for the r-5 catchment with a comprehensive physics-based model. *Water Resour. Res.* **2001**, *37*, 999–1013. [CrossRef]

27. VanderKwaak, J.E. Numerical Simulation of Flow and Chemical Transport in Integrated Surface-Subsurface Hydrologic Systems. Ph.D. Thesis, University of Waterloo, Waterloo, ON, Canada, 1999.

28. Ran, Q.; Heppner, C.S.; Vanderkwaak, J.E.; Loague, K. Further testing of the integrated hydrology model (inhm): Multiple-species sediment transport. *Hydrol. Process.* **2007**, *21*, 1522–1531. [CrossRef]

29. Ran, Q.; Hong, Y.; Li, W.; Gao, J. A modelling study of rainfall-induced shallow landslide mechanisms under different rainfall characteristics. *J. Hydrol.* **2018**, *563*, 790–801. [CrossRef]

30. Heppner, C.S.; Ran, Q.; Vanderkwaak, J.E.; Loague, K. Adding sediment transport to the integrated hydrology model (inhm): Development and testing. *Adv. Water Resour.* **2006**, *29*, 930–943. [CrossRef]

31. Johansen, M.P.; Hakonson, T.E.; Breshears, D.D. Post-fire runoff and erosion from rainfall simulation: Contrasting forests with shrublands and grasslands. *Hydrol. Process.* **2001**, *15*, 2953–2965. [CrossRef]

32. Van Genuchten, M.T. A closed-form equation for predicting the hydraulic conductivity of unsaturated soils. *Soil Sci. Soc. Am. J.* **1980**, *44*, 892–898. [CrossRef]

33. Gao, J.; Kirkby, M.; Holden, J. The effect of interactions between rainfall patterns and land-cover change on flood peaks in upland peatlands. *J. Hydrol.* **2018**, *567*, 546–559. [CrossRef]

34. Flanagan, D.C.; Foster, G.R.; Moldenhauer, W.C. Storm pattern effect on infiltration, runoff, and erosion. *Trans. ASAE* **1988**, *31*, 546–559. [CrossRef]

35. Wang, B.; Steiner, J.; Zheng, F.; Gowda, P. Impact of rainfall pattern on interrill erosion process. *Earth Surf. Process. Landf.* **2017**, *42*, 1833–1846. [CrossRef]

36. Xue, J.; Gavin, K. Effect of rainfall intensity on infiltration into partly saturated slopes. *Geotech. Geol. Eng.* **2008**, *26*, 199–209. [CrossRef]

37. Wu, S.; Yu, M.; Chen, L. Nonmonotonic and spatial-temporal dynamic slope effects on soil erosion during rainfall-runoff processes. *Water Resour. Res.* **2017**, *53*, 1369–1389. [CrossRef]

38. Liu, Q.Q.; Xiang, H.; Singh, V.P. A simulation model for unified interrill erosion and rill erosion on hillslopes. *Hydrol. Process.* **2006**, *20*, 469–486. [CrossRef]

39. Dunkerley, D. Stemflow production and intrastorm rainfall intensity variation: An experimental analysis using laboratory rainfall simulation. *Earth Surf. Process. Landf.* **2014**, *39*, 1741–1752. [CrossRef]

40. Fang, H.; Sun, L.; Tang, Z. Effects of rainfall and slope on runoff, soil erosion and rill development: An experimental study using two loess soils. *Hydrol. Process.* **2015**, *29*, 2649–2658. [CrossRef]

41. Wirtz, S.; Seeger, M.; Ries, J.B. Field experiments for understanding and quantification of rill erosion processes. *Catena* **2012**, *91*, 21–34. [CrossRef]

Article

Physics-Based Simulation of Hydrologic Response and Sediment Transport in a Hilly-Gully Catchment with a Check Dam System on the Loess Plateau, China

Honglei Tang [1], Qihua Ran [1] and Jihui Gao [2,*]

[1] Institute of Hydrology and Water Resources, Zhejiang University; Hangzhou 310058, China; tanghonglei@zju.edu.cn (H.T.); ranqihua@zju.edu.cn (Q.R.)
[2] State Key Laboratory of Hydraulics and Mountain River Engineering, Sichuan University, Chengdu 610065, China
* Correspondence: jgao@scu.edu.cn

Received: 7 May 2019; Accepted: 29 May 2019; Published: 2 June 2019

Abstract: Check dams are among of the most widespread and effective engineering structures for conserving water and soil in the Loess Plateau since the 1950s, and have significantly modified the local hydrologic responses and landforms. A representative small catchment was chosen as an example to study the influences of check dams. A physics-based distributed model, the Integrated Hydrology Model (InHM), was employed to simulate the impacts of check dam systems considering four scenarios (pre-dam, single-dam, early dam-system, current dam-system). The results showed that check dams significantly alter the water redistribution in the catchment and influence the groundwater table in different periods. It was also shown that gully erosion can be alleviated indirectly due to the formation of the expanding sedimentary areas. The simulated residual deposition heights (Δh) matched reasonably well with the observed values, demonstrating that physics-based simulation can help to better understand the hydrologic impacts as well as predicting changes in sediment transport caused by check dams in the Loess Plateau.

Keywords: check dam; hydrologic response; sediment transport; InHM; Loess Plateau

---

1. Introduction

The Chinese Loess Plateau has suffered from severe water and soil loss for decades [1,2]. Many measures, including artificial forestation, terraced farming, and check dam construction, have been implemented to conserve soil and water since the 1950s. Since the first check dam appeared in the Loess Plateau 400 years ago, the effective engineering structure has prevailed all over the Loess Plateau, especially in the Loess Mesa Ravine Region and the Loess Hill Ravine Region, to create productive farmlands and conserve soil and water [3]. There were 122,028 check dams in the Loess Plateau at the end of 2005, which held 2.1×10^{10} m^3 of sediments and formed 3340 km^2 of dam farmlands [4,5]. The number of check dams and the area of dam-farmlands is expected to double by 2020, with the completion of check dam systems for all the main tributaries of the Yellow River in the Loess Plateau.

Check dams have been shown to be an effective engineering structure to reduce water discharge [6,7] and sediment yields [8–10] at the basin scale. For example, Xu et al. [6] applied Soil and Water Assessment Tool (SWAT) in a 7725 km^2 watershed with 6572 check dams and found that the annual runoff was reduced by 14.3%, comparing to when there were very few dams.

Ran et al. [9] compared the sediment retention by check dams in five typical watersheds in the Hekou-Longmen section of the Yellow River and found that the average sediment reduction ratio can reach 60% when the percentage of the basin area above check dams in the catchments reached 3.0%. Previous numerical simulations studying the impacts of check dams mainly focused on water discharges and sediment yields at the outlet. However, other hydrologic features such as groundwater table and subsurface storage changes and the sedimentary processes behind the dams may be also influenced by check dams, and sometimes, much more important (e.g., the primary drinking water source for local residents is the groundwater in many catchments). Hydrologic-response change phenomena such as near-surface K_{sat} increase [11], hillslope-channel decoupling [12], groundwater recharge increase [13] were reported in many catchments with check dams around the world. Water resources in semi-arid areas are precious and the hydrologic-response changes induced by check dams should be better understood. Huang et al. [14] evaluated the impacts of a 30-year-old check dam on water redistribution and the results showed that infiltration was enhanced in the sedimentary field.

Check dams are often constructed as a system in which individual dam is operated in different purposes, and it is important to assess the impacts of a check dam system on the environment. Considering that hydraulic erosion plays an important role in most parts of the Loess Plateau, simulating hydrologic response and sediment transport simultaneously is essential to further understand the influences of check dams. A large amount of sediment eroded from slopes is deposited along gullies due to the interception of check dams. However, the useful lifetime of many small-size check dams was shortened by rapid sedimentary processes behind dams. Rapid sedimentation is expected when a check dam is mainly used as a productive dam to quickly create farmlands but should be avoided for check dams used for preventing floods. The sedimentary processes in dam-controlled gullies, though seldom reported, are important because local people's lives and property are often threatened by dam-break events caused by overtopping floods. In fact, a large number of check dams were destroyed by floods due to inappropriate position or rapid deposition in the Loess Plateau [6]. Thus, an appropriate forecast of the sedimentary processes behind check dams is necessary for future check dam planning and management.

The objectives of this simulation-based study were to capture, as best as possible, the phenomena of hydrologic-response changes and the sedimentary processes caused by a check dam system and to demonstrate that physics-based simulation can be a useful tool for predicting the sedimentary processes induced by the widely launched soil and water conservation measures in future planning and management. The Integrated Hydrology Model (InHM), a physics-based distributed hydrologic model with sediment-transport capabilities, was employed to "revisit" what had happened in the first few years after the construction of check dams and compare the water table changes and sedimentary processes of the current dam-system with early dam-system. This work focused on a small gully catchment with a developed 5-dam system, located in Loess Plateau. Annual continuous simulations were conducted to capture the changes induced by check dams.

2. Materials and Methods

2.1. Study Site and Data

Wangmaogou catchment (WMG) is a loess hilly catchment located near the outlet of Wudinghe Basin, China (Figure 1). The area of the catchment is 5.97 km^2, and the elevation ranges from 940 to 1188 m a.s.l. Under the impacts of severe soil and water erosion, landscape featured as interlaced gullies are formed gradually, with a gully density of 4.3 km/km^2. The geological structure of WMG is relatively simple: (1) the uppermost layer is 20~30 m deep homogeneous loess soil; (2) the second layer is 50~100 m deep red soil, most of which emerges in the gully head; (3) the underlying bedrock is Triassic sandy shale. WMG catchment has a typical semiarid continental climate. The annual potential evapotranspiration is around 800 mm, while the mean annual precipitation is 513 mm, more than 70% of which is received during the rainy season from June to September [15].

Engineering measures focusing on gully erosion control (i.e., check dams) were carried out in the catchment to alleviate soil and water loss since 1953, before which the average erosion rate was 18,000 t km^{-2} per year. 42 check dams were constructed from 1953 to 1959, but most of the dams were destroyed by heavy storms due to low design criteria. Figure 1b shows the 22 currently existing check dams. These check dams have been in operation for more than 50 years under various current conditions (i.e., some have been fully deposited, some have been partially destroyed, the others still remain in good condition). The 42 check dams effectively prevented sediments being transported downstream into Wudinghe River and further into the Yellow River by surface runoff. Other water and soil conservation measures aiming at slope erosion control such as terraced farmlands and woods planting were started in 1962. Figure 2 compares the land use conditions of 1962 and 2010. The primary land use type 50 years ago was slope farmlands (Figure 2a), which was a major erosion source of the catchment. After 50 years, 60% of the slope farmlands were turned into terraced farmlands (Figure 2b), which have a higher water erosion resisting performance. Large areas of grasslands were transformed into sparse woods (*Robinta pseudoacacia* and *Platycladus orientalis*) or orchards (apple tree and peach tree) by local farmers.

Table 1 summarizes the data set used in this study. Detailed field measurements were conducted from April to September 2014. The soil sample tests provide supplemental soil information including saturated hydraulic conductivities (K_{sat}) and soil water characteristic curves of various land use types at different sampling sites (Figure 1b). Two wells were monitored for water table data set, which was used for calibrating the initial groundwater table. The residual deposition heights (Δh) behind each dam, defined as the height between dam crest and the surface elevation behind the dam, were also carefully measured to present the sedimentary processes. The field data provides a reference for how the catchment functions currently, and serves as validation for the simulated responses. Ten rainfall-runoff events recorded by the local Suide Soil and Water Conservation Station from 1962 to 1964 were used for model calibration. Two topography maps of WMG catchment were used to construct the computational meshes.

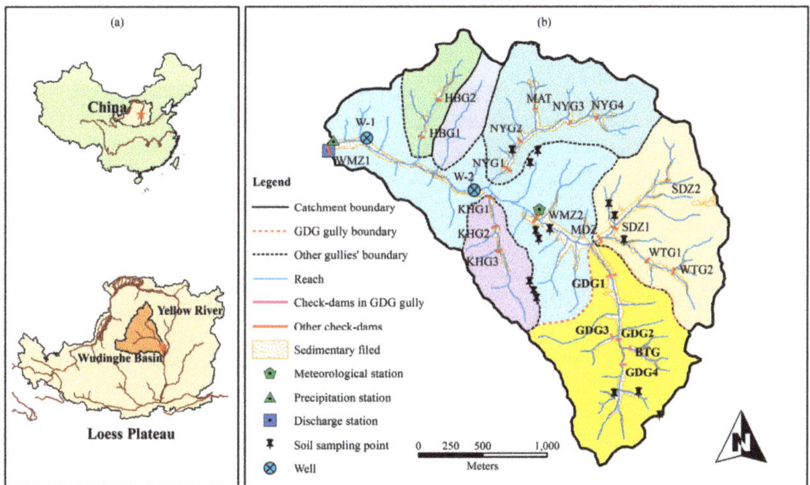

Figure 1. (**a**) Location of WMG catchment (the red star); (**b**) check dam distributions and measurement locations of soil characteristics and hydrologic data. Polygons with different colors represent different gullies and the red dash line depicts the boundary of GDG catchment.

Guangdigou catchment (GDG) is a headwater catchment locating in the south of WMG catchment, with a relatively constant dam-system and less land use changes from the 1950s to now. There are five check dams, and Table 2 shows the characteristics of them.

As shown in Figure 1b, GDG1 is located near the outlet of the main gully and installed in series with GDG2 and GDG4 along the main gully. GDG3 and BTG are located on the outlet of two tributary gullies. The five dams are still in good condition. Compared to other sub-catchments in WMG, the landuse change of GDG was relatively simple during the 50 years, with a large area of slope farmlands being replaced by terraced farmlands.

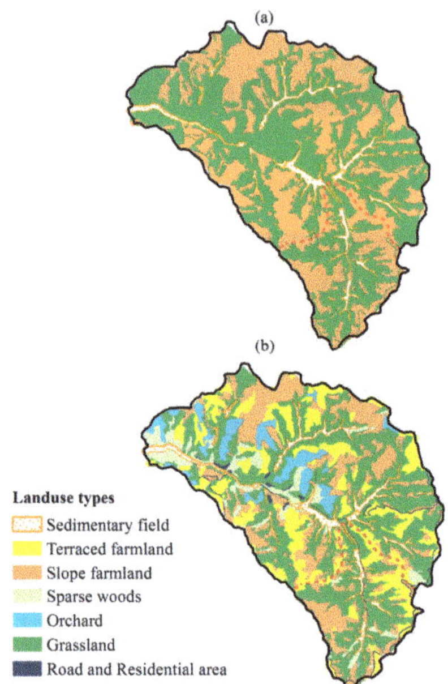

Figure 2. Landuse types of WMG catchment: (**a**) 1962; (**b**) 2010. The red dash line presents the boundary of GDG catchment.

Table 1. Data set used in this study.

Data Type	Year	Resolution	Data Source	Use in This Study
Land use type	1962, 2010	20-m	Literature [15]	Model construction
Soil data	2014		Field survey	Model construction
Groundwater data	2014	Semi-monthly	Field survey	Model construction
Rainfall-runoff event	10 events during 1962–1964	5-min to 1-hour	Suide Soil and Water Conservation Station	Model calibration
Topography data	1953, 2010	25-m for 1953 map; 5-m for 2010 map	Shaanxi Administration of Surveying [16]	Model construction
Check dam information	1953–1964, 2014		Suide Soil and Water Conservation Station; field survey	Model construction

Table 2. Characteristics of the 5 check dams in GDG catchment.

Name	Year [1]	Height (m)	Length (m)	Control Area (km^2)	Sedimentary Area ($\times 10^4$ m^2)	Δh [2] (m)
GDG1	1955 (1959) [3]	10 (+13) [3]	66	1.14	2.81	7.00
GDG2	1959	8	28	0.10	0.24	1.00
GDG3	1959	12	36	0.05	0.24	1.10
GDG4	1959	8	46	0.40	1.00	2.90
BTG	1959	10	45	0.20	0.94	0.80

[1] Completion year, all dams were completed before the rainy season.; [2] Δh is the residual deposition height, measured in April 2014.; [3] GDG1 dam was heightened by 13 m in 1959.

2.2. The Integrated Hydrology Model (InHM)

Based on the blueprint of the distributed physics-based hydrological model proposed by Freeze and Harlan [17], InHM was developed to quantitatively simulate, via a fully coupled approach, 3-dimensional variably saturated flow in soil and 2-D flow and sediment transport across land surface [18–21]. The 3-D Richards' equation was implemented to describe variably saturated flow in soil, while the 2-D diffusion-wave equation coupled with depth-integrated multiple-species sediment transport was applied to describe surface flow movement and sediment transportation. Those surface and subsurface governing equations were discretized in space using the control volume finite element method and coupled in one coherent framework using physics-based first-order flux relationship driven by pressure head gradients. Newton iteration was used to implicitly solve each coupled system of nonlinear equations. More details of InHM can be found in Appendix A.

With the most important and innovative characteristic (i.e., no priori assumption of specific runoff-generation mechanisms), InHM has been successfully employed in many different catchments across the world for event-based or continuous hydrological-response simulations [22,23] as well as hydrologically-driven sediment transport simulations [24,25]. In the Loess Plateau catchments of Wangmaogou and its sub-catchment Guangdigou, InHM is capable of simulating rainfall-runoff processes dominated by the infiltration-excess surface flow mechanism and also rainsplash and hydraulic erosion processes in flood events. Spatially distributed information (e.g., surface flow velocity and sediment flux) of these processes can be provided by the model.

2.3. Scenario Setting and Modelling Procedure

Two simulation stages were designed in this study. The first stage is calibration and validation simulations, to obtain the actual values of important and sensitive parameters. In the second stage, which was the focus of the study, annual continuous simulations were conducted to evaluate the hydrologic effects and sedimentary processes of a check dam system. The effects of check dam operation in the early period (1955–1962) and in the current period (2010–2013) were both studied. In the second stage, the model took us to "revisit" what had happened in the first few years after the construction of check dams and compared the water table changes and sedimentary processes of the current dam-system with the early dam-system.

2.3.1. Calibration and Validation Simulations in WMG Catchment

Calibration and validation simulations were conducted in WMG catchment, using the data of ten rainfall-runoff events recorded at the discharge station (Figure 1b) from 1961 to 1965, which were the only available observation data (Table 1). Four events were used for calibration and the other six events were used for validation. For a specified flood event, check dams were manually added/removed from the mesh according to their status (already existing or destroyed) during the event.

Simulated water discharges and sediment discharges were compared with observed values and the preliminary results were improved by adjusting surface and subsurface parameters of InHM. Parameters, related to infiltration and runoff-generation (i.e., saturated hydraulic conductivity and Manning's roughness coefficient) and erosion (i.e., rainsplash coefficient and surface erodibility), were carefully adjusted to improve the simulated results in the course of calibration. The Nash and Sutcliffe modelling efficiency (EF) [26], given by Equation (1), was employed here to evaluate the model performance.

$$EF = \left[\sum_{i=1}^{n} (O_i - \overline{O})^2 - \sum_{i=1}^{n} (S_i - O_i)^2 \right] / \sum_{i=1}^{n} (O_i - \overline{O})^2 \tag{1}$$

where O_i was the observed value (water discharge or sediment discharge), $\overline{O}$ was the average value of the observed values, and S_i was the simulated value and n was the number of samples. Considering the lack of available measurements to provide land surface information of WMG catchment from 1955 to 1962, these calibration and validation efforts helped to obtain important parameters (e.g., calibrated

K_{sat} for various soil types) to construct a more realistic boundary-value problem for the following simulations.

2.3.2. Annual Continuous Simulations in GDG Catchment

Using the calibrated parameters but different meteorological and land use data, annual continuous simulations were carried out for research period 1953–1962 and 2010–2013. It should be noticed that we had to use event-based calibrated parameters to conduct annual continuous simulations because of the lack of long-term observation data. Considering the fact that most of the observed runoff events in a year only occurred in several rainfall storms in the catchment and InHM performed reasonably well in the ten rainfall-runoff events with various rainfall characteristics (rainfall intensity, rainfall amount and duration, see details in Section 3.1), it was technically sound to conduct the following annual continuous simulations using event-based calibrated parameters.

A WMG-catchment boundary-value problem (BVP) was first built to conduct the calibration simulations in the first stage because the discharge station was located at the outlet of WMG catchment. Another BVP was needed to simulate the influence of check dams because: (1) it was difficult to evaluate the hydrologic effects and sedimentary processes of a complicated check dam system which contained 42 check dams 50 years ago and only 22 now; (2) different land use types of WMG catchment after the construction of check dams also increased the degree of complexity. To simplify the problem, the GDG sub-catchment, within WMG with a relatively constant dam-system and less land use changes, was chosen as an example to study the effects of check dams. As mentioned above, the geological structure of WMG was relatively simple. The soil types and features of GDG gully were the same as those of the rest part of WMG catchment. The parameters calibrated in the WMG BVP were mainly related to soil characteristics (e.g., saturated hydraulic conductivity, rainsplash coefficient) and could be directly used in GDG BVP. It was technically reasonable to conduct the GDG BVP simulations using parameters from WMG BVP simulations.

Four scenarios were designed in this study to evaluate the impacts of check dams on hydrologic response and the sedimentary processes (Table 3). Pre-dam scenario (PD) represented the situation before the construction of the first check dam (i.e., base case), with a two-year simulation because the available precipitation data started in 1953. Single-dam scenario (SD) represented the situation after the construction of GDG1 dam in the downstream. Early dam-system scenario (EDS) represented the conditions after four extra dams (i.e., GDG2-4, BTG) were constructed in the upstream of the gully. Current dam-system scenario (CDS) represented the current conditions including large areas of terraced farmlands and a more than 50-year-old check dam system.

Table 3. Description of the four scenarios in the study.

Scenario Name	Simulation Time Range	Check Dam Involved
Pre-dam (PD)	1953–1954	none
Single-dam (SD)	1955–1958	GDG1
Early dam-system (EDS)	1959–1962	GDG1–4, BTG
Current dam-system (CDS)	2010–2013	GDG1–4, BTG

The impacts of check dams on water redistribution were compared among the four scenarios, based on water balance calculation. Changes in groundwater table of channel reach A0-A2 (Figure 3) were analyzed. A series of observation nodes, located on/near the A0-A2 profile at a 5-m (near check dam) and 20 m interval, were set. Pressure head and soil water content were calculated in InHM for each observation nodes at every time step. The surface and subsurface nodes with zero pressure head values together outlined the groundwater table profile of A0-A2 reach. Soil erosion of the GDG catchment in the four scenarios was also evaluated by calculating the eroded sediment mass of different surface zones (Table 4), which was calculated in InHM by the integration of sediment flux through the boundary we set for each surface zone.

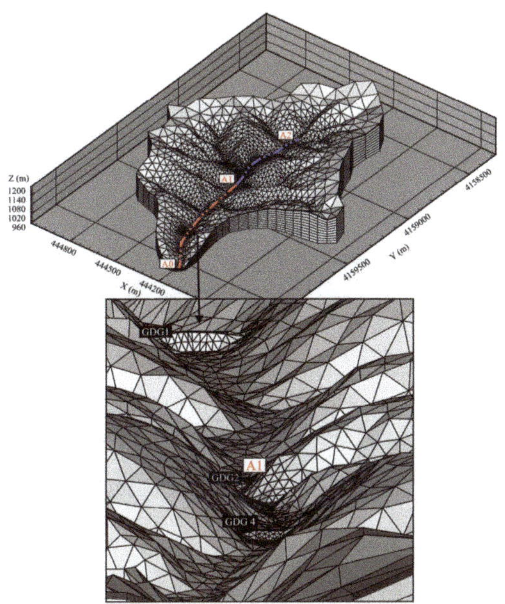

Figure 3. 3D mesh for EDS scenario, the inset is a downstream view from position A2.

Table 4. Surface parameters used in InHM for the four scenarios.

| Scenario | Surface Zone [a] | Hydrologic Response | | | | | Sediment Transport | | | |
		n [b] (-)	ψ_{im} [c] (m)	S_{sr} [d] (-)	H_t [e] (m)	c_f [f] (s m^{-1})$^{0.6}$	c_d [g] (1/m)	b [h] (-)	ξ [i] (-)	(m^{-1})
PD	Slope_G	0.10	0.0005	0.01	0.001	4.000	600	1.6	0.25	0.050
	Slope_F	0.12				8.000				0.050
	Gully_G	0.05				4.000				0.050
SD & EDS	Slope_G	0.10				4.000				0.050
	Slope_F	0.12				8.000				0.050
	Gully_G	0.05				4.000				0.050
	Gully_D	0.01				0.001				0.050
	Gully_S	0.05~0.12 k				0.100				0.050
CDS	Slope_G	0.10				4.000				0.038
	Slope_F	0.12				8.000				0.037
	Slope_T	0.50				0.500				0.038
	Gully_G	0.05				4.000				0.040
	Gully_D	0.01				0.001				0.010
	Gully_S	0.05~0.12 k				0.100				0.040

[a] The letter behind the underline represents different landuse types: G represents grasslands; F represents slope farmlands; D represents check dams; S represents sedimentary fields caused by dams; T represents terraced farmlands; [b] Manning's roughness coefficient [20], calibrated from the first stage.; [c] Immobile water depth [20], identical for all surface zones.; [d] Surface residual saturation [20], identical for all surface zones.; [e] Average height of non-discretized micro-topography [20], identical for all surface zones.; [f] Rainsplash coefficient [18], calibrated from the first stage.; [g] Rainsplash depth dampening factor [27], identical for all surface zones.; [h] Rain intensity exponent [27], identical for all surface zones.; [i] Rain-induced turbulence coefficient [27], identical for all surface zones.; [j] Surface erodibility coefficient [27], calibrated from the first stage. The values for CDS scenario were derived from Gao et al. [15].; [k] Sedimentary fields are usually used as productive farmlands after averagely 2-year deposition, increasing manning's roughness coefficient to 0.12.

2.4. Model Settings and Parameters

2.4.1. Boundary Conditions and Initial Conditions

The 3D meshes for the four scenarios were all constructed by adding layers to 2D surface meshes. The 2D surface meshes of GDG gully for the first three scenarios were constructed based upon the topographic map surveyed in 1953. A surface mesh with 2165 nodes and 4252 triangular elements was first generated for PD scenario. GDG1 dam was then added to the first surface mesh according to characteristics of GDG1 in Table 2 to generate the second surface mesh for the SD scenario, which has 2243 nodes and 4408 triangular elements. The other four dams were then added on the second mesh to generate the third surface mesh for EDS scenario (Figure 3), which consists of 2499 nodes and 4920 triangular elements. The fourth surface mesh for CDS scenario, which has 2451 nodes and 4799 triangular elements, was generated using the DEM in 2010 with 5 m horizontal resolution. The discretization of all the surface meshes varied from 50 m along the boundary to 20 m along the gullies and 5 m on the five check dams. 30 subsurface layers were added below the surface mesh. Considering that the 0.5 m deep soil near surface is the most sensitive to land use changes and important for surface hydrology, a constant thickness of 0.05 m was assigned to the first ten sublayers. The second ten sublayers (layer 11 to layer 20) had a uniform thickness of 0.5 m. A base elevation of 976.5 m (the elevation of GDG gully ranges from 992 to 1188 m a.s.l) was set as the bottom of the 3D mesh, creating variable thickness for the third ten layers (layer 21 to layer 30).

The vertical discretization for layer 21 to layer 30 ranged from 1.0 m near gully's outlet to 20.5 m at the boundary upland of the gully. This discretization method both ensured a fine resolution in the hydrologically active areas (i.e., the channel, sedimentary land, and the near-surface soil) and simultaneously saved computation resources in the relatively inactive areas (i.e., the headwater regions and deep spaces).

Using a similar method as Heppner and Loague [24], the initial 3D pressure-head distribution of GDG-gully for PD scenario was generated by conducting a one-year quick-drainage simulation starting from the following condition:

$$\psi_{t=0} = \begin{cases} 0.992 \times z_{surf} - z, & z_{surf} \leq 1060 \\ -0.34 \times z_{surf} - z + 1411.92, & z_{surf} > 1060 \end{cases} \tag{2}$$

where $\psi_{t=0}[L]$ was the initial pressure head of the simulation for a certain node, $z_{surf}[L]$ referred to the surface elevation directly above the node and $z[L]$ was the elevation of the node. 1060 m-contour was the typical dividing line of gully and slope. The choice of the initial condition for the one-year simulation was motivated by the fact that the groundwater table of the north-western Loess Plateau is around 5~10 m below surface in gullies and nearly 50~150 m below surface for slopes. This quick-drainage simulation used a synthetic rainfall time-series which only contained several small rainfall events to represent a dry year before 1953. A self-consistent head distribution, obtained from the quick-drainage simulation, was assigned to the first GDG-gully BVP (i.e., PD scenario). The initial conditions of SD scenario and EDS scenario were gleaned from the simulation results of PD scenario and SD scenario, respectively. The initial pressure head values for newly added nodes in SD scenario and EDS scenario were determined as the weighted average values of the nearest eight nodes.

To generate the initial 3D pressure-head distribution for CDS scenario, another quick-drainage simulation for the whole WMG catchment was conducted starting from the following condition:

$$\psi_{t=0} = \begin{cases} 0.999 \times z_{surf} - z, & z_{surf} \leq 1060 \\ -0.34 \times z_{surf} - z + 1419.34, & z_{surf} > 1060 \end{cases} \tag{3}$$

The choice of 0.999 in the first part of Equation (3) generated a high water-table shape along gullies at the beginning of the simulation. Then, the quick-drainage simulation started with no flux applied at the surface and ended when the simulated water table depths matched with observed average water table depths at the well-1 and well-2 (i.e., 8.0 m and 5.0 m, respectively) in Figure 1b. The pressure head values of all subsurface nodes for GDG catchment were extracted from the drainage simulation and assigned to the fourth GDG gully-BVP (i.e., CDS scenario) as initial subsurface conditions.

Three subsurface boundary conditions were assigned to the 3D boundary-value problems: (1) impermeable for each lateral face; (2) leaking at the saturated hydraulic conductivity for the basal boundary; (3) a local sink (i.e., head-dependent flux) at the down-gradient face. Specified fluxes for precipitation and evapotranspiration and a critical depth (d = 0 m) at the gully outlet were the three surface boundary conditions. The rainfall time-series spanning from 1953 to 1962 for the former 3 scenarios and from 2010 to 2013 for the CDS scenario were obtained from the precipitation station (Figure 1b). Figure 4 shows the cumulative rainfall and annual potential evapotranspiration (ET_0) for all the simulation years. The FAO56 recommended revised-Penman-Monteith method was used to estimate the daily potential evapotranspiration, using meteorological data such as daily temperature, daily vapor pressure, daily atmospheric pressure, daily solar radiation and wind speed from the meteorological station. The calculated daily potential evapotranspiration was then incorporated into InHM as actual evapotranspiration (ET) using a set of sink functions [22]:

$$Q_b^E = \alpha(\psi)q_{max}^E A_b \qquad (4)$$

where $Q_b^E[L^3T^{-1}]$ represented the volumetric evapotranspiration rate, $\alpha(\psi)[-]$ was a response function of the saturation of the porous medium and the degree of ponding at the land surface, $q_{max}^E[LT^{-1}]$ was the potential evapotranspiration rate per unit area estimated by the revised Penman-Monteith method, $\psi[L]$ was the pressure head of the subsurface nodes or water depth of the surface nodes, and $A_b[L^2]$ was the area associated with the surface water equation.

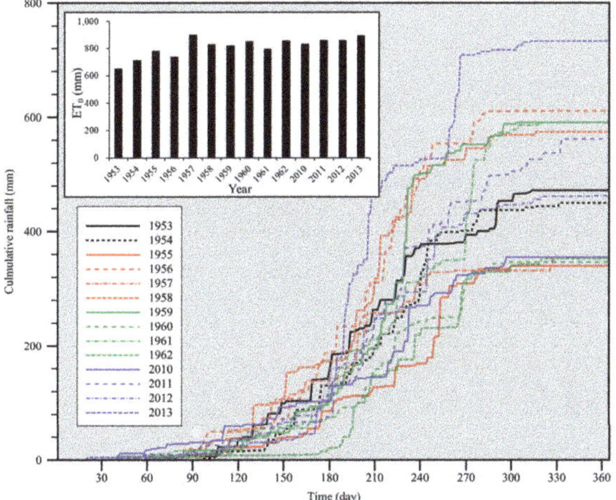

Figure 4. Measured cumulative rainfall for the 14 simulating years. The inset is the annual potential evapotranspiration (ET_0) calculated by the revised Penman-Monteith method.

2.4.2. Soil Parameters

Soils were classified as one layer of surface soil (0~20 cm) and two layers of subsurface soil (20~50 cm and below 50 cm). The surface soil layer was furtherly divided into six types according to land covers. Several soil parameters influencing hydrologic response and sediment transport were determined by field measurements or derived from the literature (Table 4). For example, the saturated hydraulic conductivity (K_{sat}) values for CDS scenario were measured in April 2014, while K_{sat} values for the other three scenarios were obtained via model calibration. Derived from the previous studies based on soil texture, the damping coefficient, raindrop turbulence factor, and rainfall intensity exponent were, respectively, set to 600 m^{-1}, 0.25, 1.6 [25,28]. Manning's roughness coefficient and rainsplash coefficient were obtained from model calibration. Surface erodibility coefficient for the first 3 scenarios was calibrated to 0.050. The surface erodibility coefficients for CDS scenario, lower than calibrated value, were derived from the work by Gao et al. [15] to represent the current surface condition. van Genuchten [29] equation was employed to describe the soil water characteristics and the parameters of the equation (Table 5) for loess soils were derived from infiltration experiments conducted in WMG catchment. According to the soil sample data in the catchment, the median diameter of the soil was set to 0.05 mm to represent the uppermost homogeneous loess soil and a single species particle with a particle density of 2650 kg·m^{-3} was used for all soil layers for the sediment transport simulation. The soil cohesion coefficient was 0.30 [24] for all surface soil except that of the check dam body, which was assigned a larger cohesion coefficient (i.e., 0.60) due to compaction.

Table 5. Soil parameters used in InHM for the four scenarios.

Scenario	Subsurface Zone	K_{sat} (m s^{-1})	Porosity (-)	Sr [a] (-)	α [b] (m^{-1})	n [c] (-)
PD	Slope_G [d]	6.20×10^{-7}	0.4800	0.0900	1.29	1.630
	Slope_F [d]	3.40×10^{-6}	0.4500	0.1000	1.27	1.710
	Gully_G [d]	3.50×10^{-6}	0.4800	0.0900	1.40	1.650
	20~50 cm [e]	5.00×10^{-6}	0.4500	0.0976	1.20	1.503
	Below 50 cm [e]	3.00×10^{-6}	0.4000	0.1000	0.92	1.542
SD & EDS	Slope_G [d]	6.20×10^{-7}	0.4800	0.0900	1.29	1.630
	Slope_F [d]	3.40×10^{-6}	0.4500	0.1000	1.27	1.710
	Gully_G [d]	3.50×10^{-6}	0.4800	0.0900	1.40	1.650
	Gully_D [e]	3.23×10^{-8}	0.2500	0.0599	1.01	1.374
	Gully_S [e]	3.19×10^{-5}	0.5000	0.0931	3.05	1.667
	20~50 cm [e]	5.00×10^{-6}	0.4500	0.0976	1.20	1.503
	Below 50 cm [e]	3.00×10^{-6}	0.4000	0.1000	0.92	1.542
CDS	Slope_G [e]	8.45×10^{-7}	0.5068	0.1176	1.74	1.579
	Slope_F [e]	2.18×10^{-6}	0.4268	0.0929	1.50	1.601
	Slope_T [e]	7.81×10^{-6}	0.4819	0.0833	2.19	1.735
	Gully_G [e]	3.96×10^{-6}	0.4640	0.0888	1.63	1.704
	Gully_D [e]	3.23×10^{-8}	0.2500	0.0599	1.01	1.374
	Gully_S [e]	3.19×10^{-5}	0.5221	0.0931	3.05	1.667
	20~50 cm [e]	5.00×10^{-6}	0.4500	0.0976	1.20	1.503
	Below 50 cm [e]	3.00×10^{-6}	0.4000	0.1000	0.92	1.542

[a] Residual soil-water content.; [b] Parameter related to the inverse of the air-entry pressure [29].; [c] Parameter related to the pore-size distribution [29].; [d] Values were all calibrated in WMG BVP.; [e] Values were all measured in April 2014.

3. Results and Discussion

3.1. Results of Calibration and Validation in WMG Catchment

Table 6 compares the observed and simulated peak discharges and the time to peak discharges of water and sediment of the ten rainfall-runoff events. EF values for water discharge and sediment discharge were calculated, separately. The EF values in the four calibration events were all higher than 0.70, and the model produced the best simulation results in event 4, which was characterized as low rainfall intensity and rainfall amount with low water and sediment discharges. The average EF values of water discharge and sediment discharge for the six validation events were all higher than 0.55, illustrating that model performances were acceptable both in water discharge simulation and sediment discharge simulation. The observed and simulated hydrographs (Figure 5) and sedigraphs (Figure 6) for the 10 events matched reasonably well, also indicating a good representation of the hydrologically-driven sediment transport processes.

Table 6. Comparisons of observed versus simulated results for 10 rainfall-runoff events in the first simulation stage.

N [a]	Date	P (mm)	Q_{pk} (m³/s) [b] O	S	T_{pk} (h) [c] O	S	EF_h [d]	$Q_{s,\,pk}$ (Kg/s) [b] O	S	$T_{s,\,pk}$ (h) [c] O	S	EF_s [d]
1	1962/7/15	29.3	1.90	1.62	5.8	5.4	0.78	1276	890	5.8	5.7	0.71
2	1963/7/05	64.8	0.58	0.48	6.9	7.1	0.79	37	25	6.9	7.3	0.72
3	1964/9/07	26.8	1.60	1.59	11.5	11.6	0.71	133	120	11.5	11.6	0.72
4	1965/7/07	19.0	0.20	0.15	2.5	2.6	0.93	7	8	2.3	2.5	0.88
5	1961/7/05	46.3	1.50	1.32	5.7	6.0	0.59	93	74	5.7	6.0	0.60
6	1961/8/01	54.7	21.00	12.12	2.8	4.0	0.71	9240	4863	2.8	5.0	0.41
7	1961/9/27	35.2	1.53	1.10	4.9	5.3	0.68	137	91	4.9	4.9	0.62
8	1964/7/05	133.1	7.07	4.61	16.1	16.0	0.60	4242	3726	16.1	16.3	0.50
9	1964/7/12	24.6	1.63	1.43	5.2	5.3	0.56	904	681	5.2	5.1	0.69
10	1964/9/11	51.4	1.82	1.63	8.1	8.0	0.65	820	777	8.1	8.0	0.45

[a] Event number. Events 1 to 4 were used for calibration, and events 5 to 10 were used for validation.; [b] Q_{pk} referred to the peak discharge, and $Q_{s,\,pk}$ was the peak sediment discharge.; [c] T_{pk} referred to the time to Q_{pk} from the start of the event, and $T_{s,\,pk}$ was the time to $Q_{s,\,pk}$ from the start of the event.; [d] EF_h referred to the EF value for hydrologic-response simulation, and EF_s was the EF value for sediment-transport simulation.

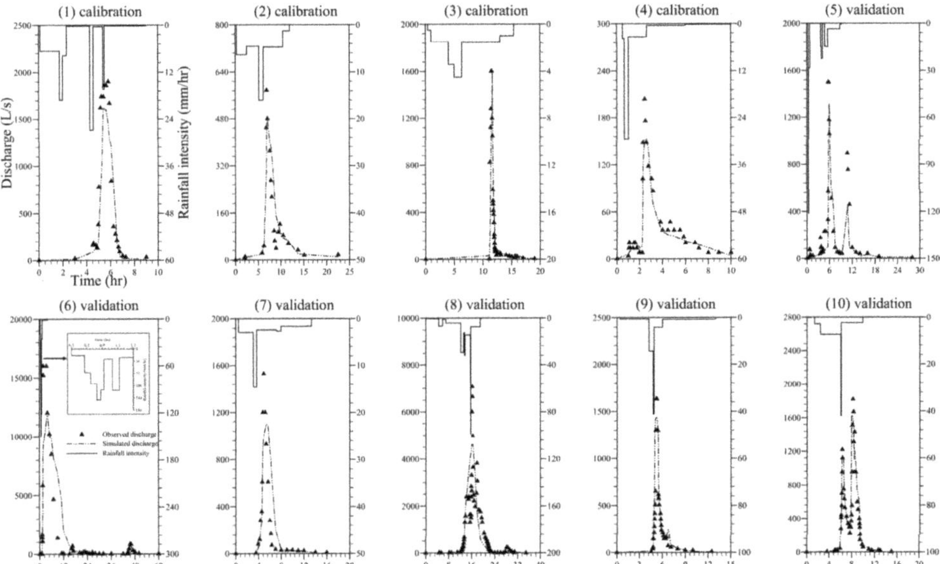

Figure 5. Observed and simulated hydrographs for the ten events in the first simulation stage. **(1)**–**(4)** were events for calibration, **(6)**–**(10)** were events for validation.

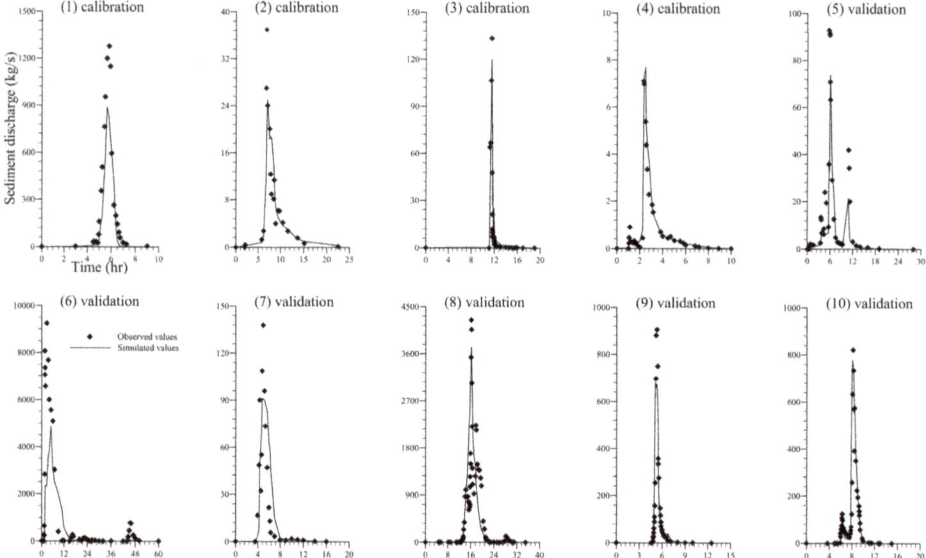

Figure 6. Observed and simulated sedigraphs for the ten events in the first simulation stage. (**1**)–(**4**) were events for calibration, (**6**)–(**10**) were events for validation.

3.2. Water Redistribution in Long-Term Simulations of GDG Catchment

Table 7 provides the InHM simulated GDG water-balance components for each year of the four scenarios. Precipitation of each year was redistributed into three water-balance components (i.e., outflow, storage change, and evapotranspiration). Surface outflow referred to the surface runoff at the outlet of GDG gully, while subsurface outflow included lateral flow leaving the downstream outlet and vertical drainage. Inspection of Table 7 showed that the percentage of each component in the four scenarios was obviously different:

- Surface runoff decreased significantly after the construction of GDG1 dam (i.e., averagely reduced by 12.74%) and remained a low percentage in the following 12 simulation years.
- Surface storage responded in two different ways in the four scenarios. Surface storage dramatically increased by 8.94% and 18.37% immediately at the end of the first year in SD and EDS scenarios, indicating a large amount of stormwater during the rainy seasons of 1955 and 1959 was retained behind check dams. Comparing to the large growth in 1955 and 1959, small increases of surface storage occurred in the following years of SD and EDS scenarios (i.e., 1956, 1957, 1958, 1960, and 1962). PD and CDS scenarios had similarly slight surface storage increases (i.e., average changes were +0.90% and +1.53%, respectively) for different reasons: (1) surface runoff left the gully without the interception of dams in PD scenario; (2) surface water was retained behind dams and infiltrated into subsurface quickly in CDS scenario. (See the expanding permeable sedimentary layers with relatively high infiltration rate in Figure 7)
- Subsurface outflow increased to 7.17% after the construction of GDG1 dam. The mean annual subsurface outflow was 33.5 mm (7.17%) and 54.0 mm (11.49%) for SD and EDS scenarios, increasing by 4.32% after the construction of four more dams. The subsurface outflow remained stable in CDS scenario, fluctuating slightly between wet years and dry years.
- Subsurface storage started increasing significantly in 1956, and fluctuating slightly between dry years and wet years. The check dams' role in transforming from a surface reservoir to subsurface reservoir was obvious when comparing the storage-change tendencies of the surface and subsurface.

- Evapotranspiration, the largest term in GDG water balance, showed a slight decline from PD to CDS scenarios. It could be explained by the enhancement of infiltration in the sedimentary areas, which reduced the residence time of rainfall as surface water. In SD and EDS scenarios, dry years showed higher ET proportions than wet years.

Table 7. Water redistribution in long-term simulations of GDG catchment.

Scenario	Year	Precipitation (mm)	Outflow (%)		Storage Change (%)		ET (%)
			Surface	Subsurface	Surface	Subsurface	
PD	1953	476.50	15.00	5.00	0.80	1.70	77.50
	1954	449.30	14.32	3.94	1.00	−0.63	81.37
	Average	462.90	14.66	4.47	0.90	0.54	79.43
SD	1955	326.90	1.01	4.51	8.94	4.93	80.61
	1956	610.88	2.11	7.35	3.20	11.01	76.33
	1957	354.70	1.36	7.54	3.18	8.69	79.23
	1958	574.50	3.20	9.27	2.24	15.10	70.19
	Average	466.75	1.92	7.17	4.39	9.93	76.59
EDS	1959	590.80	4.39	8.74	18.37	11.09	57.41
	1960	346.70	1.03	12.31	4.87	14.99	66.80
	1961	591.11	2.57	10.74	8.96	12.38	65.35
	1962	353.00	0.97	14.17	3.01	6.74	75.11
	Average	470.40	2.24	11.49	8.80	11.30	66.17
CDS	2010	355.00	1.12	9.79	1.64	10.24	77.21
	2011	562.30	3.71	11.62	1.02	12.12	71.53
	2012	462.70	3.58	11.69	0.67	9.10	74.96
	2013	733.70	3.08	13.15	2.79	8.88	72.10
	Average	646.03	2.90	11.56	1.53	10.09	73.92

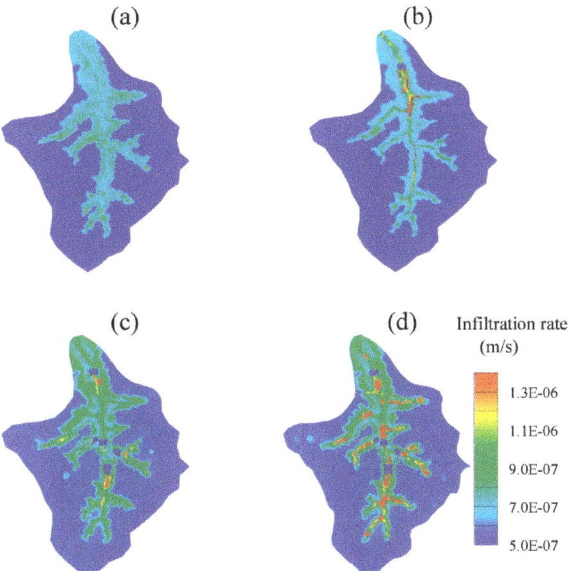

Figure 7. Snapshots of infiltration rate at the time of peak discharge of four events with similar rainfall characteristics extracted from the four scenarios, respectively. (**a**) PD scenario; (**b**) SD scenario; (**c**) EDS scenario; (**d**) CDS scenario.

3.3. Impacts of Check Dams on Groundwater in Long-Term Simulations of GDG Catchment

To further capture the influences of check dams on groundwater and the sedimentary processes, the water table profiles and surface elevations along the gully of each year were compared (Figures 8–11). The impacts of check dams on the GDG1 dam-controlled reach (i.e., A0-A1 reach in Figure 3) in different stages (i.e., SD, EDS, and CDS) are compared in Figures 8–10. Figure 11 compared the water table changes and sedimentary processes of the two-dam reach (i.e., A1-A2 reach in Figure 3) in EDS and CDS scenarios. Perusal of the water table changes lead to the following results:

- The water table rose sharply at 200 m upstream from the dam heel of GDG1 and dropped sharply at the dam toe at the end of 1955 (see the green dash line with square symbols in Figure 8), indicating the formation a "subsurface reservoir".
- Surface ponding was found in the first year of SD scenario (also see the green line in Figure 8) and disappeared in the following years.
- The position where water table started rising moved upstream to the location 320 m away from the dam heel of GDG1 at the end of 1958 (see the orange dash line with diamond symbols in Figure 8) and remained there in EDS scenario (Figure 9), indicating the expansion of "subsurface reservoir" referred above.
- The water table downstream the dam averagely increased by 2.0 m and 0.6 m from the 1st year to the 4th year in SD and EDS scenarios, respectively. However, the water table in the upstream reach remained stable in the SD scenario but averagely rose by 1.0 m in the EDS scenario (see Figures 8 and 9).
- Another difference in water table profiles between SD scenario and EDS scenario was the sensitivity to climate. Comparing to the SD scenario, the water table behind GDG1 showed less extreme changes between wet years and dry years in the EDS scenario (see Figures 8 and 9).
- The water table profiles in the CDS scenario (Figure 10) were different from those in the former two scenarios. First, the dam's impact that caused the water table rising sharply disappeared in CDS scenario. Second, the water table along the gully had risen by 3~5 m both in the upstream reach and in the downstream reach. Third, the water table at the dam toe was relatively high and might expose as springs, which was actually observed during the field survey in April 2014.
- The water table in front of GDG2 and GDG4 in EDS scenario showed similar profile to the water table in front of GDG1 in SD scenario (see the green dash lines with square symbols in Figures 8 and 11). However, the intermediate reach of the two dams (i.e., the reach ranging from 880 m to 980 m away from the outlet) experienced a higher water table lifting (3.0 m averagely), indicating a promotion effect of water table rising by two adjacent check dams (Figure 11).

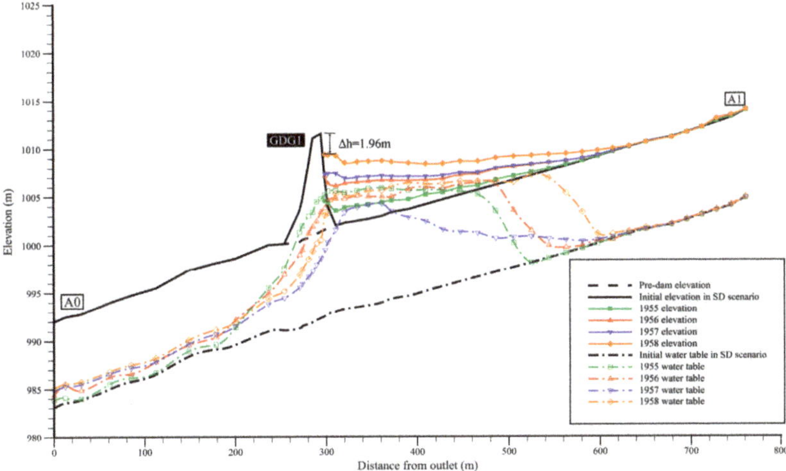

Figure 8. Surface elevation changes and water table changes of GDG1 dam-controlled reach during SD scenario. Note that the simulation results in the end of month 11 of each year were used for comparison.

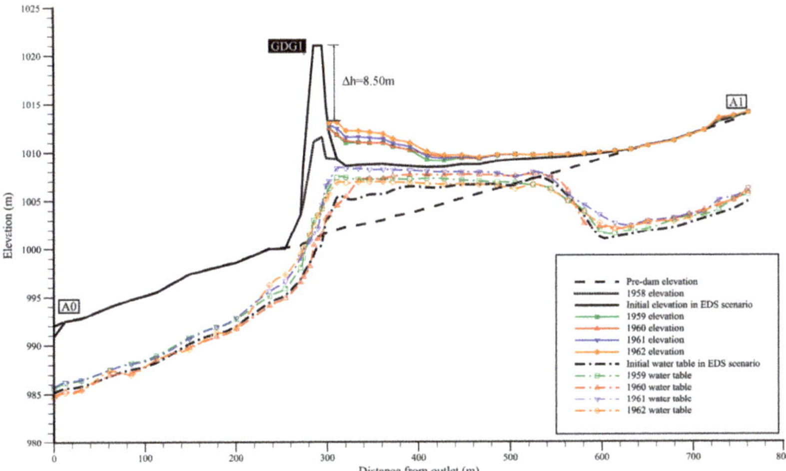

Figure 9. Surface elevation changes and water table changes of GDG1 dam-controlled reach during EDS scenario. Note that the simulation results in the end of month 11 of each year were used for comparison.

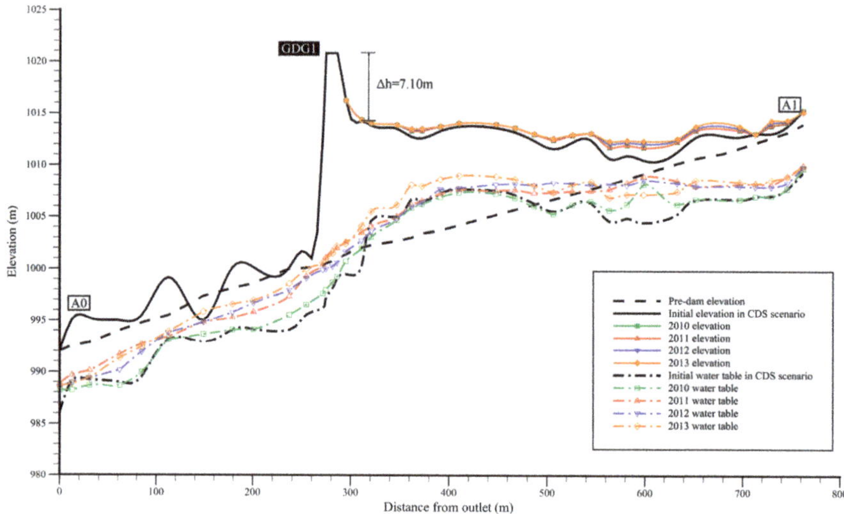

Figure 10. Surface elevation changes and water table changes of GDG1 dam-controlled reach during CDS scenario. Note that the simulation results in the end of month 11 of each year were used for comparison.

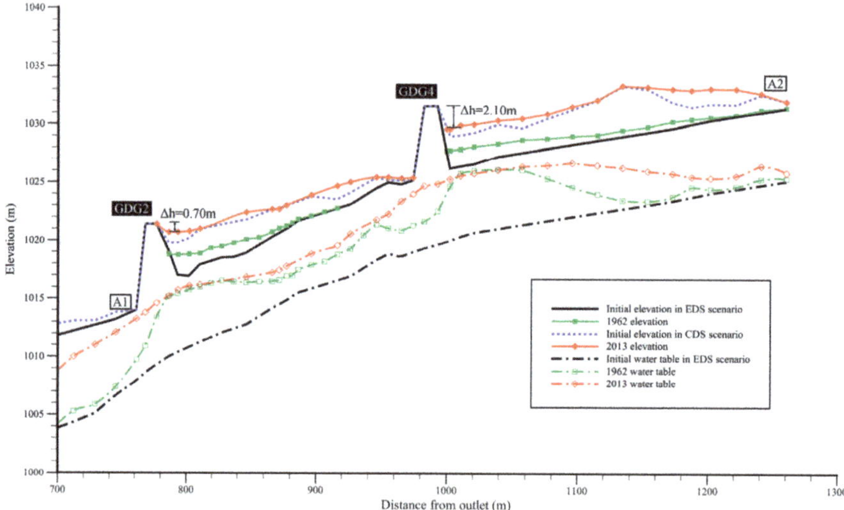

Figure 11. Surface elevation changes and water table changes of GDG2-GDG4 dam-controlled reach in EDS scenario and CDS scenario. Note that the simulation results in the end of month 11 of each year were used for comparison.

3.4. Impacts of Check Dams on Sediment Transport in Long-Term Simulations of GDG Catchment

Table 8 summarizes the eroded sediment mass of each year and the proportion of different zones. The mean erosion moduli, calculated by dividing the eroded mass to the area of GDG gully, was 233.95 t·ha^{-1}·a^{-1}, 226.18 t·ha^{-1}·a^{-1}, 206.77 t·ha^{-1}·a^{-1} and 121.74 t·ha^{-1}·a^{-1} for the four scenarios, respectively. Table 8 shows that the eroded sediment mass decreased apparently in CDS scenario. The main reason for the decrease was that slope erosion was directly alleviated by terraced farming, which replaced nearly 60% of slope farmlands with terraced farmlands. For example, 1.17 × 10^4 t sediment (39%) was eroded from slope farmlands in 1957 (SD scenario with 354.70 mm precipitation). Compared to a similar rainfall condition in 2010 (CDS scenario with 355.00 mm precipitation), only 0.40×10^4 t sediment was eroded from the remaining slope farmlands and the terraced farmlands. Another reason was that gully erosion was indirectly alleviated by the existence of check dams, which formed expanding and elevating sedimentary fields in the channel. For example, the total amount of eroded sediment mass from the two gully-zones (Gully_G, Gully_S) decreased from PD scenario to CDS scenario (Table 8).

Table 8. Eroded sediment mass of the four simulation scenarios.

Scenario	Year	Eroded Mass ($\times 10^4$ t)	Eroded Mass of Different Zones ($\times 10^4$ t) (%)				
			Slope_G	Slope_F	Slope_T	Gully_G	Gully_S
PD	1953	3.26	0.85 (26)	1.08 (33)	0.00 (0)	1.34 (41)	0.00 (0)
	1954	3.37	0.94 (28)	1.25 (37)	0.00 (0)	1.18 (35)	0.00 (0)
SD	1955	2.13	0.60 (28)	0.75 (35)	0.00 (0)	0.77 (36)	0.02 (1)
	1956	4.75	1.19 (25)	1.95 (41)	0.00 (0)	1.57 (33)	0.05 (1)
	1957	3.01	0.69 (23)	1.17 (39)	0.00 (0)	1.08 (36)	0.06 (2)
	1958	2.93	1.23 (42)	0.88 (30)	0.00 (0)	0.79 (27)	0.03 (1)
EDS	1959	3.27	0.98 (30)	1.21 (37)	0.00 (0)	0.98 (30)	0.10 (3)
	1960	3.36	0.97 (29)	1.34 (40)	0.00 (0)	0.84 (25)	0.20 (6)
	1961	3.12	0.94 (30)	1.12 (36)	0.00 (0)	0.81 (26)	0.25 (8)
	1962	1.97	0.59 (30)	0.81 (41)	0.00 (0)	0.47 (24)	0.10 (5)
CDS	2010	1.18	0.46 (39)	0.28 (24)	0.12 (10)	0.24 (20)	0.08 (7)
	2011	1.37	0.60 (44)	0.29 (21)	0.16 (12)	0.26 (19)	0.05 (4)
	2012	1.42	0.58 (41)	0.23 (16)	0.26 (18)	0.31 (22)	0.04 (3)
	2013	2.93	1.00 (34)	0.67 (23)	0.41 (14)	0.59 (20)	0.26 (9)

Table 9 compares the simulated and measured residual deposition heights (Δh) in different stages. Inspection of the surface elevation changes in Figures 8–11 and Table 9 helped to revisit the sedimentary processes of GDG gully:

- Deposition in front of GDG1 occurred quickly in SD scenario (1955–1958). The 10 m high check dam was nearly fully deposited and facing the risk of dam-break at the end of 1958 (Figure 8).
- Being heightened by 13 m before the rainy season of 1959, GDG1 was prevented from being over-deposited during the rainy season of 1959 (Figure 9). Less sediment silted in front of GDG1 in EDS scenario due to the existence of the four newly-built dams.
- GDG2 and GDG4, both of which were 8 m high, experienced quick deposition in EDS scenario (see the green solid line with square symbols in Figure 11). GDG3 and BTG, both of which located on the outlet of tributary gullies, also intercepted large amounts of sediment in EDS scenario (Table 9).
- Comparing to PD scenario, a sedimentary layer which has an average thickness of 8 m formed in front of GDG1 dam at the beginning of CDS scenario (see the black solid line in Figure 10). The flat sedimentary area with relatively high soil water content was turned into productive croplands. Long-term alternative erosion and sedimentary formed the undulating terrain in the downstream reach of GDG1. More sediment deposited in the upper reach 600 m away from the outlet rather than at the dam heel for GDG1.
- GDG1, which formed productive croplands, had a residual deposition height (Δh) of 7.10 m at the end of CDS scenario. However, the other four check dams all had been almost fully-filled and were vulnerable to floods.
- Table 9 also shows that most of the simulated Δh values were higher than the observed values (except for GDG3 in 1962, GDG2 and GDG4 in 2013), indicating that the sediment volumes deposited in front of dams were underestimated in most simulations.

Table 9. Observed versus simulated residual deposition height (Δh).

Name	1958			1962			2013		
	O [a] (m)	S [b] (m)	D [c] (%)	O [a] (m)	S [b] (m)	D [c] (%)	O [a] (m)	S [b] (m)	D [c] (%)
GDG1	0.50	1.96	292.0	7.90	8.50	7.6	7.00	7.10	1.4
GDG2	-	-	-	3.60	3.90	7.7	1.00	0.70	−30.0
GDG3	-	-	-	6.93	6.20	−10.5	1.10	2.10	90.9
GDG4	-	-	-	4.20	5.10	21.4	2.90	2.10	−27.6
BTG	-	-	-	4.70	5.65	20.2	0.80	1.05	31.3

[a] O refers to the observed Δh.; [b] S refers to the simulated Δh.; [c] D refers to Difference = [(Simulated-Observed)/Observed] × 100.

3.5. Model Performance Evaluation: Influence of Gravity Erosion

Figures 5 and 6 and Table 6 show underestimations of peak sediment discharges, which were most obvious in event 6 with long-duration high rainfall intensities (i.e., 120–150 mm/h). Most of the simulated residual deposition heights (Δh) were larger than the observed ones (8 out of 11), indicating an underestimation of deposited sediment behind check dams (Table 9). For example, the simulated Δh (2.10 m) was nearly two times the observed one (1.10 m) for GDG3 dam in 2013. The most likely reason is that serious collapses occurred on the hillslopes near the two dams. Soils used for constructing check dams were obtained from nearby hillslopes on the two sides of check dams, making the slopes steeper and easier to induce collapse or landslide. According to the inquiries from local farmers, there were several collapses in WMG catchment during the large floods in 27 June 2013 (sediment yield 16,318 tons), one of which occurred on the hillslope near GDG3 dam. Another reason might be the construction of the road on the dam crest. Extra soil from dam crest was poured into the sedimentary area during road construction.

Except for hydraulic erosion, gravity erosion, which usually occurred in the form of landslide or collapse, could dramatically increase sediment yields in a single storm. Since the sediment transport component of InHM was first developed to apply in the hydraulic-erosion-dominating catchments, the sediment transport processes induced by landslides or collapses are not supported yet. Inspiringly, several studies have recently shown the potential of physics-based simulations to forecast landslides and collapses from hydrogeological perspectives (i.e., subsurface fluid-pressures changes in failure-prone locations) [30–32]. Future works will be focused on combining the two erosion processes, both of which are related to hydrologic responses, to more accurately predict sediment yields and estimate the sedimentary processes in the Loess Plateau.

4. Conclusions

A comprehensive physics-based model InHM, after model calibration and validation, was employed in a small gully catchment with a more than 50-year-old check dam system. The simulated residual deposition heights reasonably match with the observed values, indicating the ability of InHM in check dam planning and management. The impacts of check dams on the hydrological response and landforms were investigated and the results are summarized as follows.

(1) Check dams do change the water redistribution in catchment-scale. GDG1 check dam near the gully outlet can effectively reduce surface outflow and increase subsurface outflow. The four other check dams located in the upstream can protect GDG1 from floods risk and promote the redistribution of water.

(2) The construction of check dam can significantly change the water table profile along the gully. A surface reservoir behind the dam will be formed in the first few years and gradually transformed into a subsurface reservoir, resulting in relatively high soil water content in the sedimentary areas. After more than 50 years of operation, the water table along the gully has been averagely elevated by 3–5 m. Adjacent check dams have a promoting effect on water table rising.

(3)   Check dams intercept surface water and force sediment in the water to deposit. Gully erosion can be alleviated indirectly due to the formation of expanding sedimentary areas.

The simulations reported herein demonstrate that physics-based simulation can provide a framework for better understanding the impacts (both on hydrological response and landform evolution) of check dams in the Loess Plateau. The study is like a "revisit" to the hydrologic and geomorphic changes that occurred after the construction of these dams, and a prediction of what will happen after long-term operation of the dam system, which could be a useful reference in guiding future check dam construction and management. However, as with previous InHM simulations studying the impacts of human activities on the environment, this study also suffered from the difficulty of validation because of lacking observation data that are not only accurate and credible, but also of the correct kind for distributed simulations [22,24]. Although the residual deposition height (Δh) can be used as a reference to prove the simulated sedimentary processes, more detailed measurements on catchment-scale erosion/deposition are needed to aid future physics-based distributed simulations of hydrologic-response-driven geomorphic evolution processes. Furthermore, gravity erosion should be considered in future InHM simulations and more integrated long-term continuous observations such as the groundwater table distribution along the gully and subsurface outflow rates are needed to further guide the search for a comprehensive understanding of hydrologic responses.

Author Contributions: Conceptualization, H.T.; methodology, J.G.; software, Q.R.; validation, H.T.; formal analysis, H.T., J.G.; investigation, H.T.; writing—original draft preparation, H.T.; writing—review and editing, J.G.; supervision, Q.R., J.G.; funding acquisition, Q.R.

Funding: This study was funded by the National Key Research and Development Program of China (2016YFC0402404), and the National Natural Science Foundation of China (51679209, 51379184).

Acknowledgments: Special thanks were given to Suide Soil and Water Conservation station for providing essential field data.

Conflicts of Interest: The authors declare no conflict of interest.

Appendix A The Integrated Hydrologic Model (InHM)

Appendix A.1 Hydrologic-Response Module

The 3D subsurface flow in variably saturated porous medium is estimated in InHM by:

$$\nabla \cdot f^a \vec{q} \pm q^b \pm q^e_{ps} = f^v \frac{\partial \phi S_w}{\partial t} \tag{A1}$$

$$\vec{q} = -k_{rw} \frac{\rho_w g}{\mu_w} \vec{k} \nabla (\psi_p + z) \tag{A2}$$

where $f^a[-]$ is the area fraction related to each continuum, $\vec{q}\,[LT^{-1}]$ is the Darcy flux, $q^b[T^{-1}]$ is a specified rate source/sink, $q^e_{ps}[T^{-1}]$ (equal to $-q^e_{sp}$) is the rate of water exchange between the porous medium and surface continua, $f^v[-]$ is the volume fraction associated with each continuum, $\phi[L^3L^{-3}]$ is porosity, $S_w[L^3L^{-3}]$ is the water saturation, $t[T]$ is time, $k_{rw}[-]$ is the relative permeability, $\rho_w[ML^{-3}]$ is the density of water, $g[LT^{-2}]$ is the gravitational acceleration, $\mu_w[ML^{-1}T^{-1}]$ is the dynamic viscosity of water, $\vec{k}[L^2]$ is the intrinsic permeability vector, $\psi_p[L]$ is the pressure head, and $z[L]$ is the elevation head.

The diffusion wave approximation to the depth-integrated shallow wave equations is employed in InHM to estimate the 2D transient surface flow on the land surface (both overland and open channel), with the conservation of water on the land surface expressed by:

$$\nabla \cdot \psi_s^{mobile} \vec{q}_s \pm a_s q^b \pm a_s q^e_{sp} = \frac{\partial (S_{w_s} H_t + \psi_{im})}{\partial t} \tag{A3}$$

$$\vec{q}_s = -\frac{(\psi_s^{mobile})^{2/3}}{\vec{n}\,\Phi^{1/2}}\nabla(\psi_s + z) \tag{A4}$$

where $\psi_s[L]$ is the surface water depth, $\vec{q}_s[LT^{-1}]$ is the surface water velocity calculated by a two-dimensional form of the Manning's equation, $a_s[L]$ is the surface coupling length scale, $q^b[T^{-1}]$ is the source/sink rate (i.e., rainfall/evaporation), $q_{sp}^e[T^{-1}]$ is the rate of water exchange between the surface continua and porous medium, $H_t[L]$ is the average height of non-discretized surface microtopography, $\vec{n}[TL^{-1/3}]$ is the Manning's surface roughness tensor, and $\Phi[-]$ is the friction (or energy) slope; $\psi_s^{mobile}[L]$ and $\psi_{im}[L]$ refer to the water depth exceeding and held in depression storage, respectively.

The calculated daily potential evapotranspiration was then incorporated into InHM as actual evapotranspiration (ET) using a set of sink functions [21]:

$$Q_b^E = \alpha(\psi)q_{max}^E A_b \tag{A5}$$

where $Q_b^E[L^3T^{-1}]$ represents the volumetric evapotranspiration rate, $\alpha(\psi)[-]$ is a response function of the saturation of the porous medium and the degree of ponding at the land surface, $q_{max}^E[LT^{-1}]$ is the potential evapotranspiration rate per unit area estimated by the revised Penman-Monteith method, $\psi[L]$ is the pressure head of the subsurface nodes or water depth of the surface nodes, and $A_b[L^2]$ is the area associated with the surface water equation.

The first-order coupling between the surface and subsurface continua, driven by pressure head gradients, occurs via a thin soil layer of thickness, a_s in Equation (A3). The control volume finite-element method is employed to discretize the equations in space. Each coupled system of nonlinear equations is solved implicitly using Newton iteration. A more detailed description of the hydrologic-response module of InHM can be found in [19].

Appendix A.2 Sediment-Transport Module

Depth-integrated multiple-species sediment transport, restricted to the surface continuum, is calculated for each sediment species by:

$$\frac{\partial C_{sed}}{\partial t} = -\nabla \cdot \vec{q}_s C_{sed} + \frac{1}{V_w}\left(e_{sed} + \sum_{j=1}^{BC} q_{s_j}^b C_{sed_j}^*\right) \tag{A6}$$

$$e_{sed} = e_s + e_h \tag{A7}$$

where $C_{sed}[L^3L^{-3}]$ is volumetric sediment concentration, $\vec{q}_s[LT^{-1}]$ is the depth-averaged surface water velocity, $V_w[L^3]$ is the volume of water at the node, $e_{sed}[L^3T^{-1}]$ is the volumetric rate of soil erosion and/or deposition, $q_{s_j}^b[L^3T^{-1}]$ is the rate of water added/removed via the jth boundary condition, $C_{sed_j}^*[L^3L^{-3}]$ is the sediment concentration of the water added/removed via the jth boundary condition, BC is the total number of boundary conditions. The net erosion rate is the sum of the rainsplash erosion rate $e_s[L^3T^{-1}]$ and the hydraulic erosion rate $e_h[L^3T^{-1}]$. The rainsplash erosion rate of each species in Equation (A7) is calculated as:

$$e_{s_i} = \begin{cases} \sigma_i c_f F(\psi_s)(\cos(\theta)\cdot r)^b A_{3D} & , \quad q > 0 \\ 0 & , \quad q < 0 \end{cases} \tag{A8}$$

$$F(\psi_s) = \exp(-c_d\psi_s) \tag{A9}$$

where $\sigma_i[-]$ is the source fraction of species i, $c_f[(TL^{-1})^{b-1}]$ is the rainsplash coefficient, $b[-]$ is the rainfall intensity exponent, $\theta[-]$ is the angle of the element from horizontal, $r[LT^{-1}]$ is the rainfall intensity, $A_{3D}[L^2]$ is the three-dimensional area associated with the node, and $q[LT^{-1}]$ is the sum of rainfall intensity and infiltration intensity; $F(\psi_s)[-]$ is a damping function, related to surface water depth $\psi_s[L]$ and the surface water damping-effectiveness coefficient $c_d[L^{-1}]$, to represent the reduction in splash erosion with increasing surface water depth. The hydraulic erosion rate in Equation (A7) estimated as:

$$e_{h_i} = \alpha_{sed_i}\left(C_{sed_{max_i}}\sigma_i - C_{sed_i}\right) \tag{A10}$$

$$C_{sed_{max_i}} = 0.05\frac{\vec{q}_s q_*^3}{g^2 d_{sed_i}\psi_s\left(\gamma_{sed_i}-1\right)^2} \tag{A11}$$

$$\alpha_{sed_i} = A\begin{cases} 2v_{sed_i}\xi\, , & C_{sed} > C_{sed_{max_i}}\ (erosion) \\ \varphi\vec{q}_s\psi_s\chi_i\, , & C_{sed} < C_{sed_{max_i}}\ (deposition) \end{cases} \tag{A12}$$

where $\alpha_{sed_i}[L^3T^{-1}]$ is the hydraulic erosion transfer coefficient for species i, $C_{sed_{max_i}}[L^3L^{-3}]$ is the concentration at equilibrium transport capacity for species i, $q_*[LT^{-1}]$ is the local shear velocity, $d_{sed_i}[L]$ and $\gamma_{sed_i}[-]$ are the particle diameter and specific gravity, respectively; $A[L^2]$ is the area associated with the node in Equation (A12), $v_{sed_i}[LT^{-1}]$ is the particle settling velocity, $\xi[-]$ is a coefficient related to turbulence in the surface water due to raindrop impact, $\varphi[L^{-1}]$ is an erodibility coefficient related to surface properties and texture, and $\chi_i[-]$ is the particle erodibility factor (ranging from zero to one).

References

1. Shi, H.; Shao, M. Soil and water loss from the Loess Plateau in China. *J. Arid Environ.* **2000**, *45*, 9–20. [CrossRef]
2. Xin, Z.; Xu, J.; Zheng, W. Spatiotemporal variations of vegetation cover on the Chinese Loess Plateau (1981–2006): Impacts of climate changes and human activities. *Sci. China Ser. D Earth Sci.* **2008**, *51*, 67–78. [CrossRef]
3. Xu, X.; Zhang, H.; Zhang, O. Development of check dam systems in gullies on the Loess Plateau, China. *Environ. Sci. Policy* **2004**, *7*, 79–86.
4. Jin, Z.; Cui, B.; Song, Y.; Shi, W.; Wang, K.; Wang, Y.; Liang, J. How Many Check Dams Do We Need To Build on the Loess Plateau? *Environ. Sci. Technol.* **2012**, *46*, 8527–8528. [CrossRef] [PubMed]
5. Wang, Y.F.; Fu, B.J.; Chen, L.D.; Lu, Y.H.; Gao, Y. Check Dam in the Loess Plateau of China: Engineering for Environmental Services and Food Security. *Environ. Sci. Technol.* **2011**, *45*, 10298–10299. [CrossRef] [PubMed]
6. Xu, Y.; Fu, B.; He, C. Assessing the hydrological effect of the check dams in the Loess Plateau, China, by model simulations. *Hydrol. Earth Syst. Sci.* **2013**, *17*, 2185–2193. [CrossRef]
7. Shi, H.; Li, T.; Wang, K.; Zhang, A.; Wang, G.; Fu, X. Physically-based simulation of the streamflow decrease caused by sediment-trapping dams in the middle Yellow River. *Hydrol. Process.* **2015**, *30*, 783–794. [CrossRef]
8. Boix-Fayos, C.; de Vente, J.; Martínez-Mena, M.; Barberá, G.G.; Castillo, V. The impact of land use change and check dams on catchment sediment yield. *Hydrol. Process.* **2008**, *22*, 4922–4935. [CrossRef]
9. Ran, D.-C.; Luo, Q.-H.; Zhou, Z.-H.; Wang, G.-Q.; Zhang, X.-H. Sediment retention by check dams in the Hekouzhen-Longmen Section of the Yellow River. *Int. J. Sediment Res.* **2008**, *23*, 159–166. [CrossRef]
10. Shi, H.; Wang, G. Impacts of climate change and hydraulic structures on runoff and sediment discharge in the middle Yellow River. *Hydrol. Process.* **2015**, *29*, 3236–3246. [CrossRef]
11. Zhao, P.; Shao, M.; Wang, T. Spatial distributions of soil surface-layer saturated hydraulic conductivity and controlling factors on dam farmlands. *Water Resour. Manag.* **2010**, *24*, 2247–2266. [CrossRef]
12. Callow, J.N.; Smettem, K.R.J. The effect of farm dams and constructed banks on hydrologic connectivity and runoff estimation in agricultural landscapes. *Environ. Model. Softw.* **2009**, *24*, 959–968. [CrossRef]
13. Parimalarenganayaki, S.; Elango, L. Assessment of effect of recharge from a check dam as a method of Managed Aquifer Recharge by hydrogeological investigations. *Environ. Earth. Sci.* **2015**, *73*, 5349–5361. [CrossRef]
14. Huang, J.; Hinokidani, O.; Yasuda, H.; Ojha, C.P.; Kajikawa, Y.; Li, S. Effects of the check dam system on water redistribution in the Chinese Loess Plateau. *J. Hydrol. Eng.* **2013**, *18*, 929–940. [CrossRef]
15. Gao, H. Hydro-Ecological Impact of the Gully Erosion Control Works in Loess Hilly-Gully Region. Ph.D. Thesis, The University of Chinese Academy of Sciences, Beijing, China, 2013.

16. Shaanxi Administration of Surveying, Mapping and Geoinformation. Available online: http://www.shasm.gov.cn/ (accessed on 4 April 2014).
17. Freeze, R.A.; Harlan, R.L. Blueprint for a physically-based, digitally-simulated hydrologic response model. *J. Hydrol.* **1969**, *9*, 237–258. [CrossRef]
18. Heppner, C.S.; Ran, Q.; VanderKwaak, J.E.; Loague, K. Adding sediment transport to the integrated hydrology model (InHM): Development and testing. *Adv. Water Resour.* **2006**, *29*, 930–943. [CrossRef]
19. Ran, Q.; Heppner, C.S.; VanderKwaak, J.E.; Loague, K. Further testing of the integrated hydrology model (InHM): Multiple-species sediment transport. *Hydrol. Process.* **2007**, *21*, 1522–1531. [CrossRef]
20. VanderKwaak, J.E. Numerical Simulation of Flow and Chemical Transport in Integrated Surface-Subsurface Hydrologic Systems. Ph.D. Thesis, University of Waterloo, Waterloo, ON, Canada, 1999.
21. VanderKwaak, J.E.; Loague, K. Hydrologic-response simulations for the R-5 catchment with a comprehensive physics-based model. *Water Resour. Res.* **2001**, *37*, 999–1013. [CrossRef]
22. Carr, A.E.; Loague, K.; VanderKwaak, J.E. Hydrologic-response simulations for the North Fork of Caspar Creek: Second-growth, clear-cut, new-growth, and cumulative watershed effect scenarios. *Hydrol. Process.* **2014**, *28*, 1476–1494. [CrossRef]
23. Mirus, B.B.; Ebel, B.A.; Loague, K.; Wemple, B.C. Simulated effect of a forest road on near-surface hydrologic response: Redux. *Earth Surf. Process. Landf.* **2007**, *32*, 126–142. [CrossRef]
24. Heppner, C.S.; Loague, K. A dam problem: Simulated upstream impacts for a Searsville-like watershed. *Ecohydrology* **2008**, *1*, 408–424. [CrossRef]
25. Ran, Q.; Loague, K.; VanderKwaak, J.E. Hydrologic-response-driven sediment transport at a regional scale, process-based simulation. *Hydrol. Process.* **2012**, *26*, 159–167. [CrossRef]
26. Nash, J.E.; Sutcliffe, J.V. River flow forecasting through conceptual models part 1-a discussion of principles. *J. Hydrol.* **1970**, *10*, 282–290. [CrossRef]
27. Ran, Q. Regional Scale Landscape Evolution: Physics-Based Simulation of Hydrologically-Driven Surface Erosion. Ph.D. Thesis, Stanford University, Stanford, CA, USA, 2006.
28. Gabet, E.J.; Dunne, T. Sediment detachment by rain power. *Water Resour. Res.* **2003**, *39*, 1002. [CrossRef]
29. Van Genuchten, M.T. A closed-form equation for predicting the hydraulic conductivity of unsaturated soils. *Soil Sci. Soc. Am. Proc.* **1980**, *44*, 892–898. [CrossRef]
30. Mirus, B.B.; Ebel, B.A.; Heppner, C.S.; Loague, K. Assessing the detail needed to capture rainfall-runoff dynamics with physics-based hydrologic response simulation. *Water Resour. Res.* **2011**, *47*, W00H10. [CrossRef]
31. BeVille, S.H.; Mirus, B.B.; Ebel, B.A.; Mader, G.G.; Loague, K. Using simulated hydrologic response to revisit the 1973 Lerida Court landslide. *Environ. Earth Sci.* **2010**, *61*, 1249–1257. [CrossRef]
32. Thomas, M.A.; Loague, K. Hydrogeologic insights for a Devil's Slide-like system. *Water Resour. Res.* **2014**, *50*, 6628–6645. [CrossRef]

 *water*

Article

Impacts of Filled Check Dams with Different Deployment Strategies on the Flood and Sediment Transport Processes in a Loess Plateau Catchment

Honglei Tang, Hailong Pan and Qihua Ran *

Institute of Hydrology and Water Resources, Zhejiang University; Hangzhou 310058, China; tanghonglei@zju.edu.cn (H.T.); panhailong@zju.edu.cn (H.P.)
* Correspondence: ranqihua@zju.edu.cn

Received: 2 April 2020; Accepted: 3 May 2020; Published: 7 May 2020

Abstract: As one of the most widespread engineering structures for conserving water and soil, check dams have significantly modified the local landform and hydrologic responses. However, the influences of sedimentary lands caused by filled up check dams on the runoff and sediment transport processes were seldom studied. Employing an integrated hydrologic-response and sediment transport model, this study investigated the influences of filled check dams with different deployment strategies in a Loess Plateau catchment. Six hypothetical deployment strategies of check dams were compared with no-dam scenario and the reality scenario. Results showed that filled check dams were still able to reduce Flood peak (Q_p) by 31% to 93% under different deployment strategies. Considerable delays of peak time and decreases were also found in scenarios, which were characterized as having larger and more connective sedimentary lands on the main channel. Reduction rates of Sediment yield (SY) and the total mass of Eroded sediment (ES) ranged from 4% to 52% and 2% to 16%, respectively, indicating that proper distributions of check dams can promote sediment deposition in the channel and reduce soil erosion. The results of this study indicate that (1) check dam systems could still be useful in flood attenuation and sediment control even when they were filled, and (2) optimizing the deployment strategies of check dams can help reduce erosion.

Keywords: check dam system; sedimentary land; flood control; sediment transport; Loess Plateau

1. Introduction

Building check dams is an effective engineering measure for flood attenuation in mountainous streams that experience torrents [1,2] and soil erosion control in arid and semiarid regions with large sediment yields [3–5]. Because of their high efficiency in sediment reduction and relatively low building investment, check dams are constructed in many countries, such as Mexico [6], Spain [7], Italy [8], Iran [9], and China [10]. The Loess Plateau in Northwest China has suffered severe soil erosion due to its combination of highly-erodible soil layers and considerable human activities [11]. The erosion rate in the region ranged from 2000 to 20,000 t km^{-2} yr^{-1} in the 1950s, resulting in the riverbed uplift of the downstream Yellow River [12]. To reduce the large catchment sediment yield caused by soil erosion, more than 110,000 check dams have been constructed in the Loess Plateau catchments over the past 50 years (especially in the 1970s) to prevent sediment from being transported to the Yellow River [13]. Significant decreases in annual runoff and sediment yields of different catchments due to the construction of check dams have been widely observed and investigated in the Loess Plateau [14–16]. However, most of the check dams have been filled quickly due to the large amount of sediment discharge during floods, and a large number of check dams were destroyed due to low design criteria and unscientific site location [17].

As an engineering structure chronically changing the landforms of a natural channel, the influences of filled check dams on the flood and sediment transport processes have seldom been quantitatively studied. In the early stage of a check dam (i.e., the design storage capacity has not been consumed), channel flow and sediment can be easily blocked by the dam body. When a check dam is filled, the block effect of dam body no longer exists, but a large sedimentary land is formed behind the check dam. The sedimentary land is normally a flat platform with relatively higher soil water content and much lower slope gradient, which is widely used as farmlands in the Loess Plateau. The hydrological impacts of sedimentary land, such as infiltration promotion [18], evaporation increase [19], and enhancement of groundwater recharge [20], have been reported. According to the low slope gradient of sedimentary land and its impacts on hydrologic response, it may influence the flood and sediment transport processes in the long term [21], even as the check dam is filled. In fact, approximately 320,000 ha of sedimentary lands behind check dams were formed during the 1950s to 2000s in the Loess Plateau [13], and the area of sedimentary lands are expected to continue increasing as more and more check dams are about to be fully filled. Considering that all check dams are going to be filled one day and the sedimentary lands formed behind check dams will exist in a long term, it is necessary to quantitatively study their influences on the flood and sediment transport processes.

On the other hand, improving the performance of check dams in flood attenuation and sediment control by optimizing the deployment strategies for treated catchments is still a challenge, given the fact that many of the check dams built in the past 50 years were prone to dam break [17]. A recent study has paid attention to the optimization of check dams in a Loess Plateau catchment [22]. Employing a distributed erosion model (WaTEM/SEDEM) based on the Revised Universal Soil Loss Equation, the life expectancy of check dams under different deployment strategies and their impacts on sediment delivery ratio of the Shejiagou catchment were compared in [22]. Several insights on extending the life expectancy of check dams and simultaneously keeping the system effective in sediment reduction were obtained in this simulation-based study. However, limited by the deficiency of WaTEM/SEDEM, which does not simulate the flow and sediment movement on the sedimentary land, the hydrologic responses of sedimentary land such as infiltration improvement and flow-velocity decrease were not considered in this study. Besides, including the studies introduced above, most of the studies on the optimization of check dam systems focused on improving the sediment trapping efficiency of the system during its lifespan [22–24]. Few studies considered how to optimize the system in advance so as to achieve a better performance of sedimentary lands in flood and sediment control when all the check dams are filled.

Driven by these twofold challenges, the study reported herein investigated the influences of filled check dams with different deployment strategies on the flood and sediment transport processes. Following the work in [22], the Integrated Hydrology Model (InHM), which is fully physics-based, was employed to the 4.26 km^2 Shejiagou catchment. The objectives were (1) to quantitatively understand whether and how much filled check dams with their sedimentary lands could attenuate flood processes and reduce sediment output, and (2) to compare the influences caused by filled check dams under different deployment strategies on the flood and sediment transport processes.

2. Data and Methods

2.1. Study Site and Data Sources

The Shejiagou catchment (109°58′ E, 37°42′ N) is a small catchment belonging to Wudinghe Basin in the Loess Plateau, China (Figure 1a). The catchment is 4.26 km^2 with an elevation ranging from 920 m to 1143 m (22.8° mean slope, Figure 1b). The catchment is covered by a thick loess layer with an average depth of over 100 m. Being highly eroded, the hillslopes of the catchment are deeply incised by gullies. The land cover of the catchment was mainly sparse grassland, bare land, and cultivated land in the 1960s (Figure 1c) [25,26]. The Shejiagou catchment has a semiarid continental climate with a mean annual precipitation of 500 mm and a mean annual potential evaporation of 1570 mm [27]. Seventy percent of the precipitation falls in the wet seasons from June to September. Infiltration-excess surface flow dominates runoff generation during floods in the catchment, and most eroded sediment is carried downstream by surface flow in heavy storms. There is normally no flow in the stream channels and gullies during inter-storm periods. The mean annual specified sediment yield was 12,800 t km^{-2} yr^{-1} based on records from 1959 to 1969 (2100 t km^{-2} yr^{-1} to 24,400 t km^{-2} yr^{-1}). According to the Strahler stream order, the river network is relatively simple and mainly consists of one third-order stream (i.e., the main channel) and several first-order streams (i.e., gullies), which symmetrically located in the two sides of the main channel (Figure 1d). Because the left-side gullies were more incised, the left side of the catchment has a steeper slope distribution and contributes more sediment than the right side. According to the model simulation in [22], major erosion areas distributed on hillslopes with slope gradients greater than 30° (Figure 1b), and the headwater areas of the catchment and the fifth gully on the right side of the main channel (Figure 1d) contribute the largest amount of sediment (see the Supplementary Materials).

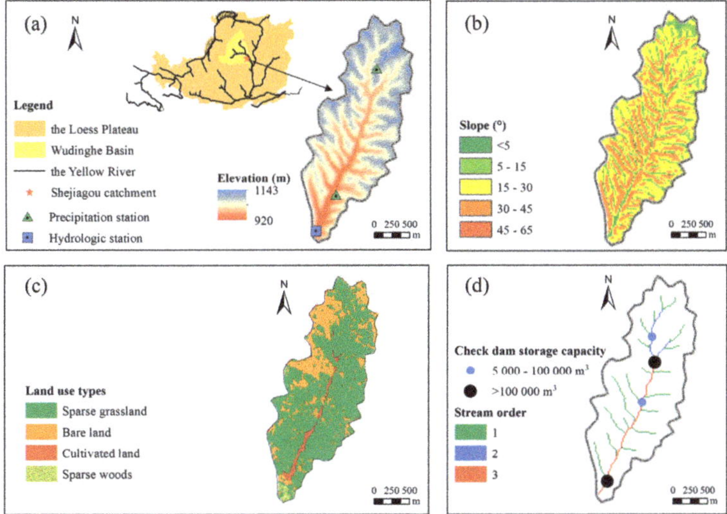

Figure 1. Information of the Shejiagou catchment: (**a**) location and Digital Elevation Model (DEM); (**b**) slope distribution; (**c**) land use types; (**d**) stream network classification and check dam distribution.

Table 1 summarizes the data set used in this study. The 20-m resolution Digital Elevation Model (DEM) of Shejiagou catchment was obtained from the State Bureau of Surveying and Mapping. Infiltration experiments and soil sampling were conducted by the Zizhou Experimental Station in Tuanshangou, a sub-catchment of Shejiagou, to obtain data of soil water characteristics during 1960–1962. By using these soil data, the parameters of van Genuchten Model [28], which was used to describe the soil-water retention and permeability functions, were determined. Runoff, sediment, and precipitation data during floods were obtained from the Experimental Station of Soil and Water Conservation of Zizhou County, which belongs to the Yellow River Conservancy Commission of the Ministry of Water Resources (YRCC). The Shejiagou hydrologic station was set at the outlet of Shejiagou catchment to monitor the runoff and sediment data including water level (m), discharge (m^3 s^{-1}), sediment concentration (kg m^{-3}), and sediment discharge (kg s^{-1}) during 1959–1969. During this monitoring period, the hydrographs and sedigraphs of 31 rainfall-runoff events were recorded. The time intervals of the flow and sediment discharge data ranged from 5 min around flow peaks to 1 h for low-flow periods. According to three field surveys of check dams, four check dams were constructed sequentially from 1957 to 1973, holding a total storage capacity of 465,100 m^3 (Figure 1d).

The Shejiagou catchment was chosen in this study mainly because (1) the data and information of the catchment were sufficient for model construction and (2) the land cover of the Shejiagou catchment was relatively simple in the 1960s, which was beneficial to model parameterization.

Table 1. Data set used in this study.

Data Type	Year	Resoloution	Data Source	Used in This Study
Topography data	1975	20 m	The State Bureau of Surveying and Mapping	Finite-element mesh generation
Soil data	1960–1962	Plot scale	Data Sharing Infrastructure of Loess Plateau [29]	Model parameterization
Flood events	1959–1969	5 min to 1 h	Data Sharing Infrastructure of Loess Plateau [29]	Model calibration
Precipitation	1959–1969	5 min to 1 h	Data Sharing Infrastructure of Loess Plateau [29]	Model calibration
Land use types	1975, 1986, 2006	30 m	Data Sharing Infrastructure of Loess Plateau [30]	Model parameterization
Check dam information	1978, 1993, 2001		Check dam survey of Chabagou watershed	Model parameterization

2.2. The Integrated Hydrology Model (InHM)

Based on the blueprint of the distributed physics-based hydrological model proposed by Freeze and Harlan [31], InHM was developed to quantitatively simulate, via a fully coupled approach, 3-D variably saturated flow in soil and 2-D flow and sediment transport across the land surface [32–34]. The 3-D Richards' equation was implemented to describe variably saturated flow in soil, while the 2-D diffusion-wave equation coupled with depth-integrated multiple-species sediment transport was applied to describe surface flow movement and sediment transportation. They were discretized in space using the control-volume finite element method and coupled in one coherent framework using a physics-based first-order flux relationship driven by pressure head gradients [35]. Newton iteration was used to implicitly solve each coupled system of nonlinear equations. More details of InHM can be found in the Appendix A.

As a research tool with robust, general and efficient solution methods [36], InHM can run event-based or continuous simulations in various catchments with different types of topography and climate [37]. Compared to many process-based models [38–41], InHM has the following advantages in this study. (1) The time step is adaptive, and the output information is spatially distributed. This feature enables the model to represent vital variables (e.g., flow velocity, water depth, and sediment concentration) in a certain area (e.g., the sedimentary land) and at a specified time step. (2) The triangle spatial elements used in InHM are flexible and accurate to construct different shapes of sedimentary lands under various check dam deployment strategies. Hydrologically active areas such as gully mouths and tail-end areas can be refined to capture small variations in flow and sediment. The model is more dynamic compared to the methods in which the check dam is generalized as a sediment-trapping function [22] and more physics-based compared to the approaches that calculate flow and sediment output of a check dam by a simple water depth-discharge relationship [42].

The model has been utilized to examine the environmental impacts of many engineering structures, such as dam removal [43] and forest roads [44]. In the Loess Plateau catchment of Shejiagou, InHM is capable of simulating rainfall-runoff processes dominated by the infiltration-excess overland flow mechanism and rainsplash erosion as well as hydraulic erosion processes in flood events [26]. These successful applications indicated that InHM is a competent tool to study the impacts of filled check dams under different deployment strategies in the Shejiagou catchment.

2.3. Modelling Scenario

Three variables, namely, number, size, and site location of check dams, were usually considered to compare the runoff-sediment reduction efficiencies of different deployment strategies [23]. Apart from increasing number and/or size of check dams, which inevitably increases the deployment costs, appropriate choice of site locations may also improve the sediment reduction rate of a check dam system [22]. To study the impacts on runoff and sediment processes of filled check dam systems under different deployment strategies, three steps were conducted to set up modelling scenarios: (1) add different check dam systems into the model, (2) fill the check dams via rising the elevations of the channels influenced by each check dam to represent a filled check dam, and (3) conduct rainfall-runoff events in each modelling scenarios.

In this study, eight scenarios were designed (Figure 2, Table 2). Scenario-0 and Scenario-7 were cases which represent the no-check dam situation and the current situation of Shejiagou catchment, respectively. Scenarios 1-6 were cases considering different numbers, sizes and site locations of check dams. Although in all scenarios check dams were all filled (i.e., the design storage capacity was used up), the sizes still mattered because the sedimentary land formed by the check dam was different between small check dams and large check dams. For example, the total areas of sedimentary lands in scenario-2 and scenario-3 were 11.21×10^4 m^2 and 5.93×10^4 m^2, respectively. As for the determination of site locations, whether the check dams were constructed on gullies (the first or second order streams) or on the main channel (i.e., the third order stream) was considered in this study. For example, the second check dam was constructed on a first-order stream, which receives a high sediment load (2000–3600 t yr^{-1}, see the Supplementary Materials) in scenario-5, while the second check dam was on the main channel which receives a much higher cumulative sediment load (9000–37,521 t yr^{-1}) in scenario-6. The six scenarios (i.e., scenarios 1–6) were conceived to compare the effects of deployment strategies with different number (e.g., scenarios 1, 2, 4), size (e.g., scenarios 5, 6), and site location (e.g., scenarios 2, 3) of check dams. The six scenarios were also compared with scenario-0 and scenario-7, which provides indications of how the performance of a filled check dam system could be improved.

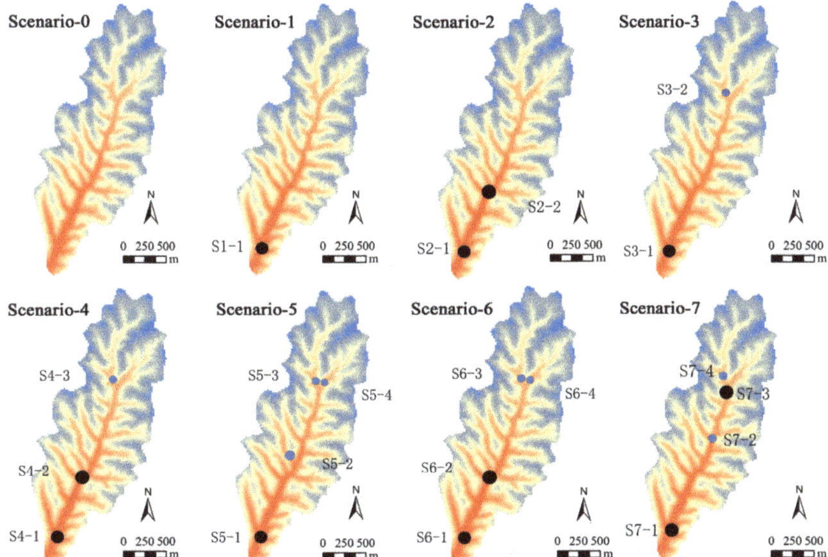

Figure 2. Check dam systems under different deployment strategies. Note that scenario-7 is the current deployment strategy of Shejiagou catchment.

Table 2. Description of the eight scenarios in the study.

Scenario Number	Number of Filled Check Dams	Size Combination	Total Area of Sedimentary Lands ($\times 10^4$ m^2)	Main Feature of the Scenario
0	0	(0,0)	0	No check dam in the catchment
1	1	(1,0)	4.69	One large check dam downstream
2	2	(2,0)	11.21	Two large check dams connected
3	2	(1,1)	5.93	Two check dams with different sizes
4	3	(2,1)	12.45	Three check dams with different sizes
5	4	(1,3)	9.00	One large check dam on main the channel and targeted treatment to erosion-prone gully
6	4	(2,2)	14.18	Two large check dams on main the channel and targeted treatment to erosion-prone gullies
7	4	(2,2)	8.92	Current deployment strategy

2.4. Model Settings and Parameters

2.4.1. Finite-Element Mesh

Eight 3-D meshes for the scenarios were constructed by adding subsurface layers to 2-D triangular surface meshes. The major differences of the eight 3-D meshes were the different combination of check dams and the sedimentary lands formed behind check dams. Figure 3a shows the 3-D mesh that contained 10,062 nodes and 19,918 triangular elements in scenario-1, in which only one filled check dam was considered. The cell dimensions of all the surface meshes varied from 180 m near the headwater boundary to 20 m along the gullies, and the meshes of the check dam were refined to 3 to 10 m. Sixteen subsurface meshes were added under the surface mesh with varying thickness from 0.10 m near the surface to 39.30 m near the bedrock with a constant basal elevation of 900 m (see Figure 3b). This setting of mesh size ensured a fine resolution in the hydrologically active areas (i.e., the channel, sedimentary land, and the near-surface soil) and simultaneously saved computation resources in the relatively inactive areas (i.e., the headwater regions and deep spaces). Figure 3c shows the check dam (i.e., S1-1) and its sedimentary land in the finite-element mesh. To model a filled check dam, the sedimentary land was elevated to the same elevation of the spillway, indicating that the dam body can no longer block surface flow. Based on the fact that the sedimentary lands normally have low slope gradients (usually less than 2%) when compared to the original channel [45,46], it was assumed that, for each filled check dam, the nodes on the sedimentary land have the same elevations to simplify the model construction.

2.4.2. Boundary Conditions and Initial Conditions

Two subsurface boundary conditions were assigned to each 3D mesh: (1) impermeable for each lateral face and the bottom face (A-E-D-C-B and F-J-I-H-G in Figure 3a); and (2) a local sink (i.e., head-dependent flux) at the down-gradient face [47], subsurface flowing out through A-B-G-F in Figure 3b. The surface boundary conditions were driven by the given precipitation to the whole catchment and constrained by a critical water depth boundary condition (no dam at the outlet, so the critical water depth = 0 m) at the catchment outlet (A–B in Figure 3b). A one-hour uniform rainfall with an intensity of 90 mm hr^{-1} was chosen to conduct the rainfall-runoff simulations in this study because this rainfall intensity was common in the Shejiagou catchment to induce floods, according to the 31 recorded rainfall-runoff events during 1959–1969. The evapotranspiration process was not considered in these storm event simulations.

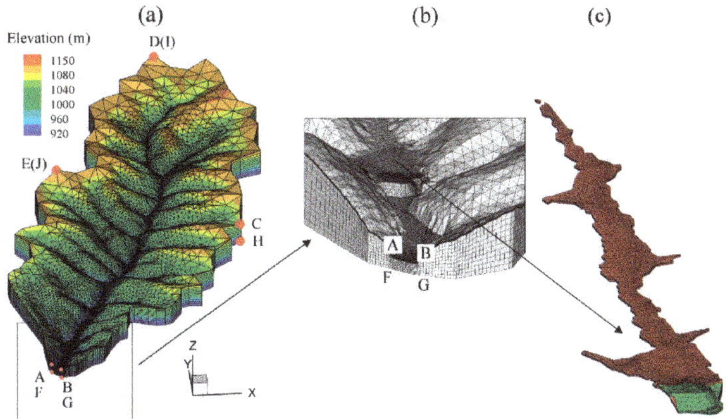

Figure 3. 3D mesh used in scenario-1: (**a**) the whole catchment and boundary conditions; (**b**) the outlet area and the outflow boundary; (**c**) the surface mesh representing the filled check dam (S1-1) and its sedimentary land. Note that the elevation of sedimentary land was the same as that of the spillway on the right side of the check dam, indicating the check dam was filled.

Before all the scenario runs, a drainage run with no rainfall input was started to drive the whole catchment from a fully saturated state to a realistic state, and it stopped when the simulated discharge at the catchment outlet became stable and matched the average no-rainfall discharge of the catchment at the beginning of wet seasons (i.e., 0.006 m s^{-1}). This drainage simulation was performed to obtain a continuous initial distribution of the hydraulic head that was close to the reality of the catchment. Using this initial condition, a one-week warming-up stage with 0.75 mm hr^{-1} rainfall was conducted at the beginning of each scenario run, and then the one-hour 90 mm hr^{-1} rainfall for the scenario run was input to the model. Each simulation ended 42 h after the rainfall, lasting for 2 days total.

2.4.3. Model Parameters

A near-surface loess layer (0–0.5 m) and a deep loess layer (0.5 m to the bedrock) constituted the soil profile in the model. The near-surface soil layer was classified into three types, i.e., the normal surface loess soil (labeled as "SJG-soil (0–0.5 m)"), the silted soil of the sedimentary land (labeled as "Sedimentary land"), and the compacted soil of the check dam (labeled as "Check dam").

Table 3 summarizes the values of parameters that represent the soil hydraulic characteristics. The saturated hydraulic conductivities (K_{sat}) of soils were calibrated (see Section 2.4.4), while the porosity values and van Genuchten parameters of the soils were derived from the experiment data by the Zizhou Experimental Station. Table 4 summarizes the land surface parameters representing the hydraulic characteristics and erosion/deposition capabilities. For the hydrologic-response module in InHM, Manning's roughness coefficients (n) were obtained by calibration (see Section 2.4.4). For the sediment-transport module, the rainsplash coefficient (c_f) and surface erodibility coefficient (φ), the two important parameters in InHM to describe the rainsplash erosion and hydraulic erosion processes, were determined by model calibration (see Section 2.4.4). Derived from previous studies based on soil texture [48,49], the damping coefficient (c_d), rainfall intensity exponent (b), and raindrop turbulence factor (ξ) were set to 600 m^{-1}, 1.6, and 0.25, respectively. According to the soil sample data in the Shejiagou catchment, the median diameter of the soil was set to 0.05 mm to represent the uppermost homogeneous loess soil for the sediment transport simulation. The soil cohesion coefficient was 0.30 [43] for all surface soils except that of the check dam body, which was assigned a larger cohesion coefficient (i.e., 0.60) due to compaction.

Table 3. Soil parameters used in InHM for the eight scenarios.

Soil Types	K_{sat} [a] (m s^{-1})	Porosity (-)	α [b] (m^{-1})	n [c] (-)	S_r [d] (-)
SJG-soil (0–0.5 m)	2.90×10^{-6}	0.42	1.38	1.74	0.08
SJG-soil (below 0.5 m)	2.70×10^{-6}	0.40	1.35	1.83	0.11
Check dam	4.00×10^{-10}	0.35	1.52	1.29	0.04
Sedimentary land	4.00×10^{-6}	0.45	0.37	1.19	0.13
Bedrock	1.00×10^{-9}	0.20	4.30	1.25	0.08
Parameter source	Calibration	The Zizhou Experimental Station			

[a] Saturated hydraulic conductivity.; [b] Parameter related to the inverse of the air-entry pressure.; [c] Parameter related to the pore-size distribution.; [d] Residual soil-water content [28].

Table 4. Surface parameters used in InHM for the four scenarios.

Surface Zone [a]	Hydrologic Response				Sediment Transport				
	n [a] (s m$^{-1/3}$)	ψ_{im} [b] (m)	S_{sr} [c] (-)	H_t [d] (m)	c_d [e] (m^{-1})	b [f] (-)	ξ [g] (-)	c_f [h] (s m^{-1})$^{0.6}$	φ [i] (m^{-1})
Hillslope	0.010	0.0015	0.01	0.0015	600	1.6	0.25	50.00	0.005
Gully and Channel	0.008							50.00	0.005
Sedimentary Land	0.008							0.05	0.001
Check Dam	0.004							5.00	0.001
Parameter Source	Calibration			Literature				Calibration	

[a] Manning's roughness coefficient. [b] Immobile water depth, identical for all surface zones. [c] Surface residual saturation, identical for all surface zones. [d] Average height of non-discretized micro-topography, identical for all surface zones. [e] Rainsplash depth dampening factor, identical for all surface zones. [f] Rain intensity exponent, identical for all surface zones. [g] Rain-induced turbulence coefficient, identical for all surface zones. [h] Rainsplash coefficient. [i] Surface erodibility coefficient.

2.4.4. Model Calibration and Validation

Two rainfall-runoff events were selected for the model calibration (event of 1964/7/14) and validation (event of 1964/8/2). The two events were chosen among the 31 recorded events because they were two of the largest events with the most detailed (i.e., complete and fine in temporal resolution) precipitation data. The saturated hydraulic conductivity and Manning's roughness coefficient were calibrated to fit the observed discharge at the Shejiagou hydrologic station. After that, the rainsplash coefficient and the surface erodibility coefficient were calibrated to fit the observed sediment discharge. The Nash-Sutcliffe efficiency [50] was used as the criterion to evaluate the model performance:

$$NSE = \left[\sum_{i=1}^{n} \left(O_i - \overline{O} \right)^2 - \sum_{i=1}^{n} \left(S_i - O_i \right)^2 \right] / \sum_{i=1}^{n} \left(O_i - \overline{O} \right)^2 \tag{1}$$

where O_i was the observed value (water discharge or sediment discharge), $\overline{O}$ was the average value of the observed values, and S_i was the simulated value and n was the number of samples. The Nash-Sutcliffe coefficients of the calibration were 0.763 and 0.738 for river flow and sediment discharge, and those of the validation were 0.781 and 0.732, respectively (Figure 4). These results showed good performance of the InHM with the optimized parameter set in the Shejiagou catchment. This parameter set (Tables 3 and 4) was then used in the scenario modelling runs.

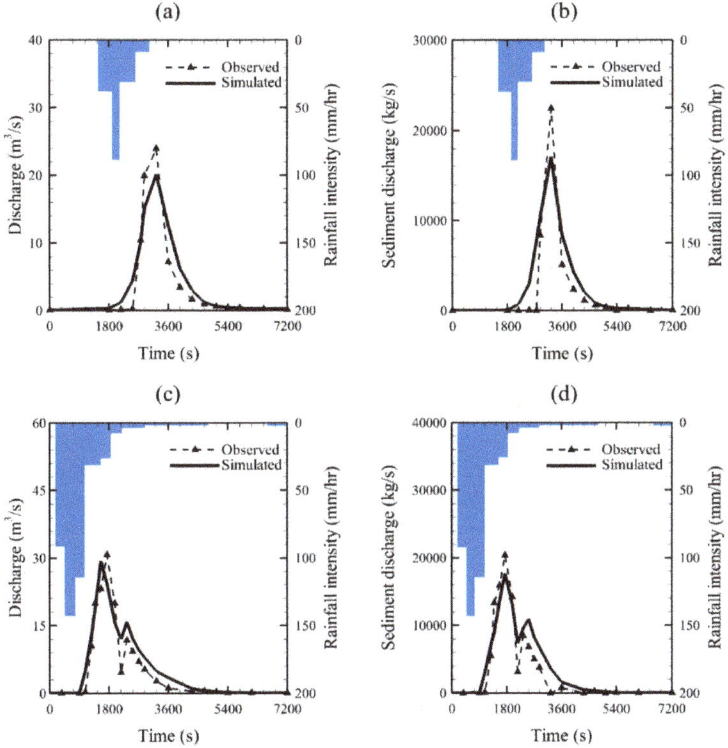

Figure 4. Hydrographs and sedigraphs of the model calibration ((**a**) and (**b**), event of 14 July 1964) and validation ((**c**) and (**d**), event of 2 August 1964). Note that the blue bars represent the rainfall intensity.

The Manning's roughness coefficient values were calibrated to be 0.010 s m$^{-1/3}$ for the hillslopes and 0.008 s m$^{-1/3}$ for the gullies (including the main channel and sedimentary land) and 0.004 s m$^{-1/3}$ for the dam body in the catchment (Table 4). The calibrated Manning's roughness coefficient values were small but reasonable due to the low vegetation coverage and the large area of bare land and cultivated land in the Shejiagou catchment during the simulation period (i.e., 1964) [25,27]. Similar values have been measured on bare land and cultivated land through field experiments in the Loess Plateau in a few studies e.g., [51,52].

2.4.5. Evaluation Criteria

To evaluate the influences on flood processes under different deployment strategies, the hydrograph and sedigraph at the catchment outlet were compared among the eight scenarios. Flood peak discharge (Q_p), peak time (T_p), and flood volume (Q_v) of each scenario were derived from the hydrographs. The runoff ratio (RR), which reflected the influences of check dams on the runoff-precipitation relationship in this study, was also calculated through dividing the runoff depth by the rainfall depth. To compare the efficiencies of sediment reduction under different deployment strategies, indicators such as $Q_{s,p}$ (sediment discharge peak), SY (sediment yield observed at the catchment outlet), and ES (the total mass of eroded sediment in the catchment) were derived from the modelling results. Besides, the sediment delivery ratio (SDR) was calculated as follows:

$$\text{SDR} = \frac{\text{SY}}{\text{ES}} \times 100\% \tag{2}$$

SDR has multi-fold environmental implications both in evaluating the soil erosion processes of a catchment and the effectiveness of soil and water conservation measures conducted in a catchment [53]. Variations of SDR of a specified catchment can be induced either by changes in SY or in ES. A comprehensively analysis of the changes in SDR, SY, and ES under different scenarios led to a more complete understanding of the changes in sediment transport processes caused by filled check dams. Moreover, the final distribution of eroded or deposited sediment in the catchment was also analyzed to furtherly understand the influences of filled check dams on sediment transport processes.

3. Results

3.1. Variation Characteristics of Flood Processes under Different Deployment Strategies

The flood processes were significantly changed by filled check dams under different deployment strategies (Table 5, Figure 5). Compared to scenario-0 (i.e., no check dams), the existence of check dams and their sedimentary lands reduced and delayed the flood peak. The efficiencies of flood-peak reduction and delay under the 7 deployment strategies were scenario-6 > scenario-4 > scenario-2 > scenario-7 > scenario-5 > scenario-3 > scenario-1. The hydrographs at the catchment outlet showed two different patterns: (1) the flood discharge rose and dropped sharply with a large peak discharge in scenario-0, scenario-1, scenario-3 and scenario-5; (2) the rising limb of hydrographs were delayed and more gentle, and the recession limb of hydrographs were longer in scenario-2, scenario-4, scenario-6, and scenario-7. The major difference between scenarios 2,4,6 and scenarios 1,3,5 was the existence of the large check dam (e.g., S2-2) on the middle reach of the main channel (Figure 2), which indicated that two large check dams constructed in series on the main channel could be more beneficial in flood peak attenuation than scenarios with only one large check dam. For scenarios with only one large check dam (i.e., scenarios 1,3,5), increasing the number of small check dams on the gullies led to significant reduction of flood peak and insignificant delay of peak time. For scenarios with two large check dams (e.g., scenarios 2,4,6), the delay of flood peak was longer when adding more small check dams in the catchment. Besides, proper site selection of large check dams on the main channel also made a difference in flood peak attenuation (e.g., scenario-6 versus scenario-7).

Table 5. Flood characteristics under different scenarios.

Scenario	Q_p (m^3 s^{-1})	Reduction of Q_p (%)	T_p (s)	Q_v (m^3)	Reduction of Q_v (%)	Runoff Coefficient (%)
Scenario-0	34.85	-	7300	58,984.10	-	15.38
Scenario-1	24.21	30.53	7512	56,768.51	3.76	14.81
Scenario-2	8.06	76.87	11,300	52,481.28	11.02	13.69
Scenario-3	18.04	48.24	7548	55,854.77	5.31	14.57
Scenario-4	4.04	88.41	13,300	50,505.12	14.38	13.17
Scenario-5	10.76	69.12	7520	53,163.99	9.87	13.87
Scenario-6	2.32	93.34	15,800	38,871.10	34.10	10.14
Scenario-7	9.95	71.45	8950	47,187.28	20.00	12.31

Different from unfilled check dam systems, which not only reduce flood peak but also reduce flood volume significantly [24], the reductions of flood volume by filled check dam systems were smaller (Table 5). For unfilled check dams, surface flow is intercepted by the dam body until the water depth is higher than the inlet of spillway, which forces a large amount of water to retain behind the check dam. For filled check dams, the elevation superiority of the spillway over the upstream channel no longer existed, and the flood volumes were mainly reduced by the promoted infiltration processes and the slowdown of flood velocity on the wider channel (i.e., sedimentary lands). Even so, the flood reduction rate can still be as much as 34.10% in scenario-6, in which the size combination was two large check dams and two small check dams. The efficiencies of flood-volume reduction under the 7 deployment strategies were scenario-6 > scenario-7 > scenario-4 > scenario-2 > scenario-5 >

scenario-3 > scenario-1. Different from the flood peak, the flood-volume reduction rate of scenario-7 was larger than scenario-3 and scenario-4. A possible explanation for this result may be that there were more check dams and sedimentary lands in scenario-7 in the headwater areas, where more runoff was generated compared to other areas [22]. In scenario-7, the sedimentary lands of the three upstream check dams were connected and became a larger infiltration zone for the upstream runoff than those in scenario-2 and scenario-4. Under the same rainfall condition (i.e., 90 mm), changes of the runoff coefficient under different deployment strategies were consistent with those of the flood volume observed at the catchment outlet.

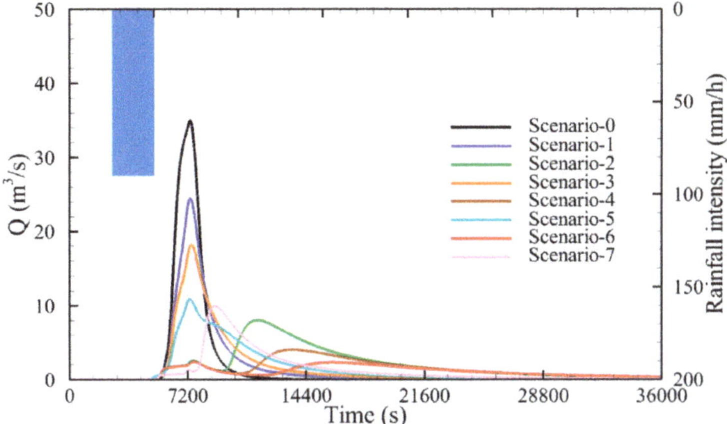

Figure 5. Hydrographs at the catchment outlet under different scenarios (Q: discharge).

3.2. Variation Characteristics of Sediment Transport Processes under Different Deployment Strategies

The changes of sediment transport processes at the catchment outlet were generally consistent with the variation characteristics of flood processes under different deployment strategies (see Figures 5 and 6). Along with the reduction and delay in flood peaks, the sediment discharge peaks (i.e., $Q_{s,p}$) were also reduced and delayed. The efficiencies of $Q_{s,p}$ reduction under the 7 deployment strategies were scenario-6 > scenario-4 > scenario-2 > scenario-5 > scenario-7 > scenario-3 > scenario-1 (Table 6). This result highlighted the importance of adding a large check dam on the middle reach because there was a large check dam on the middle reach of the catchment in scenarios 6,4,2 and no large check dam on the same position in scenarios 5,7,3,1 (see Figure 2). Differently, the reduction rate of $Q_{s,p}$ in scenario-5 (i.e., 67.43%) was larger than that in scenario-7 (i.e., 59.02%), while the reduction rate of Q_p in scenario-5 (i.e., 69.12%) was slightly smaller than that in scenario-7 (i.e., 71.45%). This was because that the gully controlled by check dam S5-2 contributed a large amount of eroded sediment into the main channel (see Figure 6 in [22]), and there was no check dam on the heavily eroded gully in scenario-7, indicating that targeted treatment to heavily eroded gullies could lead to a better performance of the check dam system in sediment reduction.

The reduction rates of sediment yield at the catchment outlet were basically higher than those of flood volume except for scenario-2 and scenario-7 (Tables 5 and 6), indicating that the deposition-promotion performance of sedimentary lands was normally better than their infiltration-promotion performance under most of the deployment strategies. The efficiencies of sediment-yield reduction under the 7 deployment strategies were scenario-6 > scenario-4 > scenario-5 > scenario-7 > scenario-2 > scenario-3 > scenario-1, implying that an increase in the number of check dams and targeted treatment to heavily eroded gullies could lead to a better performance in reducing sediment output of the catchment. Table 6 also showed that the total amount of eroded sediment in the catchment after a flood decreased due to the existence of filled check dams, and the reduction efficiencies were scenario-6 > scenario-4 > scenario-7 > scenario-2 > scenario-5 > scenario-3 > scenario-1. The reduction of total eroded sediment might be attributed to the lifting of erosion base of channels and hillslopes around the sedimentary lands, which led to decrease of runoff erosion power [24,54]. Large decreases of the sediment delivery ratio (SDR) were found in scenario-6 and scenario-4, which were consistent with the results of sediment yield at the catchment outlet. However, for scenarios with both small decreases of sediment yield and eroded sediment mass (e.g., scenario-1, scenario-2, scenario-7), their benefits for sediment reduction were hard to distinguish from SDR.

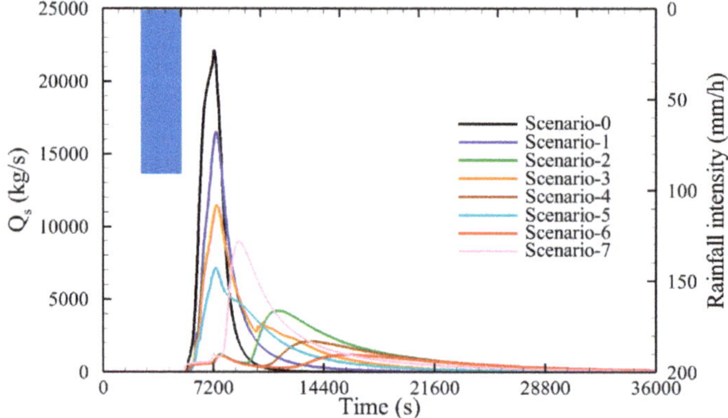

Figure 6. Sedigraphs at the catchment outlet under different scenarios (Q_s: sediment discharge).

Table 6. Sediment transport characteristics under different scenarios.

Scenario	$Q_{s,p}$ (kg s^{-1})	Reduction of $Q_{s,p}$ (%)	SY (×10^4 t)	Reduction of SY (%)	ES (×10^4 t)	Reduction of ES (%)	SDR (%)
Scenario-0	21,932.03	-	3.82	-	4.08	-	93.56
Scenario-1	16,508.84	24.73	3.66	4.14	3.99	2.18	91.78
Scenario-2	4232.80	80.70	3.53	7.55	3.71	8.98	95.12
Scenario-3	11,432.89	47.87	3.55	6.96	3.97	2.60	89.45
Scenario-4	2103.80	90.41	2.47	35.23	3.56	12.78	69.54
Scenario-5	7144.15	67.43	3.21	16.03	3.90	4.48	82.32
Scenario-6	1184.30	94.60	1.83	52.09	3.43	15.98	53.41
Scenario-7	8986.66	59.02	3.44	9.90	3.61	11.38	95.21

3.3. Erosion/Deposition Distribution in the Catchment under Different Deployment Strategies

The final distributions of erosion and deposition areas under the eight scenarios were shown in Figure 7. When there was no check dam in the catchment (i.e., scenario-0), erosion processes occurred both on hillslopes and in channels, while deposition areas were sparsely scattered in gully-head areas (Figure 7a). However, the existence of check dams, even when they were filled, could transform a large area of channels and gullies into deposition sinks (Figure 7b–h). By comparing the deposition depths behind each check dam (i.e., in the sedimentary land), we found that more sediment was deposited on the upstream sedimentary lands (i.e., the tail-end areas of the sedimentary land) than on the areas that were close to the check dams (e.g., Figure 7b). This deposition pattern was different from that of an unfilled check dam. For an unfilled check dam, flood processes were stopped at the near-dam area, and consequently, deposition processes occurred there. Therefore, the thickness of the deposition layer was thicker at the near-dam area than at the upstream area [55]. For a filled check dam with a large and flat sedimentary land, which was considered in this study, flood processes were weakened as soon as the surface flow entered the tail-end areas of sedimentary lands due to the decrease of slope gradient [21], consequently promoting sediment to deposit on these areas.

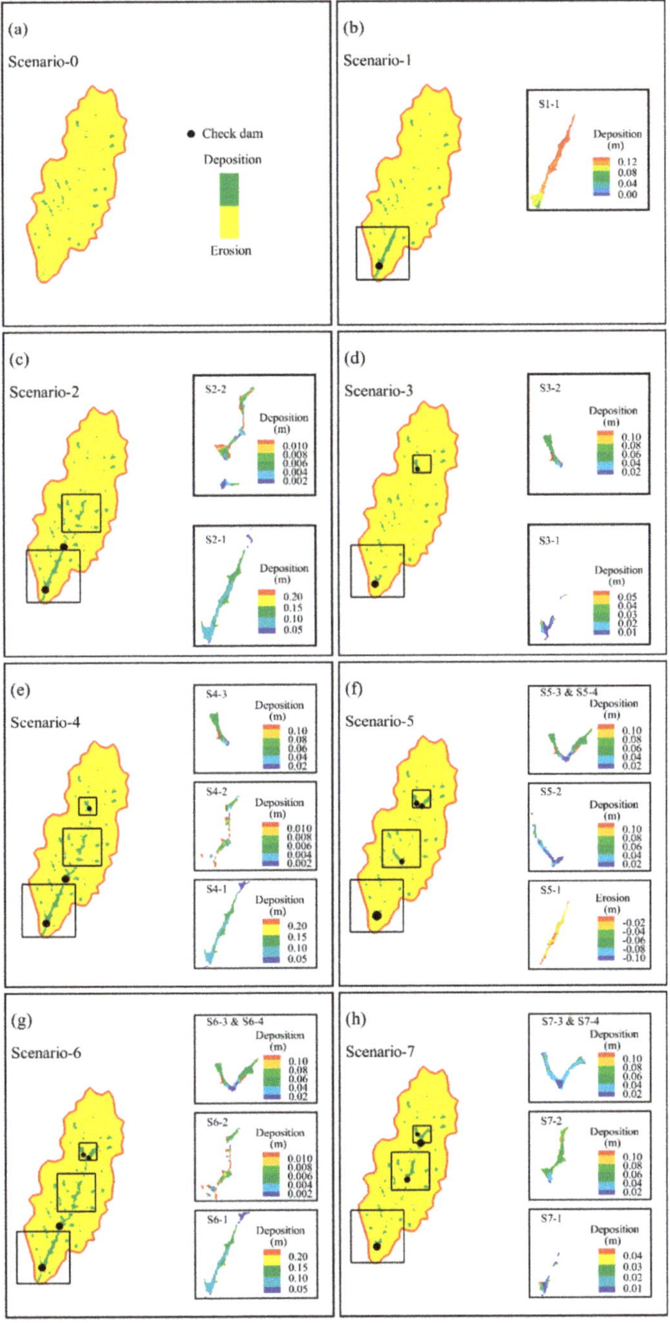

Figure 7. The final distributions of erosion and deposition areas in scenario-0 (**a**); scenario-1 (**b**); scenario-2 (**c**); scenario-3 (**d**); scenario-4 (**e**); scenario-5 (**f**); scenario-6 (**g**); and scenario-7 (**h**). Note that the inset pictures in each figure show the deposition depths on the sedimentary lands, and the areas with deposition depth below zero (i.e., erosion) were blanked for the consideration of readability.

Perusal of the deposition distributions under different deployment strategies led to the following implications:

- Increasing the number of check dams, especially large check dams, can significantly promote sediment deposition in the channel (e.g., Figure 7a–c,e,g). More check dams mean more elevated sedimentary lands with low channel gradients, creating a larger buffer zone for flood attenuation.
- The location and size of check dams do matter in sediment deposition. A large check dam on the main channel will create a larger deposition area than a small check dam (e.g., comparison of scenario-4 and scenario-5 in Figure 7e,f).
- Deployment strategies that can connect the sedimentary lands formed by different check dams lead to a better performance in flood attenuation and sediment reduction. For example, the performances of scenarios 6, 4, and 2 in the reduction of flood peak, sediment yield and eroded sediment were generally better than the performances of scenarios 3, 5, 7.

Deducing from those implications, in the situation where all check dams were filled, the best deployment strategies of check dams were those with large and connective sedimentary lands on the main channel and targeted treatments to heavily eroded gullies (e.g., scenario-6 in this study).

4. Discussion

4.1. Potential Benefits of Filled Check Dams in Flood and Erosion Control

Unlike check dams with large remaining storage capacity (i.e., low fill rate), filled check dams influence the flood processes mainly by the sedimentary land formed behind the check dams. Characterized as having a relatively high permeability [20] and a low slope gradient [46], the sedimentary land is able to attenuate flood velocity, and locally promote water infiltration and sediment deposition, consequently reducing and delaying the flood processes (Figure 5). Compared to the original channel, the sedimentary land is less susceptible to hydraulic erosion due to the attenuation of flood and the relatively large size of surface particles on the sedimentary land [56]. Therefore, the upstream expansion of sedimentary land will protect the soils of some channel areas and slope areas from being eroded, indirectly reducing the erodible area of the catchment. For example, the sedimentary land of S1-1 (i.e., the large check dam in scenario-1) protected approximate 950 m of the original channel reach and covered around 2.33×10^4 m^2 of hillslopes around the channel, indirectly reducing by 900 t (2.18%) eroded sediment. This indirect benefit of a filled check dam, which has not been quantitatively analyzed by previous studies, was termed as the lifting of erosion base by [54].

The increase of sedimentary lands resulted in considerable reductions in flood peak discharge (Q_p) and sediment peak discharge ($Q_{s,p}$), and the reduction rates responded linearly to the area of sedimentary land (Figure 8), which implies the long-term potential benefits of the large amount of check dams in the Loess Plateau. Although the remaining storage capacity of them will be totally consumed in the future, the sedimentary lands behind check dams will still be effective in flood control. As the area of sedimentary land increases, the reduction rates of flood volume (Q_v), sediment yield (SY), and eroded sediment (ES) were smaller than those of Q_p and $Q_{s,p}$ (Figure 8), because these indicators were also sensitive to the size combination and site selection of check dams (Tables 5 and 6).

Check dams, especially small check dams, are filled quickly in the catchment with severe erosion problems (e.g., in the Loess Plateau). For example, 444 check dams (including 39 large check dams) were constructed during 1953–1977 in the 200 km^2 Chabagou watershed and more than 60% of them were filled according to the check dam survey in 1978 (Figure 9). Several studies have reported that the flood processes were obviously attenuated and the runoff and sediment yield out of the Chabagou watershed were largely decreased due to check dam construction and hillslope treatments (e.g., terrace farming, afforestation) [57–59]. Based on the results of this study, it can be inferred that these filled check dams and their sedimentary lands also had a nonnegligible contribution to the reduction of runoff and sediment yield. Quantitatively modelling the cumulative effect of flood attenuation by the large number of filled check dams will help to obtain a more comprehensive understanding of the potential benefits of check dams in the long-term.

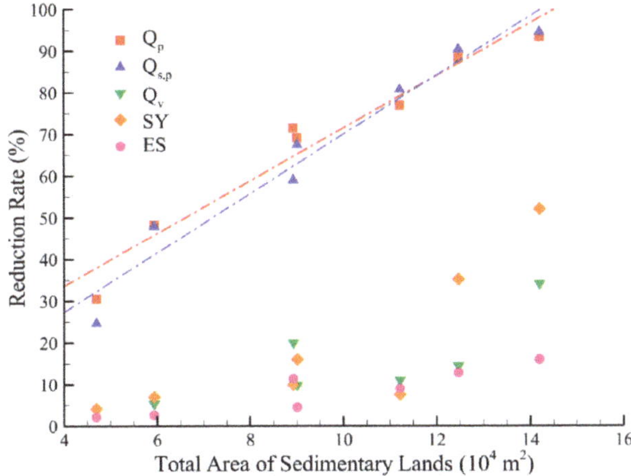

Figure 8. Relationship of the total area of sedimentary lands and reduction rate of flood processes and sediment transport processes.

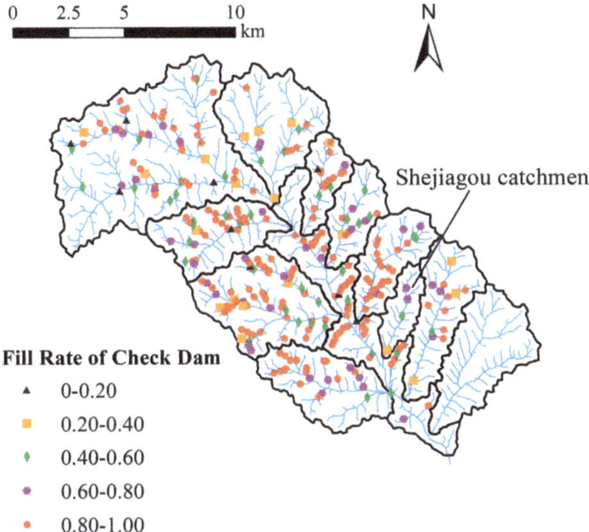

Figure 9. Fill rate of check dams in the Chabagou watershed. Note that when the check dam is filled, its fill rate equals to 1.

4.2. Implications of Future Check Dam Deployment

The quick-fill of check dams due to the large sediment yields during wet seasons and frequent dam breaks caused by flood overtopping often shorten the lifespan of check dams in the Loess Plateau [17]. For example, the two upstream check dams in the Shejiagou catchment (i.e., S7-3 and S7-4 in scenario-7) were totally filled after ten year's operation according to the check dam survey in 1978 [22]. There were technically two possible solutions to this problem: (1) dam maintenance such as regular sediment removal behind check dams can create new sediment storage capacities for filled check dams; (2) construction of open check dams, which have a permanent outlet and can let more sediment be transported downstream [60]. However, the first solution seems to be unfeasible in most of the Loess Plateau catchments due to the large amount of check dams on the Loess Plateau and the high expense for large excavators to transport and work in the gullies. On the other hand, the sediment reduction rate of open check dams is much lower than the check dam studied in this study, which cannot meet the urgent needs of sediment reduction for the heavily eroded Loess Plateau catchment. Besides, the construction materials of the dam body for open check dams are usually stone or concrete, which increase the construction expense in the hilly-gullied Loess Plateau catchment. Instead, check dams in the Loess Plateau catchment are usually gravity dams constructed by the local earth using materials from the nearby hillslopes, which largely reduces the construction expenses (see the Supplementary Materials).

Therefore, a scientific deployment strategy of check dams is necessary in order to obtain the largest benefits of flood attenuation and sediment interception when the storage capacities of check dams are consumed in the future. According to Table 7, scenarios with two large check dams (e.g., scenarios 2,4,6,7) performed better than scenarios with only one check dam (e.g., scenarios 1,3,5) in the reduction of Q_p and Q_v. Except for scenario-5, scenarios with two large check dams (i.e., scenarios 2,4,6,7) also performed better than scenarios with only one large check dam (i.e., scenarios 1,3) in the reduction of $Q_{s,p}$, SY, and ES. Apart from the number of large check dams in the system, the locations and number of small check dams also made a difference in the reduction performance. For example, scenario-5 performed better than scenario-7 in the reduction of $Q_{s,p}$ and SY because there was targeted treatment (i.e., check dam S5-2) to a heavily eroded gully in scenario-5. In the aspect of flood control, constructing large check dams on the main channel and making the sedimentary lands of the large check dams as connective as possible will obtain the best performance in the reduction and delay of flood peak (Table 7). Similar results were reported in [20], which showed that the combination of more than one large check dams on the main channel led to a considerable increase of groundwater recharge. In the aspect of sediment control, targeted treatment to specified gullies combined with large and connective sediment sinks in the main channel will largely promote sediment deposition in the channel and reduce soil erosion of hillslopes and channels (Table 7). Compared the current deployment strategy (i.e., scenario-7), the deployment strategy in the hypothetical scenario-6 which was more in line with the description above indeed performed better both in flood and sediment control.

Table 7. Performance ranking in flood and sediment control among different scenarios.

Scenario Number [a]	Performance Ranking						
	(1) [b]	(2)	(3)	(4)	(5)	(6)	(7) [b]
Reduction of Q_p	6	4	2	7	5	3	1
Reduction of Q_v	6	7	4	2	5	3	1
Reduction of $Q_{s,p}$	6	4	2	5	7	3	1
Reduction of SY	6	4	5	7	2	3	1
Reduction of ES	6	4	7	2	5	3	1

[a] Scenarios with different check dam systems (i.e., scenarios 1–7) were compared; [b] **(1)** represents the best reduction performance among all scenarios, **(7)** represents the worst reduction performance among all scenarios. For example, the best performance in the reduction of Q_p was scenario-6.

4.3. Limitations and Future Work

There were two simplifications in the parameterization of sedimentary land to simplify the problem in this study: (1) the elevations of surface nodes on the same sedimentary land were the same (i.e., the slope gradient of sedimentary land was zero) and (2) no croplands were planted on the sedimentary land. In reality, the slope of a sedimentary land in a check dam system is normally less than 2%. The first simplification might lead to a slight overestimation of the flood attenuation effects of sedimentary lands. However, most sedimentary lands in the Loess Plateau are planted with crops (e.g., corn), which increase surface roughness and local infiltration. This led to a small underestimation of the decrease in flow velocity and the sediment deposition. The two simplifications only had small influences on the flood processes, and their influences might offset each other [61].

While the implications on deployment strategies of check dams were specified to Shejiagou catchment, we believe that similar findings may be obtained for larger catchments with more complex check dam systems (e.g., other sub-catchments of Chabagou watershed in Figure 9). For this reason, future research will focus on larger sites with more complicated stream networks and check dam systems, to validate the implications and to provide a more comprehensive guidelines for the optimization of check dam deployment strategies. Besides, the influence of sedimentary land on catchment hydrologic responses and sediment transport processes in a larger time scale instead of event-scale need to be studied in order to aid the prediction of future changes in runoff and sediment yield in the Loess Plateau.

5. Conclusions

Employing the InHM, this study investigated the influences of filled check dams with different deployment strategies on the flood and sediment transport processes in a Loess Plateau catchment, which is a 4.26 km^2 erosion-prone area and currently has four check dams. Six hypothetical scenarios representing different deployment strategies were compared with no-dam scenario and the scenario that is currently applied in this catchment. Event-scale rainfall-runoff simulations were conducted with a 90 mm hr^{-1} rainfall intensity for the eight scenarios. Indicators such as flood peak discharge (Q_p), peak time (T_p), flood volume (Q_v), sediment peak discharge ($Q_{s,p}$), sediment yield (SY), and eroded sediment (ES) were used to compare the performances of different deployment strategies in flood and sediment control.

Filled check dams were still able to reduce flood peak by 31% to 93% under different deployment strategies. Considerable delays of peak time and decreases were also found in scenarios 2,4,6, which were characterized as having larger and more connective sedimentary lands on the main channel. Reduction rates of sediment yield and eroded sediment ranged from 4% to 52% and 2% to 16%, respectively, indicating that a proper distribution of check dams was beneficial to promote sediment deposition in the channel and reduce soil erosion of hillslopes and channels.

Apart from showing the potential benefits of filled check dams, this modelling exercise provided the following insights on the deployment strategy of check dams, which led to a better performance in flood and sediment control when check dams are filled:

- The number of large check dams on the main channel appears to be the most important factor influencing the flood and sediment transport processes.
- Connecting the sedimentary lands of large check dams via proper site selection is helpful for flood attenuation and sediment reduction.
- The area of sedimentary lands dominates the reduction rates of water/sediment peak discharge, while the size combination and site location can make a difference in the reduction of flood volume, sediment yield, and the total eroded sediment.
- Targeted treatment to heavily eroded gullies (e.g., scenario-5) will improve the sediment-reduction performance of the system, which requires a sediment-contribution analysis of each gullies via modelling practices or field observation to aid the site selection.

Compared to the current deployment strategy (i.e., scenario-7), a more scientific deployment strategy for the selected catchment, which is more in line with the insights above, was proposed (i.e., scenario-6).

The simulations reported herein took a first step to explore the potential benefits of filled check dams and demonstrated that check dams could have a long-term influence on the flood and sediment transport processes via forming sedimentary lands that changes the catchment landforms permanently.

Supplementary Materials: The following are available online at http://www.mdpi.com/2073-4441/12/5/1319/s1, Figure S1: (a) Sediment erosion (negative value) and deposition (positive value) rates for Shejiagou catchment; (b) Sediment inflow to stream segments; (c) Cumulative flow without check dams (Debasish et al., 2018), Figure S2. Open check dams. Beam dam (left), silt dam (right). Derived from Armanini and Larcher (2001), Figure S3. Characteristic components of a sediment trap with an open check dam: (a) inlet structure: solid body dam; (b) scour protection; (c) basin; (d) lateral dikes; (e) maintenance access; (f) open check dam; (g) counter dam. Derived from Piton and Recking (2015), Figure S4. Examples of check dams in the Loess Plateau. (a) check dam LDG-1 in Liudaogou catchment; (b) check dam MDZ-1 in Wangmaogou catchment; (c) check dam NYG-2 in Nianyangou catchment; (d) check dams system in Shejiagou catchment.

Author Contributions: Conceptualization, H.T.; methodology, H.T.; software, Q.R.; validation, H.T.; formal analysis, H.T.; investigation, H.P.; writing—original draft preparation, H.T.; writing—review and editing, Q.R.; supervision, Q.R.; funding acquisition, Q.R. All authors have read and agreed to the published version of the manuscript.

Funding: This study was funded by the China Major Science and Technology Program for Water Pollution Control and Treatment (2017ZX07206-001) and the National Natural Science Foundation of PR China (Grant no. 51379184 and no. 51328901).

Acknowledgments: We thank the Zizhou Experimental Station of YRCC and the National Earth System Science Data Center, National Science & Technology Infrastructure of China (http://www.geodata.cn) for providing essential observation data. We also thank the two anonymous reviewers for providing constructive suggestions. Special thanks to Amplin for helping improve the language of the manuscript.

Conflicts of Interest: The authors declare no conflict of interest.

Appendix A. The Integrated Hydrologic Model (InHM)

Appendix A.1. Hydrologic-Response Module

The 3D subsurface flow in variably saturated porous medium is estimated in InHM by:

$$\nabla \cdot f^a \vec{q} \pm q^b \pm q^e_{ps} = f^v \frac{\partial \phi S_w}{\partial t} \tag{A1}$$

$$\vec{q} = -k_{rw} \frac{\rho_w g}{\mu_w} \vec{k} \nabla(\psi_p + z) \tag{A2}$$

where $f^a[-]$ is the area fraction related to each continuum, $\vec{q}\,[LT^{-1}]$ is the Darcy flux, $q^b[T^{-1}]$ is a specified rate source/sink, $q^e_{ps}[T^{-1}]$ (equal to $-q^e_{sp}$) is the rate of water exchange between the porous medium and surface continua, $f^v[-]$ is the volume fraction associated with each continuum, $\phi[L^3L^{-3}]$ is porosity, $S_w[L^3L^{-3}]$ is the water saturation, $t[T]$ is time, $k_{rw}[-]$ is the relative permeability, $\rho_w[ML^{-3}]$ is the density of water, $g[LT^{-2}]$ is the gravitational acceleration, $\mu_w\,[ML^{-1}T^{-1}]$ is the dynamic viscosity of water, $\vec{k}\,[L^2]$ is the intrinsic permeability vector, $\psi_p[L]$ is the pressure head, and $z[L]$ is the elevation head.

The diffusion wave approximation to the depth-integrated shallow wave equations is employed in InHM to estimate the 2D transient surface flow on the land surface (both overland and open channel), with the conservation of water on the land surface expressed by:

$$\nabla \cdot \psi_s^{mobile} \vec{q}_s \pm a_s q^b \pm a_s q^e_{sp} = \frac{\partial(S_{w_s} H_t + \psi_{im})}{\partial t} \tag{A3}$$

$$\vec{q}_s = -\frac{(\psi_s^{mobile})^{2/3}}{\vec{n}\,\Phi^{1/2}} \nabla(\psi_s + z) \tag{A4}$$

where $\psi_s[L]$ is the surface water depth, $\vec{q}_s[LT^{-1}]$ is the surface water velocity calculated by a two-dimensional form of the Manning's equation, $a_s[L]$ is the surface coupling length scale, $q^b[T^{-1}]$ is the source/sink rate (i.e., rainfall/evaporation), $q^e_{sp}[T^{-1}]$ is the rate of water exchange between the surface continua and porous medium, $H_t[L]$ is the average height of non-discretized surface microtopography, $\vec{n}\,[TL^{-1/3}]$ is the Manning's surface roughness tensor, and $\Phi[-]$ is the friction (or energy) slope; $\psi_s^{mobile}[L]$ and $\psi_{im}[L]$ refer to the water depth exceeding and held in depression storage, respectively.

The calculated daily potential evapotranspiration was then incorporated into InHM as actual evapotranspiration (ET) using a set of sink functions:

$$Q_b^E = \alpha(\psi) q_{max}^E A_b \tag{A5}$$

where $Q_b^E[L^3 T^{-1}]$ represents the volumetric evapotranspiration rate, $\alpha(\psi)[-]$ is a response function of the saturation of the porous medium and the degree of ponding at the land surface, $q_{max}^E[LT^{-1}]$ is the potential evapotranspiration rate per unit area estimated by the revised Penman–Monteith method, $\psi[L]$ is the pressure head of the subsurface nodes or water depth of the surface nodes, and $A_b[L^2]$ is the area associated with the surface water equation

The first-order coupling between the surface and subsurface continua, driven by pressure head gradients, occurs via a thin soil layer of thickness, a_s in Equation (A3). The control volume finite-element method is employed to discretize the equations in space. Each coupled system of nonlinear equations is solved implicitly using Newton iteration.

Appendix A.2. Sediment-Transport Module

Depth-integrated multiple-species sediment transport, restricted to the surface continuum, is calculated for each sediment species by:

$$\frac{\partial C_{sed}}{\partial t} = -\nabla \cdot \vec{q}_s C_{sed} + \frac{1}{V_w}\left(e_{sed} + \sum_{j=1}^{BC} q_{s_j}^b C_{sed_j}^*\right) \tag{A6}$$

$$e_{sed} = e_s + e_h \tag{A7}$$

where $C_{sed}[L^3 L^{-3}]$ is volumetric sediment concentration, $\vec{q}_s[LT^{-1}]$ is the depth-averaged surface water velocity, $V_w[L^3]$ is the volume of water at the node, $e_{sed}[L^3 T^{-1}]$ is the volumetric rate of soil erosion and/or deposition, $q_{s_j}^b$ $[L^3 T^{-1}]$ is the rate of water added/removed via the jth boundary condition, $C_{sed_j}^*$ $[L^3 L^{-3}]$ is the sediment concentration of the water added/removed via the jth boundary condition, BC is the total number of boundary conditions. The net erosion rate is the sum of the rainsplash erosion rate $e_s[L^3 T^{-1}]$ and the hydraulic erosion rate $e_h[L^3 T^{-1}]$. The rainsplash erosion rate of each species in Equation (A7) is calculated as:

$$e_{s_i} = \begin{cases} \sigma_i c_f F(\psi_s)(\cos(\theta) \cdot r)^b A_{3D}, & q > 0 \\ 0 & , & q < 0 \end{cases} \tag{A8}$$

$$F(\psi_s) = \exp(-c_d \psi_s) \tag{A9}$$

where $\sigma_i[-]$ is the source fraction of species i, $c_f[(TL^{-1})^{b-1}]$ is the rainsplash coefficient, b $[-]$ is the rainfall intensity exponent, $\theta[-]$ is the angle of the element from horizontal, $r[LT^{-1}]$ is the rainfall intensity, $A_{3D}[L^2]$ is the three-dimensional area associated with the node, and $q[LT^{-1}]$ is the sum of rainfall intensity and infiltration intensity; $F(\psi_s)[-]$ is a damping function, related to surface water depth $\psi_s[L]$ and the surface water damping-effectiveness coefficient $c_d[L^{-1}]$, to represent the reduction in splash erosion with increasing surface water depth. The hydraulic erosion rate in Equation (A7) is estimated as:

$$e_{h_i} = \alpha_{sed_i}(C_{sed_{max_i}} \sigma_i - C_{sed_i}) \tag{A10}$$

$$C_{sed_{max_i}} = 0.05 \frac{\vec{q}_s q_*^3}{g^2 d_{sed_i} \psi_s (\gamma_{sed_i} - 1)^2} \tag{A11}$$

$$\alpha_{sed_i} = A \begin{cases} 2v_{sed_i}\xi & , & C_{sed} > C_{sed_{max_i}} \quad (erosion) \\ \varphi \vec{q}_s \psi_s \chi_i & , & C_{sed} < C_{sed_{max_i}} \quad (deposition) \end{cases} \tag{A12}$$

where $\alpha_{sed_i}[L^3 T^{-1}]$ is the hydraulic erosion transfer coefficient for species i, $C_{sed_{max_i}}[L^3 L^{-3}]$ is the concentration at equilibrium transport capacity for species i, $q_*[LT^{-1}]$ is the local shear velocity, $d_{sed_i}[L]$ and $\gamma_{sed_i}[-]$ are the particle diameter and specific gravity, respectively; $A[L^2]$ is the area associated with the node in Equation (A12), $v_{sed_i}[LT^{-1}]$ is the particle settling velocity, $\xi[-]$ is a coefficient related to turbulence in the surface water due to raindrop impact, $\varphi[L^{-1}]$ is an erodibility coefficient related to surface properties and texture, and $\chi_i[-]$ is the particle erodibility factor (ranging from zero to one).

References

1. Conesa-García, C.; López-Bermúdez, F.; García-Lorenzo, R. Bed stability variations after check dam construction in torrential channels (South-East Spain). *Earth Surf. Process. Landf.* **2007**, *32*, 2165–2184. [CrossRef]
2. Garcia, C.; Mario, M. *Check Dams, Morphological Adjustments and Erosion Control in Torrential Streams*; Nova Science Publishers: New York, NY, USA, 2010; pp. 8–15.
3. Abbasi, N.A.; Xu, X.Z.; Lucas-Borja, M.E.; Dang, W.Q.; Liu, B. The use of check dams in watershed management projects: Examples from around the world. *Sci. Total Environ.* **2019**, *676*, 683–691. [CrossRef] [PubMed]
4. Lucas-Borja, M.E.; Piton, G.; Nichols, M.; Castillo, C.; Yang, Y.; Zema, D.A. The use of check dams for soil restoration at watershed level: A century of history and perspectives. *Sci. Total Environ.* **2019**, *692*, 37–38. [CrossRef] [PubMed]
5. Piton, G.; Recking, A. Design of Sediment Traps with Open Check Dams. I: Hydraulic and Deposition Processes. *J. Hydraul. Eng.* **2016**, *142*, 16. [CrossRef]
6. Lucas-Borja, M.E.; Zema, D.A.; Hinojosa Guzman, M.D.; Yang, Y.; Hernández, A.C.; Xiangzhou, X.; Carrà, B.G.; Nichols, M.; Cerdá, A. Exploring the influence of vegetation cover, sediment storage capacity and channel dimensions on stone check dam conditions and effectiveness in a large regulated river in México. *Ecol. Eng.* **2018**, *122*, 39–47. [CrossRef]
7. Díaz-Gutiérrez, V.; Mongil-Manso, J.; Navarro-Hevia, J.; Ramos-Díez, I. Check dams and sediment control: Final results of a case study in the upper Corneja River (Central Spain). *J. Soils Sediments* **2019**, *19*, 16. [CrossRef]
8. Catella, M.; Paris, E.; Solari, L. Case study: Efficiency of slit-check dams in the mountain region of Versilia Basin. *J. Hydraul. Eng.* **2005**, *131*, 145–152. [CrossRef]
9. Hassanli, A.M.; Nameghi, A.E.; Beecham, S. Evaluation of the effect of porous check dam location on fine sediment retention (a case study). *Environ. Monit. Assess.* **2009**, *152*, 319. [CrossRef]
10. Xu, X.Z.; Zhang, H.W.; Zhang, O.Y. Development of check-dam systems in gullies on the Loess Plateau, China. *Environ. Sci. Policy* **2004**, *7*, 79–86.
11. Chen, Y.; Wang, K.; Lin, Y.; Shi, W.; Song, Y.; He, X. Balancing green and grain trade. *Nat. Geosci.* **2015**, *8*, 739–741. [CrossRef]
12. Xin, Z.B.; Yu, B.F.; Han, Y.G. Spatiotemporal Variations in Annual Sediment Yield from the Middle Yellow River, China, 1950–2010. *J. Hydrol. Eng.* **2015**, *20*, 15.
13. Jin, Z.; Cui, B.L.; Song, Y.; Shi, W.Y.; Wang, K.B.; Wang, Y.; Liang, J. How Many Check Dams Do We Need To Build on the Loess Plateau? *Environ. Sci. Technol.* **2012**, *46*, 8527–8528. [CrossRef] [PubMed]
14. Ran, D.; Luo, Q.; Zhou, Z.; Wang, G.; Zhang, X. Sediment retention by check dams in the Hekouzhen-Longmen Section of the Yellow River. *Int. J. Sediment Res.* **2008**, *23*, 159–166. [CrossRef]
15. Rustomji, P.; Zhang, X.P.; Hairsine, P.B.; Zhang, L.; Zhao, J. River sediment load and concentration responses to changes in hydrology and catchment management in the Loess Plateau region of China. *Water Resour. Res.* **2008**, *44*, 17. [CrossRef]
16. Shi, H.; Wang, G. Impacts of climate change and hydraulic structures on runoff and sediment discharge in the middle Yellow River. *Hydrol. Process.* **2015**, *29*, 3236–3246. [CrossRef]
17. Xu, Y.; Fu, B.; He, C. Assessing the hydrological effect of the check dams in the Loess Plateau, China, by model simulations. *Hydrol. Earth Syst. Sci.* **2013**, *17*, 2185–2193. [CrossRef]
18. Zhao, P.; Shao, M.; Wang, T. Spatial Distributions of Soil Surface-Layer Saturated Hydraulic Conductivity and Controlling Factors on Dam Farmlands. *Water Resour. Manag.* **2010**, *24*, 2247–2266. [CrossRef]

19. Huang, J.; Hinokidani, O.; Yasuda, H.; Ojha, C.S.P.; Kajikawa, Y.; Li, S. Effects of the Check Dam System on Water Redistribution in the Chinese Loess Plateau. *J. Hydrol. Eng.* **2013**, *18*, 929–940. [CrossRef]

20. Tang, H.; Ran, Q.; Gao, J. Physics-Based Simulation of Hydrologic Response and Sediment Transport in a Hilly-Gully Catchment with a Check Dam System on the Loess Plateau. *Water* **2019**, *11*, 1161. [CrossRef]

21. Guyassa, E.; Frankl, A.; Zenebe, A.; Poesen, J.; Nyssen, J. Effects of check dams on runoff characteristics along gully reaches, the case of Northern Ethiopia. *J. Hydrol.* **2017**, *545*, 299–309. [CrossRef]

22. Pal, D.; Galelli, S.; Tang, H.; Ran, Q. Toward improved design of check dam systems: A case study in the Loess Plateau, China. *J. Hydrol.* **2018**, *559*, 762–773. [CrossRef]

23. Pal, D.; Galelli, S. A numerical framework for the multi-objective optimal design of check dam systems in erosion-prone areas. *Environ. Model. Softw.* **2019**, *119*, 21–31. [CrossRef]

24. Yuan, S.; Li, Z.; Li, P.; Xu, G.; Gao, H.; Xiao, L.; Wang, F.; Wang, T. Influence of Check Dams on Flood and Erosion Dynamic Processes of a Small Watershed in the Loss Plateau. *Water* **2019**, *11*, 834. [CrossRef]

25. Fang, H.; Li, Q.; Cai, Q.; Liao, Y. Spatial scale dependence of sediment dynamics in a gullied rolling loess region on the Loess Plateau in China. *Environ. Earth Sci.* **2011**, *64*, 13. [CrossRef]

26. Ran, Q.; Hong, Y.; Chen, X.; Gao, J.; Ye, S. Impact of soil properties on water and sediment transport: A case study at a small catchment in the Loess Plateau. *J. Hydrol.* **2019**, *574*, 211–225. [CrossRef]

27. Zhang, L.; Li, Z.; Wang, H.; Xiao, J. Influence of intra-event-based flood regime on sediment flow behavior from a typical agro-catchment of the Chinese Loess Plateau. *J. Hydrol.* **2016**, *538*, 71–81. [CrossRef]

28. van Genuchten, M.T. A closed-form equation for predicting the hydraulic conductivity of unsaturated soils. *Soil Sci. Soc. Am. J.* **1980**, *44*, 892–898. [CrossRef]

29. YRCC. Dataset of Hydrological Experiments at Zizhou Station in the Yellow River Basin (1959–1969); Data Sharing Infrastructure of Earth System Science_Data Sharing Infrastructure of Loess Plateau; YRCC. 2012. Available online: http://www.geodata.cn/data/datadetails.html?dataguid=190520812859307&docId=9284 (accessed on 7 May 2020).

30. Yang, Q. Land Use Map (30 m Resolution) of the Area with Abundant and Coarse Sediment of the Loess Plateau; Data Sharing Infrastructure of Earth System Science_Data Sharing Infrastructure of Loess Plateau. 2011. Available online: http://www.geodata.cn/data/datadetails.html?dataguid=69574164479925&docId=9166 (accessed on 7 May 2020).

31. Freeze, R.A.; Harlan, R.L. Blueprint for a physically-based, digitally-simulated hydrologic response model. *J. Hydrol.* **1969**, *9*, 237–258. [CrossRef]

32. VanderKwaak, J.E. Numerical Simulation of Flow and Chemical Transport in Integrated Surface-Subsurface Hydrologic Systems. Ph.D. Thesis, University of Waterloo, Waterloo, ON, Canada, 1999.

33. Heppner, C.S.; Ran, Q.H.; VanderKwaak, J.E.; Loague, K. Adding sediment transport to the integrated hydrology model (InHM): Development and testing. *Adv. Water Resour.* **2006**, *29*, 930–943. [CrossRef]

34. Ran, Q.; Heppner, C.S.; VanderKwaak, J.E.; Loague, K. Further testing of the integrated hydrotogy modet (InHM): Multipte-species sediment transport. *Hydrol. Process.* **2007**, *21*, 1522–1531. [CrossRef]

35. Ebel, B.A.; Mirus, B.B.; Heppner, C.S.; VanderKwaak, J.E.; Loague, K. First-order exchange coefficient coupling for simulating surface water–groundwater interactions: Parameter sensitivity and consistency with a physics-based approach. *Hydrol. Process.* **2009**, *23*, 1949–1959. [CrossRef]

36. VanderKwaak, J.E.; Loague, K. Hydrologic-response simulations for the R-5 catchment with a comprehensive physics-based model. *Water Resour. Res.* **2001**, *37*, 999–1013. [CrossRef]

37. Mirus, B.B.; Loague, K. How runoff begins (and ends): Characterizing hydrologic response at the catchment scale. *Water Resour. Res.* **2013**, *49*, 2987–3006. [CrossRef]

38. Fatichi, S.; Vivoni, E.R.; Ogden, F.L.; Ivanov, V.Y.; Mirus, B.; Gochis, D.; Downer, C.W.; Camporese, M.; Davison, J.H.; Ebel, B.; et al. An overview of current applications, challenges, and future trends in distributed process-based models in hydrology. *J. Hydrol.* **2016**, *537*, 45–60. [CrossRef]

39. Li, P.; Mu, X.; Holden, J.; Wu, Y.; Irvine, B.; Wang, F.; Gao, P.; Zhao, G.; Sun, W. Comparison of soil erosion models used to study the Chinese Loess Plateau. *Earth-Sci. Rev.* **2017**, *170*, 17–30. [CrossRef]

40. Ran, Q.; Tong, J.; Shao, S.; Fu, X.; Xu, Y. Incompressible SPH scour model for movable bed dam break flows. *Adv. Water Resour.* **2015**, *82*, 39–50. [CrossRef]

41. Zheng, X.; Shao, S.; Khayyer, A.; Duan, W.; Ma, Q.; Liao, K. Corrected First-Order Derivative ISPH in Water Wave Simulations. *Coast. Eng. J.* **2017**, *59*, 1750010. [CrossRef]

42. Ni, G.; Liu, Z.; Lei, Z.; Yang, D.; Wang, L. Continuous simulation of water and soil erosion in a small watershed of the Loess Plateau with a distributed model. *J. Hydrol. Eng.* **2008**, *13*, 392–399. [CrossRef]

43. Heppner, C.S.; Loague, K. A dam problem: Simulated upstream impacts for a Searsville-like watershed. *Ecohydrology* **2008**, *1*, 408–424. [CrossRef]

44. Mirus, B.B.; Ebel, B.A.; Loague, K.; Wemple, B.C. Simulated effect of a forest road on near-surface hydrologic response: Redux. *Earth Surf. Process. Landf.* **2007**, *32*, 126–142. [CrossRef]

45. Wang, Y.; Fu, B.; Chen, L.; Lu, Y.; Gao, Y. Check Dam in the Loess Plateau of China: Engineering for Environmental Services and Food Security. *Environ. Sci. Technol.* **2011**, *45*, 10298–10299. [CrossRef] [PubMed]

46. Porto, P.; Gessler, J. Ultimate bed slope in calabrian streams upstream of check dams: Field study. *J. Hydraul. Eng.* **1999**, *125*, 1231–1242. [CrossRef]

47. Heppner, C.S.; Loague, K.; VanderKwaak, J.E. Long-term InHM simulations of hydrologic response and sediment transport for the R-5 catchment. *Earth Surf. Process. Landf.* **2007**, *32*, 1273–1292. [CrossRef]

48. Ran, Q.; Loague, K.; VanderKwaak, J.E. Hydrologic-response-driven sediment transport at a regional scale, process-based simulation. *Hydrol. Process.* **2012**, *26*, 159–167. [CrossRef]

49. Gabet, E.J.; Dunne, T. Sediment detachment by rain power. *Water Resour. Res.* **2003**, *39*, 1002. [CrossRef]

50. Nash, J.E.; Sutcliffe, J.V. River flow forecasting through conceptual models part I—A discussion of principles. *J. Hydrol.* **1970**, *10*, 282–290. [CrossRef]

51. Wu, B.; Wang, Z.L.; Zhang, Q.W.; Shen, N.; Liu, J. Modelling sheet erosion on steep slopes in the loess region of China. *J. Hydrol.* **2017**, *553*, 549–558. [CrossRef]

52. Xia, L.; Song, X.; Fu, N.; Cui, S.; Li, L.; Li, H.; Li, Y. Effects of rock fragment cover on hydrological processes under rainfall simulation in a semi-arid region of China. *Hydrol. Process.* **2018**, *32*, 792–804. [CrossRef]

53. Wang, L.; Yao, W.; Tang, J.; Wang, W.; Hou, X. Identifying the driving factors of sediment delivery ratio on individual flood events in a long-term monitoring headwater basin. *J. Mt. Sci.* **2018**, *15*, 1825–1835. [CrossRef]

54. Hu, C.; Zhang, X. Loess Plateau soil erosion governance and runoff-sediment variation of Yellow River. *Water Resour. Hydropower Eng.* **2020**, *51*, 1–11. (In Chinese)

55. Kang, R.S.; Chacko, E.; Kaur, D.; Viadero, R. Silting patterns in the reservoirs of small-and medium-sized earthen check dams in humid subtropical monsoon regions. *Earth Surf. Process. Landf.* **2019**, *44*, 2638–2648. [CrossRef]

56. Li, X.; Wei, X.; Wei, N. Correlating check dam sedimentation and rainstorm characteristics on the Loess Plateau, China. *Geomorphology* **2016**, *265*, 84–97. [CrossRef]

57. Fang, H.Y.; Cai, Q.G.; Chen, H.; Li, Q.Y. Temporal changes in suspended sediment transport in a gullied loess basin: The lower Chabagou Creek on the Loess Plateau in China. *Earth Surf. Process. Landf.* **2008**, *33*, 1977–1992. [CrossRef]

58. Yu, G.Q.; Zhang, M.S.; Li, Z.B.; Li, P.; Zhang, X.; Cheng, S.D. Piecewise prediction model for watershed-scale erosion and sediment yield of individual rainfall events on the Loess Plateau, China. *Hydrol. Process.* **2014**, *28*, 5322–5336.

59. Fu, S.; Yang, Y.; Liu, B.; Liu, H.; Liu, J.; Liu, L.; Li, P. Peak flow rate response to vegetation and terraces under extreme rainstorms. *Agric. Ecosyst. Environ.* **2020**, *288*, 106714. [CrossRef]

60. Armanini, A.; Larcher, M. Rational criterion for designing opening of slit-check dam. *J. Hydraul. Eng.* **2001**, *127*, 94–104. [CrossRef]

61. Tang, H. Analysis of Runoff/Sediment Reduction Efficiency of Check-Dam in Its Lifespan. Ph.D. Thesis, Zhejiang University, Hangzhou, China, 2019.

Article

Internal Stability Evaluation of Soils

Qingfeng Feng [1], Hao-Che Ho [2], Teng Man [3], Jiaming Wen [1], Yuxin Jie [1] and Xudong Fu [1,*]

[1] State Key Laboratory of Hydroscience and Engineering, Department of Hydraulic Engineering, Tsinghua University, Beijing 100084, China
[2] Ocean College, Zhejiang University, Hangzhou 310027, China
[3] Department of Civil, Environmental, and Geo-Engineering, University of Minnesota, Minneapolis, MN 55455-0116, USA
* Correspondence: xdfu@tsinghua.edu.cn; Tel.: +86-(0)10-6279-7071; Fax: +86-(0)10-6277-3576

Received: 18 May 2019; Accepted: 9 July 2019; Published: 12 July 2019

Abstract: Suffusion constitutes a major threat to the foundation of a dam, and the likelihood of suffusion is always determined by the internal stability of soils. It has been verified that internal stability is closely related to the grain size distribution (GSD) of soils. In this study, a numerical model is developed to simulate the suffusion process. The model takes the combined effects of GSD and porosity (n) into account, as well as Wilcock and Crowe's theory, which is also adopted to quantify the inception and transport of soils. This proposed model is validated with the experimental data and shows satisfactory performance in simulating the process of suffusion. By analyzing the simulation results of the model, the mechanism is disclosed on how soils with specific GSD behaving internally unstable. Moreover, the internal stability of soils can be evaluated through the model. Results show that it is able to distinguish the internal stability of 30 runs out of 36, indicating a 83.33% of accuracy, which is higher than the traditional GSD-based approaches.

Keywords: suffusion; internal stability; grain size distribution (GSD)

1. Introduction

Suffusion is the mass movement of fine particles through the pore space of a coarser matrix driven by seepage forces [1]. It is one of the main mechanisms initiating internal erosion within levees, earth dams and foundations [2–4] as well as watershed hillslopes [5,6].

A lot of experimental [7–14] and numerical (CFD) studies [15,16] were launched and expected to reveal the underlying mechanisms corresponding to suffusion. Most of these studies were macroscopic and under the important assumption that the movement of soil is a continuous process. This assumption has never been verified through or connected directly to a particle-scale study. However, some of following studies in granular system start to show their potential to support this assumption. For instance, as DEM models [17–19] and CFD+DEM models [20] are developed, the suspension, collision or spin of particles have been described in detail, but still seem to be irrelevant to the internal structure fracture (based on regime transition research [21]) or massive movement (based on rheology research [22–24]). Since then, macroscopic studies on suffusion are still mainstream, especially when focusing on engineering practices.

Suffusion can take on many forms, but it is particularly insidious when it occurs under a relatively low hydraulic gradient. Hydraulic engineers determine the internal stability by examining the critical gradient of a soil which directly affects the likelihood of diffusion. Generally, the soil is internally stable if $i_c < 0.7$ [7].

The internal stability of soils is closely related to its grain size distribution (GSD). As pointed out by Wan and Fell [1], internally unstable soils are usually broadly graded soils with particles from silt (or clay) to gravel size, whose GSD curves are concave upward (or known as gap-graded soils).

GSD-based assessment approaches, based on experimental evidence, have been used extensively in practice [8–14]. Most of them distinguish coarse and fine particles within the soil and incorporate characteristic sizes into an assessment index or indices.

Early studies accounted for two particles sizes with an additional division size. For example, Kézdi [25] divided soils into coarse and fine components. The maximum value of D_{15}/d_{85}, which depends on the division size, is used as the index of internal stability. D_{15} (d_{85}) is the size for which 85% (15%) by weight is finer in the coarse (fine) components, respectively. The soil is evaluated as unstable if $D_{15}/d_{85} > 4$. Similarly, Kenney and Lau [26] introduced a "shape curve" to identify an internally unstable soil. They defined F as the percentage of particles that are finer than an adjustable size d and $H + F$ as the percentage of particles finer than the size of $4d$. The material is assessed as internally unstable when the minimum value of H/F is less than 1.3 originally, [26] and a modified lower value of 1.0 later [27]. Different from the usage of an adjustable particle size, Burenkova [28] considered three characteristic sizes and evaluated the soils unstable when d_{90}/d_{60} is greater than $1.86\log(d_{90}/d_{15}) + 1$ or less than $0.76\log(d_{90}/d_{15})$, where d_x ($x = 15$, 60 or 90) hereafter is the diameter for which $x\%$ by weight of the soil particles are finer.

In recent decades, the idea of a more precise representation of the GSD curve has been further developed. Wan and Fell [1] extracted four sizes from the GSD curve. Two are for coarse particles and the other two for the fine counterpart of the soils. They defined two variables, $s_1 = 15/\log(d_{20}/d_{05})$ and $s_2 = 30/\log(d_{90}/d_{60})$, for each soil sample. The well-known assessment criterion is that a soil sample is internally unstable when $s_1 < 22$ and $s_2 > 80$. That is, an internally unstable soil is featured as $d_{20}/d_{05} > 4.8$ and $d_{90}/d_{60} < 2.4$. Also, its accuracy is higher than any other GSD-based approaches mentioned above.

The accuracy of the GSD-based approaches has been improving since more and more GSD information has been added in. In terms of seepage flow dynamics, however, these approaches are still empirical and with ambiguous meaning. Early research suggests that the gap-graded soils are apt to being internally unstable since fine particles easily pass through, but do not fill in the voids between the coarse particles if subjected to seepage [9]. The filtration of detached fine particles has therefore been well accepted as the mechanism responsible for the internal stability. Overall, suffusion appears as the combination of detachment, transport, and filtration of fine particles. They may depend on the excessive shear stress related to particle inception [15,29], sediment transport capacity if definable, and the size of the voids, respectively. In addition to the filtration effect, the other mechanisms may dominate in some cases, but have received less attention so far. Such incomplete understanding hampers the explanation of other important factors such as porosity [30,31] and loading history [32].

This paper presents a numerical simulation of fine particle dynamics subjected to seepage flow, within an extended numerical framework of Fujisawa's study [15]. We take into account the combined effects of GSD and porosity. Wilcock and Crowe's theory [33] is adopted to quantify the inception and transport of each size group of soil particles. We reproduce the phenomena observed in seepage tests and explain why the soils with specific GSD behave more internally unstable, and we predict the internal stability of soil through the model.

2. Numerical Model

Numerical models have been successfully applied to analyze seepage flow through embankments, dikes, and soil foundations. With a reasonable model for particle movement within soil matrix, numerical models may be able to evaluate internal stability of soils with a certain GSD. Our new model is an alternative approach for exploring internal stability.

2.1. Governing Equations

In terms of representing the dynamics of fine particles subjected to seepage forces, numerical simulation at the microscopic scale provides a potential approach. In 2010, Fujisawa et al. described the process of suffusion (or the process of porosity n change with time t) within a unit volume of seepage field by using $\partial n/\partial t = EA$ [15]; The involved parameters E and A are the erosion rate defined as the volume of eroded particles per erodible area per time, and the erodible area per unit volume.

Taking the GSD into consideration, as shown in Figure 1, we assume that the particles from the soil skeleton would be eroded into the pore flow and increase its concentration, considering the two-dimensional seepage flow in the directions 1 and 2. During the time period of dt, the volume of eroded soil particles is $\sum E_k A_k dVdt$, where E_k denotes the erosion rate of kth size group section defined as the volume of eroded particles per erodible area per time, A_k denotes the erodible area per unit volume of kth size group, $dV = dx_1 dx_2 dL$, and dx_1, dx_2, and dL are the dimensions of the unit volume, respectively.

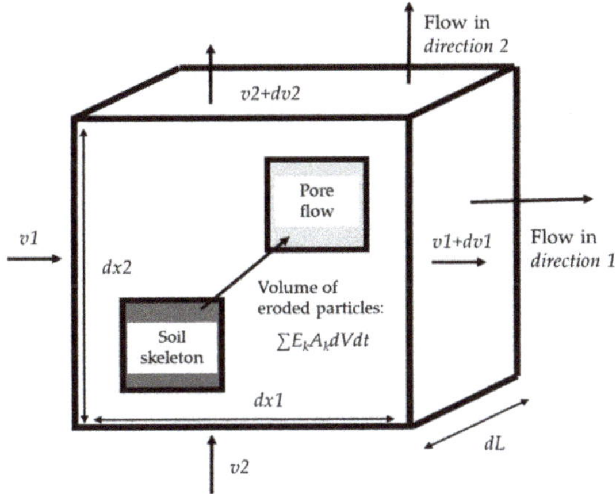

Figure 1. Unit volume of seepage field as the modeling framework.

The governing equations of mass and momentum conservation are as follows:

$$\frac{\partial n}{\partial t} = \sum E_k A_k \tag{1}$$

$$\frac{\partial nC}{\partial t} + \frac{\partial v_i C}{\partial x_i} = \sum E_k A_k \tag{2}$$

$$n\frac{\partial S_r}{\partial t} + \frac{\partial v_i}{\partial x_i} = 0 \tag{3}$$

$$v_i = -k_s \frac{\partial h}{\partial x_i} \tag{4}$$

where n denotes the porosity of soil mass, C denotes the solid concentration of seepage flow, v_i denotes the seepage velocity in the two directions ($i = 1, 2$), S_r denotes the saturation of soils, h denotes the local water head, k_s denotes the hydraulic conductivity, x_i denotes the distance along the ith direction ($i = 1, 2$), and t denotes the time passing by. Details about the closure equations will be introduced in Section 2.2.

Item $\sum E_k A_k$ differs from EA in the study of Fujisawa et al. [15] since the dependence of E_k on the size of kth group is taken into account. If the dependence is neglected, i.e., $E_k = E$ holds for each size group, the formulation by Fujisawa et al. [15] will be recovered.

2.2. Closures

Closures of items erosion rate E_k, erodible area A_k. and hydraulic conductivity k_s are listed below:

2.2.1. The Erosion Rate E_k

For suffusion, a linear relationship has been used between the erosion rate E and the excessive shear stress τ acting on eroded particles. However, this assumption does not distinguish the difference in the inception of particles with different sizes. Here, we adopt Wilcock and Crowe 's theory [33] for mixed sediments which found wide applications in geomorphic science and river engineering. That is, the erosion rate E_k is evaluated taking into account GSD, i.e.,

$$E_k = \begin{cases} \alpha(\tau - \tau_{ck}) & \text{for } \tau \geq \tau_{ck} \\ 0 & \text{for } \tau < \tau_{ck} \end{cases} \tag{5}$$

where τ is shear stress exerted on the soil particles, E_k and τ_{ck} are the erosion rate and critical shear stress for the kth size group, respectively, provided that the soil particles can be divided into several size groups (group number $\geq k$) and the particles within the same size group have the same critical shear stress.

The selection of α is based on calibration. More details will be introduced in Section 3.3.

The shear stress τ that is exerted on soil particles, is generated by the seepage flow with the soil fabric. In 1980, Hillel [34] assumed that the seepage field within a unit volume (see Figure 2a) can be treated as equivalent tubes in a length of L and radius of $R = D/2$ (see Figure 2b).

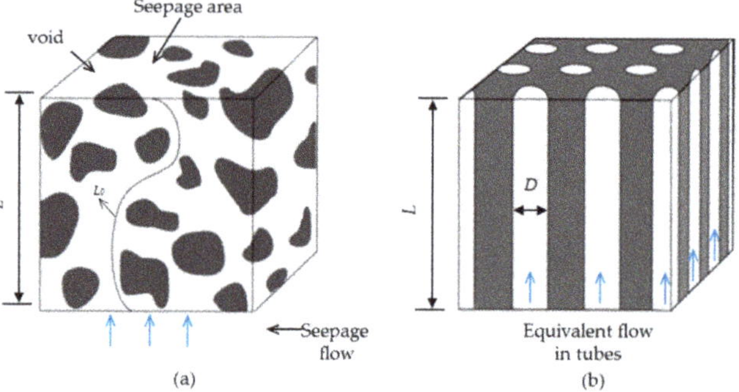

Figure 2. (**a**) Seepage field and (**b**) its equivalent tubes within a unit volume.

As shown in Figure 2, the equivalence guarantees the same discharge due to seepage flow under the same hydraulic gradient as well as the same porosity in average. The shear stress τ is then defined at the wall of tube. Using the Poiseuille law, it can be obtained by the following:

$$\tau = I\sqrt{2\rho g \frac{k_s \mu}{n}} \tag{6}$$

where $I = h/L$ denotes the local hydraulic gradient; g denotes the acceleration of gravity; $\rho = C\rho_s + (1 - C)\rho_w$ is the density of seepage flow; ρ_s denotes the density of soil particles; ρ_w denotes the density of pure water; $\mu = \mu_w(1 + 2.5C)$ is the dynamic viscosity of seepage flow with suspended sediments, which is corrected by the suspension concentration; and μ_w denotes the viscosity of pure water. It is noted that Equation (6) quantifies an equivalent shear stress within the unit volume, which balances the pressure drop driving the seepage flow.

The equivalent shear stress τ in the model is a macroscopic measure of the microscopic seepage shear stress exerted on the soil particles in the unit volume, and the potential difference between these two items is compensated through the calibration of erodible coefficient α in Equation (5). Meanwhile, the erosion of the fine particles is assumed to be equivalent to the process of sediment transportation, such an idea is also introduced when calculating the critical shear stress τ_c.

In previous studies [15,35], the critical shear stress τ_c was assumed to be a constant. However, since the GSD cannot be ignored, coarse particles will protect the fine ones from being eroded (known as the "Hiding Function"). Thus, the critical shear stress varies when particle size differs (see Figure 3). By integrating Wilcock and Crowe's theory [33] for the inception and transport of non-uniform sediments, for the kth group containing the soil particles of size between d_k and d_{k+1}, the critical shear stress τ_{ck} is evaluated based on Equations (7) and (8):

$$\tau_{ck} = \tau_{csm}\left(\frac{d_k}{d_{sm}}\right)^b \tag{7}$$

$$\frac{\tau_{csm}}{(\rho_s - \rho_w)gd_{sm}} = A + B\exp(-20F_s) \tag{8}$$

where d_{sm} denotes the medium size of soil particles; τ_{csm} is the corresponding critical shear stress; b denotes an exponent that may differ from the erosion in open channel flows and is calibrated in the simulation; F_s denotes the content ratio of fine particles (diameter $d < 2$ mm); and A and B are the two empirical parameters, and can also be calibrated during the simulation.

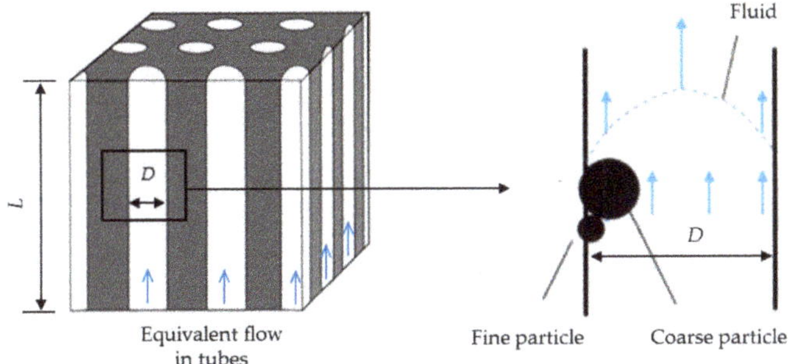

Figure 3. Effect of GSD in suffusion process.

Parameter A in Equation (8) is usually related to the cohesion of the soil. When there are cohesive particles in the soil, the value of A will be relatively higher [33]; parameter B in Equation (8) is usually related to fluid properties [33]. Due to the nature differences between the Poiseuille flow in our model and the open channel flow in the sediment transport process in Wilcock and Crowe's study [33], the value of B may also be different. In our simulation, these two parameters are obtained through calibration. More details will be introduced in Section 3.3.

2.2.2. The Erodible Area A_k

Within a volume of dV, the A_k for the kth size group can be quantified with the total projective area of the given size d_k, that is, the product of the projective area of a particle, πd_k^2, and the number of such particles per unit volume, $6(1-n)(P_{k+1} - P_k)/\pi d_k^3$. Therefore, the A_k reads,

$$A_k = 6(1-n)(P_{k+1} - P_k)/d_k \tag{9}$$

where P_{k+1} and P_k denote the percentage passing by weight of soil particles for the size finer than d_{k+1} and d_k respectively.

2.2.3. The Hydraulic Conductivity k_s

The hydraulic conductivity k_s is related to the saturation, porosity, and GSD of the soil. A relationship for the hydraulic conductivity of the saturated soil is proposed, and it reads

$$k_s = \Gamma d_{10}^{1.565} \frac{n^{2.3475}}{(1-n)^{1.565}} \tag{10}$$

where d_{10} denotes the diameter of 10% for which soil particles by weight are finer, and Γ is a constant (calibrated in Section 3.3). Such a relationship is based on Chapuis' study [10], as well as the selection of exponents of each item.

2.2.4. Update of d_{10} in Equation (10)

In the case of suffusion driven by a seepage flow, the moving out of eroded particles will change the local porosity and GSD. As a result, the hydraulic conductivity k_s as well as the critical shear stress τ_{ck} has a feature of dynamic adjustment during the erosion process. In order to capture the dynamic features, it is necessary to formulate the temporal change of P_k corresponding to the particle size d_k, and then to specify d_{10} in Equation (11). For this purpose, one can quantify the $P_k(t + dt)$ after a time period dt as

$$P_k(t + dt) = \frac{P_k(t)(1-n) - \sum\limits_{i=1}^{k} E_i A_i dt}{(1-n) - EAdt} \tag{11}$$

A rearrangement of Equation (11) without considering the second-order term results in

$$\frac{dP_k(t)}{dt} = \frac{P_k(t)EA - \sum\limits_{i=1}^{k} E_i A_i}{1-n} \tag{12}$$

Accordingly, the size of d_{10} corresponding to 10% for which soil particles by weight are fine after the time period dt can be quantified approximately as

$$d_{10} = (d_x + d_{x+1})/2 \tag{13}$$

where d_x and d_{x+1} can be obtained through exhaustive method among all soil size groups, under the condition that item $(P_x - 10\%) + (P_{x+1} - 10\%)$ reaches the smallest absolute value.

2.3. Numerical Method and Solution Procedure

The numerical model is solved by using a central scheme of the finite volume method (FVM) built on the OpenFOAM platform (version 2.2.2, https://openfoam.org/release/2-2-2/). More details when assigning basic variables (seepage velocity v_i and porosity n) and governing equations are shown in Appendix A.

The flow chart indicating the solution procedure is illustrated in Figure 4.

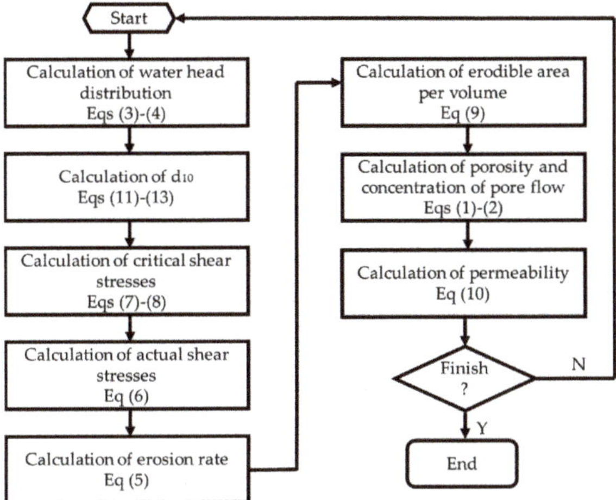

Figure 4. Flow chart of simulation program (Eq(s): Equation(s)).

Variables in closures are updated in every cell of simulation domain, and calculations above are repeated at following time steps.

3. Model Verification

In this section, the proposed model is examined by previous experiments (studies by Skempton and Brogan in 1994 [9] and Wan and Fell in 2008 [1]). The experimental works have detailed records of hydraulic gradient (or eroded soil mass) at different stage of seepage flow and are therefore used for model calibration and validation.

Furthermore, in the next section, more experiments (studies by Lau in 1984 [36]; Lafleur et al. in 1989 [8]; and Chapuis in 1996 [10]) are also included when examining the model's capability in predicting internal stability.

3.1. Overview of Experiments

3.1.1. Devices

The designs in experimental devices between these studies were similar: their experiments were conducted using acrylic cylinders filled up with fully mixed soils (see Figure 5). The upper end of the cylinder was the outlet of the seepage field, which allowed free outflow (also known as being downstream of the seepage field), and the discharge of the seepage flow was measured when running the experiment. The lower end was the inlet of the seepage field (also known as being upstream of seepage field), which connected the filter layer and the water pipe. The filter layer was designed to make the inflow distribution more stable, and the water pipe was used to connect the soil device with the water tank.

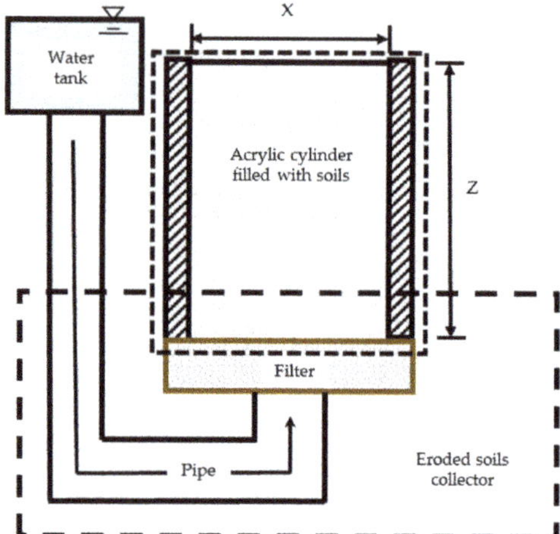

Figure 5. Designs of experimental devices.

3.1.2. Observed Stages of Suffusion Process

With the increase of hydraulic gradient, Skempton and Brogan [9] observed the seepage flow and identified three stages of suffusion: (1) the "dancing-like movement of particles" stage, during which initial signal demonstrating that the suffusion had begun; (2) the "slight but general movement of particles" stage, during which suffusion kept developing and formed the shape of the "pipe" started stretching towards the upstream, and (3) the "strong general piping of particles" stage, where the soil particles were strongly moving out of the soil fabric.

Wan and Fell [1] gave similar characterizations. They identified the flow at the outlet as "clear", "cloudy", and "very cloudy" by manually observing and recording, because the degree of turbidity indicated the degree of suffusion, just like the judgment shown in Skempton and Brogan [9]. The "clear" flow indicated no suffusion and other conditions meant erosion at different stages. The soil samples were evaluated as internally unstable when the stage of "slight but general movement of particles" or "cloudy" flow was reached below the hydraulic gradient of a manually fixed value which was set as the critical one (e.g., 0.7 mentioned in Section 1).

3.1.3. Soil Samples

Skempton and Brogan [9] used three different samples of non-cohesive soils and reported three runs of data, namely Runs A, B, C, respectively. In comparison, Wan and Fell [1] presented 18 runs of experimental data with 14 soil samples (upward flow tests only), their soils contained a portion of cohesive particles (diameter $d < 0.02$ mm). The maximum soil diameter was $d = 30$ mm. Both gap-graded soils and continuous-graded soils were used in the experiments. (see Figure 6).

The definition of gap-graded soils varies in different studies. However, among them, most gap-graded soils have irregular shapes. For instance, a horizontal part of the curve could be graded from 0.02 mm to 2 mm. Otherwise the soils are defined as continuous-graded.

For the runs that lacked complete curves (e.g., Runs 2 and 7), we assume that the diameter of particles in the missing part is the same size as the finest soil particles shown in curves, so that we can perform simulations for these runs.

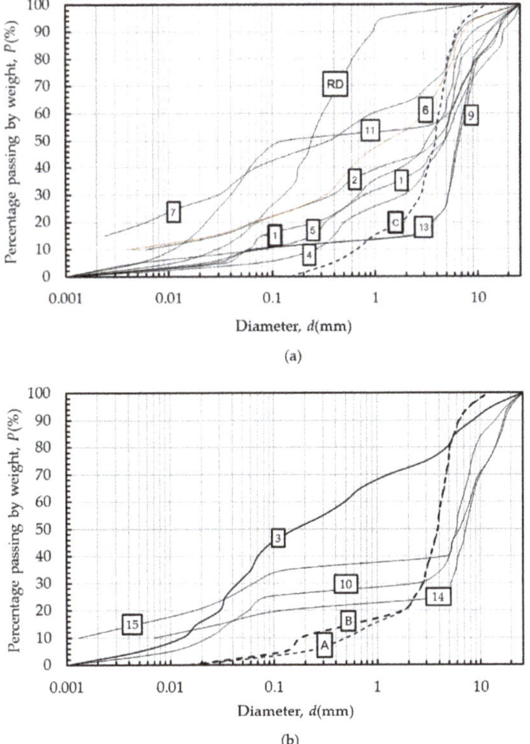

Figure 6. GSD curves of soils in (**a**) internal stable runs; (**b**) internal unstable runs in the studies of Skempton and Brogan [9] (dashed lines) and Wan and Fell [1] (solid lines).

3.2. Simulation Domain and Boundary Settings

The simulation domain is proposed considering the circular symmetry of the seepage field, and assuming a two-dimensional flow at the vertical cross-section plane. As shown in Figure 7a, the domain covers a rectangle cross-sectional area. The size of the area is corresponding to the actual size of experimental device. Therefore, for the runs in Skempton and Brogan [9] the height of the domain is set as 155 mm and the width is set as 139 mm; and for the runs in Wan and Fell [1] the height is set as 300 mm and the width is set as 300 mm. The inlet boundary is imposed with a stable, consistent water head at each stage of hydraulic gradient. Taking Run B in Skempton and Brogan [9] as an example, the loading history is shown in Figure 7b.

In 2017, Rochim et al. [32] noted that the loading history might affect the experimental results of critical hydraulic gradient as well as the eroded mass. They found that if they set a longer duration time under a specific hydraulic gradient, they would record a lower critical hydraulic gradient and more eroded mass.

Such a phenomenon can be explained: when the duration time is not long enough under a specific hydraulic gradient, the erosion process may not reach a steady/stable status, which will lead to the unreliable judgment of critical hydraulic gradient.

To eliminate the effect of different loading histories, in our simulation, a long enough duration time and corresponding stable results are used to validate our model.

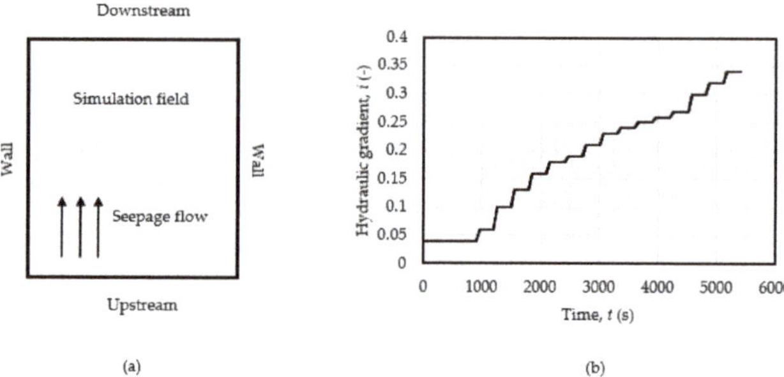

Figure 7. (**a**) Simulation domain and (**b**) loading history of hydraulic gradient of Run B by Skempton and Brogan [9].

3.3. Parameter Calibration

During the simulation, the pure water density ρ_w is set as 1000 kg/m^3, and the viscosity coefficient μ_w is set as 10^{-6} Pa·s. Besides GSD, other initial conditions of the soils (including initial porosity n_0, initial permeability coefficient k_{s0}, and density of soil particles ρ_s) are also reproduced according to the experimental records.

There are two ways to obtain the initial porosity n_0 of the soil: (1) directly from the experimental records, and (2) by calculating $n_0 = 1 - V_{s0}/V$, where V_{s0} denotes the initial soil volume and V denotes the initial volume of seepage field.

Parameters need to be calibrated include α in Equation (5), b in Equation (7), A and B in Equation (8), and Γ in Equation (10). The calibrated parameters are presented in Table 1. These parameters are calibrated using a part of experimental data, i.e., Runs A and B from Skempton and Brogan [9] (for non-cohesive soils) and Runs 2 and 3 from Wan and Fell [1] (for cohesive cases).

Table 1. Details of parameter calibration.

Parameters	Calibrated Values		Notes
	Runs in Skempton and Brogan [9] (for Non-Cohesive Soils)	Runs in Wan and Fell [1] (for Cohesive Soils)	
A	0.021	0.053	Following Wilcock and Crowe [25]
B	0.12	0.12	
b	0.6	0.6	
α	0.005	0.005	Following Fujisawa et al. [20]
Γ	Based on Equation (14)	Based on Equation (14)	

Based on these runs, A, B, b and α are calibrated by trial and error. $A_0 = 0.021$, $B_0 = 0.015$, and $b_0 = 0.067$, as suggested by Wilcock and Crowe [33] are used as initial values for A, B, and b, and $\alpha_0 = 0.0000021$, as suggested by Fujisawa et al. [15], is used as the initial value of α.

It should be noted again that Wan and Fell [1] did not record the amount of soil erosion, and α ranges from 0.0001 to 0.01 would not affect the calculation results in Section 3.4.1. However, Skempton and Brogan [9] did record the amount of soil erosion during the experiment, the model can reasonably reproduce the q-i relationship as well as the eroded mass when $\alpha = 0.005$ is subtracted. Therefore, α is fixed as 0.005 in our model. Value of Γ is calculated based on Equation (14):

$$\Gamma = \frac{k_{s0}}{\left(d_{10}^{1.565} \frac{n_0^{2.3475}}{(1-n_0)^{1.565}} \right)} \tag{14}$$

3.4. Calibration Results

3.4.1. *q-i* Relationship

Figure 8 presents the calibration results for runs in Section 3.3:

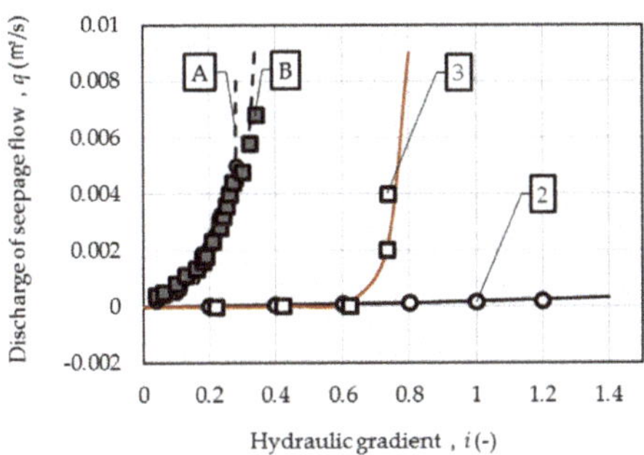

Figure 8. Calibration: dashed lines for computation results of runs in Skempton and Brogan [9]; solid lines for computation results of runs in Wan and Fell [1]; Solid points for test results of runs in Skempton and Brogan [9]; and hollow points for runs in Wan and Fell [1].

It shows that the computed *q-i* relationships agrees well with measured ones. It should be noted that ±10% change of these parameters will not affect the calibration results.

3.4.2. Eroded Mass

The computed eroded mass in Run A (duration time, 80 min; last hydraulic gradient, $i = 0.28$) and Run B (duration time, 90 min; last hydraulic gradient, $i = 0.34$) in Skempton and Brogan [9] are 220.40 g and 77.16 g, which are close to the measured mass 225 g and 75 g, with relative errors less than 3%.

3.5. Model Verification

3.5.1. *q-i* Relationship in Other Runs

Using the calibrated parameters, we conducted the simulation with the data from other runs, the results of representative runs are shown in Figure 9.

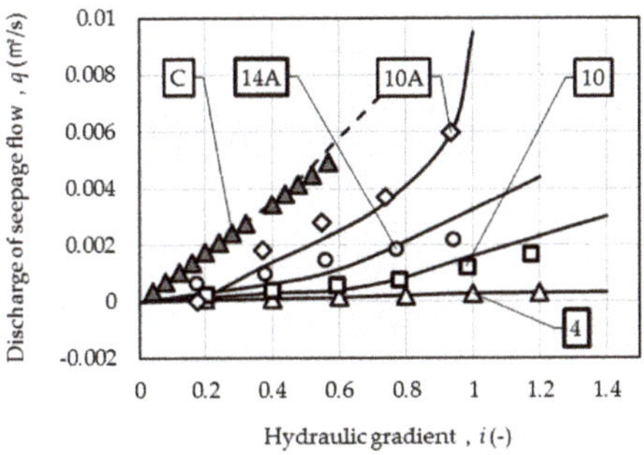

Figure 9. Verification: dashed lines for computation results of runs in Skempton and Brogan [9]; solid lines for computation results of runs in Wan and Fell [1]; solid points for test results of runs in Skempton and Brogan [9]; and hollow points for runs in Wan and Fell [1].

The internally unstable runs (e.g., Runs A and B) and internally stable runs (e.g., Run C) in Skempton and Brogan [9] show obvious difference in shape of q-i curves. For example, when $i = 0.28$ and 0.34, curves of Runs A and B show a sudden rise, respectively, whereas the curve of Run C shows a stable trend all over the experiment. The explanation is as follows: for Runs A and B, as particles are massively eroded by the seepage flow under a specific hydraulic gradient i, the porosity n and the hydraulic conductivity k_s increases, as well as the discharge q; while for Run C, less soil particles are eroded under such a hydraulic gradient which triggers massive suffusion in Runs A and B, or an even higher hydraulic gradient. The simulation result of the data by Wan and Fell [1] show a similar phenomenon.

3.5.2. Eroded Mass for Run C in Skempton and Brogan

The computed eroded mass for Run C in Skempton and Brogan [9] is 57.4 g, close to the measurement of 55 g. The relative error is less than 5%.

3.6. Determination of Critical Hydraulic Gradient

When i reaches critical hydraulic gradient i_c, it can be inferred that the soil has undergone a sudden, irreversible, and largely destructive adjustment due to suffusion. Run 10 in Wan and Fell [1] is used as an example to reveal the reason. In Figure 10, the suffusion process can be divided into three different stages according to the different slopes of the q-i curve:

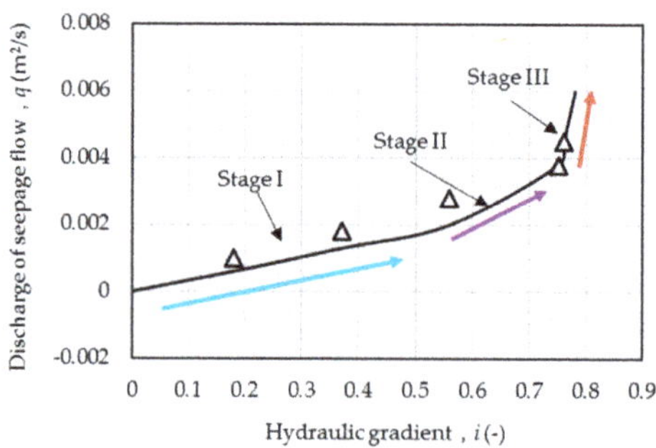

Figure 10. Three stages of suffusion process for Run 10 in Wan and Fell [1].

In stage I ($i < 0.55$), discharge of seepage flow q grows linearly with the hydraulic gradient i, during this stage the out flow is "clear" according to the corresponding experimental record; and in stage II ($0.55 < i < 0.76$), q grows in a nonlinear, but controlled, trend with i, during this stage the outflow is "cloudy" according to the corresponding experimental record; and in stage III ($i > 0.76 = i_c$), one may notice an abrupt change of q-i curve, and this abrupt change may indicate the failure of soil fabrics.

When $i < i_c$, the suffusion behavior has already taken place in the formation of losses of fine particles in limited scale, and such losses will not affect the stability of the whole soil fabrics. When $i = i_c$ or $i > i_c$, more particles are lost until the failure of whole structure occurs.

Following the justification, i_c is evaluated by observing the boundary between stage II and stage III for all the runs in the three studies. Figure 11 compares the computed critical gradients with the measured gradients. 18 of the total 21 runs are reproduced satisfactorily (relative error <10%). The three exceptions are Runs 5, 11, 13 in Wan and Fell [1].

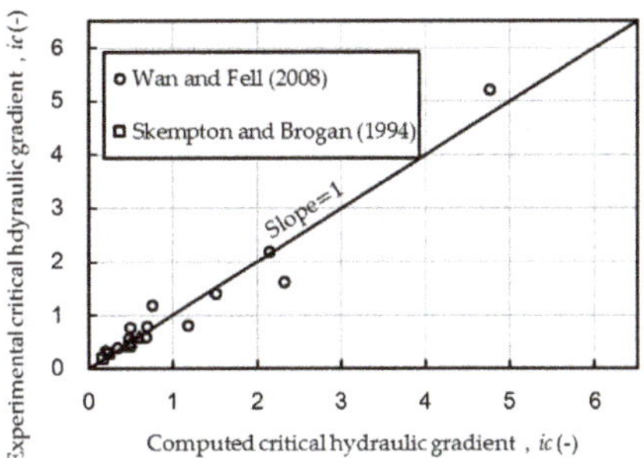

Figure 11. Comparison between computed critical hydraulic gradient and experimental critical hydraulic gradient.

3.7. Porosity Distribution in Different Suffusion Stages

In this study, taking Run A in the study by Skempton and Brogan [9] as an example in order to describe the suffusion process more directly and concretely, we present the spatial distribution of porosity corresponding to stage I, II, and III in 3.6 (see Figures 12 and 13), respectively:

Figure 12a denotes the initial porosity distribution (corresponding to stage I), Figure 12b,c denotes the porosity distribution when it reaches a stable one under hydraulic gradient $i = 0.1$ and $i = 0.2$ respectively (corresponding to stage II since $i < i_c = 0.28$). When $i = 0.1$, the high-porosity zone only appears limitedly in the local region near the outlet, which means that there is no significant erosion in the seepage field. When $i = 0.2$, the high-porosity zone expands for a while, however, it is still limited to the local region near the outlet.

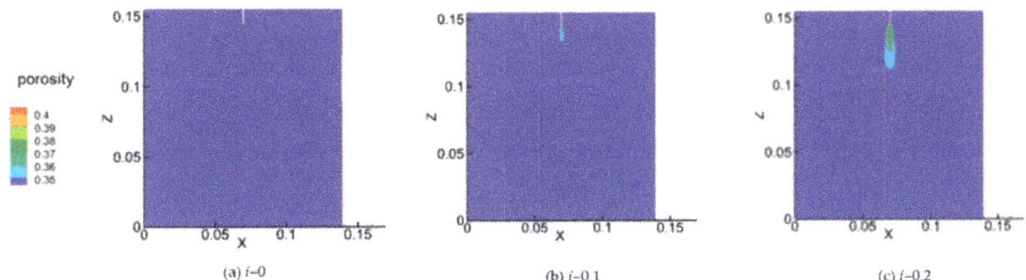

Figure 12. Porosity distribution for Run A in Skempton and Brogan [9] when $i < i_c = 0.28$.

Figure 13a–c denotes porosity distribution when $i = i_c = 0.28$ (corresponding to stage III), during the loading time $t = 3000$ s, 4000 s and 5000 s. The high-porosity zone grows so rapidly and intensely that it connects the upstream to the downstream. Such a phenomenon indicates an unsteady status under the massive suffusion behavior.

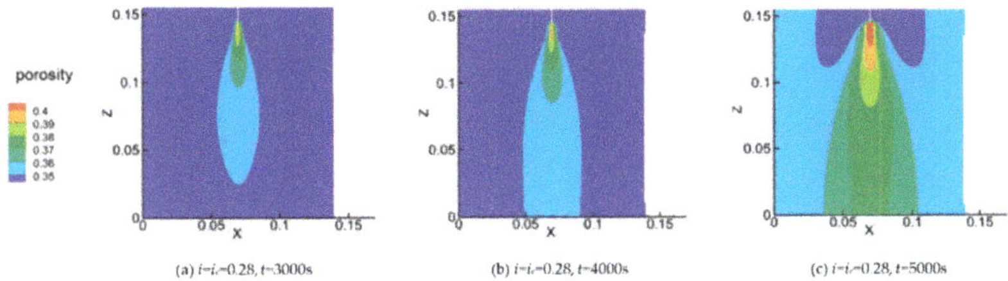

Figure 13. Porosity distribution for Run A in Skempton and Brogan [9] when $i = i_c = 0.28$.

4. Predicting Internal Stability

In Section 3, we describe the suffusion process numerically and evaluate the critical hydraulic gradient i_c, and in this section we attempt to give a further investigation in predicting internal stability.

4.1. Shear Stress Comparison Approach

Averagely, suffusion will be triggered when $\tau > \tau_{cr}$, where τ_{cr} denotes the critical shear stress of the finest soil particles (particle size d_r) remaining in soil fabrics. According to Equation (7), $\tau_{cr} = \min(\tau_{ck})$ as it has been already known that $d_r = \min(d_k)$, and it will increase when suffusion proceeds. Meanwhile, as the particles are continuously eroded out of the seepage field, the porosity ρ and hydraulic conductivity k_s of the soils become higher, and based on Equation (10), τ will also increase. When $i = i_c$, τ will keep exceeding τ_{cr} as the suffusion continues. The suffusion will develop inevitably until the whole fabrics are destroyed.

For each of the soils, there must exist a hydraulic gradient $i_k < i_c$, when $i = i_k$, if the loading time is long enough, and all the particles with sizes $d < d_k$ will be eroded out of the soil fabrics. Under this assumed situation, $d_r = d_k$, and $\tau_{cr} = \tau_{ck}$. At the moment, the volume of the soils V_s will be decreased by $k\%$ (ρ_s is a constant), and the porosity of the soil will be $n = n_0 + (1 - n_0)k\%$.

Therefore, when $n = n_0 + (1 - n_0)k\%$ is given, $\tau_{cr} = \tau_{ck}$; where τ_{ck} is determined by Equations (7) and (8). Since then, a quantitative τ_{cr}-n relationship can be computed. A τ-n relationship can also be obtained through Equation (6).

Here, an approach to predict S (stable) and U (unstable) based on the comparison of the relationship between τ_{cr}-n and τ-n is illustrated: when i is given, by comparing the relative locations of the τ_{cr}-n curve and the τ-n curve in the same coordinate system, we can judge whether suffusion starts, stops or proceeds irreversibly.

4.2. Evaluation Results

Taking Run A in Skempton and Brogan [9] as an example (see Figure 14), under the hydraulic gradient $i = 0.28$, when $n < 0.405$, $\tau > \tau_c$, which means suffusion is triggered and developed; when $n > 0.405$, the τ-n curve almost coincides with the τ_c-n curve, which means that the suffusion process is irreversible. When $i = 0.30$, the τ-n curve is always above the τ_c-n curve and there is no intersection between the two curves. This indicates that under this hydraulic gradient, the suffusion process of the soil is still irreversible.

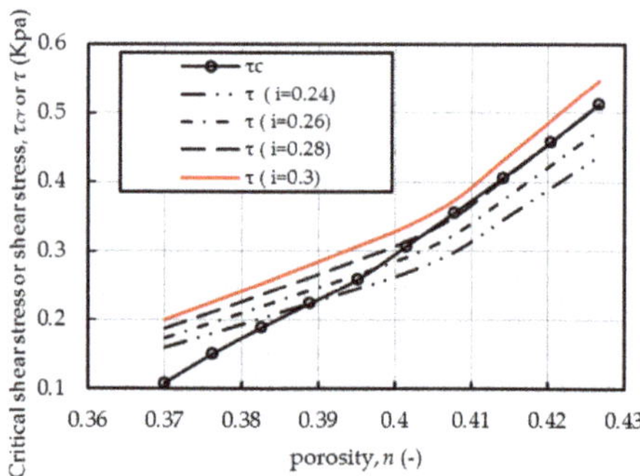

Figure 14. Change of shear stress and critical shear stress with porosity (τ-n and τ_c-n curves).

According to the results shown in Figure 14, the critical hydraulic gradient is $i_c = 0.28$, which is in accordance to the experimental data of Skempton and Brogan [9], and since $i_c < 0.7$, according to the study by Hillel [34], the internal stability of soils in this run is U. It should be noted that α does not involve with the analysis process, which indicates that it does not affect the results of internal stability prediction.

To verify the approach, experimental data from Lau [36], Lafleur et al. [8], and Chapuis et al. [10] are also studied. The GSD curves are presented in Appendix B. This approach is able to distinguish the internal stability of 30 runs out of 36, indicating 83.33% of accuracy, while the most accepted traditional GSD-based approach based on the study by Wan and Fell [1] is 75%.

5. Conclusions

In this study, a new numerical model taking combined effects of GSD and porosity into account is developed to describe the suffusion process. The theory by Wilcock and Crowe [33] in field of sediment transportation is adopted to quantitively describe the inception and transport of finer soil particles. During the simulation, all of the related parameters are calibrated in a solid way. The numerical results showed satisfactory performance when validating with the experimental data of Skempton and Brogan [9] and Wan and Fell [1].

The model can reproduce the *q-i* relationships of experiments as well as critical hydraulic gradients i_c of runs, and describe the suffusion process by showing porosity spatial distribution at different stages.

Inspired by the modeling work, an approach to predicting internal stability is also proposed. Disclosing the relationships among the factors, we found that the soils with specific GSD and porosity are prone to induce the suffusion. The model demonstrates better performance in the judgments of the internal stability (83.33% of correctness) of 36 experimental runs. The accuracy is higher than most of traditional GSD-based approaches. The results provide useful guidelines in macroscopic engineering practice.

Author Contributions: Conceptualization, Q.F. and X.F.; methodology, Q.F. and H.-C.H.; software, Q.F. and T.M.; validation, Q.F., Y.J. and J.W.; numerical runs, Q.F. and X.F.; investigation, Q.F.; data curation, H.-C.H. and J.W.; writing—original draft preparation, Q.F.; writing—review and editing, Q.F., Y.J. and X.F.; visualization, Q.F. and T.M.; supervision, X.F.; project administration, X.F.; funding acquisition, X.F. and Y.J.

Funding: This research is funded by The National Key Research and Development Program of China (Grant No. 2017YFC0804602), and The National Natural Science Foundation of China (Grant No. 51525901). The second author acknowledges the support of open fund of State Key Laboratory of Hydroscience and Engineering, Tsinghua University.

Conflicts of Interest: The authors declare no conflict of interest.

Appendix A

The numerical method for assigning basic variables (seepage velocity v_i and porosity n) within each cell are listed as:

$$v_i^{l,t} = \frac{\int_{\Omega^l} v_i^t dV^l}{V^l} \tag{A1}$$

$$n^{l,t} = \frac{\int_{\Omega^l} n^t dV^l}{V^l} \tag{A2}$$

where Ω^l and V^l denote the region and volume of the *l*th cell, respectively. $v_i^{l,t}$ and $n^{l,t}$ denotes the seepage velocity and porosity within *l*th during *t*th time step, respectively.

Based on Equations (A1) and (A2), the unknown variable porosity $n^{l,t+1}$ in governing Equation (1) can be solved numerically as:

$$n^{l,t+1} = n^{l,t} + dt \sum_k E_k^t A_k^t \tag{A3}$$

where E_k^t denotes the erosion rate of *k*th size group section at *t*th time step, and A_k^t denotes the erodible area per unit volume of *k*th size group at *t*th time step.

Adopting the derivation by Fujisawa et al. [21], the unknown variable concentration $C^{l,t+1}$ in governing Equation (2) can be solved numerically as:

$$V^l \frac{n^{l,t+1}C^{l,t+1} - n^{l,t}C^{l,t}}{dt} + q_i^t t^l_{ij}\Delta s^l_j = V^l \frac{n^{l,t+1} - n^{l,t}}{dt} \tag{A4}$$

As shown in Figure A1, t^l_{ij} and Δs^l_j denotes the unit normal vector and length of the jth ($j = 1, 2, 3$ and 4) boundary segment of the lth cell. Also, item $q_i^t\, t^l_{ij}$ can be calculated through Equation (A5) using the FVS scheme:

$$q_i^t t^l_{ij} = C^{l,t} v^{l,t} t^l_{ij} + C^{r,t} v^{r,t} t^l_{ij} \tag{A5}$$

where superscript $r = r_1, r_2, r_3$, and r_4 denotes the neighboring cells (shearing the boundary) of lth cell.

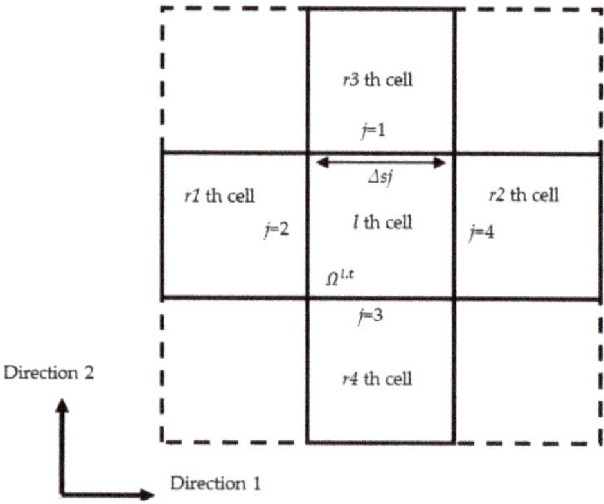

Figure A1. Geometry of the cells.

In the simulation, item $\partial Sr/\partial t$ is always 0 since the soil are fully saturated all the time. Similar to Equations (A4) and (A5), the unknown variable water head $h^{l,t}$ in Equations (3) and (4) can be solved numerically based on the following Equation:

$$k_s^{l,t} \frac{\mathrm{grad}h_1^{r_1,t} - 2\mathrm{grad}h_1^{l,t} + \mathrm{grad}h_1^{r_2,t}}{\Delta s_1^l} + k_s^{l,t} \frac{\mathrm{grad}h_2^{r_3,t} - 2\mathrm{grad}h_2^{l,t} + \mathrm{grad}h_2^{r_4,t}}{\Delta s_1^l} = 0 \tag{A6}$$

where,

$$\mathrm{grad}h_1^l = \frac{h^{r_1,t} - 2h^{l,t} + h^{r_2,t}}{\Delta s_1^l} \tag{A7}$$

$$\mathrm{grad}h_2^l = \frac{h^{r_3,t} - 2h^{l,t} + h^{r_4,t}}{\Delta s_2^l} \tag{A8}$$

Appendix B

The GSD curve of soils in Lau [36], Lafleur et al. [8], and Chapuis et al. [10] are shown in Figure A2:

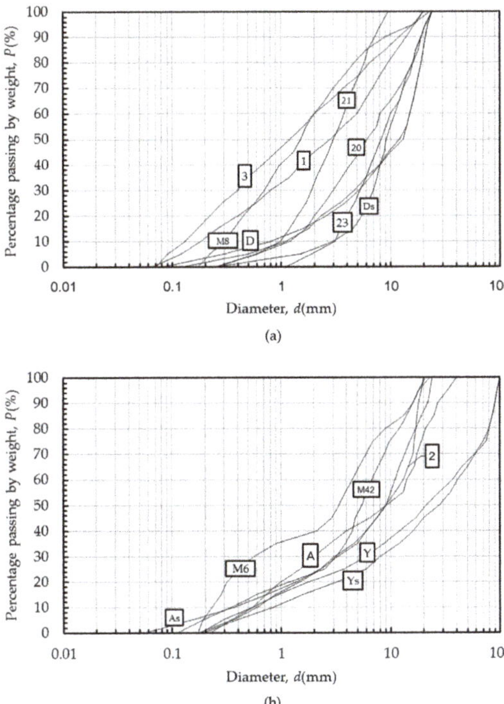

Figure A2. GSD curves of soils in (**a**) internal stable runs; (**b**) internal unstable runs in studies by Lau [36], Lafleur et al. [8], and Chapuis et al. [10].

References

1. Wan, C.F.; Fell, R. Assessing the potential of internal instability and suffusion in embankment dams and their foundations. *J. Geotech. Geoenviron.* **2008**, *134*, 401–407. [CrossRef]
2. Wan, C.F.; Fell, R. Investigation of rate of erosion of soils in embankment dams. *J. Geotech. Geoenviron.* **2004**, *130*, 373–380. [CrossRef]
3. Fell, R.; Fry, J.J. State of the art on the likelihood of internal erosion of dams and levees by means of testing. In *Erosion in Geomechanics Applied to Dams and Levees*, 1st ed.; Bonelli, S., Nicot, F., Eds.; John Wiley & Sons Inc: London, UK, 2013; Volume 1.
4. Fleshman, M.S.; Rice, J.D. Laboratory Modeling of the mechanisms of piping erosion initiation. *J. Geotech. Geoenviron.* **2014**, *140*, 04014017. [CrossRef]
5. Abrams, D.M.; Lobkovsky, A.E. Growth laws for channel networks incised by groundwater flow. *Nat. Geosci.* **2009**, *2*, 193–196. [CrossRef]
6. Howard, A.D.; McLane, C.F. Erosion of cohesionless sediment by groundwater seepage. *Water Resour. Res.* **1998**, *24*, 1659–1674. [CrossRef]
7. Adel, H.D.; Bakker, K.J. Internal stability of minestone. In Proceedings of the International Symposium on Modelling Soil–Water-Structure Interactions, Rotterdam, The Netherlands, 29 August–2 September 1988; pp. 225–231.
8. Lafleur, J.; Mlynarek, J. Filtration of broadly graded cohesionless soils. *J. Geotech. Eng.* **1989**, *115*, 1747–1768. [CrossRef]
9. Skempton, A.W.; Brogan, J.M. Experiments on piping in sandy gravels. *Geotechnique* **1994**, *44*, 449–460. [CrossRef]
10. Chapuis, R.P.; Contant, A. Migration of fines in 0–20 mm crushed base during placement, compaction, and seepage under laboratory conditions. *Can. Geotech. J.* **1996**, *33*, 168–176. [CrossRef]

11. Chang, D.S.; Zhang, L.M. A Stress-controlled erosion apparatus for studying internal erosion in soils. *Geotech. Test. J.* **2011**, *34*, 579–589.

12. Chang, D.S.; Zhang, L.M. Critical hydraulic gradients of internal erosion under complex stress states. *J. Geotech. Geoenviron.* **2013**, *139*, 1454–1467. [CrossRef]

13. Fleshman, M.S.; Rice, J.D. Constant gradient piping test apparatus for evaluation of critical hydraulic conditions for the initiation of piping. *Geotech. Test. J.* **2013**, *36*, 834–846. [CrossRef]

14. Rönnqvist, H.; Viklander, P. On the Kenney-Lau approach to internal stability evaluation of soils. *Geomaterials* **2014**, *4*, 129. [CrossRef]

15. Fujisawa, K.; Murakami, A. Numerical analysis of the erosion and the transport of fine particles within soils leading to the piping phenomenon. *Soils Found.* **2010**, *50*, 471–482. [CrossRef]

16. Li, M.X.; Fannin, R.J. Capillary tube model for internal stability of cohesionless soil. *J. Geotech. Geoenviron.* **2013**, *139*, 831–834. [CrossRef]

17. Guo, N.; Zhao, J. A coupled FEM/DEM approach for hierarchical multiscale modelling of granular media. *Int. J. Numer. Methods Eng.* **2014**, *99*, 789–818. [CrossRef]

18. Wagner, G.J.; Gregory, J. Coupling of atomistic and continuum simulations using a bridging scale decomposition. *J. Comput. Phys.* **2003**, *190*, 249–274. [CrossRef]

19. Li, X.; Ke, W. A bridging scale method for granular materials with discrete particle assembly—Cosserat continuum modeling. *Comput. Geotech.* **2011**, *38*, 1052–1068. [CrossRef]

20. Lominé, F.; Scholtes, L. Modeling of fluid–solid interaction in granular media with coupled lattice Boltzmann/discrete element methods: Application to piping erosion. *Int. J. Numer. Anal. Methods Geomech.* **2013**, *37*, 577–596. [CrossRef]

21. Bi, D.; Zhang, J. Jamming by shear. *Nature* **2011**, *480*, 355–358. [CrossRef]

22. Boyer, F.; Élisabeth, G. Unifying suspension and granular rheology. *Phys. Rev. Lett.* **2011**, *107*, 188301. [CrossRef]

23. Cassar, C.; Nicolas, M. Submarine granular flows down inclined planes. *Phys. Fluids* **2005**, *17*, 103301. [CrossRef]

24. Yohannes, B.; Hill, K.M. Rheology of dense granular mixtures: Particle-size distributions, boundary conditions, and collisional time scales. *Phys. Rev. E* **2010**, *82*, 061301. [CrossRef] [PubMed]

25. Kézdi, A. *Soil Physics-Selected Topics-Developments in Geotechnical Engineering-25*, 1st ed.; Elsevier Publishing Company, Limited: Essex, UK, 1979; Volume 1.

26. Kenney, T.C.; Lau, D. Internal stability of granular filters. *Can. Geotech. J.* **1985**, *22*, 215–225. [CrossRef]

27. Kenney, T.C.; Lau, D. Internal stability of granular filters: Reply. *Can. Geotech. J.* **1986**, *23*, 420–423. [CrossRef]

28. Burenkova, V. Assessment of suffusion in non-cohesive and graded soils. In *Title of Filters in Geotechnical and Hydraulic Engineering, Proceedings of Geo-Filters, Karlsruhe, Germany, 20–22 October 1992*; A A Balkema: Rotterdam, The Netherlands, 1993; pp. 357–360.

29. Reddi, L.N.; Lee, I.M. Comparison of internal and surface erosion using flow pump tests on a sand-kaolinite mixture. *Geotech. Test. J.* **2000**, *23*, 116–122.

30. Marot, D.; Rochim, A. Assessing the susceptibility of gap-graded soils to internal erosion: Proposition of a new experimental methodology. *Nat. Hazards* **2016**, *83*, 365–388. [CrossRef]

31. Hieu, D.M.; Kawamura, S. Internal erosion of volcanic coarse grained soils and its evaluation. *Int. J. Geomate* **2017**, *13*, 165–172. [CrossRef]

32. Rochim, A.; Marot, D. Effects of hydraulic loading history on suffusion susceptibility of cohesionless soils. *J. Geotech. Geoenviron.* **2017**, *143*, 04017025. [CrossRef]

33. Wilcock, P.R.; Crowe, J.C. Surface-based transport model for mixed-size sediment. *J. Hydraul. Eng.* **2003**, *129*, 120–128. [CrossRef]

34. Hillel, D. *Fundamentals of Soil Physics*, 1st ed.; Academic Press: San Diego, CA, USA, 1980.

35. Khilar, K.C.; Fogler, H.S. Model for piping-plugging in earthen structures. *J. Geotech. Eng.* **1985**, *111*, 833–846. [CrossRef]

36. Lau, D. Stability of Particle Grading of Compacted Granular Filters. Ph.D. Thesis, University of Toronto, Toronto, ON, Canada, 1984.

Review

SPH Modeling of Water-Related Natural Hazards

Sauro Manenti [1], **Dong Wang** [2], **José M. Domínguez** [3], **Shaowu Li** [4], **Andrea Amicarelli** [5] and **Raffaele Albano** [6,*]

[1]   Department of Civil Engineering and Architecture, University of Pavia, via Ferrata 3, 27100 Pavia, Italy
[2]   Department of Civil & Environmental Engineering, National University of Singapore,
      21 Lower Kent Ridge Road, Singapore 119077, Singapore
[3]   EPHYSLAB Environmental Physics Laboratory, Universidade de Vigo, AS LAGOAS-32004, Ourense, Spain
[4]   State Key Laboratory of Hydraulic Engineering Simulation and Safety, Tianjin University,
      Tianjin 300072, China
[5]   Department SFE, Ricerca sul Sistema Energetico—RSE SpA, via Rubattino 54, 20134 Milan, Italy
[6]   School of Engineering, University of Basilicata, 85100 Potenza, Italy
*   Correspondence: raffaele.albano@unibas.it; Tel.: +39-097-120-5157

Received: 4 July 2019; Accepted: 6 September 2019; Published: 9 September 2019

Abstract: This paper collects some recent smoothed particle hydrodynamic (SPH) applications in the field of natural hazards connected to rapidly varied flows of both water and dense granular mixtures including sediment erosion and bed load transport. The paper gathers together and outlines the basic aspects of some relevant works dealing with flooding on complex topography, sediment scouring, fast landslide dynamics, and induced surge wave. Additionally, the preliminary results of a new study regarding the post-failure dynamics of rainfall-induced shallow landslide are presented. The paper also shows the latest advances in the use of high performance computing (HPC) techniques to accelerate computational fluid dynamic (CFD) codes through the efficient use of current computational resources. This aspect is extremely important when simulating complex three-dimensional problems that require a high computational cost and are generally involved in the modeling of water-related natural hazards of practical interest. The paper provides an overview of some widespread SPH free open source software (FOSS) codes applied to multiphase problems of theoretical and practical interest in the field of hydraulic engineering. The paper aims to provide insight into the SPH modeling of some relevant physical aspects involved in water-related natural hazards (e.g., sediment erosion and non-Newtonian rheology). The future perspectives of SPH in this application field are finally pointed out.

Keywords: SPH (Smoothed Particle Hydrodynamics); water-related natural hazards; sediment scouring; dense granular flow; fast landslide; surge wave; flooding on complex topography; HPC (High Performance Computing); FOSS (Free Open Source Software)

---

1. Introduction

Thanks to the availability of high-performance computers, in the last few years, computational fluid dynamics (CFD) has been widely applied to simulate natural hazards in the field of hydraulic engineering. Due to the fast and large deformations characterizing the problems in this research field, meshless techniques allow for some intrinsic limitations of traditional grid-based methods (e.g., mesh deformation and cracking; free-surface, and interface treatment) to be overcome. Among the meshless techniques, the smoothed particle hydrodynamics (SPH) method has proven to have advantages over other methods. Following a Lagrangian approach, each continuum is discretized through a discrete set of material particles that lack connective mesh and follow the deformation undergone by the material. The dynamics of material particles obeys Newton's laws of motion and the discretized form of the governing equations (i.e., momentum and mass balance) is obtained through particle approximation using an interpolant kernel function (i.e., a central function of the particles' relative distance). Based on the solution algorithm of the discretized governing equations, two different approaches can usually be defined: weakly compressible SPH (WCSPH), if the continuum is assumed to be slightly compressible (governing equations can be decoupled), and incompressible SPH (ISPH).

Even if SPH was originally developed for astrophysical problems, it has been subsequently extended to the analysis of free-surface flows and several kinds of multiphase flows including non-Newtonian fluid and post-failure dynamics of rainfall induced landslide. Many works have shown that the multiphase coupled dynamics of water and non-cohesive sediment mixture may be simulated following a fluidic approach [1–3]. Through the proper definition of appropriate yielding criteria and non-Newtonian rheological models, the triggering and propagation of multiphase flow as well as dense rapid granular flow can be simulated through the SPH approach with a suitable degree of accuracy in reproducing the experimental results, showing that this approach can be conveniently applied to real case studies. In this way, it is possible to simulate scouring around complex structures that may cause hazards related to both structural instability and threats to the safe management of hydro-power plants [4–7]. Another important branch of non-cohesive sediment dynamics is related to two-phase flows on complex 3D topography such as the sediment eroded by a dam-break wave propagating on a mountain valley [8].

In the field of dense granular flows, fast landslide dynamics can be properly simulated by accounting for the interactions with water (both pore water and stored water) that influences landslide deformability and run-out characteristics. Thus, complex multiphase flows can be simulated for investigating natural hazard related to a tsunami wave generated by a quick landslide at the slope of a reservoir [9–13]. Additionally, a fast shallow landslide triggered by intense rainfall at the slope of a hill represents the most common natural hazards in some areas of the world [14,15] and can be simulated with proper numerical modeling. Furthermore, SPH can be successfully applied to the analysis of slope stability, with the detection of failure initiation and propagation leading to delineation of the potential slip surface [16].

Computational methods are also successful in simulating complex nonlinear flows and transport processes involved in natural hazards related to the flooding risk, as in the case of dam and dyke failures. Water wave propagation in complex 3D topography can be properly predicted [17–20] including urban areas where the complexity of the 3D flow pattern is increased by sudden changes in the propagation direction due to buildings as well as different types of floating bodies such as cars, trucks, etc., as in the works in [21,22].

Even if mathematical models have been widely established to simulate the relevant hydraulic and hydro-geologic processes involved in the simulation of water-related natural hazards [23–25], the relevant complexity (especially from a geometric point of view) of practical applications in the field of applied engineering poses some difficulties in the applications of the computer codes that implement the above-mentioned mathematical models. The numerical modeling of a complex 3D problem frequently requires the discretization of large domains with a high resolution, which implies simulating a very large number of elements (usually at least some millions), thus increasing the required computational time and resources in a non-linear way. Fortunately, the parallel computational power of the computer hardware has increased significantly in recent years, especially in the case of graphics processing units (GPUs) which are now several times more powerful than traditional central processing units (CPUs). However, it is important to note that the GPU is more powerful only for properly posed problems that are parallelizable on GPUs, and not necessarily for general computing. Furthermore, it is imperative that the models are implemented using high performance computing (HPC) techniques to take advantage of the power of current hardware (i.e., multi- cored CPUs, GPUs, or other hardware accelerators).

Another limitation that computational modeling suffers is the inherent uncertainties of some modeling parameters that influence (in some cases even significantly) the model response and lower the reliability of the results. Despite this intrinsic limitation, such deterministic models may be useful anyhow to support risk analysis and mitigation through the development of fast-running numerical models that can help to create probabilistic maps of risk variables. In this way, a multi-disciplinary decision support system for natural hazard risk reduction and management could be developed that allows for the exploration of several scenarios including potential risk-reduction options [26].

In this context, the aim of the paper was to provide an overview of some recent applications of the SPH method to the modeling of water-related natural hazards. The selected works illustrate some of the relevant problems of practical and theoretical interest in the field of hydraulic engineering that could be useful to provide an introduction to the SPH method as applied to the analysis and mitigation of water-related natural hazards.

Additionally, some widespread free and open-source software (FOSS) CFD codes for multiphase engineering applications are reviewed as well as their capabilities in modeling some of the problems described in this paper.

2. Two-Phase Coupled Dynamics

This section illustrates some engineering applications of the SPH method in the simulation of two-phase flows involved in hydraulic engineering problems of practical interest in the field of water-related natural hazards.

When dealing with the numerical simulation of two-phase flows of immiscible fluids, a central aspect to be considered is the interface treatment. Concerning incompressible SPH (ISPH) simulations of gravity currents induced by the inflow of a certain fluid into another fluid with a density difference, two different approaches were proposed in [27] and classified as a coupled model and a decoupled model, respectively. The coupled model is based on the assumption that standard ISPH governing equations can be applied to every particle in the computational domain regardless of the phase it belongs to. Therefore, when an interface particle is considered, the kernel interpolation is extended over all the heterogeneous neighbors in its interaction domain, which is the kernel compact support (circular in 2D problems or spherical in 3D problems) centered on the considered interface particle. In contrast, the decoupled model assumes that standard ISPH governing equations (including the Poisson equation) should be solved separately for each phase by identifying the free-surface. Therefore, if an interface particle is considered to be the kernel, interpolation is restricted to its homogeneous neighbors solely within the kernel support, thus kernel truncation occurs. Proper interface boundary conditions should be satisfied to couple both phases: these conditions can be represented by the continuity of both normal and tangential stresses between heterogeneous particles at the interface. According to the results in [27], the decoupled model also provides higher accuracy than the coupled model in the case of multi-fluid flow with a high-density ratio.

A modified formulation of the standard weakly compressible SPH (WCSPH) governing equations were presented in [28] for modeling rapid free-surface multiphase flows with a high-density ratio involving violent impact against a rigid wall. This formulation, which is based on the coupled approach, allows for the numerical instability induced by the discontinuity of the fluid properties across the interface to be eliminated without excluding the heterogeneous particle during kernel interpolation. Fairly accurate results were achieved with this coupled model simulating some flows with continuous pressure across the interface. Other good examples of multi-phase SPH formulations (both WCSPH and ISPH) that deal with the interface between fluids under certain constraints are described in the works in [29–32].

Among the variety of problems involving the numerical simulation of multi-phase flows, two topics of concern in the field of water-related natural hazards have been examined in the following: (i) scour with non-cohesive sediment transport induced by rapidly varied free-surface flow and erosion around complex structures; and (ii) fast dynamics of dense granular flows and landslides. Some relevant works on these topics are discussed, pointing out the peculiar aspects of numerical modeling.

2.1. Scouring and Sediment Transport

An important aspect of the design and maintenance of the bottom-founded submerged structure is non-cohesive sediment scouring. Bottom sediment erosion around the structure is caused by a complex flow pattern induced by the presence of the structure that strongly modifies the upstream undisturbed flow field [33]. The scouring evolution over time must be properly analyzed and mitigated in order to avoid worsening of structural stability due to foundation exposure.

This topic is widely investigated in fluvial hydraulics, in coastal areas, and in the marine environment. When dealing with a safety assessment of a hydraulic structure placed in the riverbed (e.g., bridges, barrages, etc.), the erosive action of the river stream must be reliably evaluated [34]. Several empirical formulas to predict final scouring are available, but the phenomenon is time-dependent and is affected by several uncertainties (related to both sediment and flow characteristics) as well as stochastic factors influencing the flow evolution such as transport and deposition of wood debris [5,35,36].

Additionally, in the design of highly demanding marine structures, the scour around the foundations should be dependably predicted [37,38]. For instance, a continuous one-way current due to the overtopping flow of an extreme tsunami wave may cause long-term erosion of the foundation on the rear side of the coastal protection structures [39,40]. Scour pit evolution behind the seawall induces the formation of eddies with increasing size to adapt the growing dimension of the scour cavity. Excessive sediment erosion directly results in the instability, and even destruction, of these hydraulic structures [41–44]. Most empirical relations have been established to predict the bulk and time-averaged sediment transport rate, but their application requires experimental data for calibration. Furthermore, these empirical relations were obtained from small scale laboratory experiments and their extension to a real scale problem may lead to significant erosion estimate errors [45].

An assessment of the mechanics of detailed temporal erosion processes as well as for the reliable design and assessment of these structures requires both physical investigations and advanced numerical modeling that accounts for the effects of the stochastic variables on the scouring process. Of course, numerical modeling is generally less expensive than physical experimentation, but, in some cases, experimental data may be available from technical literature. As with all engineering problems, any numerical tool must be properly validated against experimental data. After the model has been validated, its parameters require tuning based on the model/experiment results.

Many sediment modeling techniques use mesh-based approaches such as finite difference method and finite volume method to simulate the erosion and fluid-sediment coupled dynamics. However, these mesh-based methods suffer from some intrinsic limitations due to the fixed grid system, which leads to some difficulties in effectively simulating the bed-grain interactions, fluid-sediment momentum transmissions, and the dynamics within the deposits because of their fixed grid system. Furthermore, accurately tracking the free surface and fluid-sediment interface is also a big challenge for these approaches. Lagrangian meshfree particle methods overcome these limitations by intrinsically capturing the free-surface and tracking the particles. These methods have been widely used in recent years for the analysis of erosion processes.

The SPH technique has been proven to be effective in complex multiphase applications [46] such as water–gas flows or bubbly flow simulations [47,48]. In addition, when simulating non-cohesive bottom sediment scouring by rapid water flow, both weakly compressible (WCSPH) and incompressible (ISPH) formulations allow for a reliable description to be obtained. The basic idea is to model the sediment dynamics, likewise a pseudo-fluid, once the onset of sediment particle motion is attained. According to this approach, a criterion should be defined that represents the critical condition for the incipient motion of sediment; this can be done in terms of either critical velocity [49,50], or critical shear stress. The first approach may lead to some problems if critical velocity is evaluated as the depth-averaged value that in some cases may not be representative of the scouring, as for the continuous tsunami overflow behind a seawall that induces rapid water depth variation inside the scour pit [2]. The second approach based on the critical shear stress has been widely adopted to analyze non-cohesive sediment scouring by rapid water flows.

The authors in [7] implemented two different criteria in a WCSPH model to simulate the erosion process of non-cohesive bed sediment due to constant bottom water discharge from 2D tank (flushing). These criteria are based on Mohr–Coulomb (M–C) yielding theory and on Shields (SH) theory, respectively. The M–C critical condition defining sediment failure and the onset of erosion requires the introduction of a maximum viscosity as the product between sediment viscosity, μ_s, and a magnification factor, η, representing the numerical parameters to be tuned. When the apparent viscosity is lower than the maximum viscosity, the sediment is treated as a non-Newtonian fluid of Bingham type and solid particles are set in motion with constant viscosity μ_s (green curve in Figure 1). The strategy of introducing an upper viscosity limit for the sediment was also followed in [51] in the WCSPH simulation of complex problems in the field of marine engineering; below this maximum limit, the work in [51] adopted a variable apparent viscosity calculated through the M–C theory for the soil phase. The SH critical condition does not require the introduction of a numerical threshold for the viscosity of the solid phase. However, in [7], both the M–C and SH approaches require tuning of the mechanical parameters of the bottom sediment such as the angle of internal friction, φ, and sediment viscosity, μ_s, that became numerical parameters to fit the experimental eroded profile. This may be not be practical when calibration data are not available for the investigated problem.

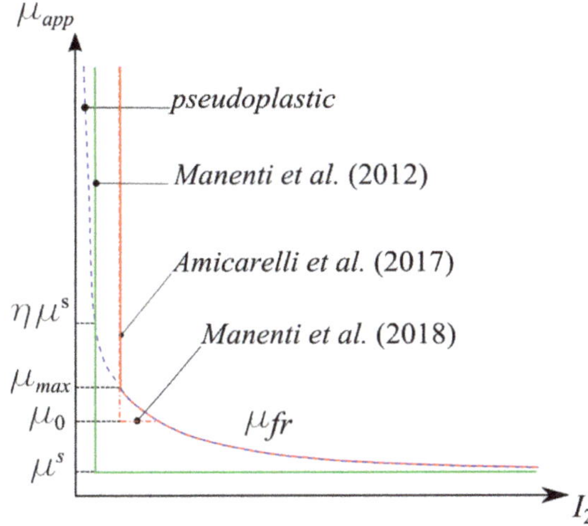

Figure 1. Rheological models for non-cohesive sediment erosion with bed load transport and for dense granular flows. I_2: second invariant of the rate of deformation tensor; μ_{app}: apparent viscosity.

In order to overcome this limitation, [8] proposed a WCSPH formulation of a mixture model for the analysis of dense granular flows consistent with the kinetic theory of granular flow (KTGF). This mixture model, which avoids the use of an erosion criterion, has been integrated into the FOSS code SPHERA v.9.0.0 (RSE SpA). The relevant physical properties (i.e., density and velocity) of the mixture of pure fluid and granular material are expressed as a function of the volume fraction ε, occupied by each phase at a material point. The balance equations for the mixture are defined accordingly and discretized consistently with the WCSPH approach. The frictional regime of the mixture dynamics is represented under the packing limit of the KTGF, which holds for the volume fraction ε_s of the solid (granular) phase and is characteristic of bed-load transport and fast landslides (see also Section 2.2). In the frictional regime, the mixture (or apparent) viscosity, μ, is calculated as a weighted sum of the pure fluid viscosity μ_f, and the frictional viscosity μ_{fr}, the latter being evaluated on the basis of the mean effective stress σ'_m, angle of internal friction φ, and the second invariant I_2 of the rate of the deformation tensor of the sediment. The frictional viscosity increases as the shear rate tends to zero, in accordance with the pseudo-plastic rheological behavior (dashed blue curve in Figure 1).

To avoid the unbounded growth of apparent viscosity of the mixture, a threshold (or maximum) viscosity $\mu_{\max}$ is introduced with a physical meaning. Threshold viscosity acts when approaching the zero shear rate; mixture particles with an apparent viscosity higher than the threshold viscosity are considered in the elastic–plastic regime of soil deformation where the kinetic energy of solid particles is relatively small and the frictional regime of the packing limit in the KTGF does not apply. For these reasons, the threshold viscosity is assigned to those particles that are excluded from the SPH computation (continuous red curve in Figure 1; below $\mu_{\max}$, the red curve coincides with the dashed blue curve of the pseudoplastic model). The excluded particles represent a fixed boundary with suitable values of the relevant physical properties and are included in the neighbor list of the nearby moving particle. The value of the threshold viscosity does not require tuning or calibration, but it should be selected for the specific problem; the value assigned to the threshold viscosity is the higher value that does not influence the numerical results significantly. Further increase of the value assigned to the threshold viscosity only affects the computational time because it determines an increase in the maximum value that can be assumed by the apparent mixture viscosity during the computation. The relationship between the mixture viscosity and the time step value that assures the stability of the adopted explicit time-stepping scheme is given by Equation 2.33 in [8]. When the numerical stability of the time integration scheme is dominated by the viscous criterion, the threshold viscosity reduces the computational time.

In [8], some experiments of erosional dam breaks adopting the physical values for the mechanical parameters of the sediments including the angle of internal friction were simulated with good reliability. The 2D experimental tests, involving rapidly varied mixture flows with erosion and the transport of bed sediment, were simulated in [7] and [8]. Moreover, in [6], mixture flow computation was performed with the open-source DualSPHysics solver [52] accelerated with a graphic processing unit (GPU). The adopted WCSPH algorithm, representing the improvement of the model in [53], reproduces the dynamics of the bottom sediment phase combining two erosion criteria with the non-Newtonian rheological model. The adopted rheological model is based on the Bingham-type Herschel–Bulkley–Papanastasiou (HBP) model providing the apparent viscosity μ_{HBP} of the sediment:

$$\mu_{HBP} = \frac{|\tau_c|}{\sqrt{I_2}}\left(1 - e^{-m\,\sqrt{I_2}}\right) + 2\mu\left(2\,\sqrt{I_2}\right)^{n-1} \tag{1}$$

This model allows for the transition between un-yielded (fixed) and yielded (mobile) granular material to be described without introducing a maximum value of the viscosity for the solid phase. Proper selection of the HBP model's parameters, m and n, allows for the reproduction of the required rheological behavior as the Newtonian or the pseudoplastic model. The yield stress parameter τ_c in the HBP model was evaluated in two different calibration procedures depending on the erosion criterion that holds for modeling the yielding mechanism of sediment. The qualitative representation of Equation (1) with $n < 1$ is denoted by the blue dashed curve in Figure 1.

If a solid particle is identified as being near the sediment–water interface by means of the criteria described in [6], then the Shields criterion determines the onset of bed erosion at the sediment–water interface and provides the critical bed shear stress τ_{bcr}, replacing the yield stress parameter τ_c, in the HBP model.

If the particle is near the water–sediment interface, but Shields erosion criterion is not satisfied or if the particle is far from the water–sediment interface, then the yielding mechanism is modeled through the Drucker–Prager criterion that allows the critical shear stress τ_y, replacing the yield stress parameter, τ_c, in the HBP model to be defined to determine the apparent viscosity for the sediment.

As a result, three distinct regions may be defined, starting from the water–sediment interface and going downward: (a) eroded sediment exceeding the Shields critical bed shear stress (bed load transport); (b) yielded sediment (plastic deformation with slow kinematics); and (c) un-yielded sediment (high viscosity, static condition).

The model proposed in [6] also accounts for suspended transport. Following [51], the identification of the suspension layer is obtained through a non-dimensional concentration, c_v, computed for an interface particle as the ratio of the sediment particle volume to the total particle volume within the interaction domain. The onset of suspended transport is determined by c_v falling below the threshold value of 0.3 and the suspended sediment viscosity is computed through [54] the experimental colloidal equation, which is more simple to implement with respect to the piece-wise function adopted in [51]. The density of suspended particles is computed by solving the mass balance equation. Even if, in some cases, the percentage gap between the experimental and numerically predicted maximum scouring depth is significant, it can be seen that the scour process is affected by several random factors and therefore reliable predictions of scouring effects are quite difficult to obtain, even with experimental modeling. The sediment dynamics models based on a synthetic rheological law (e.g., [6]) assume the same rheological behavior for the bed-load transport (frictional regime of KTGF), suspension for dense granular flows (kinetic-collisional regime of KTGF),a and suspension for diluted granular flows (kinetic regime of KTGF). This feature provides advantages and drawbacks with respect to KTGF-based sediment dynamics models (e.g., [8]). The model of [6] can reproduce several sediment transport regimes (not only bed-load transport), but is not coherent with KTGF, some parameters require tuning procedures, and non-transported granular material (e.g., landslides) is not treated.

The work in [2] investigated by means of ISPH the water-induced 2D sediment scouring where the erosion process is mainly related to loose sediment particles suspended in the water flow, as in the case of scouring behind a seawall produced by continuous tsunami overtopping. In contrast to the physics of sediment flushing, where sediment moves as bed load at very high density, in the case of overtopping erosion, the solid particles move more loosely and the density of turbid water is significantly lower than the sediment density. In such situations, the erosion dynamics is controlled by the balance between the suspension effect due to turbulent mixing and settling of suspended solid particles owing to gravity force. This process is strongly affected by the size of numerical particles, which is usually far bigger than the size of a real sediment grain in practical problems. For the reasons explained above, adoption of real sediment density in the computation may lead to an unreliable representation of the erosion. The model proposed in [2] introduces a simplification due to the size (and hence number) of the particles required to properly model the turbulent mixing. This model is based on the concepts of numerical turbid water particle (TWP) and clear water particle (CWP) to simulate the sediment-entrained flow in cases where the granular particles of sediment move more loosely and sediment is washed away, mainly in the form of a suspended load. Due to the size considerations discussed above, numerical particles should be treated as a combination of clear water and turbid water particles if a suspended load is simulated; therefore, eroded solid particles represent a water–sediment mixture whose density is calculated based on the integral interpolation theory, thus accounting for density reduction as solid particles are suspended and mixed with water. The value of 1250 kg/m^3 has been suggested for the reasonable initial density of the TWPs based on studies of bottom sediment movement. Figure 2 shows that the proposed ISPH model can simulate the real-time processes of the 2D overflow induced scouring. The detailed comparisons between numerical and experimental data can be found in [2].

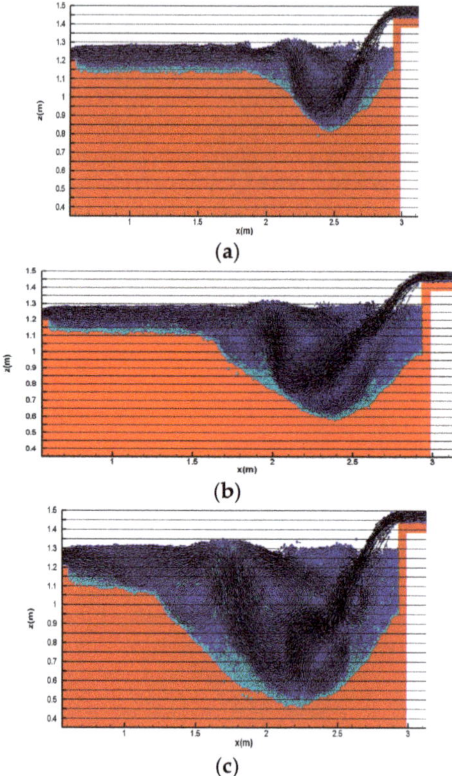

Figure 2. Snapshots of incompressible smoothed particle hydrodynamics (ISPH) calculated equilibrium scouring pit with different overflow depth: (**a**) η = 3.3 cm; (**b**) η = 4.7 cm, and (**c**) η = 6.0 cm.

In the subsequent work in [1], the ISPH model of [2] was successfully extended to the simulation of 3D local bed scour induced by clear water stationary flow passing a submerged vertical cylinder of relatively large size. Additional formulations were introduced to account for transversal and longitudinal bed slope. The erosion model was based on the turbidity water particle concept and the sediment motion was initiated when the calculated shear stress on the interface particles exceeded the critical value. The 3D ISPH erosion model was used to simulate the scouring process around a large vertical cylinder with a diameter of 60 cm in Figure 3, where the vorticity and the shape of the scour pit are illustrated at time t = 2.5 s and t = 3.5 s. The numerical results show that the proposed ISPH model could simulate the relevant features of the flow and the scour process (Figure 3). The detailed comparisons between the numerical and experimental data can be found in [1] and the scour dynamics were validated with a suitable degree of accuracy for engineering purposes. Even if a suitable representation of the complex scouring process can be obtained, the vorticity field shows some numerical noise (Figure 3a). Improvement of the calculated vorticity field could be obtained through the approach proposed in [46].

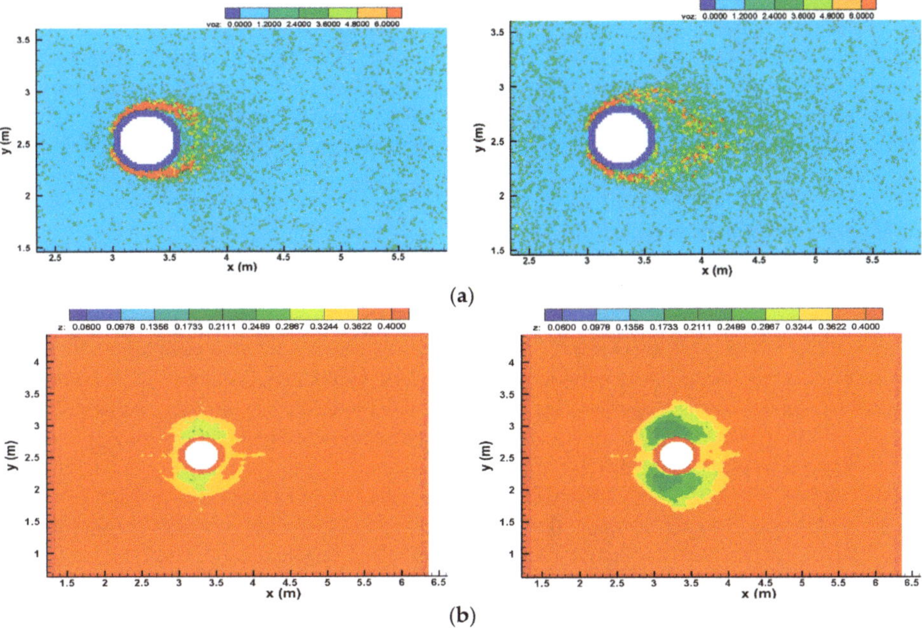

Figure 3. ISPH computed sediment bed scouring process: (**a**) snapshots of ISPH calculated vorticity in the scouring pit; (**b**) Snapshots of the ISPH calculated scouring pit.

2.2. Fast Landslides and Dense Granular Flows Interacting with Water

Numerical modeling of dense granular flows and landslides is still a challenging topic, especially when considering the interaction between the sediment and the water that may be both an internal interaction, related to pore water in landslide-prone saturated soil, and an external interaction with stored water in a basin with unstable slopes.

The interaction between pore water and soil matrix is a fundamental aspect that influences the triggering and propagation of shallow landslides induced by intense rainfall events that represent the most common natural hazards in some areas of the world [14]. Intense rainfall events induce water infiltration at slopes, leading to an increase of the volumetric water content and pore water pressure that worsen the slope stability of the soil layer close to the surface and may cause its failure. Reliable assessment of landslide susceptibility also requires a proper definition of the rainfall characteristics considering recent climate trends affecting rainfall [55,56].

If landslide triggering occurs, in the post-failure phase, the pore water content, combined with geo-mechanical properties of the soil, influence the sediment dynamics, and may induce in some cases flow-like fast earth movements classified as complex landslides because their run-out start as shallow rotational-translational failure, but changes into dense granular flows due to the large water content [15]. In this case, the fast landslide dynamics is more difficult to predict using traditional analytical models and a numerical approach could be helpful for landslide susceptibility assessment and the creation of debris–flow inundation maps.

The work in [57] proposed a combined triggering–propagation modeling approach for the evaluation of rainfall induced debris flow susceptibility. They adopted the transient rainfall infiltration and grid-based regional slope-stability model (TRIGRS) [58], which is based on the infinite slope stability approach, to obtain a map of potentially unstable cells within a study catchment under an intense rainfall event. Since not all unstable cells, in general, evolve to a debris flow, an empirical instability-to-debris-flow triggering threshold is calibrated on the basis of observed events and used to identify, among unstable cells, the so-called triggering cells that could most likely contribute to debris flow. The triggering threshold is, therefore, the key element that allows coupling between the TRIGRS slope instability model and the debris flow propagation model FLO-2D [59], a finite volume model that numerically solves the depth-integrated flow equations. The work in [57] assumed a zero excess rainfall intensity and a total friction slope depending on the Bingham-type rheological parameters as a function of the sediment concentration. The calibration of the triggering threshold with geo-morphological data of the catchment area represents a crucial step for obtaining reliable susceptibility maps in the nearby areas. Back-analysis of a catastrophic event that occurred on 1 October 2009 in the Peloritani mountain area (Italy) provided fairly good results.

In order to quantify the level of risk addressing the uncertainty inherent in landslide hazard, the susceptibility evaluation represents only one of the relevant issues involved in risk analysis. Additionally, run-out dynamics should be properly assessed in order to provide a quantitative estimate of the landslide hazard and select appropriate protective measures for risk mitigation. In order to reach such a goal, reliable predictive models should be used to obtain quantitative information on the destructive potential of the landslide. This is mainly related to the following characteristics: run-out distance; width of damage corridor; travel velocity; characteristic depth of the moving mass; and characteristic depth of the deposits [60].

The work in [61] proposed a 2D depth-integrated, coupled, SPH model for predicting the path, velocity, and depth of flow-like landslides. While the post-failure flow model of [57] assumes the heterogeneous moving mass as a single-phase continuum, [61] modeled the dense granular flow as a two-phase mixture composed of a solid skeleton with the voids filled by a liquid phase. Assuming that the shear strength of the liquid phase can be disregarded, the stress tensor within the mixture is composed of pore pressure and effective stress. Then, the mixture dynamics was described by quasi-Lagrangian depth integrated governing equations of mass and momentum balance, and the pore pressure dissipation equation that were discretized according to the standard SPH approach. Depth-integrated equations do not provide information on the vertical flow structure that is needed for evaluating shear stress on the bottom and depth integrated stress tensor. For this reason, [61] assumed that this vertical structure would be the same as the uniform steady-state flow according to the so-called model of the infinite landslide having constant depth and moving at a constant velocity on a constant slope. The work in [61] adopted the simple method proposed in [62] to obtain the bottom shear stress in a non-Newtonian fluid of Bingham-type. The model in [61] was used to simulate some catastrophic event that occurred on May 1998 in the Campania region (Italy), showing the relevant role of geotechnical parameters (especially the fluid phase and angle of internal friction) for the reliable prediction of the run-out distance, velocity, and height of the landslides. Proper selection of the value assigned to these parameters assured the best agreement with the field observations.

Some flow-like landslides are characterized by a relatively small average depth if compared with the horizontal linear dimensions, therefore the assumption of a depth-integrated model is strongly consistent with the physics of the phenomenon. Furthermore, there could be rainfall induced landslides where the initial average depth is comparable with the horizontal length and width. In these cases, significant variations of the vertical thickness and the vertical velocity profile may occur along the landslide body in the flow direction. This is illustrated in Figure 4, showing the preliminary results of an on-going study. In this new study, a narrow landslide that occurred during an intense rainfall event on April 2009 in a hilly area of the Oltrepò Pavese, named the Recoaro Valley, Northern Italy, was reproduced by means of the 2D WCSPH simulation. It can be seen that in the early phase, just after the failure, the mass portion close to the landslide front moved faster than the rear portion. Additionally, just after the impact against the downstream vertical wall of a damaged building, the landslide front began decelerating while the rear mass portion on the steep slope still maintained a relatively high average speed.

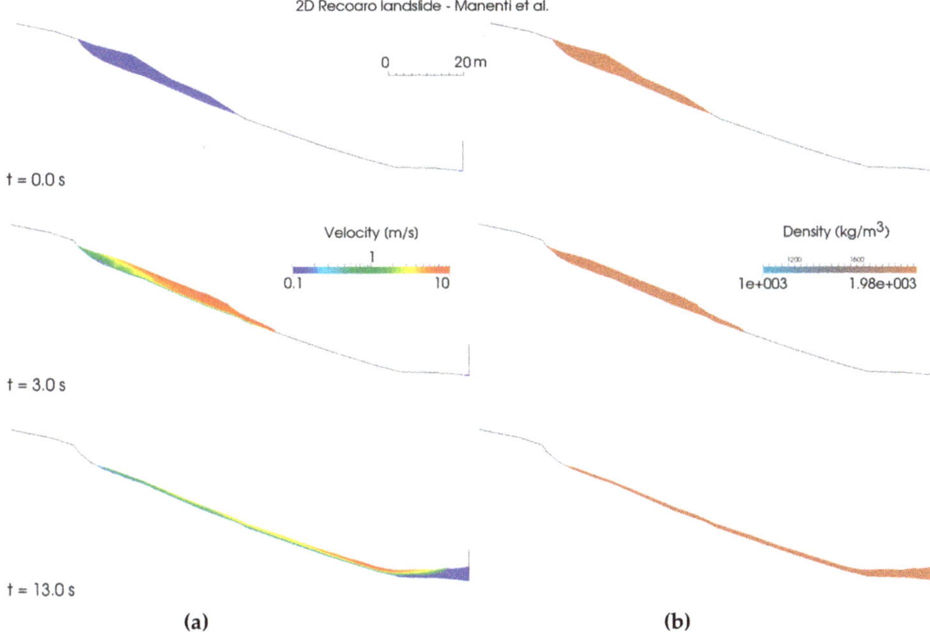

Figure 4. Frames of the 2D smoothed particle hydrodynamics (SPH) simulation of the rainfall induced shallow landslide occurred on April 2009 at Recoaro, Oltrepò Pavese (Northern Italy). (**a**) Velocity field; (**b**) Density field.

In the peculiar case described above, the modeling approach based on the infinite landslide with constant depth and moving at a constant velocity on a constant slope may be less appropriate. Therefore, the solution of the governing equation in the three-dimensional form seems more appropriate than the depth-integrated model. The simulation shown in Figure 4 was carried out with the code SPHERA v.9.0.0 adopting the mixture model for dense granular flow discussed in [8]. Even if the code has a 3D formulation, a 2D approach may be conveniently adopted in this case because the landslide is relatively narrow and the flow may be assumed to be identical on the vertical planes along the flow direction.

The work in [63] used a finite volume approach to simulate mudflows and hyper-concentrated flows characterized by suspended fine material by adopting a Bingham rheological model. Similar to [64], the Cross model was adopted for modeling the non-Newtonian flow, assuming that the constant parameter m was equal to 1, resulting in the following formulation of the apparent viscosity:

$$\mu_{eff} = \frac{\mu_0 + \mu_\infty K \dot{\gamma}}{1 + K \dot{\gamma}}$$
$$K = \frac{\mu_0}{\tau_B} \quad \mu_\infty = \mu_B$$

(2)

In Equation (2) $\dot{\gamma}$ denotes the shear rate defined through the second invariant I_2 of the rate of deformation tensor; K, μ_0, and μ_∞ are three constant parameters that can be conveniently related to common Bingham rheological parameters, namely the yield stress τ_B and viscosity μ_B. From a physical point of view, μ_0 and μ_∞ denote the viscosity at very low and very high shear rate, respectively. In order to avoid numerical divergence caused by the unbounded growth of effective viscosity as the shear rate approaches zero, μ_{eff} is limited to a suitably high threshold value, which is set to $\mu_0 = 10^3 \mu_B$ to assure convergence. The test cases simulated in work in [63] considered the flow on inclined surfaces and analyzed the role of the Froude number [65] during the propagation phase, which may be helpful in designing the control works.

Landslides occurring at the slopes of confined water bodies (e.g., artificial basin or river-valley reservoir) or at coastal regions involve complex interactions between the solid and the fluid phase. In the post-failure phase of rapid landslides generated at the slope of a water body, an impulse wave is generally induced, which is usually referred to as tsunami [66,67]. The characteristics of the generated water wave are related to the velocity and the shape (e.g., thickness and slope angle) of the landslide front, whose dynamic deformation is in turn affected by the water induced stresses at the interface while the landslide is penetrating the water body. Accurate prediction of the landslide induced wave hazard depends on reliable numerical models that can simulate the coupled dynamics.

The work in [11] proposed a WCSPH model for the analysis of impulsive wave generated in a basin by a deformable landslide. In order to properly reproduce landslide deformation during the post-failure phase, they adopted an elastic–plastic constitutive model for the soil combined with the Drucker–Prager criterion. To account for the interaction with stored water, occurring at a very small timescale due to the fast dynamics, [11] introduced a bilateral coupling model consisting of two sequential steps: the interface soil particles were initially considered as the moving deformable boundary whose velocity and position was used for solving the governing equations of the water phase; then the soil constitutive equation was solved with the corrected stress taking into account the water-induced surface force. This coupling model would require, in theory, an iterative procedure to assure the consistency of both stress and deformation at the interface at each time step. However, the sequential approach is quite suitable because the interface deformation is mainly caused by the landslide dynamics and small displacements occur within a time step. The mechanical parameters in the model of [11] have a physical meaning and the corresponding values can be deduced from conventional soil mechanic experiments. Two experiments were simulated concerning the wave generated by a slow landslide [12] and a fast landslide [68], respectively, obtaining in both cases good agreement with the experimental data.

The work in [10] proposed a hybrid model for simulating the coupled dynamics between the landslide and stored water as well as the propagation of generated surge waves. The discrete element method (DEM) was used for simulating the landslide dynamics, while WCSPH was used to solve the governing equations for water. Spurious numerical noise affecting the water pressure field was removed by applying the δ-SPH technique [69]. The interaction mechanism between the solid and fluid phase in the hybrid DEM-SPH model was based on the drag force and buoyancy. At each time step, these forces were first calculated considering (i) the initial position and velocity of both fluid and solid particles, (ii) the pressure of the fluid, and (iii) the local soil porosity, ε, evaluated at the landslide–water interface with a kernel interpolation of the solid particle volume. Based on the calculated drag force and buoyancy, the corresponding new position and velocity of both fluid and solid particles were then calculated to solve the corresponding discretized governing equations. The updated position and the velocity field were subsequently used to recalculate drag force and buoyancy. Therefore, this interaction mechanism requires an iterative process to assure convergence of the calculated forces and consistency of the displacements and velocity field of the involved phases. The hybrid DEM-SPH model was used to simulate the sliding along a 45° sloping plane of a rigid body that mimics a submarine landslide, obtaining a suitable prediction of the experimental time evolution of water surface elevation [70]. A modified simulation was additionally performed by considering the deformability of the sliding body, showing that smaller and less violent surge waves were generated owing to the landslide deformation.

A similar approach was proposed in [71] for the analysis of underwater granular collapse by coupling WCSPH (for the fluid phase) with DEM (for the solid incoherent phase). A coupling module was developed for fluid–grain interaction: the force exerted by the fluid on a solid particle was obtained by integrating the contributions on its surface; in turn, the effect of the solid particles on the fluid motion was calculated by including the neighboring DEM particles in the SPH interpolation of the governing equations for the considered liquid particle. Another attempt of coupling SPH, in particular, the DualSPHysics model, with a distributed-contact discrete-element method (DCDEM) was proposed in [72] to explicitly solve the fluid and solid phases to model a real case of an experimental debris flow. An experimental setup for stony debris flows in a slit check dam was reproduced numerically, where solid material was introduced through a hopper, assuring a constant solid discharge for the considered time interval.

The reference mixture model of [8] for the dynamics of dense granular flow was modified by introducing a numerical parameter, μ_0, referred to as limiting viscosity. The effect of limiting viscosity arises in the frictional regime at low deformation rates near the transition zone to the elastic–plastic regime: in this shear rate interval, a constant value μ_0 (lower than μ_{max}) is assigned to the mixture viscosity (see red dot-dashed curve in Figure 1), thus reducing the computational time in the case where the viscous criterion dominates the numerical stability of the time integration scheme. There are other alternative approaches to keep control of computational time in the simulation of high-viscosity flows. In the work of [73], a semi-implicit integration scheme was proposed to overcome the severe time-stepping restrictions caused by the WCSPH explicit integration scheme when simulating highly viscous fluids, as in the case of lava flow with thermal-dependent rheology. According to this approach, only the viscous part of the momentum equation is solved implicitly, thus saving computational time and obtaining an improved quality of the results with respect to the fully explicit scheme.

The mixture model of [8], modified with limiting viscosity, was successfully applied in [9] to the analysis of a fast massive landslide at the slope of an artificial reservoir. The simulated case reproduced a two-dimensional scale laboratory test carried out in 1968 at the University of Padua reproducing in Froude similitude a characteristic cross-section of the Vajont artificial basin. In 1963, a catastrophic landslide, estimated to be about 270 million cubic meters, fell into the Vajont reservoir, generating a tsunami that caused about 2000 casualties [74].

The experimental campaign of Padua aimed to evaluate the effects of both the material type and the landslide falling time on the maximum run-up of the generated wave over the opposite side of the valley. The landslide velocity was imposed through a rigid plate pulled by an engine through a steel cable. Several tests were carried out considering two types of rounded gravel (3–4 mm and 6–8 mm), crushed stone, and squared tile. For each of these material types, different values of the run-out velocity were tested by varying the plate stroke (in the range 0.5–0.8 m) and its velocity, thus resulting in a landslide falling time ranging approximately between 15 s and 500 s at full scale.

The frames in Figure 5 show the simulation results obtained with SPHERA v.9.0.0 [75]; some representative instants were selected during the acceleration phase where the vertical rigid plate pushes the landslide toward the basin, and the subsequent run-out phase of the generated wave climbs the opposite slope. Fairly good results were obtained in terms of the maximum height reached by the rising wave front. The left-hand panels show the velocity field evolution. The right-hand panels show the density field; it can be noticed that the lower landslide layer (light-brown) has a higher density because of the saturation. The effect of pore water allows for the landslide dynamics to be obtained more close to the experimental results because at $t = 1.35$ s, its front comes into contact with the opposite side of the basin. No tuning of the physical parameters of the model was necessary. The required accuracy was controlled by the proper adoption of the limiting viscosity value, μ_0, reaching a reasonable compromise with consumed computational time. This test case is provided as tutorial number 35 in the documentation of SPHERA v.9.0.0 that is freely available in [76].

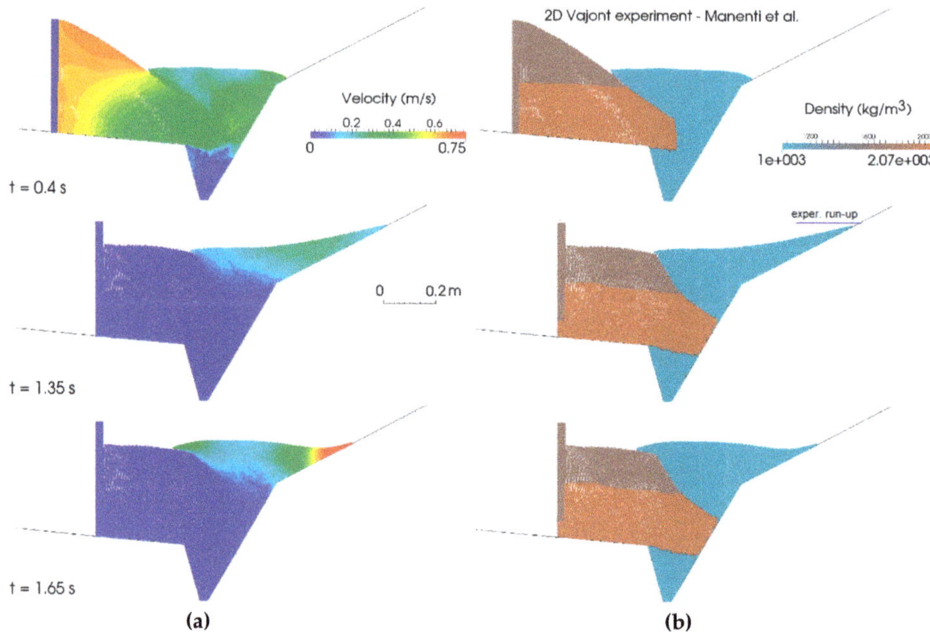

Figure 5. Frames of the simulation of the Padua experiment. (**a**) Velocity field; (**b**) Density field.

Calculation of the flow-impact induced forces on the submerged structure may be required in order to estimate landslide and water-related hazards. Using the standard WCSPH approach usually leads to high-frequency numerical noise in the pressure field [29] that may taint the calculation of the load time history acting on the solid structure. To overcome this problem and obtain the correct prediction of the hydrodynamic load for structural stability assessment, several strategies have been successfully introduced [77].

3. Flooding in Complex Topography with the Transport of Sediments

This section provides synthetic discussions on the following topics: state-of-the-art of CFD mesh-based codes for flood propagation on complex topography with sediment transport; advantages of SPH modeling for flood propagation on complex topography; state-of-the-art of SPH codes for sediment transport; example of a SPH code for flood propagation on complex topography with sediment transport; and pre-processing and post-processing tools for floods on complex topographies (i.e., real topographic surfaces or their scale models). The most popular codes used to represent erosional floods (i.e., flood propagation over granular beds) rely on mesh-based numerical methods and the shallow water equations (SWE). These are briefly recalled in the following. The work in [78] presents a finite element code for bed-load transport. A major novelty is the 3D mathematical formulation. However, the code validations only refer to 1D configurations in 2D domains where analytical solutions are available. The work in [79] introduced a Godunov-type 1D SWE model for 2D erosional dam-break floods. The work in [80] presented a 2D SWE-FVM (finite volume method) model for 3D erosional dam-break floods, which uses Exner's equation for the bed top evolution, a 1D heuristic formula for the bed-load transport rate, and a 1D spatial reconstruction scheme based on Riemann solvers suitable for multi-phase flows. The work in [81] presented a 2D SWE model for 3D erosional dam-break floods; validations refer to 1D and 2D bed-load transport configurations, whereas a 3D demonstrative configuration (still represented with a 2D code) was reported on a quake-induced erosional dam-break flood (Tangjiashan Quake Lake). The work in [82] applied a 2D SWE-FVM code to simulate an erosional flood with bed-load transport in the Yellow River. The last two examples might represent typical applications of 2D SWE mesh-based models to 3D erosional floods on complex topography with 1D schemes for sediment transport.

The above mesh-based models share the following drawbacks: 2D modeling; no consistency with the KTGF; ad-hoc tuning procedure for the viscosity of the 2-phase mixture (water and granular material); and shallow-water approximation (e.g., hydrostatic pressure profiles, velocity is uniform along the vertical).

With respect to the above state-of-the-art mesh-based models, the meshless SPH numerical method introduces several advantages, which are synthesized in the following. SPH allows the detailed 3D fluid dynamics fields within the urban canopy (urban fabric) to be simulated including fluid-structure interactions, whilst the mesh-based 2D porous models do not provide a direct modeling of the flood-building interactions. The SPH method can simulate the 3D transport of solid structures and fluid interactions with other mobile and fixed structures and can also represent 3D bed-load transport and its impact on mobile and fixed solid structures. The SPH method provides a direct estimation of the position of the free surface and the fluid and phase interfaces due to its Lagrangian nature. The direct representation of Lagrangian derivatives is also responsible for the absence of the advective non-linear terms arising in the Eulerian formulation of the balance equations. No computational mesh generation is requested, thus saving person-months, software licenses, and computational resources. Nonetheless, several drawbacks have been reported. SPH is slightly more time consuming then mesh-based methods (fixed with the same spatial resolution and accuracy) and has a narrower, but peculiar range of application fields, whose number is nonetheless elevated and relevantly involve floods. Furthermore, a SPH code can usually cover a wide range of spatial resolutions at different accuracy levels, but maintain stable algorithms. This allows the same code to be used for both preliminary analyses at a coarse spatial resolution and accurate simulations (at fine spatial resolution).

Although several SPH studies are available for 2D and 3D applications for granular flows, only a few have been dedicated to floods with sediment transport (over a simplified topography). Among these studies, [7] introduced a 2D erosion criterion to represent the sediment removal from water bodies by means of discharge channels (i.e., flushing procedures). Regardless of the application, the SPH models for granular flows were mostly restricted to 2D codes, featured by either 2-phase models or ad-hoc tuning for mixture viscosity.

Recently, a numerical mixture model for dense granular flows was presented in [8] to simulate the sediment dynamics phenomena, which typically involve the failure of earth-filled dams and dykes, bed-load transport, and fast landslides over complex topography. This model is discussed in the following. This was integrated into the FOSS SPH code SPHERA (RSE SpA) [76]. This mixture model permits the simulation of the above phenomena by solving a system of balance equations, coherent with the theoretical state-of-the-art frame represented by the KTG [83], under the conditions of the "packing limit". The model is based on mixture parameters (velocity, density, viscosity, and pressure) and phase variables (e.g., mean effective stress, frictional viscosity, and liquid phase pressure). The viscous parameters (mixture viscosity, frictional viscosity, and liquid viscosity) do not need any calibration/tuning (Section 2.1) [8]. Filtration is partly and implicitly represented; despite the absence of an explicit filtration scheme, a Lagrangian sub-scheme for saturation conditions was based on the hypotheses of stratified flows and local 1D filtration flows parallel to the local seepage [8]. A separated treatment involves the mixture particles under the elastic–plastic strain regime: they are held fixed as their velocities are negligible for applications such as bed-load transport and fast landslides [8]. The mixture model allows a high number of fluids to be simultaneously represented in the same domain, provided that they are either liquids or granular materials (both fully saturated or dry).

The model above was validated through laboratory experiments [8] and applied to a 3D erosional dam-break flood on complex topography (Figure 6). This demonstrative test case showed the applicability of the SPH method in simulating a flood on complex topography with bed-load transport. The 3D erosional dam break was triggered by an instantaneous and almost complete failure of a gravity dam, whose structure was not simulated. The water flow impacts, thrusts, or erodes, and then transports a portion of the downstream mobile bed, which is composed of a bed of granular material. Its original sedimentation was related to the presence of a weir, whose structure was then removed before the dam building. The 3D fields of the velocity vector and pressure were computed and the 2D fields of the maximum (over time) water depth and specific flow rate (i.e., flow rate per width unit) were elaborated. These quantities, requested for risk analyses, were derived from the particle and the topography heights, and the magnitude of the depth-averaged velocity vectors. The time series of the water depth, the fluid volumes (cumulated in selected sub-domains), and the flow rate were assessed at specific monitoring sections. This demonstrative test, reported in [8] and available as tutorial no. 18 of SPHERA v.9.0.0 [76], shows the potential of the SPH method in simulating a full-scale 3D flood on complex topography with sediment transport. Although validations were reported in terms of comparisons with laboratory datasets [8], these only refer to simplified topographies. A full scale validation still has to be investigated due to the lack of available measures for complex topographies with sediment transport.

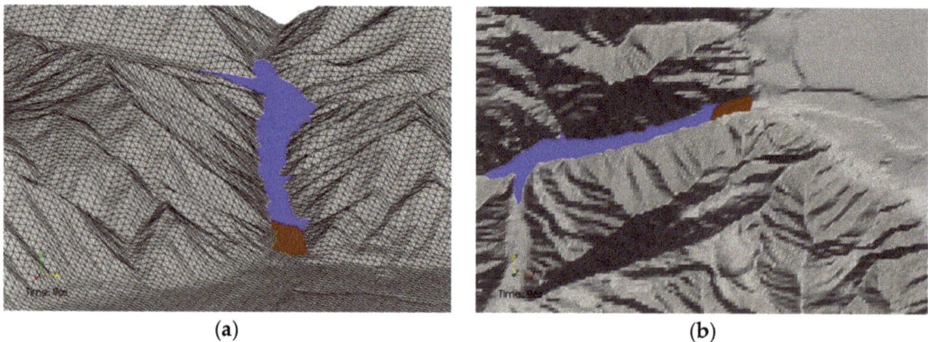

(a)   (b)

Figure 6. Demonstrative test case for the meshless SPH method applied to a 3D erosional dam-break flood on complex topography [8]. (**a**) 3D view. (**b**) Top view. Digital elevation model (grey, with a black vertex triangulation), liquid SPH particles (blue), mixture SPH particles (brown).

CFD codes for flood propagation on complex topography need a suitable numerical chain. A non-exhaustive list of pre-processing and post-processing free tools for floods on complex topography is discussed hereafter.

The dataset SRTM3 (USGS) represents the most accurate open data DEM ("digital elevation model", not to be confused with the "discrete element method" from Section 2.2) archive with a spatial resolution of 1″ (spatial resolution length scale approximately equal to 31 m) and an almost global cover. The SRTM3 files are available in the ".tif" format. The numerical tool GDAL [84] can be used to convert the DEM ".tif" file format of SRTM3 in the alternative format ".dem". DEM2xyz [85] can read the DEM file (".dem" format), convert the geographic coordinates in Cartesian coordinates over a regular grid, and write the resulting DEM on an output file (".xyz" format), possibly coarsening the spatial resolution and reconstructing bathymetry where it is not available. Paraview (Kitware) [86] can read the ".xyz" output file of DEM2xyz and elaborate a 2D Delaunay grid starting from the DEM vertices. Paraview also allows cutting the numerical domain, drawing particular regions of interest (e.g., water bodies), engineering works (e.g., dams), and monitoring elements (e.g., points, lines, surfaces). The above information, derived from Paraview, can be provided to DEM2xyz, which can be executed again to add the new elements designed with Paraview.

The 3D fluid dynamics output files of a CFD code for floods can be further elaborated by means of Grid Interpolator [87], which reads a 3D field of values from an input grid and interpolates them on an output grid with a different spatial resolution. At this point, Paraview can be used to visualize the 2D and 3D fluid dynamics fields and return the associated image files.

All the pre-processing and post-processing items above are FOSS, but SRTM3, which is "open-data" (dataset available upon public and free access) [88] can be used within the same numerical chain and can be replaced by other free or proprietary items depending on the user resources and targets.

4. High Performance Computing Solutions for Complex Hydraulic Engineering Problems

Numerical modeling is becoming more useful and practical thanks to the capability of the current computer hardware. Years ago, significant simplifications of the problem needed to be undertaken in order to make the numerical simulation feasible due to past hardware limitations. Nowadays, thanks to the continuous hardware improvement and the use of HPC techniques that allow for the advantages of the enormous calculation power of the current hardware to be taken, it is possible to now simulate complex problems with fewer simplifications, and problems that were impossible to be simulated due to their spatial and/or temporal scale can now be carried out. Therefore, several codes have been developed that use these HPC techniques to simulate complex hydraulic engineering problems of water-related natural hazards in reasonable execution times. Some examples are SPHERA for sediment transport [8] and landslides [9]; DualSPHysics to model the scouring of two-phase liquid-sediments flows [6,53]; and GPUSPH for simulating lava flows [74]. As explained in [89], due to the limitation of sequential computing and the high cost of increasing performance in sequential architectures, the era of sequential computing was replaced by the era of parallel computing. Currently, hardware performance is fundamentally increased by expanding the number of processing units, instead of increasing the power of a single processing unit. Therefore, it is mandatory to use parallel programming techniques to distribute the workload among the available processing units and synchronize their execution.

OpenMP (Open Multi-Processing) is the most common parallel programming technique for multi-core systems with shared memory because it does not involve major code changes. MPI (message passing interface) is a standard for distributed memory systems that allows for the work to be divided among the multiple calculation nodes connected to each other. On the other hand, and more recently, GPGPU (general-purpose computing on graphics processing units) has emerged as an alternative to the use of traditional CPUs (central processing units). This technique allows the code to be executed on the GPU (graphics processing units). GPUs initially developed for rendering graphics have greatly increased their parallel computing power over the last decade due to the increasing computing demands of the gaming market, and more recently due to the significant growth of markets related to AI (artificial intelligence), DL (deep learning), and DNN (deep neural networks). They are in fact now several times faster than CPUs for problems that demand a high computational cost whilst offering a high degree of parallelism, to which SPH falls under [90]. At the same time, the emergence of programming languages such as CUDA (compute unified device architecture) and OpenCL makes the development of GPU codes easier. These two reasons have encouraged many researchers to implement their models for GPU, thus starting the era of GPU computing [91].

Particle methods such as smoothed particle hydrodynamics (SPH) [92] or moving particle semi-implicit (MPS) [93] have a very high computational cost. The high computational cost of particle methods is due to the fact that each particle has to interact with a large number of neighboring particles, which, in addition, may change every time step. The way that particles are organized in memory and the neighbor search (NS) algorithm is key to maximizing performance in parallel architectures. The most common algorithms for NS are the Verlet list and Cell-linked list. The work in [94] compared both algorithms applied to SPH and, in [95], variants of these algorithms were evaluated to conclude that the Verlet list may be faster than Cell-linked list, but the memory consumption is prohibitive for a high number of particles, so the Cell-linked list was found to be more efficient. The same conclusion was confirmed in [96] for GPU simulations. Quad-tree partitions combined with the Morton space filling curve (SFC) was recommended when using variable resolution in multi-core architectures [97] and GPU devices [98]. The work in [99] proposed another algorithm based on hierarchical cell decomposition for variable resolution in distributed memory architectures using MPI. A very fast NS, where the speed is the main priority, was implemented in [100] for real-time animations, but this approach is invalid for real-physic simulations since some interactions of neighboring particles are missing.

The high computational cost of particle methods can be countered using the programming techniques discussed above. Thus, different implementations of the SPH method using OpenMP can be found [97,101,102], although the same algorithms can be applied to other particle methods. The work of [101] included a performance analysis of SPH in multi-core CPUs and studied how to avoid bottlenecks. The work in [102] implemented an optimized SPH for parallel machines with shared memory and compared the effective computing performance on multi-core CPUs and Xeon Phi MIC (many integrated core) coprocessors with 61 cores. The work in [97] evaluated the efficiency of a neighbor search algorithm based on a quad-tree partitioning on 32 processors. The SPH method has also been implemented for distributed memory systems using MPI for WCSPH [99,103–106], and MPI for ISPH [107,108]. The execution on distributed systems requires the division of the simulation domain into multiple subdomains. Each subdomain is executed by an independent process, but all processes must be synchronized for every calculation step. Therefore, load balancing between the different processes is essential to minimize the time that each process waits for the rest of the processes with a consequent loss of performance. In addition, this balancing needs to be dynamic where the workload can shift from one subdomain to another during simulation in Lagrangian methods. On the other hand, the decomposition of the domain into distributed memory systems requires the transfer of data between the nodes that process each domain, especially in the case of GPUs where transfers via PCI Express on the node itself are also required. The overhead of these memory transfers can be significant, so it is a major problem in the dynamic decomposition of the domain. In [103], the dynamic load balancing was based on the METIS package. The work in [104] implemented the code JOSEPHINE using the programming language Fortran 90 and OpenMPI, obtaining a good efficiency on 16 processors. The implementation of [99] used a hierarchical cell decomposition to allow variable resolution. The work in [105] obtained a high efficiency with 32,768 CPU cores using a decomposition algorithm based on orthogonal recursive bisection [109]. Most recently, [109] implemented dynamic domain decomposition between Voronoi subdomains and achieved a good speedup using 1024 processes. The work in [107] presented an ISPH solver where a simple spatial domain decomposition in cells was combined with the Hilbert space filling curve (SFC) and load balancing was only applied when particle imbalances were detected. This implementation reported an efficiency of 63%, simulating a still water case with 66.5 million particles on 512 processors compared to the execution time using 32 processors. The ISPH implementation of [108] achieved an efficiency close to 80% on 1536 MPI nodes, simulating a dam break with 100 million particles compared to the runtime using one node with 24 cores. Nowadays, powerful GPUs can be used instead of expensive parallel machines or clusters based on CPU processors to simulate several millions of particles. The first full GPU implementation of the SPH method was reported in [110] using shader programming. Other implementations on GPU are described in the works [52,111–115]. The work in [111] presented the SPH code, named DualSPHysics, implemented for GPU or CPU executions where a Cell-linked list [116] is used for a neighbor search. This work showed a comparison between several GPU models of different generations and several CPU models, achieving an improvement of almost two orders of magnitude. Hérault et al. [112] described another SPH solver for GPU named GPUSPH using a Verlet list. This solver allowed them to simulate lava flows on a complex topography with accuracy and efficiency, obtaining speedups of up to two orders of magnitude with respect to an equivalent CPU as claimed by the authors of [112]. The work in [52] presented an improved DualSPHysics code where the CPU and GPU optimization strategies described in [96] were applied to obtain a fair comparison between GPU and multi-core CPU executions. A multi-phase SPH model for liquid–gas simulations on GPUs based on DualSPHysics was shown in [113]. The use of GPUs is mandatory in this multi-phase approach since both fluids, water, and the surrounding air volume must be discretized in SPH particles. Therefore, the total number of particles and the execution time are usually increased several times.

The work in [114] described a 2D SPH code optimized for long hydraulic simulations using the Hilbert space filling curve to improve the data locality and increase the performance. The work in [115] presented the AQUAgpusph code using the programming language OpenCL. The use of OpenCL allows the use of NVIDIA or AMD GPUs and the same code can be executed on multi-core CPUs. GPU implementations of ISPH can be found in [117–119]. The work in [117] presented a solver written in OpenCL that could be executed on a CPU or GPU and the performance of both architectures was compared. The work in [118] showed an ISPH code to perform realistic fluid simulations in real time, so some simplifications were applied to increase the speed of execution. The work in [119] implemented an optimized GPU solver starting from [111] and achieved a speedup close to 5 against CPU simulations with 16 threads. GPU implementations of MPS can also be found in [120] for 2D and in [121] for 3D free surface simulations.

Several examples of solvers based on shallow water equations on GPU are in [98,122–126]. These allow for large-scale flood simulations to be performed using only one GPU. The work in [98] presented a SPH solver for shallow water with adaptive resolution using a quad-tree algorithm for neighbor search. The work in [122] presented a 2D flood model that allowed for the simulation of long duration floods in a few minutes. The work in [123] simulated more than one hour of the Malpasset dam failure case in less than 30 s and proved that the use of single precision was enough for that problem. The work in [124] presented a finite volume solver with explicit discretization and an efficient algorithm to deactivate dry zones improving performance. The work in [125] introduced an optimized version of the IBER solver [127] for CPU and GPU simulations. That new version of IBER solved 24 h of an extreme flash flood in less than 10 min. The work in [126] presented a shallow water model based on the finite volume approach implemented for GPU with the OpenACC language [128]. OpenACC allows GPU parallelization to be implemented automatically using compiler directives like OpenMP for CPUs. Some examples of flood simulation, possibly thanks to the computing power of a single GPU, can be found in [18,129,130]. The work in [129] used more than 1.5 million particles to simulate a complicated city layout including underground spaces with the GPUSPH model [112]. The work in [18] simulated runoff on a real terrain generated from photogrammetry information obtained by an UAV (unmanned aerial vehicle). This simulation with 7.5 million particles was performed using the DualSPHysics code [111] and took 138 h for 15 physical minutes. The work in [130] showed a large-scale urban flood performed by the GPU model implemented in [98]. This simulation with 230,000 particles took 135 h to simulate five physical hours.

One GPU can perform simulations with a high particle number in affordable execution times as demonstrated by the works mentioned before. The same results on CPU systems would require expensive CPU clusters. However, a fair comparison between GPU and CPU performance is not straightforward, since optimized solvers for both architectures and hardware from the same period should be used to avoid unreal speedups [116,131]. The work in [132] disproved the speedups of 100x or 1000x as shown in some GPU–CPU comparisons.

Although today's GPUs provide high computation power, the simulation of real cases implies huge domains with a high resolution, which implies simulating tens or hundreds of million particles and these simulations are not viable in a single GPU due to memory limitations and prohibitive execution times. The solution is to use multi-GPU machines with shared memory or GPU clusters. Examples of multi-GPU SPH solvers can be found in [133–136]. The work in [133] explored the use of MPI in the DualSPHysics code to perform multi-GPU simulation on GPU clusters. This approach allowed for the simulation of 32 million particles on four GPUs, achieving an efficiency of 80% using weak scaling by simulating eight million particles per GPU. The work in [134] described the implementation of an optimized multi-GPU version of DualSPHysics using an MPI that was able to simulate 1024 million particles on 128 GPUs, achieving almost 100% efficiency using weak scaling by simulating eight million particles per GPU. The work in [135] extended the GPUSPH solver to multi-GPU machines with shared memory using threads and domain decomposition in two dimensions. This implementation is limited by the number of GPUs hosted by one machine or computation node, typically six or eight

GPUs. The work in [136] presented a multi-GPU implementation with 3D domain decomposition that achieved 89% efficiency using strong scaling by simulating more than 30 million particles on eight GPUs using MPI.

5. Conclusions and Future Perspectives

This paper collected some recent works showing the application of CFD techniques for modeling problems of practical and theoretical interest involving complex multiphase flows relevant for the analysis and mitigation of water-related natural hazards. The paper focused on meshless techniques for the numerical modeling of fast landslides, tsunami wave, flooding in complex geometry and sediment scouring; few relevant examples have also been mentioned concerning traditional grid-based methods applied to the analysis of environmental risks related to flooding in complex topography. The peculiar features of the examined works in terms of mathematical modeling and numerical implementation have been illustrated and outline the principal differences.

Different approaches are commonly used for modeling the non-Newtonian dynamics of dense granular flows. Some of them consider the sediment as a single-phase moving mass; other approaches model the sediment as a two-phase mixture where the voids of the solid matrix are filled by some liquid phase in order to account for the pore pressure effects. In all of the works examined where the SPH method was applied to the solid, the motion of the granular phase was treated as a pseudo-fluid once the sediment particles had been mobilized. Furthermore, the criterion adopted for determining the onset of sediment motion is, in general, not uniform. In some of the considered works, a numerical threshold was used for this purpose and the involved mechanical parameters, despite their physical meaning, require proper tuning so that the model results can suitably fit the experimental data. This approach is used both in the analysis of bed sediment erosion and landslide run-out where the definition of a triggering threshold that establishes the critical condition for the onset of sediment motion is required. In these cases, the need of tuning parameters, which assume a numerical role rather than a physical meaning, may limit the applicability of the model to those situations of practical interest where calibration data are available. An alternative approach to overcome this limitation is provided by the kinetic theory of dense granular flows (KTGF) that is put into SPH formalism and has been proven to be able to simulate dense granular flows (the so-called "packing limit" of KTGF) and fast landslides with a suitable degree of accuracy.

Proper numerical modeling of the erosion processes that are related to the motion of loose sediment particles suspended in the water flow would require that the size of the numerical particles should be of the same order of the size of real sediment grains; but this task is not allowed in practical problems where the geometrical complexity may induce excessive computational cost. However, reliable results may be obtained by simulating numerical particles as a combination of clear water and turbid water particles that mimic a suspended load.

Concerning the interaction between the phases (i.e., water and sediment), different approaches have been followed. In some cases, the governing equations of motion are solved simultaneously for both phases, thus obtaining coupled dynamics. In other cases, the governing equations are solved separately for each phase and the interface conditions should be enforced to assure kinematic and dynamic continuity between phases through an iterative procedure that can, however, affect the computational time. While some models were examined in terms of the mathematical features, these were not the same methods that were examined in terms of numerical implementation.

The above-mentioned relevant aspects may exert significant influence both in the numerical representation of the multiphase flow and in the reliability of the model results. Therefore, they should be carefully evaluated in the analysis of water-related natural hazards in order to obtain a reliable representation of the investigated problem. Given the complexity (especially geometrical) that usually characterizes the problems of practical interest, these numerical models could be used to support risk analysis and mitigation if appropriate programming techniques and modern architectures for scientific computation are used to obtain fast-running computer codes. This goal is not simple to accomplish, even with the adoption of HPC techniques and parallel computing. The simulation of real cases with large domains and high resolution will probably become more affordable with the use of GPU clusters. However, this task needs better implementation of the identified useful formulations of SPH for multi-GPU multi-node. This requires overcoming the problems associated with domain decomposition and memory transfer across nodes, which are particularly difficult for the fully-coupled two-phase formulations.

Author Contributions: Conceptualization, R.A. and S.M.; Design and enhanced structure of the manuscript, R.A. and S.M.; Writing Sections 1, 2 and 5–original draft preparation, S.M.; Writing Section 2.1–original draft preparation, D.W.; Writing Section 3–original draft preparation, A.A.; Writing Section 4–original draft preparation, J.M.D.; Writing—review and editing, S.M., D.W., J.M.D., S.L., A.A., and R.A.

Funding: This work was partially financed by Xunta de Galicia (Spain) under project ED431C 2017/64 "Programa de Consolidación e Estructuración de Unidades de Investigación Competitivas (Grupos de Referencia Competitiva)" co-funded by European Regional Development Fund (FEDER), and by the Xunta de Galicia postdoctoral grant ED481B-2018/020. The work was also funded by the Ministry of Economy and Competitiveness of the Government of Spain under project "WELCOME ENE2016-75074-C2-1-R" and by the EU under the ERDF (European Regional Development Fund) through the Interreg project "MarRISK" (0262_MARRISK_1_E). The contribution of the RSE author (Section 3) was financed by the Research Fund for the Italian Electrical System (for "Ricerca di Sistema -RdS-") in compliance with the Decree of Minister of Economic Development 16 April 2018.

Acknowledgments: We acknowledge the CINECA award under the ISCRA initiative for the availability of high performance computing resources and support. HPC simulations on SPHERA refer to the following HPC research project: HPCNHLW1—High Performance Computing for the SPH Analysis of Natural Hazard related to Landslide and Water interaction (Italian National HPC Research Project); instrumental funding based on competitive calls (ISCRA-C project at CINECA, Italy); 2019; S. Manenti (Principal Investigator) et al.; 360,000 core-hours. We acknowledge the CINECA award under the ISCRA initiative, for the availability of high performance computing resources and support. HPC simulations on SPHERA (Section 3) refer to the following HPC research project: HPCEFM17—High Performance Computing for Environmental Fluid Mechanics 2017 (Italian National HPC Research Project); instrumental funding based on competitive calls (ISCRA-C project at CINECA, Italy); 2016–2017; 200,000 core-hours; Amicarelli A. (Principal Investigator) et al.

Conflicts of Interest: The authors declare no conflict of interest.

References

1. Wang, D.; Shao, S.; Li, S.; Shi, Y.; Arikawa, T.; Zhang, H. 3D ISPH erosion model for flow passing a vertical cylinder. *J. Fluids Struct.* **2018**, *78*, 374–399. [CrossRef]

2. Wang, D.; Shaowu, L.; Arikawa, T.; Gen, H. ISPH Simulation of Scour Behind Seawall Due to Continuous Tsunami Overflow. *Coast. Eng. J.* **2016**, *58*, 1650014. [CrossRef]

3. Wang, C.; Peng, C.; Meng, X. Smoothed Particle Hydrodynamics Simulation of Water-Soil Mixture Flows. *J. Hydraul. Eng.* **2016**, *142*, 04016032. [CrossRef]

4. Guandalini, R.; Agate, G.; Manenti, S.; Sibilla, S.; Gallati, M. SPH Based Approach toward the Simulation of Non-cohesive Sediment Removal by an Innovative Technique Using a Controlled Sequence of Underwater Micro-explosions. *Procedia IUTAM* **2015**, *18*, 28–39. [CrossRef]

5. Guandalini, R.; Agate, G.; Manenti, S.; Sibilla, S.; Gallati, M. Innovative numerical modeling to investigate local scouring problems induced by fluvial structures. In Proceedings of the Sixth International Conference on Bridge Maintenance, Safety and Management (IABMAS 2012), Stresa, Italy, 8–12 July 2012; pp. 3110–3116.

6. Zubeldia, E.H.; Fourtakas, G.; Rogers, B.D.; Farias, M.M. Multi-phase SPH model for simulation of erosion and scouring by means of the shields and Drucker–Prager criteria. *Adv. Water Resour.* **2018**, *117*, 98–114. [CrossRef]

7. Manenti, S.; Sibilla, S.; Gallati, M.; Agate, G.; Guandalini, R. SPH simulation of sediment flushing induced by a rapid water flow. *J. Hydraul. Eng.* **2012**, *138*, 272–284. [CrossRef]

8. Amicarelli, A.; Kocak, B.; Sibilla, S.; Grabe, J. A 3D smoothed particle hydrodynamics model for erosional dam-break floods. *Int. J. Comput. Fluid Dyn.* **2017**, *31*, 413–434. [CrossRef]

9. Manenti, S.; Amicarelli, A.; Todeschini, S. WCSPH with Limiting Viscosity for Modeling Landslide Hazard at the Slopes of Artificial Reservoir. *Water* **2018**, *10*, 515. [CrossRef]

10. Tan, H.; Chen, S. A hybrid DEM-SPH model for deformable landslide and its generated surge waves. *Adv. Water Resour.* **2017**, *108*, 256–276. [CrossRef]

11. Shi, C.; An, Y.; Wu, Q.; Liu, Q.; Cao, Z. Numerical simulation of landslide-generated waves using a soil–water coupling smoothed particle hydrodynamics model. *Adv. Water Resour.* **2016**, *92*, 130–141. [CrossRef]

12. Viroulet, S.; Sauret, A.; Kimmoun, O.; Kharif, C. Granular collapse into water: Toward tsunami landslides. *J. Vis.* **2013**, *16*, 189–191. [CrossRef]

13. Capone, T.; Panizzo, A.; Monaghan, J.J. SPH modeling of water waves generated by submarine landslides. *J. Hydraul. Res.* **2010**, *48*, 80–84. [CrossRef]

14. Bordoni, M.; Meisina, C.; Valentino, R.; Bittelli, M.; Chersich, S. Site-specific to local-scale shallow landslides triggering zones assessment using TRIGRS. *Nat. Hazards Earth Syst. Sci.* **2015**, *15*, 1025–1050. [CrossRef]

15. Zizioli, D.; Meisina, C.; Valentino, R.; Montrasio, L. Comparison between different approaches to modeling shallow landslide susceptibility: A case history in Oltrepo Pavese, Northern Italy. *Nat. Hazards Earth Syst. Sci.* **2013**, *13*, 559–573. [CrossRef]

16. Ray, R.; Deb, K.; Shaw, A. Pseudo-Spring smoothed particle hydrodynamics (SPH) based computational model for slope failure. *Eng. Anal. Bound. Elem.* **2019**, *102*, 139–148. [CrossRef]

17. Gu, S.; Zheng, X.; Ren, L.; Xie, H.; Huang, Y.; Wei, J.; Shao, S. SWE-SPHysics Simulation of Dam Break Flows at South-Gate Gorges Reservoir. *Water* **2017**, *9*, 387. [CrossRef]

18. Barreiro, A.; Domínguez, J.M.; Crespo, A.J.C.; González-Jorge, H.; Roca, D.; Gómez-Gesteira, M. Integration of UAV photogrammetry and SPH modeling of fluids to study runoff on real terrains. *PLoS ONE* **2014**, *9*, e111031. [CrossRef]

19. Vacondio, R.; Mignosa, P.; Pagani, S. 3D SPH numerical simulation of the wave generated by the Vajont rock slide. *Adv. Water Res.* **2013**, *59*, 146–156. [CrossRef]

20. Qiu, L.C.; Liu, Y.; Han, Y. A 3D Simulation of a Moving Solid in Viscous Free-Surface Flows by Coupling SPH and DEM. *Math. Probl. Eng.* **2017**, *2017*, 1–7. [CrossRef]

21. Albano, R.; Sole, A.; Mirauda, D.; Adamowski, J. Modeling large floating bodies in urban area flash-floods via a Smoothed Particle Hydrodynamics model. *J. Hydrol.* **2016**, *541*, 344–358. [CrossRef]

22. Amicarelli, A.; Albano, R.; Mirauda, D.; Agate, G.; Sole, A.; Guandalini, R. A smoothed particle hydrodynamics model for 3D solid body transport in free surface flows. *Comput. Fluids* **2015**, *116*, 205–228. [CrossRef]

23. Lu, N.; Godt, J. *Hillslope Hydrology and Stability*; Cambridge University Press: New York, NY, USA, 2013.

24. Iverson, R.M. Landslide triggering by rain infiltration. *Water Resour. Res.* **2000**, *36*, 1897–1910. [CrossRef]

25. Iverson, R.M. The physics of debris flows. *Rev. Geophys.* **1997**, *35*, 245–296. [CrossRef]

26. Inam, A.; Adamowski, J.; Prasher, S.; Halbe, J.; Malard, J.; Albano, R. Coupling of a distributed stakeholder-built system dynamics socio-economic model with SAHYSMOD for sustainable soil salinity management—Part 1: Model development. *J. Hydrol.* **2017**, *55*, 596–618. [CrossRef]

27. Shao, S. Incompressible smoothed particle hydrodynamics simulation of multi-fluid flows. *Int. J. Numer. Meth. Fluids* **2012**, *69*, 1715–1735. [CrossRef]

28. Manenti, S. Standard WCSPH for free-surface multi-phase flows with a large density ratio. *Int. J. Ocean Coast. Eng.* **2018**, *1*. [CrossRef]

29. Colagrossi, A.; Landrini, M. Numerical simulation of interfacial ows by smoothed particle hydrodynamics. *J. Comput. Phys.* **2003**, *191*, 448–475. [CrossRef]

30. Hu, X.Y.; Adams, N.A. An incompressible multi-phase SPH method. *J. Comput. Phys.* **2007**, *227*, 264–278. [CrossRef]

31. Grenier, N.; Antuono, M.; Colagrossi, A.; le Touzé, D.; Alessandrini, B. An Hamiltonian interface SPH formulation for multi-fluid and free-surface flows. *J. Comput. Phys.* **2009**, *228*, 380–393. [CrossRef]

32. Monaghan, J.J.; Ashkan, R. A simple SPH algorithm for multi-fluid ow with high density ratios. *Int. J. Numer. Methods Fluids* **2013**, *71*, 537–561. [CrossRef]

33. Bouscasse, B.; Colagrossi, A.; Marrone, S.; Souto-Iglesias, A. SPH modelling of viscous flow past a circular cylinder interacting with a free surface. *Comput. Fluids* **2017**, *146*, 190–212. [CrossRef]

34. Hamill, L. *Bridge Hydraulics*; E&FN Spon: New York, NY, USA, 1999.

35. Persi, E.; Petaccia, G.; Fenocchi, A.; Manenti, S.; Ghilardi, P.; Sibilla, S. Hydrodynamic coefficients of yawed cylinders in open-channel flow. *Flow Meas. Instrum.* **2019**, *65*, 288–296. [CrossRef]

36. Dordoni, S.; Malerba, P.; Sgambi, L.; Manenti, S. Fuzzy reliability assessment of bridge piers in presence of scouring. In Proceedings of the 5th International Conference on Bridge Maintenance, Safety and Management, Philadelphia, PA, USA, 11–15 July 2010; pp. 1388–1395.

37. Sumer, B.M.; Fredsøe, J. *The Mechanics of Scour in the Marine Environment*; Advanced Series on Ocean Engineering; World Scientific Publishing Company: Singapore, 2002; Volume 17.

38. Whitehouse, R.J.S. *Scour at Marine Structures*; Thomas Telford: London, UK, 1998.

39. Wang, D.; Arikawa, T.; Li, S.W.; Gen, H. Numerical Simulation on Scour behind Seawall due to Tsunami Overflow. In Proceedings of the Coastal Structures & Solutions to Coastal Disasters Joint Conference, Boston, MA, USA, 9–11 September 2015; ASCE: Reston, VA, USA, 2015.

40. Li, S.W.; Wang, D. Tsunami occurrences in China and numerical simulation of a supposed tsunami process in Bohai Sea. In Proceedings of the China Ocean Engineering, Dailian, China, August 2013; pp. 490–496. Available online: http://cpfd.cnki.com.cn/Article/CPFDTOTAL-HYGC201308001076.htm (accessed on 30 August 2013).

41. Sugano, T.; Nozu, A.; Kohama, E.; Shimosako, K.; Kikuchi, Y. Damage to coastal structures. *Soils Found.* **2014**, *54*, 883–901. [CrossRef]

42. Takahashi, H.; Sassa, S.; Morikawa, Y.; Takano, D.; Maruyama, K. Stability of caisson-type breakwater foundation under tsunami-induced seepage. *Soils Found.* **2014**, *54*, 789–805. [CrossRef]

43. Nakamura, T.; Mizutani, N. Sediment Transport Calculation Considering Laminar and Turbulent Resistance Forces Caused by Infiltration/Exfiltration and its Application to Tsunami-induced Local Scouring. *J. Offshore Mech. Arct. Eng.* **2013**, *136*, 011105. [CrossRef]

44. Oie, T.; Dong, W.; Takatani, T.; Araki, K.; Shaowu, L.; Goto, H.; Arikawa, T. Numerical simulation of scouring behind the seawall caused by tsunami overflow with accurate ISPH method. *Jpn. Soc. Civ. Eng.* **2015**, *71*, 253–258. (In Japanese)

45. Ettema, R.; Kirkil, G.; Muste, M. Similitude of large-scale turbulence in experiments on local scour at cylinders. *J. Hydraul. Eng.* **2006**, *132*, 33–40. [CrossRef]

46. Sun, P.N.; Colagrossi, A.; Marrone, S.; Antuono, M.; Zhang, A.M. Multi-resolution Delta-plus-SPH with tensile instability control: Towards high Reynolds number flows. *Comput. Phys. Commun.* **2018**, *224*, 63–80. [CrossRef]

47. Zhang, A.; Sun, P.; Ming, F. An SPH modeling of bubble rising and coalescing in three dimensions. *Comput. Methods Appl. Mech. Eng.* **2015**, *294*, 189–209. [CrossRef]

48. Ming, F.R.; Sun, P.N.; Zhang, A.M. Numerical investigation of rising bubbles bursting at a free surface through a multiphase SPH model. *Meccanica* **2017**, *52*, 2665–2684. [CrossRef]

49. Ran, Q.; Tong, J.; Shao, S.; Fu, X.; Xu, Y. Incompressible SPH scour model for movable bed dam break flows. *Adv. Water Resour.* **2015**, *82*, 39–50. [CrossRef]

50. Gotoh, H.; Ikari, H.; Memita, T.; Sakai, T. Lagrangian Particle Method for Simulation of Wave Overtopping on a Vertical Seawall. *Coast. Eng. J.* **2005**, *47*, 157–181. [CrossRef]

51. Ulrich, C.; Leonardi, M.; Rung, T. Multi-physics SPH simulation of complex marine-engineering hydrodynamic problems. *Ocean Eng.* **2013**, *64*, 109–121. [CrossRef]

52. Crespo, A.J.C.; Domínguez, J.M.; Rogers, B.D.; Gómez-Gesteira, M.; Longshaw, S.; Canelas, R.; Vacondio, R.; Barreiro, A.; García-Feal, O. DualSPHysics: Open-source parallel CFD solver based on smoothed particle hydrodynamics (SPH). *Comput. Phys. Commun.* **2015**, *187*, 204–216. [CrossRef]

53. Fourtakas, G.; Rogers, B.D. Modeling multi-phase liquid-sediment scour and resuspension induced by rapid flows using Smoothed Particle Hydrodynamics (SPH) accelerated with a Graphics Processing Unit (GPU). *Adv. Water Resour.* **2016**, *92*, 186–199. [CrossRef]

54. Vand, V. Viscosity of solutions and suspensions. I. Theory. *J. Phys. Colloid Chem.* **1948**, *52*, 277–299. [CrossRef] [PubMed]

55. Todeschini, S. Trends in long daily rainfall series of Lombardia (Northern Italy) affecting urban stormwater control. *Int. J. Climatol.* **2012**, *32*, 900–919. [CrossRef]

56. Todeschini, S.; Papiri, S.; Ciaponi, C. Placement Strategies and Cumulative Effects of Wet-weather Control Practices for Intermunicipal Sewerage Systems. *Water Resour. Manag.* **2018**, *32*, 2885–2900. [CrossRef]

57. Stancanelli, L.M.; Peres, D.J.; Cancelliere, A.; Foti, E. A combined triggering-propagation modeling approach for the assessment of rainfall induced debris flow susceptibility. *J. Hydrol.* **2017**, *550*, 130–143. [CrossRef]

58. Baum, R.L.; Savage, W.Z.; Godt, J.W. *TRIGRS—A FORTRAN Program for Transient Rainfall Infiltration and Grid-Based Regional Slope-Stability Analysis, Version 2.0*; U.S. Geological Survey: Reston, VA, USA, 2008.

59. O'Brien, J. *Flo-2d User's Manual*, version 2006.01; flo-2d Software. Inc.: Nutrioso, AZ, USA, 2006.

60. Dai, F.C.; Lee, C.F.; Ngai, Y.Y. Landslide risk assessment and management: An overview. *Eng. Geol.* **2002**, *64*, 65–87. [CrossRef]

61. Pastor, M.; Haddad, B.; Sorbino, G.; Cuomo, S.; Drempetic, V. A depth-integrated, coupled SPH model for flow-like landslides and related phenomena. *Int. J. Numer. Anal. Methods Geomech.* **2009**, *33*, 143–172. [CrossRef]

62. Pastor, M.; Quecedo, M.; González, E.; Herreros, I.; Merodo, J.A.F.; Mira, P. A simple approximation to bottom friction for Bingham fluid depth integrated models. *J. Hydraul. Eng.* **2004**, *130*, 149–155. [CrossRef]

63. Rendina, I.; Viccione, G.; Cascini, L. Kinematics of flow mass movements on inclined surfaces. *Theor. Comput. Fluid Dyn.* **2019**, *33*, 107–123. [CrossRef]

64. Shao, S.; Lo, E.Y.M. Incompressible SPH method for simulating Newtonian and non-Newtonian flows with a free surface. *Adv. Water Resour.* **2003**, *26*, 787–800. [CrossRef]

65. Albano, R.; Craciun, I.; Mancusi, L.; Sole, A.; Ozunu, A. Flood damage assessment and uncertainity analysis: the case study of 2006 flood in Ilisua Basin in Romania. *Carpath. J. Earth. Environ. Sci.* **2017**, *2*, 12.

66. Di Risio, M.; Bellotti, G.; Panizzo, A.; de Girolamo, P. Three-dimensional experiments on landslide generated waves at a sloping coast. *Coast. Eng.* **2009**, *56*, 659–671. [CrossRef]

67. Panizzo, A.; de Girolamo, P.; di Risio, M.; Maistri, A.; Petaccia, A. Great landslide events in Italian artificial reservoirs. *Nat. Hazards Earth Syst. Sci.* **2005**, *5*, 733–740. [CrossRef]

68. Fritz, H.M.; Hager, W.H.; Minor, H.E. Lituya Bay case: Rockslide impact and wave run-up. *Sci. Tsunami Hazards* **2001**, *19*, 3–22.

69. Marrone, S.; Antuono, M.; Colagrossi, A.; Colicchio, G.; le Touzé, D.; Graziani, G. δ-SPH model for simulating violent impact flows. *Comput. Methods Appl. Mech. Eng.* **2011**, *200*, 1526–1542. [CrossRef]

70. Heinrich, P. Nonlinear water waves generated by submarine and aerial landslides. *J. Waterw. Port Coast. Ocean Eng.* **1992**, *118*, 249–266. [CrossRef]

71. Xu, W.J.; Dong, X.Y.; Ding, W.T. Analysis of fluid-particle interaction in granular materials using coupled SPH-DEM method. *Powder Technol.* **2019**, *353*, 459–472. [CrossRef]

72. Canelas, R.B.; Dominiguez, J.M.; Crespo, A.J.C.; Gomez-Gesteira, M.; Ferreira, R.M.L. Resolved Simulation of a granular-fluid flow with a coupled SPH-DCDEM model. *J. Hydraul. Eng.* **2017**, *143*, 6017012. [CrossRef]

73. Zago, V.; Bilotta, G.; Hérault, A.; Dalrymple, R.A.; Fortuna, L.; Cappello, A.; Ganci, G.; del Negro, C. Semi-implicit 3D SPH on GPU for lava flows. *J. Comp. Phys.* **2018**, *375*, 854–870. [CrossRef]

74. Manenti, S.; Pierobon, E.; Gallati, M.; Sibilla, S.; D'Alpaos, L.; Macchi, E.; Todeschini, S. Vajont Disaster: Smoothed Particle Hydrodynamics Modeling of the Postevent 2D Experiments. *J. Hydraul. Eng.* **2016**, *142*, 5015007. [CrossRef]

75. Amicarelli, A.; Manenti, S.; Albano, R.; Agate, G.; Paggi, M.; Longoni, L.; Mirauda, D.; Ziane, L.; Viccione, G.; Todeschini, S.; et al. SPHERA v.9.0.0: A Computational Fluid Dynamics research code, based on the Smoothed Particle Hydrodynamics mesh-less method. *Comput. Phys. Commun.* **2019**. submitted.

76. SPHERA v.9.0.0 (RSE SpA). Available online: https://github.com/AndreaAmicarelliRSE/SPHERA (accessed on 21 February 2019).

77. Aristodemo, F.; Merignolo, D.D.; Groenenboom, P.; Schiavo, A.L.; Veltri, P.; Veltri, M. Assessment of Dynamic Pressures at Vertical and Perforated Breakwaters through Diffusive SPH Schemes. *Math. Probl. Eng.* **2015**, *2015*, 1–10. [CrossRef]

78. Chauchat, J.; Médale, M. A three-dimensional numerical model for incompressible two-phase flow of a granular bed submitted to a laminar shearing flow. *Comput. Methods Appl. Mech. Eng.* **2010**, *199*, 439–449. [CrossRef]

79. Fraccarollo, L.; Capart, H.; Zech, Y. A Godunov method for the computation of erosional shallow water transients. *Int. J. Numer. Meth. Fluids* **2003**, *41*, 951–976. [CrossRef]

80. Soares-Frazão, S.; Zech, Y. Experimental study of dam-break flow against an isolated obstacle. *J. Hydraul. Res.* **2007**, *45* (Suppl. 1), 27–36. [CrossRef]

81. Lin, P.; Wu, Y.; Bai, J.; Lin, Q. A numerical study of dam-break flow and sediment transport from a quake lake. *J. Earthq. Tsunami* **2011**, *5*, 401–428. [CrossRef]

82. Wu, W.; Jiang, E.; China, P.R.; Wang, S.S. Depth-averaged 2-D calculation of flow and sediment transport in the lower Yellow River. *Int. J. River Basin Manag.* **2004**, *2*, 51–59. [CrossRef]

83. Armstrong, L.; Gu, S.; Luo, K. Study of wall-to-bed heat transfer in a bubbling fluidised bed using the kinetic theory of granular flow. *Int. J. Heat Mass Transf.* **2010**, *53*, 4949–4959. [CrossRef]

84. GDAL (OSGEO). Available online: https://github.com/OSGeo/gdal (accessed on 16 April 2019).

85. DEM2xyz (RSE SpA). Available online: https://github.com/AndreaAmicarelliRSE/DEM2xyz (accessed on 16 April 2019).

86. Paraview (Kitware). Available online: https://github.com/Kitware/ParaView (accessed on 16 April 2019).

87. Grid Interpolator (RSE SpA). Available online: https://github.com/AndreaAmicarelliRSE/Grid_Interpolator (accessed on 16 April 2019).

88. SRTM3/DTED1 (USGS). Available online: http://earthexplorer.usgs.gov/ (accessed on 16 April 2019).

89. Baker, M.; Buyya, R. Cluster Computing at a Glance. In *High Performance Cluster Computing—Architectures and Systems*; Prentice Hall PTR: Upper Saddle River, NJ, USA, 1999; pp. 3–47.

90. Nickolls, J.; Buck, I.; Garland, M.; Skadron, K. Scalable Parallel Programming with CUDA. *Queue GPU Comput.* **2008**, *6*, 40–53. [CrossRef]

91. Nickolls, J.; Dally, W.J. The GPU computing era. *IEEE Micro* **2010**, *30*, 56–69. [CrossRef]

92. Monaghan, J.J. Smoothed particle hydrodynamics. *Ann. Rev. Astronom. Astrophys.* **1992**, *30*, 543–574. [CrossRef]

93. Koshizuka, S.; Oka, Y. Moving Particle Semi-Implicit Method for Fragmentation of Incompressible Fluid. *Nucl. Sci. Eng.* **1996**, *123*, 421–434. [CrossRef]

94. Viccione, G.; Bovolin, V.; Carratelli, E.P. Defining and optimizing algorithms for neighbouring particle identification in SPH fluid simulations. *Int. J. Numer. Methods Fluids* **2008**, *58*, 625–638. [CrossRef]

95. Domínguez, J.M.; Crespo, A.J.C.; Gómez-Gesteira, M.; Marongiu, J.C. Neighbour lists in smoothed particle hydrodynamics. *Int. J. Numer. Methods Fluids* **2011**, *67*, 2026–2042. [CrossRef]

96. Winkler, D.; Rezavand, M.; Rauch, W. Neighbour lists for smoothed particle hydrodynamics on GPUs. *Comput. Phys. Commun.* **2018**, *225*, 140–148. [CrossRef]

97. Wang, D.; Zhou, Y.; Shao, S. Efficient Implementation of Smoothed Particle Hydrodynamics (SPH) with Plane Sweep Algorithm. *Commun. Comput. Phys.* **2016**, *19*, 770–800. [CrossRef]

98. Xia, X.; Liang, Q. A GPU-accelerated smoothed particle hydrodynamics (SPH) model for the shallow water equations. *Environ. Model. Softw.* **2016**, *75*, 28–43. [CrossRef]

99. Gonnet, P. Efficient and Scalable Algorithms for Smoothed Particle Hydrodynamics on Hybrid Shared/Distributed-Memory Architectures. *SIAM J. Sci. Comput.* **2014**, *37*, C95–C121. [CrossRef]

100. Joselli, M.; Junior, J.R.d.; Clua, E.W.; Montenegro, A.; Lage, M.; Pagliosa, P. Neighborhood grid: A novel data structure for fluids animation with GPU computing. *J. Parallel Distrib. Comput.* **2015**, *75*, 20–28. [CrossRef]

101. Wenbo, C.; Yao, Y.; Zhang, Y. Performance analysis of parallel smoothed particle hydrodynamics on multi-core CPUs. In Proceedings of the 2014 International Conference on Cloud Computing and Internet of Things, Changchun, China, 13–14 December 201.

102. Nishiura, D.; Furuichi, M.; Sakaguchi, H. Computational performance of a smoothed particle hydrodynamics simulation for shared-memory parallel computing. *Comput. Phys. Commun.* **2015**, *194*, 18–32. [CrossRef]

103. Ferrari, A.; Dumbser, M.; Toro, E.F.; Armanini, A. A new 3D parallel SPH scheme for free surface flows. *Comput. Fluids* **2009**, *38*, 1203–1217. [CrossRef]

104. Cherfils, J.M.; Pinon, G.; Rivoalen, E. JOSEPHINE: A parallel SPH code for free-surface flows. *Comput. Phys. Commun.* **2012**, *183*, 1468–1480. [CrossRef]

105. Oger, G.; Le Touzé, D.; Guibert, D.; De Leffe, M.; Biddiscombe, J.; Soumagne, J.; Piccinali, J.-G. On distributed memory MPI-based parallelization of SPH codes in massive HPC context. *Comput. Phys. Commun.* **2016**, *200*, 1–14. [CrossRef]

106. Egorova, M.S.; Dyachkov, S.A.; Parshikov, A.N.; Zhakhovsky, V.V. Parallel SPH modeling using dynamic domain decomposition and load balancing displacement of Voronoi subdomains. *Comput. Phys. Commun.* **2019**, *234*, 112–125. [CrossRef]

107. Yeylaghi, S.; Moa, B.; Oshkai, P.; Buckham, B.; Crawford, C. ISPH modeling for hydrodynamic applications using a new MPI-based parallel approach. *J. Ocean Eng. Mar. Energy* **2017**, *3*, 35–50. [CrossRef]

108. Guo, X.; Rogers, B.D.; Lind, S.; Stansby, P.K. New massively parallel scheme for Incompressible Smoothed Particle Hydrodynamics (ISPH) for highly nonlinear and distorted flow. *Comput. Phys. Commun.* **2018**, *233*, 16–28. [CrossRef]

109. Fleissner, F.; Eberhard, P. Parallel load-balanced simulation for short-range interaction particle methods with hierarchical particle grouping based on orthogonal recursive bisection. *Int. J. Numer. Methods Eng.* **2008**, *74*, 531–553. [CrossRef]

110. Harada, T.; Koshizuka, S.; Kawaguchi, Y. Smoothed particle hydrodynamics on GPUs. In Proceedings of the Computer Graphics International Conference, Petròpolis, Brazil, 30 March 2007; pp. 63–70. Available online: https://pdfs.semanticscholar.org/a132/6b93316e7ce4d2580bd5e3928ce6ff24e386.pdf (accessed on 21 February 2019).

111. Crespo, A.C.; Dominguez, J.M.; Barreiro, A.; Gómez-Gesteira, M.; Rogers, B.D. GPUs, a new tool of acceleration in CFD: Efficiency and reliability on smoothed particle hydrodynamics methods. *PLoS ONE* **2011**, *6*, e20685. [CrossRef] [PubMed]

112. Hérault, A.; Bilotta, G.; Vicari, A.; Rustico, E.; del Negro, C. Numerical simulation of lava flow using a GPU SPH model. *Ann. Geophys.* **2011**, *54*. [CrossRef]

113. Mokos, A.; Rogers, B.D.; Stansby, P.K.; Domínguez, J.M. Multi-phase SPH modeling of violent hydrodynamics on GPUs. *Comput. Phys. Commun.* **2015**, *196*, 304–316. [CrossRef]

114. Winkler, D.; Meister, M.; Rezavand, M.; Rauch, W. gpuSPHASE—A shared memory caching implementation for 2D SPH using CUDA. *Comput. Phys. Commun.* **2017**, *213*, 165–180. [CrossRef]

115. Cercos-Pita, J.L. AQUAgpusph, a new free 3D SPH solver accelerated with OpenCL. *Comput. Phys. Commun.* **2015**, *192*, 295–312. [CrossRef]

116. Domínguez, J.M.; Crespo, A.J.C.; Gómez-Gesteira, M. Optimization strategies for CPU and GPU implementations of a smoothed particle hydrodynamics method. *Comput. Phys. Commun.* **2013**, *184*, 617–627. [CrossRef]

117. Qiu, L.C. OpenCL-Based GPU Acceleration of ISPH Simulation for Incompressible Flows. *Appl. Mech. Mater.* **2014**, *444*, 380–384. [CrossRef]

118. Nie, X.; Chen, L.; Xiang, T. Real-Time Incompressible Fluid Simulation on the GPU. *Int. J. Comput. Games Technol.* **2015**, *2015*, 417417. [CrossRef]

119. Chow, A.D.; Rogers, B.D.; Lind, S.J.; Stansby, P.K. Incompressible SPH (ISPH) with fast Poisson solver on a GPU. *Comput. Phys. Commun.* **2018**, *226*, 81–103. [CrossRef]

120. Hori, C.; Gotoh, H.; Ikari, H.; Khayyer, A. GPU-acceleration for Moving Particle Semi-Implicit method. *Comput. Fluids* **2011**, *51*, 174–183. [CrossRef]

121. Kakuda, K.; Nagashima, T.; Hayashi, Y.; Obara, S.; Toyotani, J.; Miura, S. Three-dimensional fluid flow simulations using GPU-based particle method. *Comput. Model. Eng. Sci.* **2013**, *95*, 363–376.

122. Kalyanapu, A.J.; Shankar, S.; Pardyjak, E.R.; Judi, D.R.; Burian, S.J. Assessment of GPU computational enhancement to a 2D flood model. *Environ. Model. Softw.* **2011**, *26*, 1009–1016. [CrossRef]

123. Brodtkorb, A.R.; Sætra, M.L.; Altinakar, M. Efficient shallow water simulations on GPUs: Implementation, visualization, verification, and validation. *Comput. Fluids* **2012**, *55*, 1–12. [CrossRef]

124. Vacondio, R.; Palù, A.D.; Mignosa, P. GPU-enhanced finite volume shallow water solver for fast flood simulations. *Environ. Model. Softw.* **2014**, *57*, 60–75. [CrossRef]

125. García-Feal, O.; González-Cao, J.; Gómez-Gesteira, M.; Cea, L.; Domínguez, J.M.; Formella, A. An accelerated tool for flood modeling based on Iber. *Water* **2018**, *10*, 1459. [CrossRef]

126. Liu, Q.; Qin, Y.; Li, G. Fast simulation of large-scale floods based on GPU parallel computing. *Water* **2018**, *10*, 589. [CrossRef]

127. Bladé, E.; Cea, L.; Corestein, G.; Escolano, E.; Puertas, J.; Vázquez-Cendón, E.; Dolz, J.; Coll, A. Iber: Herramienta de simulación numérica del flujo en ríos. *Rev. Int. Metod. Numer. Calc. Disen. Ing.* **2014**, *30*, 1–10. [CrossRef]

128. Wolfe, M.; Lee, S.; Kim, J.; Tian, X.; Xu, R.; Chapman, B.; Chandrasekaran, S. The OpenACC data model: Preliminary study on its major challenges and implementations. *Parallel Comput.* **2018**, *78*, 15–27. [CrossRef]

129. Hérault, A.; Zhang, H.; Dalrymple, R.A.; Yang, R.; Wu, J. Numerical modeling of dam-break flood through intricate city layouts including underground spaces using GPU-based SPH method. *J. Hydrodyn.* **2014**, *25*, 818–828.

130. Liang, Q.; Xia, X.; Hou, J. Efficient urban flood simulation using a GPU-accelerated SPH model. *Environ. Earth Sci.* **2015**, *74*, 7285–7294. [CrossRef]

131. Dickson, N.G.; Karimi, K.; Hamze, F. Importance of explicit vectorization for CPU and GPU software performance. *J. Comput. Phys.* **2011**, *230*, 5383–5398. [CrossRef]

132. Lee, V.W.; Kim, C.; Chhugani, J.; Deisher, M. Debunking the 100X GPU vs. CPU Myth: An Evaluation of Throughput Computing on CPU and GPU. *ACM SIGARCH Comput. Arch. News* **2010**, *38*, 451–460. [CrossRef]

133. Valdez-Balderas, D.; Domínguez, J.M.; Rogers, B.D.; Crespo, A.J.C. Towards accelerating smoothed particle hydrodynamics simulations for free-surface flows on multi-GPU clusters. *J. Parallel Distrib. Comput.* **2013**, *73*, 1483–1493. [CrossRef]

134. Domínguez, J.M.; Crespo, A.J.C.; Valdez-Balderas, D.; Rogers, B.D.; Gómez-Gesteira, M. New multi-GPU implementation for smoothed particle hydrodynamics on heterogeneous clusters. *Comput. Phys. Commun.* **2013**, *184*, 1848–1860. [CrossRef]

135. Rustico, E.; Bilotta, G.; Herault, A.; del Negro, C.; Gallo, G. Advances in multi-GPU smoothed particle hydrodynamics simulations. *IEEE Trans. Parallel Distrib. Syst.* **2014**, *25*, 43–52. [CrossRef]

136. Ji, Z.; Xu, F.; Takahashi, A.; Sun, Y. Large scale water entry simulation with smoothed particle hydrodynamics on single- and multi-GPU systems. *Comput. Phys. Commun.* **2016**, *209*, 1–12. [CrossRef]

Article

A New Parallel Framework of SPH-SWE for Dam Break Simulation Based on OpenMP

Yushuai Wu [1], Lirong Tian [1], Matteo Rubinato [2], Shenglong Gu [1,3,*], Teng Yu [1], Zhongliang Xu [4], Peng Cao [5,6], Xuhao Wang [6,7] and Qinxia Zhao [1]

[1]   School of Water Resources and Electric Power, Qinghai University, Xining 810016, China; ys.wu@qhu.edu.cn (Y.W.); LirongTianqhu@hotmail.com (L.T.); 2017990061@qhu.edu.cn (T.Y.); 2014990032@qhu.edu.cn (Q.Z.)

[2]   School of Energy, Construction and Environment & Centre for Agroecology, Water and Resilience, Coventry University, Coventry CV1 5FB, UK; matteo.rubinato@coventry.ac.uk

[3]   State Key Laboratory of Plateau Ecology and Agriculture, Qinghai University, Xining 810016, China

[4]   Transportation Bureau, Haiyan County, Jiaxing 812200, China; 1998990003@qhu.edu.cn

[5]   College of Architecture and Civil Engineering, Beijing University of Technology, Beijing 100124, China; 2017630009@qhu.edu.cn

[6]   Qinghai University-Tsinghua University, Sanjiangyuan University, Sanjiangyuan Research Institute, Qinghai University, Xinning 810016, China; wangxh@chd.edu.cn

[7]   School of Highway, Chang'an University, Xi'an 710064, China

*   Correspondence: sl.gu@qhu.edu.cn

Received: 7 April 2020; Accepted: 8 May 2020; Published: 14 May 2020

Abstract: Due to its Lagrangian nature, Smoothed Particle Hydrodynamics (SPH) has been used to solve a variety of fluid-dynamic processes with highly nonlinear deformation such as debris flows, wave breaking and impact, multi-phase mixing processes, jet impact, flooding and tsunami inundation, and fluid–structure interactions. In this study, the SPH method is applied to solve the two-dimensional Shallow Water Equations (SWEs), and the solution proposed was validated against two open-source case studies of a 2-D dry-bed dam break with particle splitting and a 2-D dam break with a rectangular obstacle downstream. In addition to the improvement and optimization of the existing algorithm, the CPU-OpenMP parallel computing was also implemented, and it was proven that the CPU-OpenMP parallel computing enhanced the performance for solving the SPH-SWE model, after testing it against three large sets of particles involved in the computational process. The free surface and velocities of the experimental flows were simulated accurately by the numerical model proposed, showing the ability of the SPH model to predict the behavior of debris flows induced by dam-breaks. This validation of the model is crucial to confirm its use in predicting landslides' behavior in field case studies so that it will be possible to reduce the damage that they cause. All the changes made in the SPH-SWEs method are made open-source in this paper so that more researchers can benefit from the results of this research and understand the characteristics and advantages of the solution proposed.

Keywords: dam break; SWE; SPH; openMP; numerical modelling; computational time

---

1. Introduction

The Smooth Particle Hydrodynamics (SPH) is a meshless method [1] very commonly used nowadays [2–12]. Gingold and Monaghan [13] were the first to propose this method to solve astrophysical simulations, using statistical techniques to recover analytical expressions for the physical variables from a known distribution of fluid elements. The SPH method is typically used for solving the equations of hydrodynamics in which Lagrangian discretized mass elements are followed [14].

Compared with the limitations of the Eulerian grid method [15–17], the SPH method has unique advantages in dealing with free water surface and moving boundary conditions [18–21]. In fact, the SPH method strictly runs in accordance with the law of conservation of mass and can deal with free surface and moving boundary flexibly; hence, it is very suitable for simulating dam break flows [22]. Flooding due to dam break has potentially disastrous consequences, and multiple studies were conducted to numerically replicate the hydrodynamics of this phenomenon [23–28]. In most cases, the dam break flow occurs in a wide area and lasts for a long time. Therefore, the Shallow Water wave Equations (SWEs) have become the main application formulae for this specific dam break problem.

In 1999, Wang and Shen [29] applied the SPH method to SWEs for the first time. Dam break flows are unsteady open channel flows that can be described by the St. Venant equations, which are equations that can be used for flows with strong shocks [29]. The study conducted by Wang and Shen [29] has demonstrated that the method developed based on the SPH is capable of providing accurate simulations for mixed flow regimes with strong shocks. In 2005, Ata and Soulaimani [30] tried to reduce the difficulties that were associated with the treatment of the solid boundary conditions, especially with irregular boundaries. Ata and Soulaimani [30] derived a new artificial viscosity term by using an analogy with an approximate Riemann solver, and the several numerical tests conducted have confirmed that the stabilization proposed provides more accurate results than the standard artificial viscosity introduced by Monaghan [22]. However, Ata and Soulaimani [30] found that it was difficult to implement the Dirichlet boundary conditions for bounded domains. Different techniques such as symmetrization and ghost particles were implemented; nevertheless, results obtained for irregular boundaries or in presence of shocks were not satisfactory, and there was a need to make SPH method more competitive with standard approaches [30].

A new method with good stability was then needed for the SPH numerical simulation of shallow water equations to deal with dam break flows, flood waters, debris flows, avalanches, and tidal waves.

De Leffe et al. [31] proposed an improved calculation method based on a two-dimensional SPH solid wall boundary condition. By introducing a periodic redistribution of the particles and using a kernel function with variable smoothing length, this modification was tested and validated against dam break flows on a flat dry bottom in 1D and 2D. Comparisons conducted against literature results [30,32,33] have confirmed how this new approach is robust and able to simulate complex hydrodynamic situations.

However, to date, only a few studies have investigated the efficiency of solving the SPH-SWEs model. Vacondio et al. [34–37] have developed the serial code to solve the SWEs by using the SPH method and have made an open source version called SWE-SPHysics, which has been optimized and adapted based on the hydrodynamics investigated by other researchers [38–52]. Despite continuous progress, there is still a limitation related to the computational efficiency when the number of particles to simulate is very large, and this aspect still needs to be improved.

Xia and Liang [53] explored the Graphic Processing Units (GPUs) to accelerate an SPH-SWE model for wider applications such as dam breaks. Xia and Liang [53] demonstrated that the performance of the new GPU accelerated SPH-SWE model can significantly improve the calculation efficiency and verified that the quadtree neighbor searching method may reduce redundant computation when searching neighbor particles [53]. Furthermore, Liang et al. [54] developed a shock-capturing hydrodynamic model to simulate the complex rainfall-runoff and the induced flooding process in a catchment in England, Haltwhistle Burn, of 42 km^2, and implemented it on GPUs for high-performance parallel computing.

GPU has been enhanced with a Fortran programming language capability employing CUDA (Compute Unified Device Architecture), known as CUDA Fortran [55]. Although the GPU parallel computing performance is strong, the GPU price is relatively expensive and support for the CUDA FORTRAN language compiler is limited. Furthermore, if CUDA FORTRAN language is compared with other parallels (OpenMP and MPI), the program design is also more complex [56–58].

Over the years, with the development of parallel computing, the development of OpenMP and MPI in parallel methods has matured, and MPI is widely used in the field of engineering computing [59–63]. However, in this multi-machine cluster environment, memory is not shared. For global shared data operations, data must be transferred by the communication between machines [64,65].

OpenMP is based on the shared storage mode of multi-core processors, and it is commonly used in parallel processing of single workstations. Although it is limited by the processing capacity and memory capacity of a single node, it can simplify the past multi-core computing to the present multi-core computers. The program design is relatively simple, and it can secure the advantages of economic and programming optimizations [66].

Therefore, this paper adopts the parallel computing method of CPU-OpenMP that is applied to a single machine and a multi-core to calculate the new SPH-SWEs framework for the parallel computing of the test case of *2-D dry-bed dam break with particle splitting* [67,68] and a *2-D debris flow with a rectangular obstacle downstream the dam* [2]. The accuracy of the new SPH-SWEs framework was then verified by comparing the serial algorithm, and the advantages of CPU-OpenMP parallel computing were analyzed.

The paper is organized as follows: Section 2 describes the methodology adopted presenting the theoretical derivation of the numerical model applied and the governing equations. Section 3 explains the setup of the SPH-SWEs method used. Section 4 provides the results of the application tested with a discussion of the results obtained. Lastly, Section 5 produces a brief summary and concluding remarks of the whole study.

2. Methodology

Ata et al., [30] and Paz and Bonet [32] have initiated the idea of SPH-SWE model solution, which is described in Section 2.1 with all the governing equations.

2.1. Governing Equations

By ignoring the Coriolis effect and the fluid viscosity, SWEs can be written in the Lagrangian form as follows:

$$\begin{aligned}
\frac{dd}{dt} &= -d\nabla \cdot v \\
\frac{dv}{dt} &= -g\nabla d + g(\nabla b + S_f)
\end{aligned} \tag{1}$$

d represents the water depth, g the acceleration of gravity, v is velocity, b represents the riverbed elevation, and S_f represents the riverbed friction. In the SWEs, the area density is defined as:

$$\rho = \rho_w d \tag{2}$$

ρ represents the density, and ρ_w represents the density of water.

2.2. Water Depth Solutions

According to the SPH idea, the area density (i.e., water depth) of particles is solved as shown below in the implicit function:

$$\begin{aligned}
\rho_i &= \sum_j m_j W_i(x_i - x_j, h_i) \\
h_i &= h_0 \left(\frac{\rho_0}{\rho_i}\right)^{1/d_m}
\end{aligned} \tag{3}$$

xi/xj represents the particle coordinates; mj represents the particle mass of j; h_i and ρ_i represent the smooth length and the area density of the particle i; h_0 and ρ_0 represent the initial values of the smooth length and the area density, respectively; d_m represents the latitude (1 represents one dimension, 2 represents two dimensions) and W represents the kernel function.

2.3. Speed Solution

According to the Lagrangian equation of motion [69], a_i is the acceleration of the particle i, and the solution formula of each particle can be obtained as follows:

$$a_i = \frac{g + v_i \cdot k_i v_i + t_i \cdot \nabla b_i}{1 + \nabla b_i \cdot \nabla b_i} \nabla b_i - t_i + S_{f,i} \tag{4}$$

$k_i = \nabla(\nabla b_i)$ represents $b(x)$ of the curvature tensor [70]; t_i represents the acceleration caused by the internal force; ∇b_i represents the riverbed gradient of the particle i, and in order to deal with any complex terrain problem, the riverbed gradient can be modified as follows [71,72]:

$$\nabla b_i = \sum_j b_j \widetilde{\nabla} W_i \left(x_i - x_j, h_i \right) V_j \tag{5}$$

$\widetilde{\nabla} W_i$ denotes the gradient of the modified kernel function, which is modified by the correction matrix L_i, as shown below:

$$\widetilde{\nabla} W_i\left(x_i - x_j, h_i\right) = L_i \nabla W_i\left(x_i - x_j, h_i\right) L_i = \left[\sum_j \nabla W_i\left(x_i - x_j, h_i\right) \times \left(x_i - x_j\right) V_j \right]^{-1} \tag{6}$$

In order to reduce the numerical oscillation and ensure the stability of the calculation, one method is to increase the viscosity term as introduced below [73]:

$$\begin{aligned}
t_i &= \sum_j m_j \frac{g}{2\rho_w}\left[\left(\tfrac{1}{\beta_j} + \pi_{ij}\right)\nabla W_j(x_i - x_j, h_j) - \left(\tfrac{1}{\beta_i} + \pi_{ij}\right)\nabla W_i(x_j - x_i, h_i) \right] \\
\beta_i &= -\frac{1}{\rho_i d_m} \sum_j m_j r_{ij} \frac{dW_{ij}}{dr_{ij}} \\
\pi_{ij} &= \frac{\bar{c}_{ij} v_{ij} \cdot x_{ij}}{\rho_{ij} \sqrt{|x_{ij}|^2 + \zeta^2}}
\end{aligned} \tag{7}$$

β represents the correction coefficient caused by variable smooth length; r_{ij} represents the particle spacing; π_{ij} represents the numerical viscosity added to maintain stability. However, this method has the problem of numerical dissipation.

To reduce this issue, the interaction between two particles was treated as a Riemann problem [74], as follows:

$$\begin{aligned}
t_i &= \sum_j m_j p^* \left[\frac{1}{\rho_j^2 \beta_j} \nabla W_j(x_i - x_j, h_j) - \frac{1}{\rho_i^2 \beta_i} \nabla W_i(x_j - x_i, h_i) \right] \\
p^* &= 0.5 g \rho_w (d^*)^2 \\
d^* &= \frac{g_l d_l + g_r d_r + v_{l,n} - v_{r,n}}{g_l + g_r} \\
g_k &= \sqrt{0.5 \frac{g(d_0 + d_k)}{d_0 d_k}} \\
d_0 &= \frac{1}{g}\left[0.5(c_l + c_r) + 0.25(v_{l,n} - v_{r,n}) \right]^2
\end{aligned} \tag{8}$$

d_l and d_r represent the water depth on the left and right sides, respectively; $k = l$ and $k = r$ represent the left and right states, respectively; d_0 represents the initial estimated water depth; $c = \sqrt{gh}$ represents the shallow water wave velocity.

2.4. Time Integration and Boundary Processing

In order to update the particle velocity and displacement, the leap-frog time integration scheme [75] was used. By using this method, both time and space are of second-order accuracy, and the storage demand is relatively low, however the calculation efficiency is relatively high, as shown below:

$$
\begin{aligned}
v_i^{n+1/2} &= v_i^{n-1/2} + \Delta t a_i^n \\
x_i^{n+1} &= x_i^n + \Delta t v_i^{n+1/2} \\
v_i^{n+1} &= v_i^{n+1/2} + \tfrac{1}{2}\Delta t a_i^n
\end{aligned}
\tag{9}
$$

Δt represents the time step; where the time step must meet the Courant number condition [76] displayed as follows:

$$
\Delta t = CFL \min_{i=1}^{N}\left(\frac{h_i}{c_i + \|v_i\|}\right)
\tag{10}
$$

To solve the boundary problem, this study adopts the Modified Virtual Boundary Particle (MVBP) method [36]. MVBP method is an improvement of the virtual boundary particle (VBP) method [77]. The virtual particles on the boundary will neither move with the fluid particles nor interact with them but generate the virtual particles similar to the mirror image through point symmetry. This method is easier and simpler in dealing with the complex boundary.

Compared with the VBP method, the MVBP method has two improvements: (1) When a virtual boundary particle is within the range of the kernel function of a fluid particle, two layers of newly generated virtual particles can be added ($X_{k,1} = 2X_v - X_i$ and $X_{k,2} = 4X_v - X_i$). Among them, $X_{k,1}$ and $X_{k,2}$ represent the coordinates of the newly generated virtual particles, and X_v represents the virtual boundary particles; (2) When the internal angle of the boundary is less than or equal to 180°, two newly generated virtual particles are added outside the corner. Compared with the single point of the VBP method, this improvement reduces the kernel truncation error.

In order to verify the effect of different kernel functions to simulate the dam break, this paper used the SPH-SWE open source code [34–37,68] and applied different kernel functions (*B-spline, super Gauss, quadratic spline, Gauss, quartic spline, quintic and Bell*) to simulate case 2 open source scenario [68]. These were the initial conditions considered for the dam-break: (i) simulation area was 2000 m long; (ii) the river bed elevation was 0 m; (iii) the initial fluid particle area was 1000 m long; (iv) the particle spacing was 10 m; (v) the initial water depth was 10 m, (vi) and the simulation duration was 50 s. The position, the water depth, and speed of the fluid particles were obtained as an output every 10 s. After verification, the advantages and disadvantages of different kernel functions and different numerical oscillation processing methods (see Equations (7)–(8)) identified from the results are consistent. At t = 50 s, the water depth dissipated after 1500 m, and the results of different kernel functions and different numerical oscillation processing methods can be visually reflected through the graphs in Figure 1; at the same time, it also illustrates the continuity of the numerical oscillation issue. Therefore, the data of t = 50 s (Figure 1 results) was selected for analysis in this study.

It can be noticed from Figure 1 that in the numerical simulation of dam break based on SPH-SWEs approach, all kernel functions are characterized by numerical oscillation except the option where the Bell kernel function is considered. The three kernel functions that provide a more accurate estimation of the water depth and the velocity are B-spline, quadratric spline, and quartic spline (above 85.7%). Figure 2a–d shows the absolute error between the calculated water depths and velocities using these three kernel functions vs. the analytical solution. Results displayed confirm the optimal performance of the B-spline kernel function in dam break simulations.

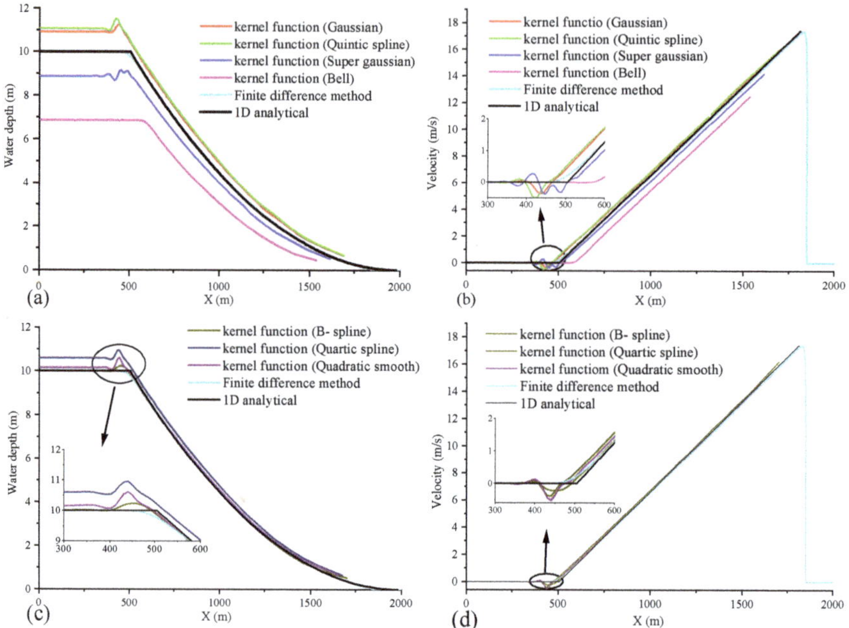

Figure 1. Simulation results of different kernel functions. (**a,c**) Results of water depths calculations. (**b,d**) Results of velocity calculations.

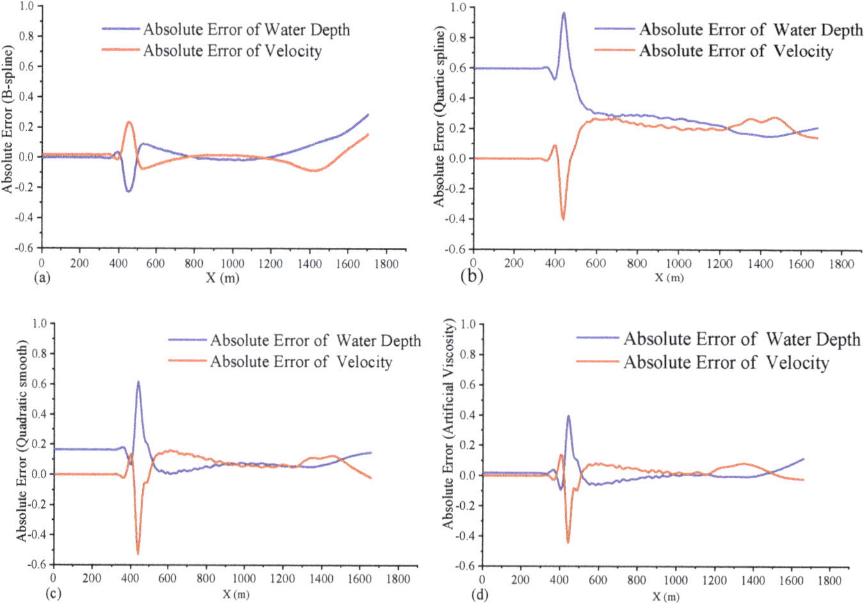

Figure 2. Absolute errors of different kernel functions and processing methods. (**a–c**) The absolute errors of water depth and velocity calculated by B-spline kernel function, quartic spline, and quadratic spline kernel function, respectively. (**d**) The absolute errors of water depth and velocity calculated by using the artificial viscosity method.

If the kernel function adopted is B-spline, the numerical oscillation is treated as adding the viscosity term (numerical viscosity, lax Friedrichs flux [78]). The results are shown in Figures 2d and 3. It is shown in Figure 3 that the Riemann solvers method has better characteristics in dealing with numerical oscillation problems when using the same kernel function. This confirms that it is possible to increase the viscosity by increasing π_{ij} in Equation (7), which helps to reduce the numerical oscillation; however, this can cause a decrease in the accuracy of the calculation results and an possible increase on computational time.

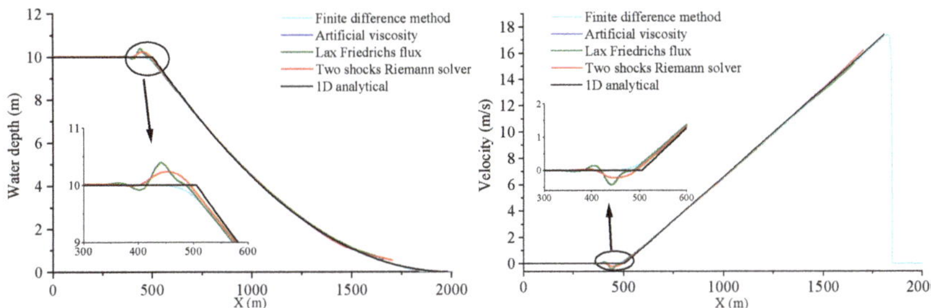

Figure 3. Simulation results of different processing methods.

According to Figure 2a,d and Figure 3, it can be found that the two-shocks Riemann solver is more advantageous in dealing with numerical oscillations when considering dam break cases. The absolute error of the solutions considered is significantly small. Table 1 displays the errors of the three methods adopted, and it can be seen that the standard deviations of the velocity and water depth (WD) of the three methods are small and, basically, of similar magnitude. However, the average relative errors of water depth and speed solved by the two-shocks Riemann solver seem to be the smallest, 0.92% and 5.86%, respectively.

Table 1. Error analysis.

	Parameters	Artificial Viscosity	Lax Friedrichs Flux	Two-Shocks Riemann Solver
Mean Absolute Error	Speed	0.0617	0.0476	0.0351
	WD	0.0515	0.0482	0.0356
Mean Relative Error	Speed	0.0718	0.0667	0.0586
	WD	0.0142	0.0089	0.0092
Standard deviation of error	Speed	0.1042	0.0728	0.0568
	WD	0.0681	0.0633	0.0484

- *WD = Water Depth.*

In order to verify the effect of the selected B-spline kernel function and the numerical oscillation processing method, a wet case was simulated (simulation range of 2000 m, initial water depth in the range of 0–1000 m is 10 m, and water depth in the range of 1000–2000 m is 5 m. The simulation time was 50 s, and the calculation results are generated every 10 s). The results are summarized in Table 2 and are shown in Figure 4.

According to the numbers displayed in Table 2, using the SPH-SWE model adopting the B-spline kernel function and the two-shocks Riemann solver method to solve the wet case, the simulation results are better, and their mean absolute errors to simulate velocity and water depth are within the 6%.

Based on these results, it was decided to select the B-spline kernel function and the two-shocks Riemann solver method to perform the two-dimensional dam-break numerical simulation to verify the computational efficiency of the new SPH-SWE model solution framework proposed.

Table 2. Error analysis for the wet case.

	Parameters	30 s	50 s
Mean Absolute Error	Speed	0.0529	0.0603
	Water Depth	0.0567	0.0613
Mean Relative Error	Speed	0.0548	0.1037
	Water Depth	0.0075	0.0081
Standard deviation of error	Speed	0.1543	0.1514
	Water Depth	0.1257	0.1238

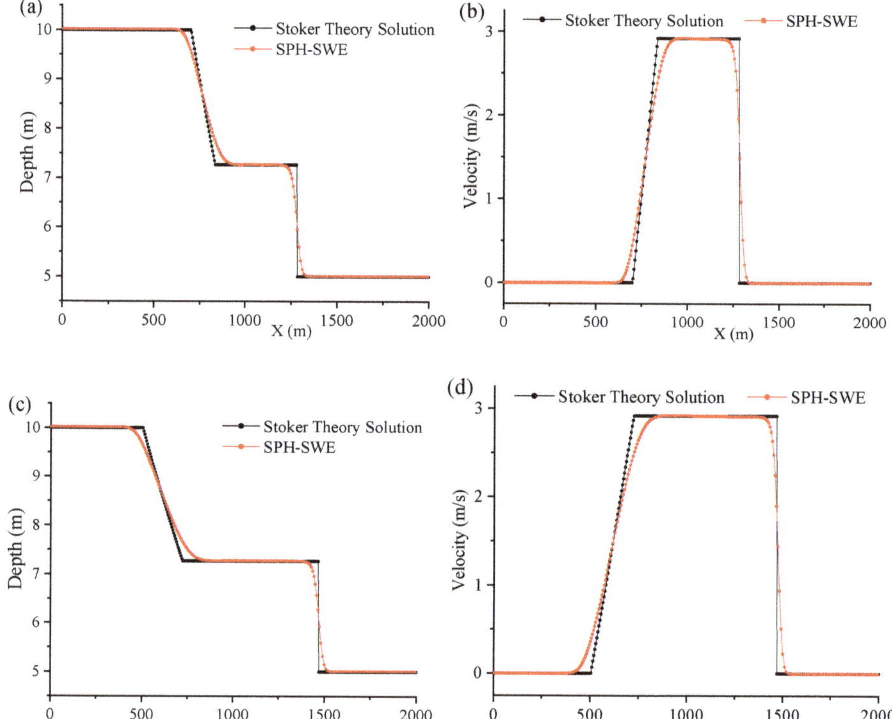

Figure 4. Simulation results of the SPH-SWE module. (**a,b**) water depth and velocity diagram at t = 30 s; (**c,d**) water depth and velocity diagram at t = 50 s.

3. SPH-SWE Model Solution Framework

When simulating dam break cases with a large number of particles involved, there is a challenge to be faced associated with low calculation efficiency. The code runs in serial steps and before each variable calculation, the mesh is divided into small grids (mesh size is *2H*, smooth length, in order to calculate the corresponding parameters of each particle), making the process more repetitive and requiring a lot of calculation time. Moreover, the code framework is complex and it is demanding to complete any modification. As the open source code solves the model with a large number of particles, it has the problem of low calculation efficiency and cannot even be calculated (the reason is that the array overflows). This paper proposes a new SPH-SWE model solution framework, which can dynamically allocate the storage space of particle information, solve the problems of repeated particle search and unsuccessful memory allocation of the array of stored particle information, and can quickly solve the large-scale SPH-SWE model. Furthermore, the model framework proposed is simple, and any modification can be made easily to this algorithm.

For this study, Algorithm 1 was developed to solve the problem of data analysis and realize the CPU-OpenMP parallel computing.

Algorithm 1. Calculation framework of the SPH-SWEs model. This algorithm is needed to read the particles data (include fluid particles/virtual particles/open boundary particle/riverbed particles).

Read parameters

Output initial data of the model

Mesh riverbed particles and calculate fluid particles and the net water depth {**Loop**   1}

Search particles {**Loop**   2}

do $t = 0 \rightarrow total\_number\_of\_timesteps$

    Step 1: *Calculate the water depth of fluid particles* {**Loop**   3}.

$$\rho_i = \sum_j m_j W_i(x_i - x_j, h_i) \qquad h_i = h_0\left(\frac{\rho_0}{\rho_i}\right)^{1/d_m}$$

    Step 2: *Calculate time water depth of fluid particles and the speed gradient* {**Loop**   4}.

$$\nabla n_i = \sum_j V_j(n_i - n_j, h_i)\nabla W_i(x_i - x_j, h_i) \qquad (n = d/u/v)$$

    Step 3: *Calculate time increments.* $\Delta t = CFL\min_{i=1}^{N}\left(\frac{h_i}{ci+\|v_i\|}\right)$

    Step 4: *Calculate accelerations of fluid particle,corrections of riverbed gradients,speeds, and displacements*

    {**Loop**   5}. $\vec{a}_i = \dfrac{g+\vec{v}_i\cdot\vec{k}_i\vec{v}_i+\vec{t}_i\cdot\nabla b_i}{1+\nabla b_i\cdot\nabla b_i}\nabla b_i - \vec{t}_i + \vec{S}_{f,i}$

    Step 5: *Calculate displacements of open boundary particles.*

    Step 6: *Fluid particle division.* $v_k = c_v\dfrac{d_N}{dk}v_N \qquad c_v = \dfrac{A_N}{\sum_{k=1}^{M} A_k}$

    Step 7: *Calculate fluid particle of the riverbed* {**Loop**   1}.

$$\begin{cases} d_i = \sum_j d_j W_i(x_i - x_j, h_i)V_j \qquad S_i = \sum_j W_i(x_i - x_j, h_i)V_j \\ shepard \quad correction : d_i = \frac{d_i}{S_i} \end{cases}$$

    Step 8: *Search partilces* {**Loop**   2}.

    end do

The calculation of each time step includes five parts, which can be summarized as follows:

1. Calculation of fluid particles in the riverbed;
2. Particle search, and calculation of the water depth;
3. Calculation of the fluid particle depth and velocity gradient;
4. Acceleration and riverbed gradient correction;
5. Calculation of velocity and displacement rate.

In this paper, multiple two-dimensional arrays were used to store the information of fluid particles, riverbed particles, virtual particles, and open boundary particles. When calculating the parameters related to the above five parts, each particle can be calculated separately, and there is no dependency between particles.

Therefore, the CPU-OpenMP parallel computing can be realized (since each cycle is for all fluid particles, the calculation amount is known, so the schedule in the cycle configuration is set to Static mode) and the SPH-SWE model with a large number of particles can also be calculated.

3.1. Fluid Particle Riverbed Calculation

When calculating the riverbed particles in each time step, considering that the bed range and the smooth length of the riverbed particles are the same and there is no relationship between them, the calculation of the riverbed fluid particles was carried as follows.

Firstly, the grid division was completed, then the calculation of the riverbed fluid particles was conducted. It was verified that the calculation efficiency has no advantage, and additional array was needed to store riverbed particle information and increase data reading time. Please see below Algorithm 2 for the calculation framework adopted to achieve this task.

Algorithm 2. Computing fluid particle riverbed.

---

1. *Stage 1:* $h\_t/sum\_h\_t = 0$, *initialize to 0*
2. *!\$OMP PARALLEL DO PRIVATE(private variable),&*
3. *!\$OMP& SHARED(shared variable), DEFAULT(none), SCHEDULE(static)*
4. *do $i = 1 \rightarrow total\_number\_of\_fluid$ particles*
5.   *if particle_i is valid then*
6.   *Calculate particles' mesh locations based on the riverbed's mesh*
7.   *!sum_h_t(i) is used to make shepard correction(CSPM)*
8.   *CALL PURE celij_hb $(i, h\_t(i), sum\_h\_t(i))$*
9.   *endif*
10. *enddo*
11. *!\$OMP END PARALEL DO*

Because the calculated variables were two arrays, *h_t and sum_h_t*, it was not difficult to realize OpenMP parallel, and multiple threads were set to calculate the riverbed fluid particles at the same time, following these formulae:

$$h\_t(i) = \sum_j hb(j) \cdot W_i(x_i - x_j, hb(j)) \cdot Vol\_b(j)$$
$$sum\_h\_t(i) = \sum_j W_i(x_i - x_j, hb(j)) \cdot Vol\_b(j) \tag{11}$$

After the calculation, the CSPM [79] was corrected, as shown in the next formula:

$$h\_t(i) = \frac{h\_t(i)}{sum\_h\_t(i)} \tag{12}$$

3.2. Particle Search

In this paper, the particle search technique [80,81] was used as a separate module to prepare for the calculation of parameters such as water depth, acceleration, and velocity.

In the particle search, the mesh was firstly divided, and then the mesh area of each particle search was calculated; in order to ensure the symmetry of particle interaction, less than $2h_i$ or $2h_j$ was used as the judgment condition of i effective particles.

The specific steps of the particle search technique adopted [82] (see Figure 5) are described as follows:

1. Before each time step, the temporary grid position was updated, and each grid was assigned to a unique number; the grid size can be set to a fixed size *dx_grid/dy_grid*;
2. According to the position of the current SPH particles, all the SPH particles were allocated to the temporary mesh space, and the particle chain in the mesh was established;

3. According to the range ($2h_i$) of the tight support region of particle i, the search of other meshes (- *xsize* to *xsize*, - *ysize* to *ysize*) was completed in the tight support region of the mesh, storing the mesh number;
4. All the SPH particles i and j in the mesh were searched (*icell-xsize* to *icell + size*, *jcell-ysize* to *jcell + ysize*) in the tight support domain.

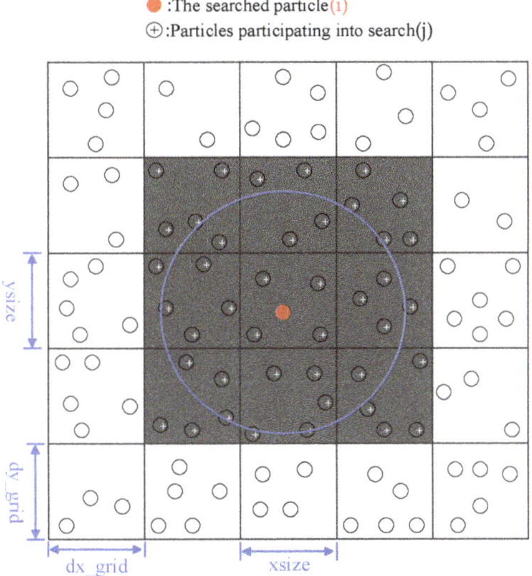

Figure 5. Framework of particle search [81].

To avoid high time consumption caused by repeated particle search in the meshless SPH-SWE model, Algorithm 3 was produced.

In the open source code, the information of particles was stored in a three-dimensional array, and the grid was divided by a maximum smooth length of *2hmax* (by adopting this, all particles can be stored into a cell, causing failure of memory allocation; however, the particle search method mentioned above solved this issue).

3.3. Water Depth Calculation

According to the Newton-iteration method, the water depth of particles was calculated, and the maximum number of iterations and the iteration-errors were taken as a criteria to terminate the iterations. In order to reduce the calculation time while ensuring the accuracy of the results, the maximum number of iterations of each particle was generally set to 50, and the iteration error was setup to 10^{-3}. Nevertheless, different approaches can be selected according to different calculation models.

Before each calculation step, the water depth and smooth length of each particle were guessed. At the same time, the smooth length h and the correction coefficient α_i^k were re-calculated and updated. In the same time step, the updated smooth length was then used for the sub-sequent SPH interpolation. The calculation framework is displayed in Algorithm 4.

Algorithm 3: The particle search. Read in the particles data (include fluid particles/virtual parti'cles/open boundary particle/riverbed particles).

---

1. *In each timestep*
2. *Mesh all particles based on fixed size **dx_grid/dy_grid**(generally select the maximum smooth length) and particles into **nc** array*
3. ***ncx/ncy:** total number of grids in x/y direction*
4. ***iboxvv/iboxff/iboxob:** store the virtual particles, fluid particles and open boundary particles within the affected region into two dimensional arrays*
5. ***!$OMP PARALLEL DO PRIVATE** (private variable),**SHARED**(shared variable),&*
6. ***!$OMP& SHARED**(shared variable),**DEFAULT**(none)*
7. ***do** i = 1 → total_number_of_fluid particles*
8.   *if particle_i is valid **then***
9.   *Calculate mesh of the particle i:**icell/jcell***
10.   *Calculate search mesh range of particle i:**xsize/yszie***
11.   ***do** row ∈ **-ysize,ysize***
12.   *irow=jcell+1*
13.   ***do** column ∈ **-xsize,xsize***
14.   *icolumn=icell+column*
15.   *Calculate number of search grid: **gridn***
16.   ***gridn=icolumn+(irow-1)*ncx***
17.   *!Search for **Virtual particles** in the scope of I particle*
18.   ***do** j ∈ nc(grindn,1)*
19.   *if particle_i and particle_j are neighbours **then***
20.   *Write particle_j to **iboxvv** array*
21.   *endif*
22.   *enddo*
23.   *!Search for **Fluid particles** in the scope of i particle*
24.   ***do** j ∈ nc(grindn,2)*
25.   *if particle_i and particle_j are neighbours **then***
26.   *Write particle_j to **iboxff** array*
27.   *endif*
28.   *enddo*
29.   *!Search for **Open boundary particles** in the scope of **i** particle*
30.   ***do** j ∈ nc(grindn,3)*
31.   *if particle_i and particle_j are neighbours **then***
32.   *Write particle_j to **iboxob** array*
33.   *endif*
34.   *enddo*
35.   *enddo*
36.   *enddo*
37.   *endif*
38. *enddo*
39. ***!$OMP END PARALLEL DO***

After each water depth calculation, each particle was judged on whether the error requirements were met, and re-calculation was then completed for the particles that did not meet them. After this iteration cycle, the water depth, speed of the water, and volume were constantly updated.

Algorithm 4: Water depth calculation.

---

1. *Stage 1: Guess for density and smoothed length*
2. *!$OMP PARALLEL DO PRIVATE(private variable),&*
3. *!$OMP& SHARED(shared variable),DEFAULT(none),SCHDULE(static)*
4. *do i = 1 → total_number_of_fluid particles*
5.   *if particle_i is valid **then***
6.   ***1a:** rhop(i) = rhop(i) + dt · rhop(i) · ar(i)*
7.   ***1b:** h_var(i) = h_var(i) − (dt/dm)·h_var(i)·ar(i)*
8.   ***endif***
9. ***enddo***
10. *!$OMP END PARALLEL DO*
11. ***CALL** particle search() %Search particles*
12. ***Stage 2: Calculate depth***
13. ***do while** ((maxval(resmax) **.gt.** Minimum error) **.and.** (Iterationtimes **.lt.** max)*
14.   *!$OMP PARALLEL DO PRIVATE(private variable),&*
15.   *!$OMP& SHARED(shared variable),DEFAULT(none),SCHEDULE(static)*
16.   ***do** i = 1 → total_number_of_fluid particles*
17.   *if particle_i is valid **then***
18.   ***CALL PURE** fluid particle(i,rhop_sum(i),alphap(i))*
19.   ***CALL PURE** virtual particle(i,rhop_sum(i),alphap(i))*
20.   ***CALL PURE** open boundary particle((i,rhop_sum(i),alphap(i))*
21.   *%Calculate next step's water depth and the smooth length*
22.
$$\varphi_i^k = rhop(i)\left[\rho_i^k\right] - rhop\_sum(i)\left[\sum_j m_j W_i(x_i - x_j, h_i)\right]$$
23.
$$\alpha_i^k = alphap(i)\left[-\frac{1}{\rho_i d_m}\sum_j m_j r_{ij}\frac{dW_i}{dr_{ij}}\right]$$
24.
$$\rho_i^{k+1} = rhop(i)\left[1 - \frac{\varphi_i^k}{\varphi_i^k + \rho_i^k \alpha_i^k}\right]$$
25.
$$h_i = h_o\left(\frac{\rho_0}{\rho_i^{k+1}}\right)^{1/d_m}$$
26.   ***endif***
27.   ***enddo***
28. *!$OMP END PARALLEL DO*
29. ***enddo***

3.4. Velocity Calculations

In order to calculate the acceleration ($\vec{a}$) caused by internal force, the gradient of velocity and water depth had to be calculated, and the kernel function was adopted to complete this task as shown below:

$$\nabla p_i = \sum_j V_j\left(p_i - p_j\right) \times \widetilde{\nabla}W_i\left(x_i - x_j, h_i\right) \quad (p = d/u/v) \tag{13}$$

where p_i is the depth/velocity of particle i. The calculation conducted for this step is displayed in Algorithm 5.

If a variable time-step was implemented, the next step was to calculate the time-step according to Equation (9). The calculation framework of the time step included a loop and no sub-routine. It was found that the speed-up of time step parallel computing was less than 2 to perform serially variable time step calculations.

Algorithm 5: Calculation of fluid particle velocity and water depth gradient.

---

1. *Stage 1: sum_f/alphap/grad_up/grad_vp/grad_dw=0, Initialize to 0*
2. *!$OMP PARALLEL DO PRIVATE(private variable),&*
3. *!$OMP& SHARED(shared variable),DEFAULT(none),SCHDULE(static)*
4. *do i* = 1 → *total_number_of_fluid particles*
5.   *if particle_i is valid* **then**
6.   *!First conduct matrix for gradient correction*
7.   *CALL PURE celij_corr(i,sum_f (I,1:4))*
8.   *CALL PURE celij_alpha(i,alphap(i),grad_dw(i,1:2),grad_up(i,1:2), grad_vp(i,1:2))*
9.   *CALL PURE celij_alpha_vir(i,alphap(i))*
10.   *CALL PURE celij_alphap_ob(i,alphap(i),grad_dw(i,1:2),grad_up(i,1:2), grad_vp(i,1:2))*
11.   *endif*
12.   *enddo*
13. *!$OMP END PARALLEL DO*

3.5. Calculation of Fluid Particle Acceleration, Riverbed Scouring, Speed, and Displacement

The acceleration $(\vec{t})$ caused by the internal force was firstly calculated, followed by the riverbed gradient and the total acceleration $(\vec{a})$. After these calculations, the velocity and the displacement of fluid particles needed to be regularly updated as shown in the calculation framework Algorithm 6.

Algorithm 6: Calculation of acceleration, velocity, and position. The Lagrangian equation of motion for a particle i is d/dt ∂L/(∂v_i)-∂L/(∂x_i)=0, where the Lagrangian functional L is defined in term of kinetic energy K and potential energy π as L = K-π, where π is a function of particles position but not velocity.

---

1. *Stage 1: Calculate $\vec{t}_i(ax(i)/ay(i))$*
2. *!$OMP PARALLEL DO PRIVATE(private variable),&*
3. *!$OMP& SHARED(shared variable),DEFAULT(none),SCHDULE(static)*
4. *do i* = 1 → *total_number_of fluid particles*
5.   *if particle_i is valid* **then**
6.   *1a. use **Riemann solution** to calculate $\vec{t}_i$*
7.   *1b. use **Numerical viscosity** to calculate $\vec{t}_i$*
8.   *! ar(i) is used to calculate depth*
9.   *CALL PURE fluid particle(i,ar(i),ay(i),ar(i))*
10.   *CALL PURE virtual particle(i,ar(i),ay(i),ar(i))*
11.   *CALL PURE open boundary particle(i,ar(i),ay(i),ar(i))*
12.   *endif*
13.   *Stage 2: Calculate* $\nabla b_i = \sum_j b_j \widetilde{\nabla} W_i\left(x_i - x_j, h_i\right) V_j$
14.   *Stage 3: Calculate* $\vec{a}_i = \dfrac{g + \vec{v}_i \cdot \vec{k}_i \vec{v}_i + \vec{t}_i \cdot \nabla b_i}{1 + \nabla b_i \cdot \nabla b_i} \nabla b_i - \vec{t}_i + \vec{S}_{f,i}$
15.   *Stage 4: Calculate velocity and position of fluid particle i*
16. *enddo*
17. *!$OMP END PARALLEL DO*

If the open boundary was adopted, the displacement was updated, and it was checked whether it became a fluid particle or a buffer zone; if the case under investigation had a particle splitting zone, the fluid particles that meet the conditions identified were split [37]. Achieved this landmark, the calculation process was then completely repeated according to Algorithm 1.

4. Applications

4.1. Validation 1: 2-D Dry Bed Dam Break with Particle Splitting

In order to test the performance of the new computing SPH-SWE framework according to the CPU-OpenMP parallel computing, the open source case *"2-D dry bed dam break with particle splitting"*, referred to as DBDB case [67,68], was considered. Initial conditions of this open source DBDB case were setup as follows: area of 2.6 m × 1.2 m, initial fluid particle layout of 0.8 m × 1.2 m, and initial water depth equal to 0.15 m, as shown in Figure 6.

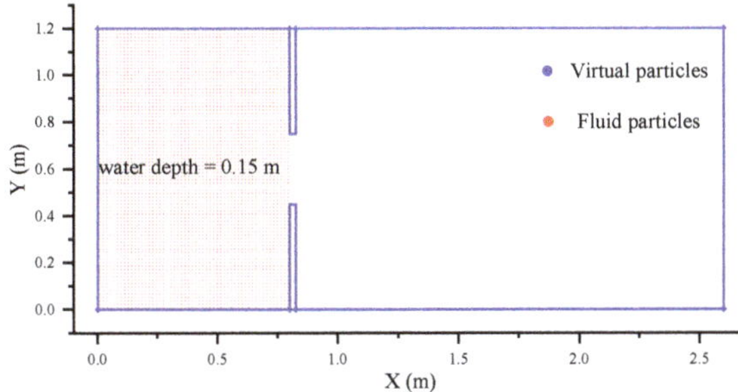

Figure 6. 2-D dry bed dam break with particle splitting (DBDB) case setup.

The spacing of the fluid particles was 0.015 m × 0.015 m, 0.01 m × 0.01 m, and 0.005 m × 0.005 m, respectively. Table 3 shows the number of fluid particles, virtual particles, and riverbed particles selected for the 3 cases tested (considering that no inflow and outflow conditions were set, the number of open boundary particles was maintained equal to 0).

Table 3. Particle numbers for each case tested (unit: pcs).

Case	Number of Fluid Particles	Number of Virtual Particles	Number of Riverbed Particles
Case 1	4374	1276	14,094
Case 2	9801	3300	31,581
Case 3	38,801	9424	125,561

The particles arrangements in the three cases were calculated using the open source code [67,68] and the CPU-OpenMP parallel code (in terms of 1000 particles per thread and 2000 particles per thread) to ensure that the calculation results were consistent and comparison of each model's performance could be completed.

For t = 0, the instantaneous burst occurs and the fluid flows downstream. Figure 7 shows flow velocities for T = 1.2 s under the three fluid particle arrangements displayed in Table 3. It can be seen from Figure 7 that the more fluid particles there were, the more water flow characteristics and velocity distribution characteristics after dam break were seen. Figure 8 shows the comparison between numerical and experimental results for the three cases tested in Table 3.

It can be concluded that with the increase of fluid particles (Case 3), the error between the numerical and the experimental results was smaller.

Hence, bigger is the number of fluid particles computed, and more valuable are the water depth and velocity calculations for each time step and location across the dam system, therefore providing more support for the formulation of dam break mitigation plans. This also reflects the superiority of SPH-SWE model in dealing with large deformation and free surface problems.

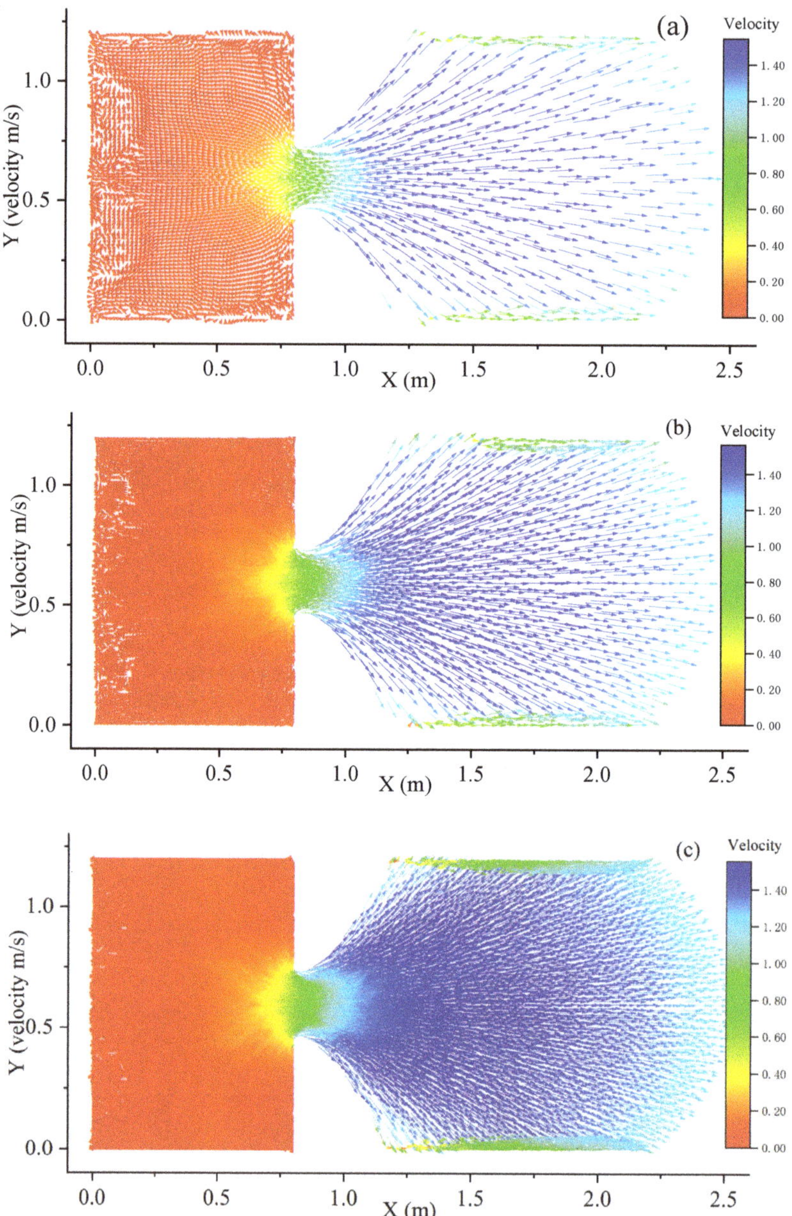

Figure 7. Results of 2-D Dry-Bed Dam Break with particle splitting at 1.2 s obtained from the simulations with: (**a**) 4374 particles; (**b**) 9801 particles; (**c**) 38,801 particles.

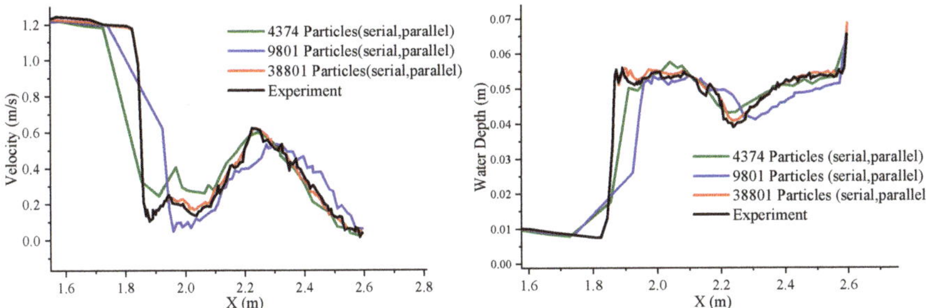

Figure 8. Numerical vs. experimental results comparison (velocities on the **left** and water depths on the **right**).

The time and speedup of total time $t(s)$ (Equation (14)) required for calculating water depth, acceleration, speed, and displacement were quantified according to open source code and CPU-OpenMP parallel code (under different thread configurations) as shown in Table 4.

$$t(s) = \frac{t(k)}{t(c)} \tag{14}$$

where $t(k)$ represents the running time of open source code and $t(c)$ represents the running time of the parallel code. In Table 4, $R(t)$ represents the particle search time; $C(t)$ represents the time to calculate the water depth; $A(t)$ represents the time to calculate the acceleration, speed, and displacement; $T(t)$ represents the total running time of the case (including $t(k)$ and $t(c)$); the time unit is always seconds (s). It can be seen from the results that the number of particles computed was higher, and the parallel calculation was larger, based on CPU-OpenMP (Figure 9).

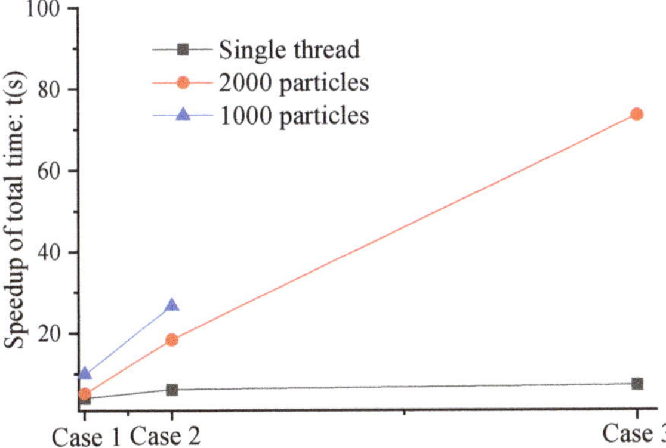

Figure 9. Speedup of total runtime for validation 1.

CPU-OpenMP allocated a thread according to 1000 fluid particles in parallel and calculated case 2 and case 3, in 26.7 s and 91.36 s, respectively.

In parallel computing, $S(p)$ (the speedup ratio) and $E(p)$ (parallel efficiency) were important indexes to evaluate the parallel effect. $S(p)$ was the ratio between the serial time and the multi-core parallel time when threads calculate (p) and solved the iteration at the same time, as follows:

$$S(p) = \frac{T_s}{T_p} \tag{15}$$

where, T_s is the time spent by a single processor in the serial mode; T_p is the time spent by threads (p) in the parallel mode. $E(p)$ (parallel efficiency) is the ratio of the acceleration ratio to the number of CPU cores used in the calculation (and $E(p) \leq 1$), indicating the average execution efficiency of each processor. When the acceleration ratio was close to the number of cores, the parallel efficiency was higher, and the utilization rate of each thread was higher, as calculated below.

$$E(p) = \frac{S(p)}{p} \tag{16}$$

Table 4. Run time in different configurations for 2-D Dry-Bed Dam Break with particle splitting.

Cases		$R(t)$	$C(t)$	$A(t)$	$T(t)$	$t(s)$
Open Source Code (Case 1)		N/A	1040.44	213.12	1253.56	1.0
Parallel Operation Code	Single Core	87.47	174.95	49.99	312.41	4.01
	2000	60.27	118.38	36.59	215.24	5.82
	1000	36.43	70.34	18.84	125.61	9.98
Open Source Code (Case 2)		N/A	5892.29	1039.82	6932.11	1.0
Parallel Operation Code	Single Core	338.853	643.8207	146.8363	1129.51	6.14
	2000	116.8514	218.6252	41.4634	376.94	18.39
	1000	75.2985	150.597	33.7545	259.65	26.70
Open Source Code (Case 3)		N/A	107,218.04	16,021.09	123,239.13	1.0
Parallel Operation Code	Single Core	5498.28	10,481.09	1202.75	17,182.12	7.17
	2000	554.20	990.83	134.35	1679.38	73.38

In order to check the parallel effect of CPU-OpenMP, case 3 was calculated with different threads. The acceleration ratio and parallel efficiency of different threads are shown in Table 5, and the performances of parallel algorithms are displayed in Figure 10.

Table 5. Case 3 speed-up and parallel efficiency under different threads.

Number of Single-Thread Particles (pcs)	p (pcs)	T_p (s)	$S(p)$ (s)	$E(p)$ (%)
2000	20	1679.38	10.23	51.16
2500	16	1759.15	9.77	61.05
3000	13	1833.58	9.37	72.08
4000	10	1935.12	8.88	88.79
5000	8	2300.79	7.47	93.35
6000	7	2579.88	6.66	95.14
7000	6	2934.02	5.86	97.60
8000	5	3455.94	4.97	99.44
10,000	4	4312.39	3.98	99.61
20,000	2	8616.82	1.99	99.70

According to Table 5 and Figure 10, as the number of enabled threads increased, the speedup ratio, parallel efficiency and calculation time were affected by the following trends: (1) the calculation time decreases with the increase of threads, but when the number of online processes exceeded 10, the time-consuming reduction speed changed from fast to slow reaching towards a balance; (2) the acceleration ratio increased all the time, but the improvements varied, and after 10 threads, the increase rate was from fast to slow; (3) the parallel efficiency decreased with the increase of threads, but the decrease rate fluctuated. When calculating the number of 10 threads, the parallel efficiency started to be less than 90%; therefore, case 3 could allocate one thread according to 5000 particles, with the acceleration ratio of 7.47 and the parallel efficiency of 93.35%.

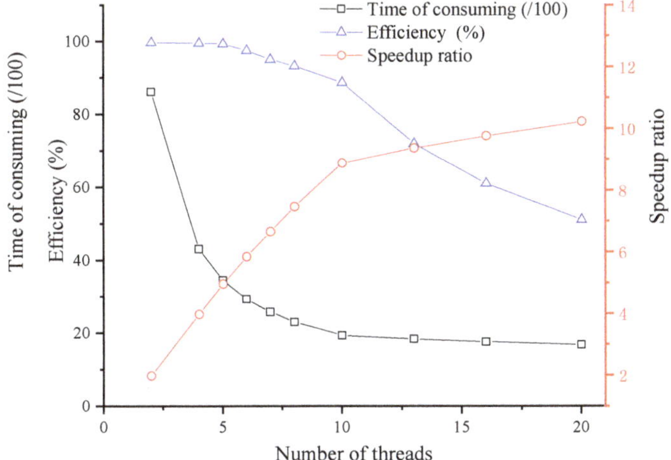

Figure 10. CPU-OpenMP model parallel computing performances.

4.2. Validation 2: 2-D Dam Break with A Rectangular Obstacle Located in the Downstream Area

In order to solidify the accuracy and advantages of the new SPH-SWE model proposed calculated by CPU-OpenMP, a second case has been considered for validation. This case involved the dam break flow with a rectangular obstacle located in the downstream area as shown in Figure 11 [2–83].

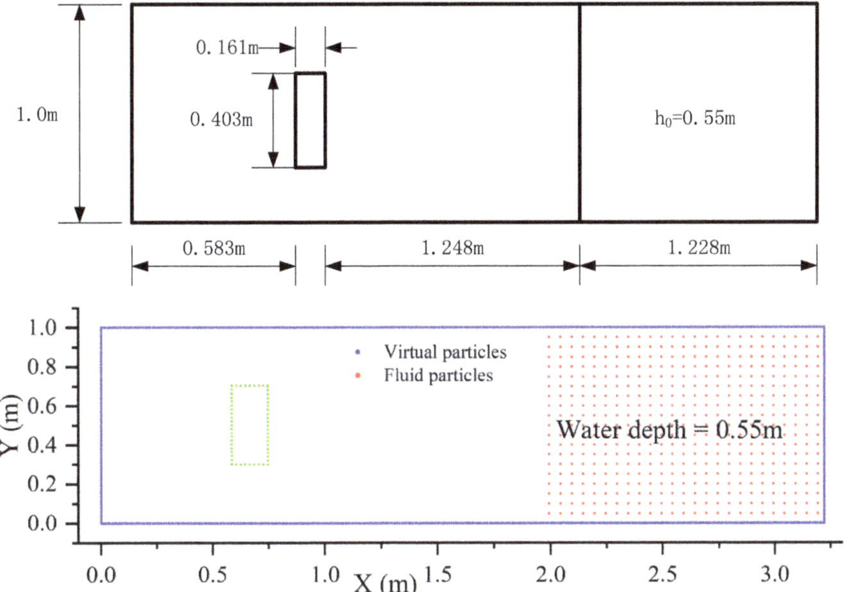

Figure 11. Scheme of the second model used for validation of the new SPH-SWE model proposed [2–83].

For this second validation scheme, the fluid particles were arranged according to the particle spacing of 0.01, 0.005, and 0.002 m. Figures 12 and 13 display the results for t = 0.74 s and t = 1.76 s for the same particle spacing used by Gu et al., [2] (0.01 m) and increasing number of fluid particles involved.

In Table 6, it is possible to check the number of particles involved in each simulation, and it can be noticed that the speed up of total time (*t*) slightly increased with the rise of particles (using 8 threads in all three processes) (Figure 14).

Table 6. Particle numbers for each case tested (unit: pcs). Remarks: In all three cases, 8 threads were used for parallel calculation.

Case	Particle Spacing	Number of Fluid Particles	Number of Virtual Particles	Number of Riverbed Particles	T_8 (s)	T (s)
Case 4	0.01	12,423	4798	129,645	1511.38	7.12
Case 5	0.005	51,858	9582	516,889	14,538.83	7.32
Case 6	0.002	323,145	23,934	807,111	108,868.42	7.46

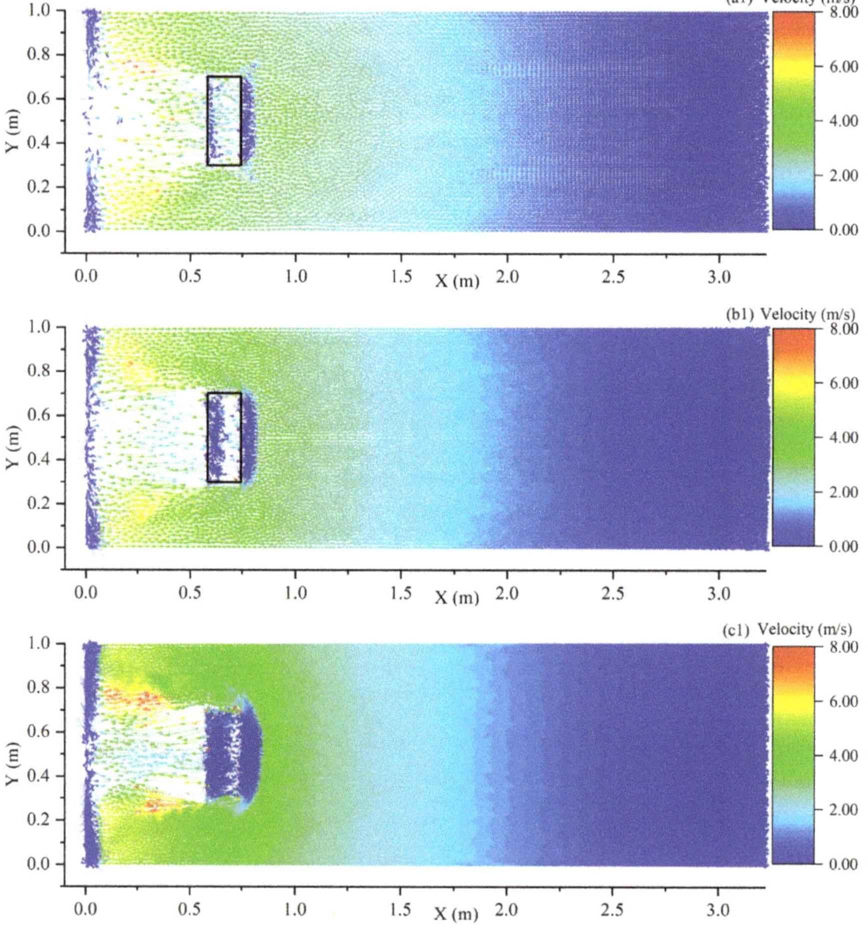

Figure 12. Velocity distribution at 0.74 s for Case 4 (**a1**), 5 (**b1**) and 6 (**c1**) displayed in Table 6.

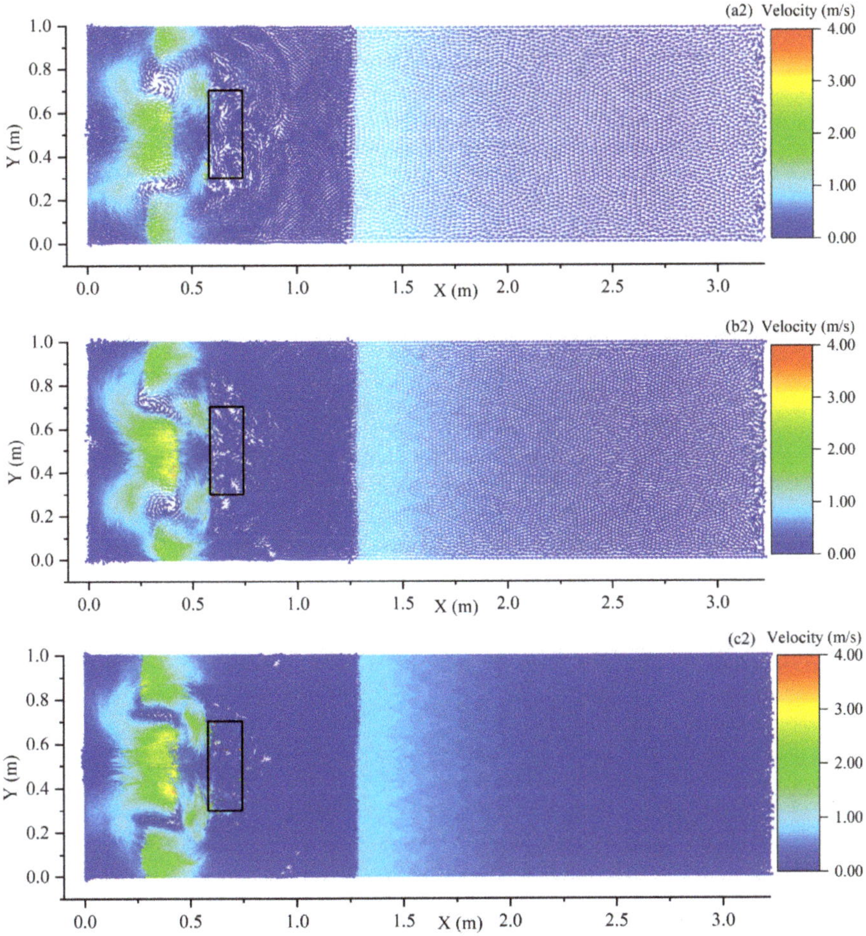

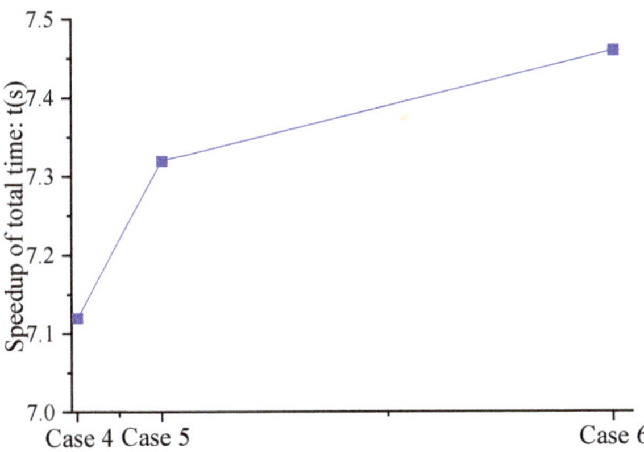

Figure 13. Velocity distribution at 1.76 s for Case 4 (**a2**), 5 (**b2**) and 6 (**c2**) displayed in Table 6.

Figure 14. Speedup of total runtime for validation 2.

Table 7 unveils values of $R(t)$, which represents the particle search time; $C(t)$, which represents the time to calculate the water depth; and $A(t)$, which represents the time to calculate the acceleration, speed, and displacement. It can be seen from the results (Figure 14) that when the number of particles computed was higher, the parallel calculation was slightly larger, based on CPU-OpenMP.

Table 7. Run time in different configurations for 2-D Dam Break with a rectangular obstacle downstream.

Cases	R (t)	C (t)	A (t)	T$_8$ (s)	t (s)
Case 4	468.53	876.60	166.25	1511.38	7.12
Case 5	4216.26	8432.52	1890.05	14,538.83	7.32
Case 6	34,837.91	66,409.72	7620.80	108,868.42	7.46

The simulation results are consistent with the previous case to demonstrate the improvement made by the proposed SPH-SWE calculation framework. When analyzing the same parameters as in the paper by Gu et al., [2], 12,423 fluid particles were involved. Results displayed in Figure 15 confirmed that the agreement between numerical results and experimental datasets improved even when simulating an increase of the number of fluid particles. However, the improvements do not involve the entire domain because in some positions (for example (a) H4 gauge, x = 1.3 m–1.8 m; (b) H2 gauge, x = 3.50 m–4.5 m) there are still minor inaccuracies (caused by the truncation error of the kernel function and the processing of boundary particles in the SPH method), even with the model with 323,145 fluid particles; hence, future work will focus on improving the algorithm to progress the calculation accuracy.

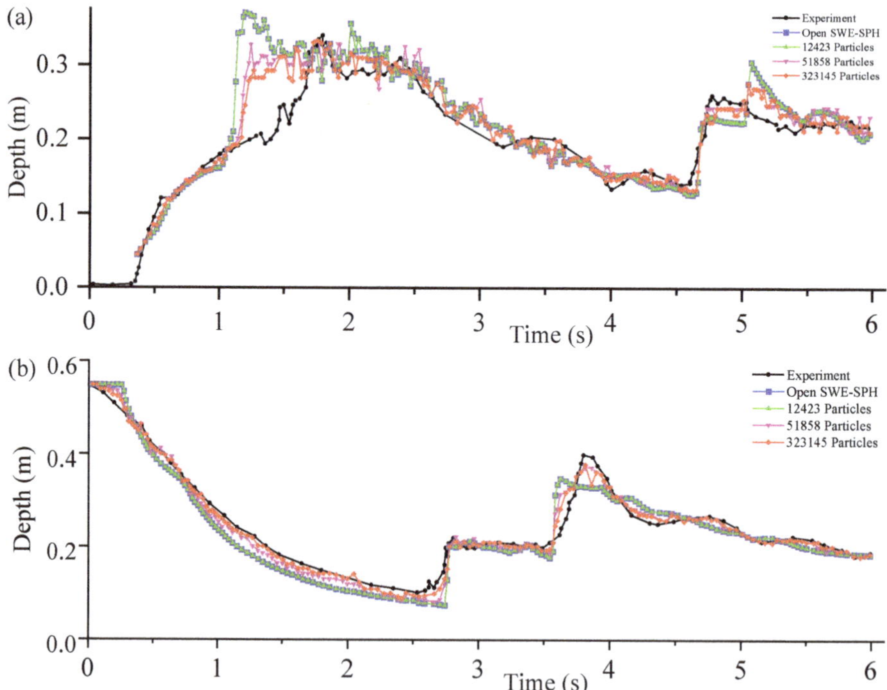

Figure 15. Comparison between experimental and numerical results water depth datasets: (a) gauge H4 in [2] and [80]; (b) gauge H2 in [2] and [80].

5. Conclusions

Vacondio et al. [34–37] made an open source version called SWE-SPHysics, which has been optimized and adapted based on the hydrodynamics investigated by other researchers during the last decade [38–47]. However, despite continuous progress, there was still a limitation related to the computational efficiency when the number of particles to simulate is very large, and this aspect still needs to be improved.

To fill this gap, in this study, a new solution to the SPH-SWE model introduced by Vacondio et al. [34–37] was proposed, and it was validated against two open source case studies of a *2-D dry-bed dam break with particle splitting* [67,68] and of a *2-D dam break with a rectangular obstacle downstream* [2,83]. To test the computing performance against the first case study, when involving large numbers of particles, three cases, involving different particles numbers, were tested (case 1—4374 particles; case 2—9801 particles; case 3—38,801 particles). Furthermore, this paper adopted the parallel computing method of CPU-OpenMP that is applied to a single machine and a multi-core to calculate the new SPH-SWEs framework.

By applying this CPU-OpenMP method, results have confirmed that the computing speeds of case 1/case 2/case 3 were increased by 4.01 times/6.14 times/7.17 times, respectively, to compute the new solution framework of the SPH-SWE model proposed to the open source case study previously mentioned [67,68]. According to the new solution framework of SPH-SWE model, case 3, characterized by the highest number of particles, was also calculated by using different threads. It was found that the speedup ratio can reach 7.47 when the parallel efficiency was more than 90%, which fully proves the good calculation performance of the CPU- OpenMP parallel new SPH-SWE model. Additionally, in the new solution framework of the SPH-SWE model proposed, particle search was used as a separate module for parallel computing, which greatly improved the computing efficiency and could replace the meshless SPH-SWE model in the open source code [67,68].

Therefore, using CPU-OpenMP parallel computing demonstrated that the SPH-SWE model new framework can accurately and timely simulate the flood evolution after a dam break.

In future works, the SPH-SWE model can be put into existing clusters to achieve more threads and further improve the calculation efficiency. Furthermore, this would enable the possibility of introducing more effective new algorithms into the SPH-SWE model (i.e., debris flow or water pollution modules) in order to expand its application. Continuous development of technology aids the improvement of new tools to design and inspect more accurate solutions, and this is an area in continuous development that needs to be addressed to support local and national authorities in making decisions to mitigate drastic effects generated by dam break.

Author Contributions: All authors contributed to the work. Conceptualization, S.G. and L.T.; methodology Y.W., M.R. and T.Y.; validation, Q.Z.; formal analysis, S.G., Y.W., M.R. and Z.X.; investigation, L.T., S.G., and M.R.; resources, S.G.; data curation, L.T. and M.R.; writing—review and editing, P.C., X.W., and M.R.; visualization, Q.Z.; supervision, S.G., M.R.; project administration, S.G.; funding acquisition, S.G. All authors have read and agreed to the published version of the manuscript.

Funding: This work was supported by the following projects: National Key R&D Program of China (No. 2017YFC0404303), National Natural Science Foundation of China (No.51869025, 51769028, 51868066), Qinghai Science and Technology Projects (No. 2018-ZJ-710), Youth Fund of Qinghai University (Grant No. 2017-QGY-7), National Key Laboratory Project for Water Sand Science and Water and Hydropower Engineering, Tsinghua University (Grant No. sklhse-2018-B-03), Beijing Institute of Structure and Environment Engineering Fund(Grant No. BQ2019001).

Conflicts of Interest: The authors declare no conflict of interest.

References

1. Chang, Y.S.; Chang, T.J. SPH simulations of solute transport in flows with steep velocity and concentration gradients. *Water* **2017**, *9*, 132. [CrossRef]
2. Gu, S.; Zheng, X.; Ren, L.; Xie, H.; Huang, Y.; Wei, J.; Shao, S. SWE-SPHysics simulation of dam break flows at South-Gate Gorges Reservoir. *Water* **2017**, *9*, 387. [CrossRef]

3. Chen, R.; Shao, S.; Liu, X.; Zhou, X. Applications of shallow water SPH model in mountainous rivers. *J. Appl. Fluid Mech.* **2015**, *8*, 863–870. [CrossRef]

4. Peng, X.; Yu, P.; Chen, G.; Xia, M.; Zhang, Y. Development of a Coupled DDA–SPH Method and its Application to Dynamic Simulation of Landslides Involving Solid–Fluid Interaction. *Rock Mech. Rock Eng.* **2020**, *53*, 113–131. [CrossRef]

5. Verbrugghe, T.; Dominguez, J.M.; Altomare, C.; Tafuni, A.; Vacondio, R.; Troch, P.; Kirtenhaus, A. Non-linear wave generation and absorption using open boundaries within DualSPHysics. *Comput. Phys. Commun.* **2019**, *240*, 46–59. [CrossRef]

6. Ni, X.; Feng, W.; Huang, S.; Zhao, X.; Li, X. Hybrid SW-NS SPH models using open boundary conditions for simulation of free-surface flows. *Ocean Eng.* **2020**, *196*, 106845. [CrossRef]

7. Gonzalez-Cao, J.; Altomare, C.; Crespo, A.J.C.; Dominguez, J.M.; Gomez-Gesteira, M.; Kisacik, D. On the accuracy of DualSPHysics to assess violent collisions with coastal structures. *Comput. Fluids* **2019**, *179*, 604–612. [CrossRef]

8. Atif, M.M.; Chi, S.W.; Grossi, E.; Shabana, A. Evaluation of breaking wave effects in liquid sloshing problems: ANCF/SPH comparative study. *Nonlinear Dyn.* **2019**, *97*, 45–62. [CrossRef]

9. Meringolo, D.D.; Marrone, S.; Colagrossi, A.; Liu, Y. A dynamic δ-SPH model: How to get rid of diffusive parameter tuning. *Comput. Fluids* **2019**, *179*, 334–355. [CrossRef]

10. Shu, A.; Wang, S.; Rubinato, M.; Wang, M.; Qin, J.; Zhu, F. Numerical Modeling of Debris Flows Induced by Dam-Break Using the Smoothed Particle Hydrodynamics (SPH) Method. *Appl. Sci.* **2020**, *10*, 2954. [CrossRef]

11. Wu, S.; Rubinato, M.; Gui, Q. SPH Simulation of interior and exterior flow field characteristics of porous media. *Water* **2020**, *12*, 918. [CrossRef]

12. Wang, S.; Shu, A.; Rubinato, M.; Wang, M.; Qin, J. Numerical Simulation of Non-Homogeneous Viscous Debris-Flows based on the Smoothed Particle Hydrodynamics (SPH) Method. *Water* **2019**, *11*, 2314. [CrossRef]

13. Gingold, R.A.; Monaghan, J.J. Smoothed particle hydrodynamics: Theory and application to non-spherical stars. *Mon. Not. R. Astron. Soc.* **1977**, *181*, 375–389. [CrossRef]

14. Hopkins, P. A general class of Lagrangian smoothed particle hydrodynamics methods and implications for fluid mixing problems. *Mon. Not. R. Astron. Soc.* **2013**, *428*, 2840–2856. [CrossRef]

15. Cremonesi, M.; Meduri, S.; Perego, U. Lagrangian-Eulerian enforcement of non-homogeneous boundary conditions in the Particle Finite Element Method. *Comput. Part. Mech.* **2020**, *7*, 41–56. [CrossRef]

16. Sugiyama, K.; Li, S.; Takeuchi, S.; Takagi, S.; Matsumoto, Y. A full Eulerian finite difference approach for solving fluid-structure coupling problems. *J. Comput. Phys.* **2011**, *230*, 596–627. [CrossRef]

17. Miller, G.H.; Colella, P. A conservative three-dimensional Eulerian method for coupled solid-fluid shock capturing. *J. Comput. Phys.* **2002**, *183*, 26–82. [CrossRef]

18. Liu, M.B.; Liu, G.R. Smoothed Particle Hydrodynamics (SPH): An Overview and Recent Developments. *Arch. Comput. Methods Eng.* **2010**, *17*, 25–76. [CrossRef]

19. Liu, G.R.; Liu, M.B. *Smoothed Particle Hydrodynamics: A Meshfree Particle Method*; World Scientific: Singapore, 2003.

20. Dalrymple, R.A.; Rogers, B.D. Numerical modeling of water waves with the SPH method. *Coast. Eng.* **2006**, *53*, 141–147. [CrossRef]

21. Huang, C.; Lei, J.M.; Peng, X.Y. A kernel gradient free (KGF) SPH method. *Int. J. Numer. Methods Fluids* **2015**, *78*. [CrossRef]

22. Monaghan, J.J.; Kocharyan, A. SPH simulation of multi-phase flow. *Comput. Phys. Commun.* **1995**, *87*, 225–235. [CrossRef]

23. Chen, A.S.; Djordjevic, S.; Leandro, J. An analysis of the combined consequences of pluvial and fluvial flooding. *Water Sci. Technol.* **2010**, *62*, 1491–1498. [CrossRef]

24. Liang, Q.; Borthwick, A.G.L.; Stelling, G. Simulation of dam and dyke break hydrodynamics on dynamically adaptive quadtree grids. *Int. J. Numer. Methods Fluids* **2004**, *46*. [CrossRef]

25. Chang, T.J.; Kao, H.M.; Chang, K.H.; Hsu, M.H. Numerical simulation of shallow water dam break flows in open channels using smoothed particle hydrodynamics. *J. Hydrol.* **2011**, *408*, 78–90. [CrossRef]

26. Kao, H.M.; Chang, T.J. Numerical modeling of dambreak-induced flood inundation using smoothed particle hydrodynamics. *J. Hydrol.* **2012**, *448–449*, 232–244. [CrossRef]

27. Colagrossi, A.; Landrini, M. Numerical simulation of interfacial flows by smoothed particle hydrodynamics. *J. Comput. Phys.* **2003**, *191*, 448–475. [CrossRef]

28. Yang, F.L.; Zhang, X.F.; Tan, G.M. One and two-dimensional coupled hydrodynamics model for dam break flow. *J. Hydrodyn.* **2007**, *19*, 769–775. [CrossRef]

29. Wang, Z.; Shen, H.T. Lagrangian simulation of one-dimensional dam-break flow. *Hydraul. Eng.* **1999**, *125*, 1217–1220. [CrossRef]

30. Ata, R.; Soulaimani, A. A stabilized SPH method for inviscid shallow water flows. *Int. J. Numer. Methods Fluids* **2005**, *47*, 139–159. [CrossRef]

31. Leffe, M.D.; Touzé, D.L.; Alessandrini, B. SPH Modeling of a shallow-water coastal flows. *Hydraul. Res.* **2010**, *48*, 118–125. [CrossRef]

32. Rodriguez-Paz, M.; Bonet, J. A corrected smooth particle hydrodynamics formulation of the shallow-water equations. *Comput. Struct.* **2005**, *83*, 1396–1410. [CrossRef]

33. Panizzo, A.; Longo, D.; Bellotti, G.; De Girolamo, P. Tsunamis early warning system. Part 3: SPH modeling of nlswe. In Proceedings of the XXX Convegno di Idraulica e Costruzioni Idrauliche, Rome, Italy, 10–15 September 2006.

34. Vacondio, R.; Rogers, B.D.; Stansby, P.K.; Mignosa, P. A correction for balancing discontinuous bed slopes in two-dimensional smoothed particle hydrodynamics shallow water modeling. *Int. J. Numer. Methods Fluids* **2013**, *71*, 850–872. [CrossRef]

35. Vacondio, R.; Rogers, B.D.; Stansby, P.K.; Mignosa, P. SPH Modeling of Shallow Flow with Open Boundaries for Practical Flood Simulation. *J. Hydraul. Eng.* **2012**, *138*, 530–541. [CrossRef]

36. Vacondio, R.; Rogers, B.D.; Stansby, P.K.; Mignosa, P. Smoothed Particle Hydrodynamics: Approximate zero-consistent 2-D boundary conditions and still shallow water tests. *Int. J. Numer. Methods Fluids* **2011**, *69*, 226–253. [CrossRef]

37. Vacondio, R.; Rogers, B.D.; Stansby, P.K. Accurate particle splitting for SPH in shallow water with shock capturing. *Int. J. Numer. Methods Fluids* **2012**, *69*, 1377–1410. [CrossRef]

38. Skillen, A.; Lind, S.J.; Stansby, P.K.; Rogers, B.D. Incompressible Smoothed Particle Hydrodynamics (SPH) with reduced temporal noise and generalised Fickian smoothing applied to body-water slam and efficient wave-body interaction. *Comput. Methods Appl. Mech. Eng.* **2013**, *265*, 163–173. [CrossRef]

39. Fourtakas, G.; Rogers, B.D.; Laurence, D.R.P. Modelling Sediment resuspension in Industrial tanks using SPH. *Houille Blanche* **2013**, *2*, 39–45. [CrossRef]

40. St-Germain, P.; Nistor, I.; Townsend, R.; Shibayama, T. Smoothed-Particle Hydrodynamics Numerical Modeling of Structures Impacted by Tsunami Bores. *J. Waterw. Port Coast. Ocean Eng.* **2014**, *140*, 66–81. [CrossRef]

41. Cunningham, L.S.; Rogers, B.D.; Pringgana, G. Tsunami wave and structure interaction: An investigation with smoothed-particle hydrodynamics. *Proc. Inst. Civ. Eng. Eng. Comput. Mech.* **2014**, *167*, 106–116. [CrossRef]

42. Aureli, F.; Dazzi, S.; Maranzoni, A.; Mignosa, P.; Vacondio, R. Experimental and numerical evaluation of the force due to the impact of a dam-break wave on a structure. *Adv. Water Resour.* **2015**, *76*, 29–42. [CrossRef]

43. Canelas, R.B.; Domínguez, J.M.; Crespo, A.J.C.; Gómez-Gesteira, M.; Ferreira, R.M.L. A Smooth Particle Hydrodynamics discretization for the modelling of free surface flows and rigid body dynamics. *Int. J. Numer. Methods Fluids* **2015**, *78*, 581–593. [CrossRef]

44. Heller, V.; Bruggemann, M.; Spinneken, J.; Rogers, B.D. Composite modelling of subaerial landslide–tsunamis in different water body geometries and novel insight into slide and wave kinematics. *Coast. Eng.* **2016**, *109*, 20–41. [CrossRef]

45. Fourtakas, G.; Rogers, B.D. Modelling multi-phase liquid-sediment scour and resuspension induced by rapid flows using Smoothed Particle Hydrodynamics (SPH) accelerated with a graphics processing unit (GPU). *Adv. Water Resour.* **2016**, *92*, 186–199. [CrossRef]

46. Mokos, A.; Rogers, B.D.; Stansby, P.K. A multi-phase particle shifting algorithm for SPH simulations of violent hydrodynamics with a large number of particles. *J. Hydraul. Res.* **2017**, *55*, 143–162. [CrossRef]

47. Alshaer, A.W.; Rogers, B.D.; Li, L. Smoothed Particle Hydrodynamics (SPH) modelling of transient heat transfer in pulsed laser ablation of Al and associated free-surface problems. *Comput. Mater. Sci.* **2017**, *127*, 161–179. [CrossRef]

48. Sun, P.N.; Colagrossi, A.; Marrone, S.; Antuono, M.; Zhang, A.M. A consistent approach to particle shifting in the δ-Plus-SPH model. *Mech. Eng.* **2019**, *348*, 912–934. [CrossRef]

49. Sun, P.; Zhang, A.M.; Marrone, S.; Ming, F. An accurate and efficient SPH modeling of the water entry of circular cylinders. *Appl. Ocean Res.* **2018**, *72*, 60–75. [CrossRef]

50. Zheng, X.; Shao, S.; Khayyer, A.; Duan, W.; Ma, Q.; Liao, K. Corrected first-order derivative ISPH in water wave simulations. *Coast. Eng. J.* **2017**, *59*. [CrossRef]

51. Luo, M.; Reeve, D.; Shao, S.; Karunarathna, H.; Lin, P.; Cai, H. Consistent Particle Method simulation of solitary wave impinging on and overtopping a seawall. *Eng. Anal. Bound. Elem.* **2019**, *103*, 160–171. [CrossRef]

52. Ran, Q.; Tong, J.; Shao, S.; Fu, X.; Xu, Y. Incompressible SPH scour model for movable bed dam break flows. *Adv. Water Resour.* **2015**, *82*, 39–50. [CrossRef]

53. Xia, X.; Liang, Q. A GPU-accelerated smoothed particle hydrodynamics (SPH) model for the shallow water equations. *Environ. Model. Softw.* **2016**, *75*, 28–43. [CrossRef]

54. Liang, Q.; Xia, X.; Hou, J. Catchment-scale High-resolution Flash Flood Simulation Using the GPU-based Technology. *Procedia Eng.* **2016**, *154*, 975–981. [CrossRef]

55. Satake, S.I.; Yoshimori, H.; Suzuki, T. Optimazations of a GPU accelerated heat conduction equation by a programming of CUDA Fortran from an analysis of a PTX file. *Comput. Phys. Commun.* **2012**, *183*, 2376–2385. [CrossRef]

56. Yang, C.T.; Huang, C.L.; Lin, C.F. Hybrid CUDA, OpenMP, and MPI parallel programming on multicore GPU clusters. *Comput. Phys. Commun.* **2011**, *182*, 266–269. [CrossRef]

57. Ohshima, S.; Hirasawa, S.; Honda, H. OMPCUDA: OpenMP Execution Framework for CUDA Based on Omni OpenMP Compiler. In *Beyond Loop Level Parallelism in OpenMP: Accelerators, Tasking and More*; Sato, M., Hanawa, T., Müller, M.S., Chapman, B.M., de Supinski, B.R., Eds.; *IWOMP* 2010. Lecture Notes in Computer Science, 6132; Springer: Berlin/Heidelberg, Germany, 2010. [CrossRef]

58. Loncar, V.; Young, S.L.E.; Skrbic, S.; Muruganandam, P.; Adhikari, S.; Balaz, A. OpenMP, OpenMP/MPI, and CUDA/MPI C programs for solving the time-dependent dipolar Gross-Pitaevskii equation. *Comput. Phys. Commun.* **2016**, *209*, 190–196. [CrossRef]

59. Bronevetsky, G.; Marques, D.; Pingali, K.; McKee, S.; Rugina, R. Compiler-enhanced incremental checkpointing for OpenMP applications. In Proceedings of the 2009 IEEE International Symposium on Parallel & Distributed Processing, Rome, Italy, 23–29 May 2009; pp. 1–12. [CrossRef]

60. Dagum, L.; Menon, R. OpenMP: An industry standard API for shared-memory programming. *IEEE Comput. Sci. Eng.* **1998**, *5*, 46–55. [CrossRef]

61. Slabaugh, G.; Boyes, R.; Yang, X. Multicore Image Processing with OpenMP [Applications Corner]. *IEEE Signal Process. Mag.* **2010**, *27*, 134–138. [CrossRef]

62. Chorley, M.J.; Walker, D.W. Performance analysis of a hybrid MPI/OpenMP application on multi-core clusters. *J. Comput. Sci.* **2010**, *1*, 168–174. [CrossRef]

63. Adhianto, L.; Chapman, B. Performance modeling of communication and computation in hybrid MPI and OpenMP applications. *Simul. Model. Pract. Theory* **2007**, *15*, 481–491. [CrossRef]

64. Wright, S.J. Parallel algorithms for banded linear systems. *Siam J. Sci. Stat. Comput.* **1991**, *12*, 824–842. [CrossRef]

65. Jiao, Y.-Y.; Zhao, Q.; Wang, L. A hybrid MPI/OpenMP parallel computing model for spherical discontinuous deformation analysis. *Comput. Geotech.* **2019**, *106*, 217–227. [CrossRef]

66. Przemysław, S. Algorithmic and language-based optimization of Marsa-LFIB4 pseudorandom number generator using OpenMP, OpenACC and CUDA. *J. Parallel Distrib. Comput.* **2020**, *137*, 238–245.

67. Vacondio, R. Shallow Water and Navier-Stokes SPH-Like Numerical Modelling of Rapidly Varying Free-Surface Flows. Ph.D. Thesis, Università degli Studi di Parma, Parma, Italy, 2010.

68. Vacondio, R.; Rodgers, B.D.; Stansby, P.K.; Mignosa, P. User Guide for the SWE-SPHysics Code. 2013. Available online: https://wiki.manchester.ac.uk/sphysics/images/SWE-SPHysics_v1.0.00.pdf (accessed on 2 April 2020).

69. Marion, J.; Thornton, S. *Classical Dynamics of Particles and Systems*; Harcourt Brace Jovanovich Inc.: San Diego, CA, USA, 1988.

70. Monaghan, J.J. Smoothed particle hydrodynamics. *Rep. Prog. Phys.* **2005**, *68*, 1703–1759. [CrossRef]

71. Bonet, J.; Lok, T.-S.L. Variational and momentum preservation aspects of Smooth Particle Hydrodynamic formulations. *Comput. Methods Appl. Mech. Eng.* **1999**, *180*, 97–115. [CrossRef]

72. Vila, J.P. On particle weighted methods and smooth particle hydrodynamics. *Math. Models Methods Appl. Sci.* **1999**, *9*, 161–209. [CrossRef]

73. Dinshaw, B.S. Von Neumann stability analysis of smoothed particle hydrodynamics—Suggestions for optimal algorithms. *J. Comput. Phys.* **1995**, *121*, 357–372.

74. Toro, E. Direct Riemann solvers for the time-dependent Euler equations. *Shock Waves* **1995**, *5*, 75–80. [CrossRef]

75. Hernquist, L.; Katz, N. TREESPH: A unification of SPH with the hierarchical tree method. *Astrophys. J. Suppl.* **1989**, *70*, 419–446. [CrossRef]

76. Toro, E. *Shock Capturing Methods for Free Surface Shallow Water Flows*; Wiley: New York, NY, USA, 1999.

77. Nikolaos, D.K. A dissipative galerkin scheme for open-channel flow. *Hydraul. Eng.* **1984**, *110*, 337–352.

78. Majda, A.; Osher, S. Numerical viscosity and the entropy condition. *Commun. Pure Appl. Math.* **1979**. [CrossRef]

79. Stranex, T.; Wheaton, S. A new corrective scheme for SPH. *Comput. Methods Appl. Mech. Eng.* **2011**, *200*, 392–402. [CrossRef]

80. Monaghan, J.J.; Gingold, R.A. Shock simulation by the particle method SPH. *J. Comput. Phys.* **1983**, *52*, 374–389. [CrossRef]

81. Monaghan, J.J. Particle methods for hydrodynamics. *Comput. Phys. Rep.* **1985**, *3*, 71–124. [CrossRef]

82. Chen, F.; Qiang, H.; Gao, W. Coupling of smoothed particle hydrodynamics and finite volume method for two-dimensional spouted beds. *Comput. Chem. Eng.* **2015**, *77*, 135–146. [CrossRef]

83. Kleefsman, K.M.T.; Fekken, G.; Veldman, A.E.P.; Iwanowski, B.; Buchner, B. A volume-of-fluid based simulation method for wave impact problems. *J. Comput. Phys.* **2005**, *206*, 363–393. [CrossRef]

Article

Numerical Simulation of Non-Homogeneous Viscous Debris-Flows Based on the Smoothed Particle Hydrodynamics (SPH) Method

Shu Wang [1], Anping Shu [1,*], Matteo Rubinato [2], Mengyao Wang [1] and Jiping Qin [1]

[1] School of Environment, Key Laboratory of Water and Sediment Sciences of MOE, Beijing Normal University, Beijing 100875, China; colinyouare@163.com (S.W.); 201721180007@mail.bnu.edu.cn (M.W.); 201821180006@mail.bnu.edu.cn (J.Q.)

[2] School of Energy, Construction and Environment & Centre for Agroecology, Water and Resilience, Coventry University, Coventry CV1 5FB, UK; matteo.rubinato@coventry.ac.uk

* Correspondence: shuap@bnu.edu.cn; Tel.: +86-010-5880-0658

Received: 11 September 2019; Accepted: 1 November 2019; Published: 5 November 2019

Abstract: Non-homogeneous viscous debris flows are characterized by high density, impact force and destructiveness, and the complexity of the materials they are made of. This has always made these flows challenging to simulate numerically, and to reproduce experimentally debris flow processes. In this study, the formation-movement process of non-homogeneous debris flow under three different soil configurations was simulated numerically by modifying the formulation of collision, friction, and yield stresses for the existing Smoothed Particle Hydrodynamics (SPH) method. The results obtained by applying this modification to the SPH model clearly demonstrated that the configuration where fine and coarse particles are fully mixed, with no specific layering, produces more fluctuations and instability of the debris flow. The kinetic and potential energies of the fluctuating particles calculated for each scenario have been shown to be affected by the water content by focusing on small local areas. Therefore, this study provides a better understanding and new insights regarding intermittent debris flows, and explains the impact of the water content on their formation and movement processes.

Keywords: non-homogeneous debris flow; viscous coefficients; intermittent debris flows; energy conversion

1. Introduction

As a frequent natural disaster, debris flow poses a serious threat to the lives and properties of people living in mountain areas [1–4]. This natural phenomenon needs to be taken into consideration when planning new developments in mountain catchments because its sudden, devastating, and extensive impacts could have strong consequences on the local economy. Debris flows can be divided into multiple categories such as viscous flow, dilatant flow, dilute flow, water-rock flow, etc. [5] based on the variety of materials and their combinations. The first category mentioned, the viscous debris flow, is particularly widely distributed in the mountainous areas of the southwest of China. This particular debris flow is characterized by high viscosity, wide particle-size distributions and uneven velocity distributions [6]. The difficulty in describing its motion process is due to the fact that the viscous debris flow is classified as a heterogeneous, non-constant, and non-Newtonian flow [7]. In previous studies, Johnson [8] attempted to describe the motion of viscous debris flow by using the Bingham Model, which is applicable to laminar viscous flows because in such fluids, the concentration of coarse particles is quite low. Bagnold [9] described it in terms of dilatant flow, and emphasized the discrete force between particles caused by collision. Based on the experiment conducted by Bagnold, Takahashi [10] proposed the idea of the presence of a vertical collision stress between particles, which supports the coarse particles not sinking into the fluid. Chen [11,12] comprehensively focused on the elastic shear force, the plastic shear force, and the laminar shear force in muddy fluids, implementing a modification of the motion resistance which could be suitable for viscoplastic fluids containing certain coarse particles. O'Brien [13,14] took into account the effects of plasticity, viscosity, collision, and turbulence on resistance forces and identified a general equation which includes the components due to cohesive yield stress, Mohr-Coulomb shear force, viscous shear stress, turbulent shear stress, and dispersive shear force, establishing a debris fluid model that could fully describe the effects due to different particle compositions and different densities.

Furthermore, the mesh-free numerical modeling techniques are being gradually introduced into this field. Dai [15] developed a three-dimensional model to simulate rapid landslide motions across 3D terrains. The artificial viscosities linearly related to the linear and quadratic terms of shear deformation were incorporated into the pressure terms in the momentum equation to dissipate energy for avoiding numerical oscillations. However, Dai [15] did not consider the effect of yield stress or the interaction between solid and liquid phases. Hosseini [16] adopted an innovative treatment similar to the one applied in Eulerian formulations to viscous terms, and hence facilitated the implementation of various inelastic non-Newtonian models. This approach utilized the exact forms of the shear strain rate tensor and its second principal invariant to calculate the shear stress tensor. Rodriguez-Paz [17] introduced a new frictional approach for the boundary conditions and modified constitutive equations in the SPH (Smoothed Particle Hydrodynamics) method to focus on the interaction between fluid particles and boundary conditions. The resulting technique was then applied for the numerical simulation of debris flows and the results were compared with those experimentally obtained and available in literature [18,19], providing good agreements.

In this paper, the SPH method was applied to estimate the movement of a non-homogeneous viscous debris-flow: the fluid under investigation was divided into solid-liquid phases. The solid phase was characterized by particles with larger size than the critical particle size, while the liquid phase was the mixture of water and particles with smaller size than the critical one. The constitutive relation for the liquid phase was characterized by yield stress, laminar viscous force, and turbulent viscous force, while the constitutive relation of the solid phase was related to support force, friction, and collision stresses. It is well known that the magnitude of shear deformation determines which force plays a dominant role in the process of the fluid movement [20]. Therefore, quantifying all the above forces, it could be possible to quantify and estimate the role of each component and further investigate how the shear sharp deformation could be reduced. However, due to the complexity of the materials composing debris flows, even the fluid with same rheological coefficients could generate effects attributable to different flow structures and characteristics. Therefore, the effect on the debris flow process of the initial vertical distribution of the two-phase solid-liquid is also considered in this study.

2. Fundamental Theories and Numerical Modeling

2.1. The SPH Method

SPH is a kind of mesh-free method based on a pure Lagrangian description, which has been widely applied in multiple engineering and science fields [21–24]. Compared with the mesh method based on continuum theory, the SPH method avoids the problem of mesh distortion in dealing with the flow issue since there is no connectivity between the particles, since it is developed on a uniform smoothed particle hydrodynamic framework. By adopting this technique, the goal is to provide accurate and stable numerical solutions for integral equations or partial differential equations (PDEs) using a series of arbitrarily distributed particles carrying field variables, such as mass, density, energy, and stress tensors [25].

2.1.1. SPH Interpolation

There are two main steps to transform PDEs equations into the SPH form, called particle approximation and kernel approximation, respectively [26]. The first step consists of representing a function in continuous form as an integral representation by using an interpolation function. In this step, the approximation of the function and its derivatives is based on the evaluation of the smooth kernel function and its derivatives. The second step involves representing the problem domain by using a set of discrete particles within the influence area of the particle at location x, and then estimating the field variables for those particles as follows:

$$\langle f(x) \rangle = \int_{\Omega} f(x')W(x-x',h)dx', \tag{1}$$

$$\langle f(x) \rangle = \sum_{j=1}^{N} m_j \frac{f(x_j)}{\rho_j} W(x-x_j,h), \tag{2}$$

where x' denotes the position of continuum in the influence domain before the discretization; W denotes the smoothing function; h is a parameter that defines the size of the kernel support, known as the smoothing length; Ω represents the problem space whose radius is taken as several times of h according to different smoothing functions; N is the total number of neighboring particles; m is the mass; and ρ is the density.

Kernel approximation is the technique of approximating the values of both the field function and the derivative of the field function. The kernels used in the SPH method approximate a δ function (the Dirac function). Monaghan [27] suggested that a suitable Kernel approximation must have a compact support in order to ensure zero interactions outside its computational range. The original calculations of Gingold and Monaghan [28] used a Gaussian Kernel. The Gaussian Kernel function is simple to use and has high accuracy. Especially for the case of disordered particle distribution, this technique generates stable and accurate approximation results. However, the Gaussian Kernel function does not have a strict compact support unless the size of the Kernel support approaches the infinity value. Additionally, further various Kernels forms with a compact support (such as spline [29], super-Gaussian [30], polynomial [31], and cosine [32]) were proposed in previous studies but the one of the most popular Kernels more commonly utilized is based on the spline functions [29] as defined by:

$$W(r,h) = \frac{10}{7\pi h^2} \times \begin{cases} 1 - 1.5q^2 + 0.75q^3 & 0 \leq q < 1 \\ 0.25(2-q)^3 & 1 \leq q < 2 \\ 0 & 2 \leq q \end{cases}, \tag{3}$$

where $q = |r|/h$ and r is the separation distance between the particles. In this study, Equation (3) has been used to approximate the values of the field under investigation. This Kernel has compact support so that its interactions are exactly zero for $r > 2h$. The smoothing distance or so called "Kernel range" h determines the degree with which a particle interacts with adjacent particles.

2.1.2. Gradient and Divergence

As a standard procedure, the gradient and divergence operators need to be formulated in a SPH algorithm if the simulation of the Navier-Stokes equations is to be attempted. In this work, the following commonly used forms are employed for gradient of a scalar A and divergence of a vector A [33]:

$$\frac{1}{\rho_a} \nabla_a A = \sum_b m_b \left(\frac{A_a}{\rho_a^2} + \frac{A_b}{\rho_b^2} \right) \nabla_a W_{ab}, \tag{4}$$

$$\frac{1}{\rho_a} \nabla_a \cdot \mathbf{A} = \sum_b m_b \left(\frac{\mathbf{A}_a}{\rho_a^2} + \frac{\mathbf{A}_b}{\rho_b^2} \right) \cdot \nabla_a W_{ab}, \tag{5}$$

where $\nabla_a W_{ab}$ is the gradient of the Kernel function $W(x - x_j, h)$ with respect to x_j, subscripts $_a$ and $_b$ represent the target particles and the particles in the influence domain, respectively, affecting the position of particle i. This choice of discretization operators ensures that an exact projection algorithm is produced. To date, there are various options to represent these operators, but only certain specific ones [34,35] have proven to be more convenient in terms of the accuracy and robustness of the method.

2.2. Governing Equations

The governing equations for transient compressible fluid flow include the conservation of mass and momentum equations. In a Lagrangian framework, these can be written as follows:

$$\frac{1}{\rho}\frac{D\rho}{Dt} + \nabla \cdot \mathbf{v} = 0, \tag{6}$$

$$\frac{D\mathbf{v}}{Dt} = \mathbf{g} + \frac{1}{\rho}\nabla \cdot \boldsymbol{\tau} - \frac{1}{\rho}\nabla P, \tag{7}$$

where t is time, $\mathbf{v}$ is the particle velocity vector, $\mathbf{g}$ is the gravitational acceleration, P stands for pressure, and D/Dt refers to the material derivative. The density ρ has been intentionally kept in the equations to be able to enforce the incompressibility of the fluid. Using an appropriate constitutive equation to model the shear stress tensor $\boldsymbol{\tau}$, Equations (6) and (7) can be used to solve both Newtonian and non-Newtonian flows.

The momentum equations include three driving force terms, i.e., body force, forces due to divergence of the stress tensor, and the pressure gradient, and these always have to be considered together with the incompressibility constraint. In a SPH formulation, the above system of governing equations must be solved for each particle at each time-step, and the order with which the force terms are incorporated into the momentum equations can be different from one algorithm to another.

2.2.1. Equation of State

SPH method can be formulated for incompressible and compressible flows. The equation of state, giving the relationship between particle density and fluid pressure, can be written as follows [28]:

$$P = P_0\left[\left(\frac{\rho}{\rho_0}\right)^7 - 1\right], \tag{8}$$

where P_0 represents a constant value of pressure, usually expressed in terms of initial pressure and ρ_0 is the reference density.

2.2.2. Viscous Terms

In the context of the SPH method, several forms of viscosity terms were introduced by Lucy [25], Gingold and Monaghan [29], Wood [34], Loewenstein and Mathews [36], and Shao and Lo [37]. As the purpose of this work was to solve non-Newtonian fluids, a new description of viscosity was developed to facilitate the modeling of such flow characteristics. Viscosity of incompressible Newtonian fluids depends only on the second principal invariant of the shear strain rate:

$$\mathbf{D} = \begin{bmatrix} 2\frac{\partial u}{\partial x} & \frac{\partial u}{\partial y} + \frac{\partial v}{\partial x} \\ \frac{\partial u}{\partial y} + \frac{\partial v}{\partial x} & 2\frac{\partial v}{\partial y} \end{bmatrix}, \tag{9}$$

In solving Equations (8) and (9), the finite difference method should firstly be used to solve the total derivative between two particles, and then decompose the results into x, y directions, following Lo and Shao [38].

$$\left(\frac{\partial u_i}{\partial x_j}\right)_a = \left(\frac{\partial u_i}{\partial r_{ab}}\right)\left(\frac{\partial r_{ab}}{\partial x_j}\right) = \frac{(u_i)_a - (u_i)_b}{r_{ab}} \frac{(x_j)_a - (x_j)_b}{r_{ab}}, \tag{10}$$

The viscous debris flow studied in this paper is a non-Newtonian fluid whose constitutive equation has the following form:

$$\text{Bingham fluid}: \ \tau = \tau_B + \mu_B \mathbf{D}, \tag{11}$$

$$\text{dilatant fluid}: \ \tau = \mu_D(|\mathbf{D}|)\mathbf{D}, \tag{12}$$

$$\text{viscoplastic fluid}: \ \tau = \frac{1}{2}\left(\frac{\mu_0}{|\mathbf{D}|} + \mu_1 + \mu_2|\mathbf{D}|\right)\mathbf{D}, \tag{13}$$

where τ_B refers to the yield stress; μ_B, μ_1 are the coefficients of friction and μ_D, μ_2 are the coefficients of collision stresses. According to the different flows considered, the symbols in Equation (13) can represent different meanings. When the object considered corresponds to solid particles, μ_0 indicates the static support force between the solid particles, μ_1 is the particle friction coefficient, and μ_2 is the particle collision coefficient. When considering a liquid phase slurry, μ_0 is the yield stress, μ_1 is the laminar viscosity coefficient, and μ_2 is the turbulent viscosity coefficient. By comparing Equations (10)–(12), Equations (10) and (11) can be regarded as a special form of Equation (12), because slurry flows consists of water and fine particles (liquid phase) and coarse particles (solid phase). In the Bingham model, since the fluid turbulence is not considered, the turbulent viscosity coefficient is $\mu_2 = 0$. In the expansion model, the inter-particle frictional force is negligible relative to the particles' collision, so the second term in Equation (11) is zero.

By substituting single components of Equations (9) and (10), the second term of right hand side in Equation (7) can be written as follows:

$$\frac{1}{\rho_a}\nabla_a \cdot \boldsymbol{\tau} = \frac{1}{\rho_a}\left[\begin{array}{c}\frac{\partial}{\partial x} \\ \frac{\partial}{\partial y}\end{array}\right] \cdot \left[\begin{array}{cc}\tau_{xx} & \tau_{xy} \\ \tau_{yx} & \tau_{yy}\end{array}\right] = \frac{1}{\rho_a}\left[\begin{array}{c}\frac{\partial}{\partial x}\cdot\tau_{xx} + \frac{\partial}{\partial y}\cdot\tau_{yx} \\ \frac{\partial}{\partial x}\cdot\tau_{xy} + \frac{\partial}{\partial y}\cdot\tau_{yy}\end{array}\right], \tag{14}$$

By substituting Equation (13) into Equation (14), the viscous term in the x, y direction can be represented by the following discretization:

$$\frac{du}{dt} = \sum_b m_b\left\{\frac{1}{\rho_a\rho_b}\cdot\frac{1}{2}\left(\frac{\mu_0}{|\mathbf{D}|} + \mu_1 + \mu_2|\mathbf{D}|\right)\left[2\frac{\partial u}{\partial x}\mathbf{i} + \left(\frac{\partial v}{\partial x} + \frac{\partial u}{\partial y}\right)\mathbf{j}\right]\right\}\cdot\left(\frac{\partial W}{\partial q}\cdot\frac{x_{ab}}{hr}\mathbf{i} + \frac{\partial W}{\partial q}\cdot\frac{y_{ab}}{hr}\mathbf{j}\right) \tag{15}$$

$$\frac{dv}{dt} = \sum_b m_b\left\{\frac{1}{\rho_a\rho_b}\cdot\frac{1}{2}\left(\frac{\mu_0}{|\mathbf{D}|} + \mu_1 + \mu_2|\mathbf{D}|\right)\left[\left(\frac{\partial v}{\partial x} + \frac{\partial u}{\partial y}\right)\mathbf{i} + 2\frac{\partial v}{\partial y}\mathbf{j}\right]\right\}\cdot\left(\frac{\partial W}{\partial q}\cdot\frac{x_{ab}}{hr}\mathbf{i} + \frac{\partial W}{\partial q}\cdot\frac{y_{ab}}{hr}\mathbf{j}\right) \tag{16}$$

It can be seen from Equation (15) that when deformation $|\mathbf{D}|$ is relatively small, yield stress (particle support force) has a greater impact on the fluid's acceleration, but there should be an upper limit to this effect in order to prevent excessive acceleration which could cause the local instability of the fluid investigated. In existing studies [16], a lower limit is usually set for $|\mathbf{D}|$. When $|\mathbf{D}|$ is less than this lower limit, the relationship between stress and $|\mathbf{D}|$ satisfies linearity:

$$\begin{aligned}|\mathbf{D}| \leq \frac{\mu_0}{\sigma} &\to \tau = \sigma\mathbf{D} \\ |\mathbf{D}| > \frac{\mu_0}{\sigma} &\to \tau = \frac{1}{2}\left(\frac{\mu_0}{|\mathbf{D}|} + \mu_1 + \mu_2|\mathbf{D}|\right)\mathbf{D}\end{aligned} \tag{17}$$

where σ is the limiting factor.

In [16], Hosseini considers that the viscosity of "solid zone" fluid is much greater than that of the main fluid (100 times). In this study, although the turbulent stress term $\frac{1}{2}\mu_2|\mathbf{D}|\mathbf{D}$ is introduced, it can be ignored in the region of low velocity. By calibrating this value against experimental results obtained for this study, $200\mu_1$ was the limited factor selected.

3. Experimental Setup and Boundary Conditions

Figure 1 shows the facility where the experiments for the numerical validation of the method previously described in Section 2 were conducted. This facility is located in the Debris Flow Observation and Research Station at Jiangjia Gully, the largest field research center in China—also known as "the debris flow museum". For this study, water is stored upstream to the material prepared under multiple configurations and water is always released by opening a steel gate at the velocity of 3 m/s. The flume can then be divided into two parts: (i) a steep upstream reach (6.0 m long and the chute slope can vary from 15° to 40° (always set up at 25° for these experiments); and (ii) a flat-bottomed downstream section (3.0 m long). The slope of the flume's bed can be manually adjusted.

Details regarding the experimental procedure conducted can be found in [6]. For this study, three slurries were testes with densities $\rho = 1400$ kg/m^3; $\rho = 1500$ kg/m^3 and $\rho = 1600$ kg/m^3. Different layers patterns were selected according to the different configurations displayed in Figure 2. By adding water to the flume, when the water level reached the height of the mixing fluid, the front-end steel gate of the mixing area was released at a speed of 3 m/s. In order to maintain the driving force of the mixtures, the water level behind the mixtures was kept at $h = 0.2$ m during the experiment.

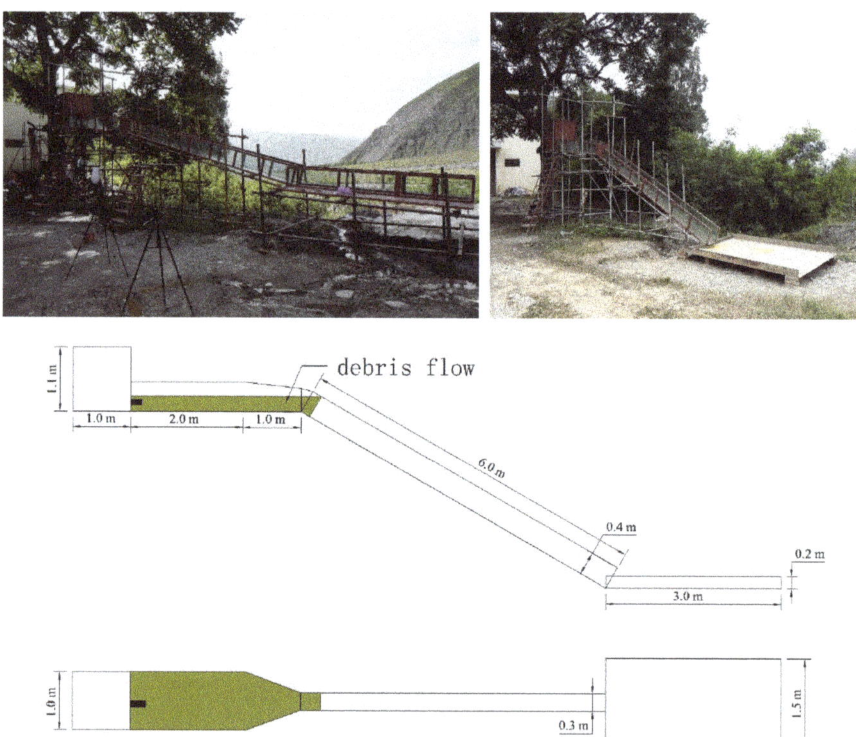

Figure 1. Experimental equipment for debris flow simulation [6].

The experimental configurations chosen to examine the influence of vertical grading patterns on the debris flow formation and propagation are displayed in Figure 2.

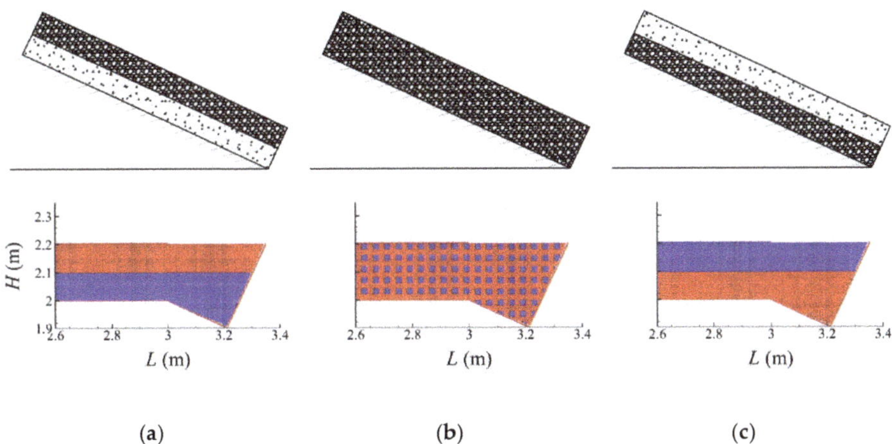

Figure 2. Schematic representation (experimental and numerical) of configurations applied for this study [6].

There are three configurations: Figure 2a shows the configuration a, with coarse particles as a top layer and fine sediment positioned below them; Figure 2b shows the configuration b, a realistic configuration with the original sediments collected from the Jiangjiagou Valley, fully mixed and distributed along the entire measurement area; Figure 2c shows the configuration c, with coarse particles distributed at the bottom and fine grains on the top of them. For this study, three types of liquid slurries were considered: (i) $\rho = 1400$ kg/m^3, (ii) $\rho = 1500$ kg/m^3, and (iii) $\rho = 1600$ kg/m^3. The slurry rheology coefficient was measured by the MCR301 advanced rotary rheometer manufactured by Anton Paar, Austria [39]. Values of viscosity μ for the fluids simulated are displayed in Table 1.

Table 1. Experimental conditions (viscosity values and vertical grading patterns) adopted for this study.

Test No.	Factors				
	ρ (kg/m^3)	Solid Phase Level	μ_0 (Pa)	μ_1 (Pa·s)	μ_2 (Pa·s^2)
1	1400	Upper			
2	1400	Mixed	0.00004	0.0048	0.0197
3	1400	Bottom			
4	1500	Upper			
5	1500	Mixed	0.00006	0.0051	0.1654
6	1500	Bottom			
7	1600	Upper			
8	1600	Mixed	0.0001	0.0034	0.8242
9	1600	Bottom			

The particle size distribution curves are shown in Figure 3, Ψ stands for the percentage of accumulated mass of particles and d denotes particle size.

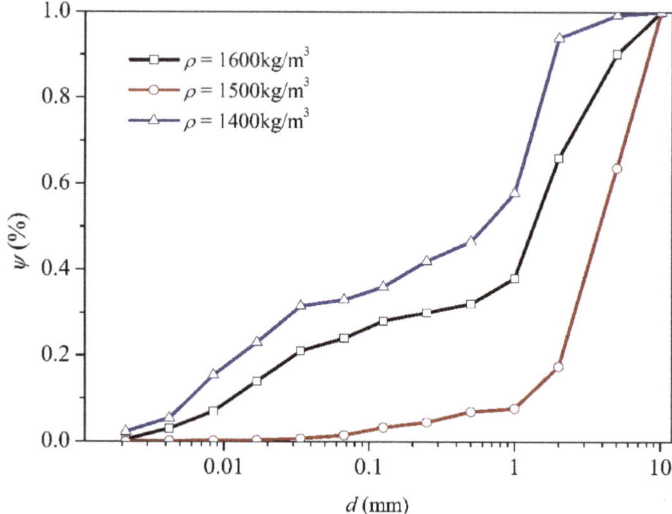

Figure 3. Particle size distribution curves.

4. Simulation and Analysis

By simulating the movement processes of different viscous debris flows, a series of experiments has been completed where free surface heights, fluid velocities, pressures, and shear deformations associated with the movement of the fluid were measured. The numerical simulation was carried out to generate the experimental conditions previously described (Figure 4). The numerical simulation replicated $N = 113{,}054$ particles, particle spacing $d_p = 0.0025$ m, solid particle density $\rho_s = 2200$ kg/m^3, thickness of solid phase $h_s = 0.1$ m and thickness of liquid phase $h_l = 0.1$ m for configuration a and c in Figure 2. Similar to the tests conducted on the experimental facility, three different viscosity coefficients for the liquid phases (as shown in Table 1) were selected in the numerical simulation. The inflow conditions were the same as those applied experimentally, and the water level as driving force of the debris flow was kept at 0.2 m for each entire simulation.

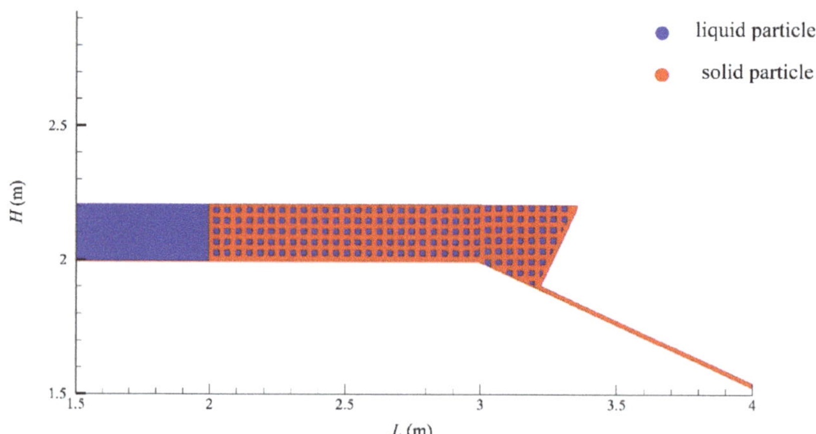

Figure 4. Initial state of debris flow numerical simulation (mixed configuration). The water level behind the mixtures was kept at $h = 0.2$ m.

Figure 5 represents the free surface values recorded at different times for each of the tests conducted in Table 1, after the debris flow initiation generated by the release of water upstream. The x-axis represents the length L of the debris flow, and the y-axis represents the height H of the debris flow. After 0.7 s, the distance reached by the debris flow (considering all the configurations) is within the range 2.48–2.66 m, and the maximum velocity range is 4.65–5.12 m/s. Authors noticed that when the debris flow has similar viscous properties but for the vertical distribution of different particles, the differences in velocities are not so significant and can considered almost negligible in most of the cases. However, the maximum velocity recorded for configuration b (Shown in Figure 2b correspond to tests 2, 5 and 8 in Figure 5) is similar to the one recorded for configuration a (Shown in Figure 2a and corresponding to tests 1, 4, and 7 in Figure 5), while configuration c (Shown in Figure 2c and corresponding to tests 3, 6, and 9 in Figure 5) was characterized by higher values of velocities and elevations measured. By comparing the shapes of head under different vertical distributions, it was found that for the tests conducted in Table 1, free surface values measured for configuration b fluctuate more than in configuration a and c, demonstrating that this scenario is typical of intermittent debris flows.

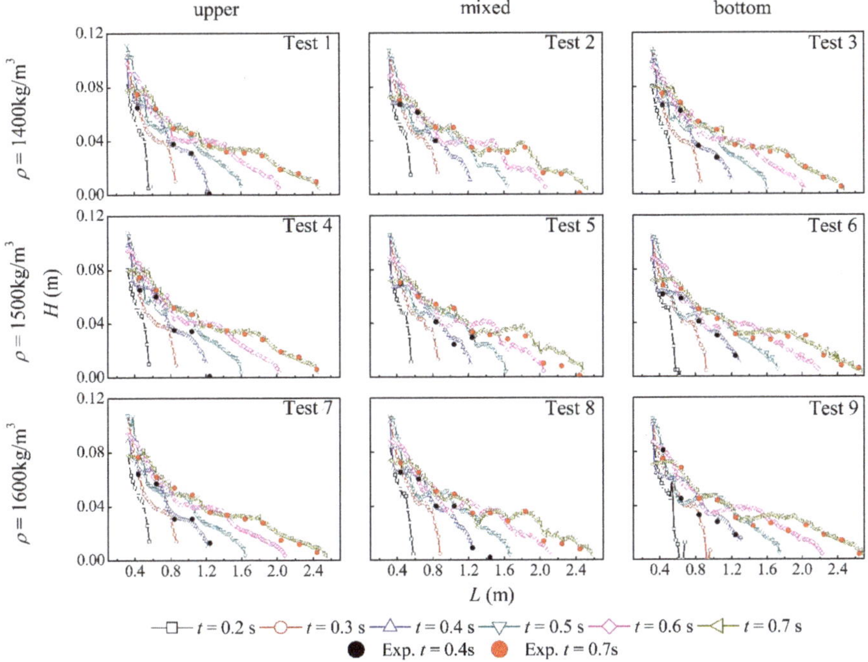

Figure 5. Debris flow free surface for tests 1, 2, 3, 4, 5, 6, 7, 8, and 9 (Table 1) and experimental results used for validation (dots), respectively.

Characterization of Intermittent Debris Flow

In order to study the causes of this phenomenon, the characteristics of the fluid movement processes associated to configuration b were analyzed. Figure 6 shows the evolution of the solid-liquid phase at different locations simulated numerically. Figure 6a shows that in the horizontal region, most of the particles still retain under the laminar form. When moving into the upstream part of the sloped section, the fluid height decreases and the liquid phase group is stretched, as shown in Figure 6b. Then, due to the slope, velocity increases while the fluid height decreases, and different layers of liquid and solid particles will appear almost as parallel mixing within the entire width of the debris flow, as shown in Figure 6c. At this stage, the altering layers interact changing continuously positions demonstrating that mixing processes are taking place and when the mixing is finally completed, the fluctuation amplitude reduce becoming more stable, while the influenced range of the fluctuations can spread over a longer length, as shown in Figure 6d. Figure 6e shows the effect of the liquid phase on the height of the debris flow. It is clear that when liquid particles accumulate due to the mixing phenomena (highlighted as circles in Figure 6e), there is a correspondent decrease of the height of the debris flow (pointed out using arrows in Figure 6e). This inverse relationship is very interesting especially because it demonstrates how the gathering and accumulation of liquid particles tends to appear towards the bottom side of the debris flow layer.

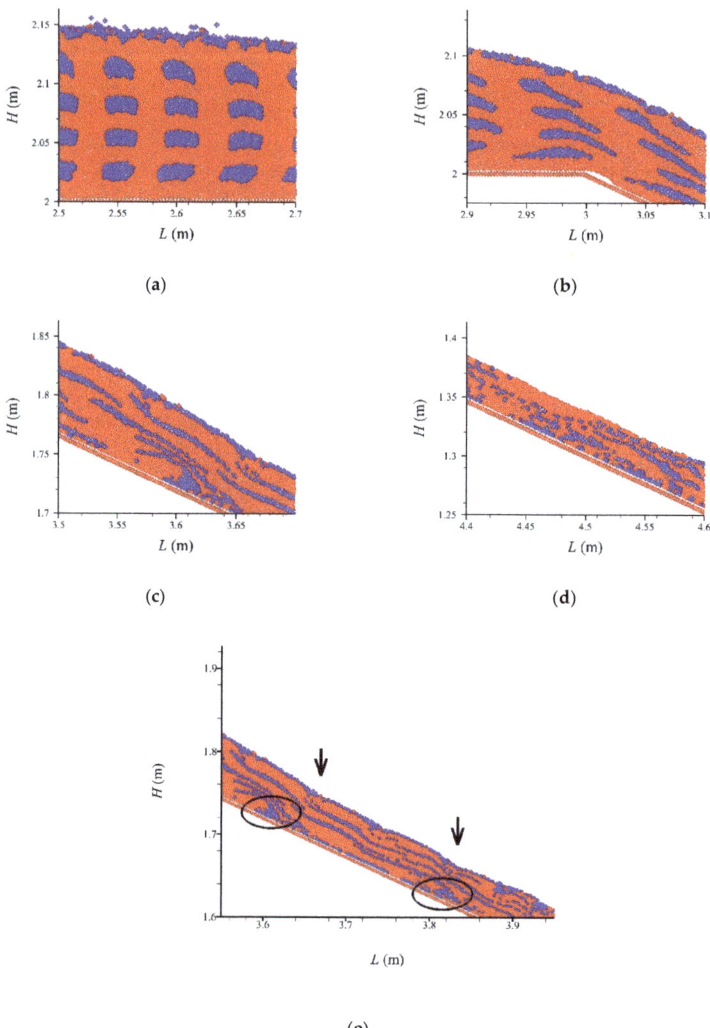

Figure 6. Analysis of the debris flow behaviors at different locations. Solid particles and liquid particles are represented by red and blue dots, respectively.

On this basis, the relationship between the moisture content ϕ (the amount of liquid particles divided by the amount of solid particles, N_l/N_s), the kinetic energy of particles E_k and the height of the free surface H (related to the potential energy of particles E_p) were calculated for the tests conducted for configuration b. As shown in Figure 7a, the height of the free surface H decreases as the debris flow develops. There is a noticeable correlation between the fluctuation of the free surface associated with the fluctuation of the moisture content. In the regions of $L = 1.00–1.38$ m and $L = 1.82–2.04$ m, the height of free surface decreases linearly, and in these two regions, the water content remains in the range 0.2–0.65. The points that obviously exceed this threshold are $L = 0.90$, $L = 1.52$, $L = 1.6$, $L = 1.74$, and $L = 1.80–1.84$, and the height of free surface is different from that of linear decline in these areas or vicinity. When the moisture content is within the range 0.20–0.65, the free surface of the debris flow is characterized by a linear change, but when the moisture content exceeds this range, it generates an impact on the free surface.

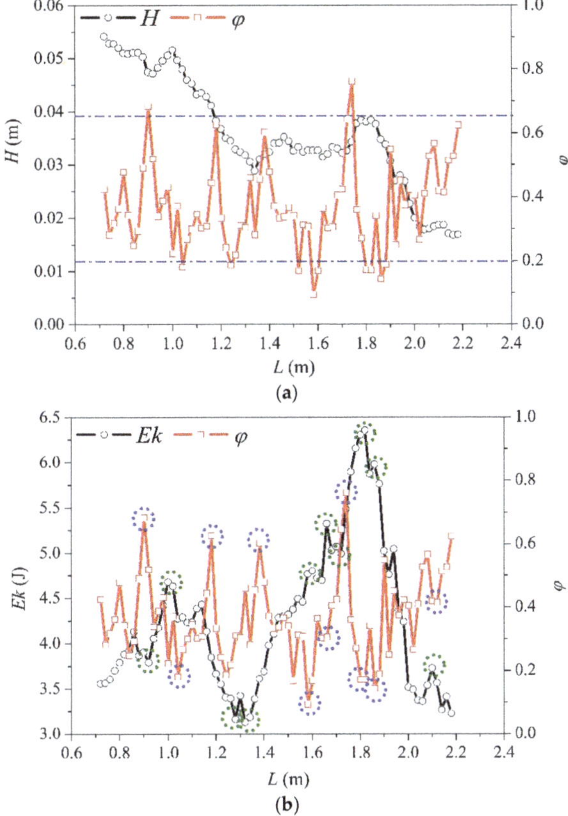

Figure 7. Relationships between the moisture content, the kinetic energy, and the height of the free surface for configuration b.

From Figure 7b, it can be seen that there is a more obvious negative correlation between the particles kinetic energy E_k and the moisture content ϕ. To almost every peak of the kinetic energy E_k (highlighted as green circles in Figure 7b) calculated corresponds a peak of the moisture content ϕ (highlighted as blue circles in Figure 7b), which indicates that kinetic energy E_k and moisture content ϕ interact directly. However, this effect can only be assigned to small-scale portions of the particle kinetic energy fluctuations. Looking at Figure 7b, at the location of $L = 1.74$ m, the moisture content value corresponds to $\phi = 0.7619$ and it is the maximum value measured in this region, and the corresponding kinetic energy E_k records its minimum value. But because of the large kinetic energy of the particles recorded in this region, the corresponding kinetic energy $E_k = 5.4848$ J is still higher than that recorded at the position of $L = 1.64$ m in the adjacent one $E_k = 0.30263$ J.

Figure 8 shows the effect of moisture content ϕ on the kinetic energy E_k and the height of the free surface H on a large scale, for configurations a and c. For both configurations, where the solid phase is located at the top and the bottom, the free surface is greatly affected by the magnitude of the moisture content ϕ, while the kinetic energy E_k is greatly affected by the derivative of moisture content along the length of the slope $\frac{d\phi}{dL}$. However, the fluctuation of the moisture content $\Delta\phi$ along the length L, especially for configuration a where the solid phase is displayed at the top of the debris flow, is relatively small. So the kinetic energy E_k and potential energy E_p curves show relatively large-scale area fluctuations and linear characteristics in comparison to the mixed distribution fluid conditions typical of configuration b.

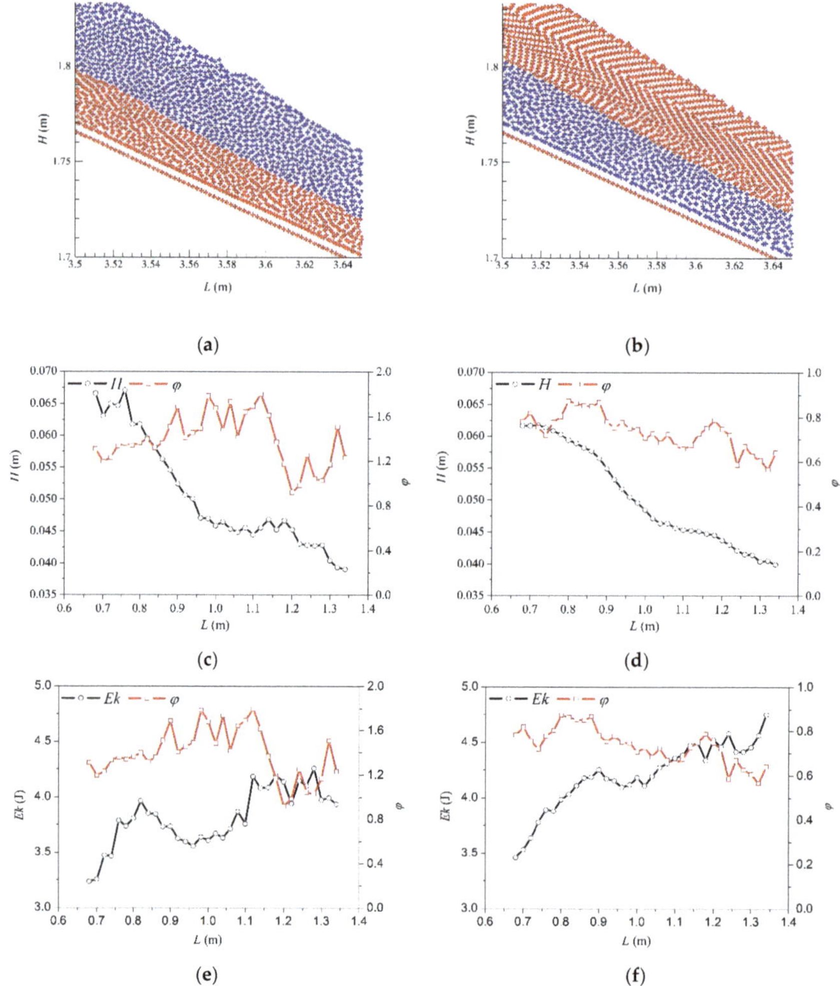

Figure 8. Relationships between the moisture content ϕ, the kinetic energy E_k, and the height of the free surface H for configuration a and c.

Finally, the energy conversion curves of fluids with different viscous coefficients were inspected and confronted, as shown in Figure 9. It was found that the gravitational potential energy ($E_p = mgH$) and the total energy ($E_0 = E_k + E_p$) of fluids decreases at a similar rate. The difference between three fluids is mainly reflected on kinetic energies. When comparing the set of fluids with the smallest density (ρ = 1400 kg/m^3, μ_0 = 0.00004, μ_1 = 0.0048, and μ_2 = 0.0197), results shows that velocity values increase from t = 0.0 s up to t = 0.6 s, reaching almost the highest values, and then the kinetic energy of the three fluids tends to be equal. As the time progresses, the same order appears again in the kinetic energy magnitude arrangement, which is $E_{k,\rho}$ = 1400 kg/m^3 > $E_{k,\rho}$ = 1500 kg/m^3 > $E_{k,\rho}$ = 1600 kg/m^3. This phenomenon is also due to the stronger fluctuation of the less dense fluids and these effects caused by different viscous fluids on debris flow array and collision, and friction forces on debris flow movement, will require further investigation in the future.

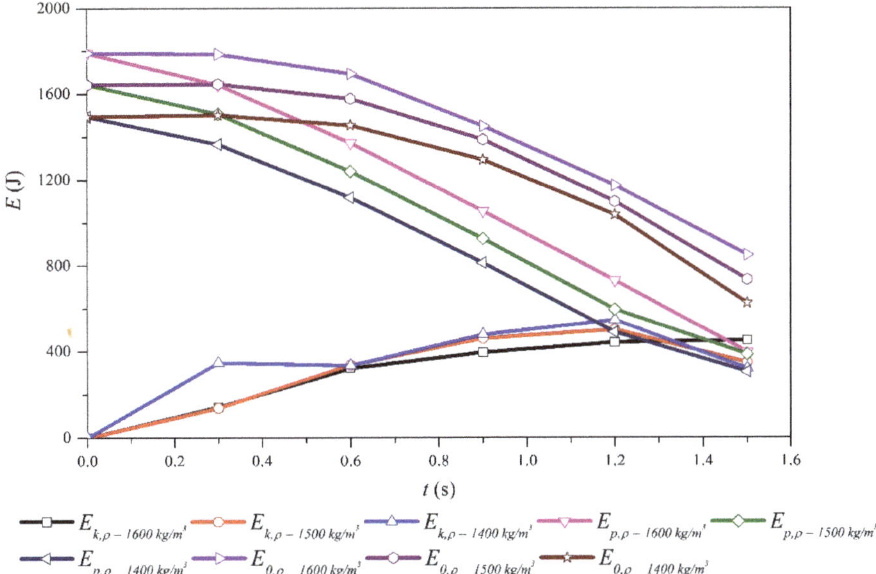

Figure 9. Energy evolution of different bulk density debris flows.

5. Summary and Conclusions

Debris flows are a natural phenomenon causing a lot of economical and human losses worldwide. Due to their nature, debris flows travel long distances at high speeds, and the time-space evolution of the relationship between soil and water content strongly affects the propagation stage. Thus, a quantitative modeling of this phenomenon is crucial to design strategies to be adopted to reduce the negative impacts. This paper contributes to this topic through the use of a SPH model investigating multiple combinations of fine and coarse particles with water content.

The SPH model was used to simulate tests conducted in the experimental facility that is located in the Debris Flow Observation and Research Station at Jiangjia Gully, the largest field research center in China—also known as "the debris flow museum". As previously demonstrated, the SPH model is capable of properly reproducing the main characteristics of debris flows (propagation height and velocity, and more importantly to correctly simulate the time-space evolution of solid and liquid particles during the whole process from initiation to propagation over an impervious/permeable bottom boundary).

Based on the theory of solid-liquid two phase flows, the viscous term in the SPH model was modified to make it suitable for nonhomogeneous viscous debris flows. It was found that the denser the fluid is, the greater are the yield stress and the turbulent viscous coefficient. However for laminar viscous coefficients, the fluid with the density of ρ = 1500 kg/m^3 has the largest values. The results obtained can be summarized as follows:

1. By comparing the shape and velocity of debris flows under different configurations, it was found that the vertical distribution of particles played a very important role in debris flow fluctuation, with a greater influence than on the viscous coefficient. The third configuration with mixed fine and coarse particles showed to fluctuate more violently, and this outcome confirmed one of the main assumptions for intermittent debris flows.

2. By analyzing the characteristics of the fluid movement processes, it was found that when the two layers (fine and coarse particles) are mixed with the water, liquid particles tended to gather towards the bottom side of the debris flow causing a correspondent decrease of height. However, this effect could only be observed at small-scale areas. The potential energy was greatly affected by the magnitude of the moisture content, while the kinetic energy was significantly affected by the derivative of moisture content in the L direction.

3. The differences of the energy conversion curves associated to different viscous coefficients were mainly noticed in kinetic energies. Fluids with smaller densities exhibited higher initiation velocities and higher fluctuations values.

The authors can also confirm that there are still some uncertainties within the results analyzed that could be reduced by the use of either novel physically-based entrainment laws or fully 3D mathematical approaches, which could surely more accurately take into consideration the variation of pore water pressures inside the propagating mass. Therefore, future research will target the development of 3D mathematical models to refine the findings and provide an even better understanding of this very complex natural phenomenon.

Author Contributions: All the authors jointly contributed to this research. A.S. was responsible for the proposition and design of the experiments, analysis of the results and conclusions of the paper; S.W. completed the numerical simulation, S.W. and M.R. analysed the experimental datasets and wrote the paper. M.W. and J.Q. performed the experiments.

Funding: The research reported in this manuscript is funded by the National key research and development plan (Grant No. 2018YFC1406404), the Interdiscipline Research Funds of Beijing Normal University and the Natural Science Foundation of China (Grant No. 11372048).

Conflicts of Interest: The authors declare no conflict of interest.

References

1. Kim, M.I.; Kwak, J.H.; Kim, B.S. Assessment of dynamic impact force of debris flow in mountain torrent based on characteristics of debris flow. *Environ. Earth Sci.* **2018**, *77*, 538. [CrossRef]

2. Zhang, S.; Zhang, L.M. Impact of the 2008 Wenchuan earthquake in China on subsequent long-term debris flow activities in the epicentral area. *Geomorphology* **2017**, *276*, 86–103. [CrossRef]

3. Silhan, K.; Tichavsky, R. Recent increase in debris flow activity in the Tatras Mountains: Results of a regional dendogeomorphic reconstruction. *CATENA* **2016**, *143*, 221–231. [CrossRef]

4. Brandinoni, F.; Mao, L.; Recking, A.; Rickenmann, D.; Turowski, J.M. Morphodynamics of steep mountain channels. *Earth Surf. Process. Landf.* **2015**, *40*, 1560–1562. [CrossRef]

5. Takahashi, T.; Das, D.K. *Debris Flow: Mechanics, Prediction and Countermeasures*; CRC Press: London, UK, 2014.

6. Shu, A.P.; Tian, L.; Wang, S.; Rubinato, M.; Zhu, F.Y.; Wang, M.Y.; Sun, J.T. Hydrodynamic characteristics of the formation processes for non-homogeneous debris-flow. *Water* **2018**, *10*, 452. [CrossRef]

7. Liu, J.J.; Li, Y. A review of study on drag reduction of viscous debris flow residual layer. *J. Sedim. Res.* **2016**, *3*, 72–80. (In Chinese)

8. Johnson, A.M. *Physical Processes in Geology*; Freeman Cooper & Company: San Francisco, CA, USA, 1970.

9. Bagnold, R.A. Experiments on a gravity-free dispersion of large solid spheres in a Newtonian fluid under shear. *Proc. R. Soc. Lond. Ser. A* **1954**, *22*, 49–63.

10. Takahashi, T. Mechanical characteristics of debris flow. *J. Hydraul. Div. Am. Soc. Civ. Eng.* **1978**, *104*, 1153–1169.

11. Chen, C.L. Generalized viscoplastic modelling of debris flow. *J. Hydraul. Div. Am. Soc. Civ. Eng.* **1988**, *114*, 237–258. [CrossRef]

12. Chen, C.L. General solutions for viscoplastic debris flow. *J. Hydraul. Div. Am. Soc. Civ. Eng.* **1988**, *114*, 259–282. [CrossRef]
13. O'Brien, J.S.; Julien, P.Y. Laboratory analysis of mudflow properties. *J. Hydraul. Div. Am. Soc. Civ. Eng.* **1988**, *114*, 877–887. [CrossRef]
14. O'Brien, J.S.; Julien, P.Y.; Fullerton, W.T. Two-dimensional water flood and mudflow simulation. *J. Hydraul. Div. Am. Soc. Civ. Eng.* **1993**, *119*, 244–261. [CrossRef]
15. Dai, Z.; Huang, Y.; Cheng, H.; Xu, Q. 3D Numerical modeling using smoothed particle hydrodynamics of flow-like landslide propagation triggered by the 2008 Wenchuan earthquake. *Eng. Geol.* **2014**, *180*, 21–33. [CrossRef]
16. Hosseini, S.M.; Manzari, M.T.; Hannani, S.K. A fully explicit three-step SPH algorithm for simulation of non-Newtonian fluid flow. *Int. J. Numer. Methods Heat Fluid Flow* **2007**, *17*, 715–735. [CrossRef]
17. Rodriguez-Paz, M.X.; Bonet, J. A corrected smooth particle hydrodynamics method for the simulation of debris flows. *Numer. Methods Part Differ. Equ.* **2004**, *20*, 140–163. [CrossRef]
18. Savage, S.B.; Hutter, K. The dynamics of avalanches of granular materials from initiation to run out. *Acta Mech. Sin.* **1991**, *86*, 201–223. [CrossRef]
19. Trunk, F.J.; Dent, J.D.; Lang, T.E. Computer modeling of large rock slides. *J. Geotech. Eng.* **1986**, *112*, 348–360. [CrossRef]
20. Iverson, R.M. The physics of debris flows. *Rev. Geophys.* **1997**, *35*, 245–296. [CrossRef]
21. Xenakis, A.M.; Lind, S.J.; Stansby, P.K.; Rogersb, B.D. An incompressible SPH scheme with improved pressure predictions for free-surface generalised Newtonian flows. *J. Non-Newton. Fluid Mech.* **2015**, *218*, 1–15. [CrossRef]
22. Springel, V.; Hernquist, L. Cosmological smoothed particle hydrodynamics simulations: A hybrid multiphase model for star formation. *Mon. Not. R. Astron. Soc.* **2003**, *339*, 289–311. [CrossRef]
23. Gómez-Gesteira, M.; Dalrymple, R.A. Using a three-dimensional Smoothed Particle Hydrodynamics method for wave impact on a tall structure. *J. Waterw Port Coast.* **2004**, *130*, 63–69. [CrossRef]
24. Flebbe, O.; Muenzel, S.; Herold, H.; Riffert, H. Smoothed Particle Hydrodynamics: Physical viscosity and the simulation of accretion disks. *Astrophys. J.* **1994**, *431*, 754–760. [CrossRef]
25. Lucy, L.B. A numerical approach to the testing of the fission hypothesis. *Astron. J.* **1977**, *82*, 1013–1024. [CrossRef]
26. Liu, G.R.; Liu, M.B. *Smoothed Particle Hydrodynamics: A Mesh-Free Particle Method*; World Scientific Publishing Company: Singapore, 2003; pp. 27–33.
27. Monaghan, J.J. Smoothed particle hydrodynamics. *Annu. Rev. Astron. Astrophys.* **1992**, *30*, 543–574. [CrossRef]
28. Gingold, R.A.; Monaghan, J.J. Smoothed particle hydrodynamics: Theory and application to non-spherical stars. *Mon. Not. R. Astron. Soc.* **1977**, *181*, 375–389. [CrossRef]
29. Gingold, R.A.; Monaghan, J.J. Kernel estimates as a basis for general particle methods in hydrodynamics. *J. Comput. Phys.* **1982**, *46*, 429–453. [CrossRef]
30. Monaghan, J.J. Particle methods for hydrodynamics. *Comput. Phys. Rep.* **1985**, *3*, 71–124. [CrossRef]
31. Fulk, D.A.; Quinn, D.W. An Analysis of 1-D Smoothed Particle Hydrodynamics kernels. *J. Comput. Phys.* **1996**, *126*, 165–180. [CrossRef]
32. Colagrossi, A.; Landrini, M. Numerical simulation of interfacial flows by Smoothed Particle Hydrodynamics. *J. Comput. Phys.* **2003**, *191*, 448–475. [CrossRef]
33. Bose, A.; Carey, G.F. Least-squares p-r finite element methods for incompressible non-Newtonian flows. *Comput. Methods Appl. Mech.* **1999**, *180*, 431–458. [CrossRef]
34. Wood, D. Collapse and fragmentation of isothermal gas clouds. *Mon. Not. R. Astron. Soc.* **1981**, *194*, 201–218. [CrossRef]
35. Bonet, J.; Lok, T.S.L. Variational and momentum preservation aspects of Smooth Particle Hydrodynamic formulations. *Comput. Method Appl. Mech.* **1999**, *180*, 97–115. [CrossRef]
36. Loewenstein, M.; Mathews, W.G. Adiabatic particle hydrodynamics in three dimensions. *J. Comput. Phys.* **1986**, *62*, 414–428. [CrossRef]
37. Shao, S.; Lo, E.Y.M. Incompressible SPH method for simulating Newtonian and non-Newtonian flows with a free surface. *Adv. Water Resour.* **2003**, *26*, 787–800. [CrossRef]

38. Lo, E.Y.M.; Shao, S.D. Simulation of near-shore solitary wave mechanics by an incompressible SPH method. *Appl. Ocean Res.* **2002**, *24*, 275–286.
39. Yang, H.J.; Wei, F.Q.; Hu, K.H. Determination of the maximum packing fraction for calculating slurry viscosity of debris flow. *J. Sediment. Res.* **2018**, *3*, 382–390.

Article

Comparative Study on Violent Sloshing with Water Jet Flows by Using the ISPH Method

Hua Jiang [1,2], **Yi You** [1], **Zhenhong Hu** [1,*], **Xing Zheng** [1] **and Qingwei Ma** [1,3]

[1] College of Shipbuilding Engineering, Harbin Engineering University, Harbin 150001, China; jiangh@gumeco.com (H.J.); youyione001@hrbeu.edu.cn (Y.Y.); zhengxing@hrbeu.edu.cn (X.Z.); q.ma@city.ac.uk (Q.M.)
[2] Guangzhou Marine Engineering Corporation, Guangzhou 510250, China
[3] School of Mathematics, Computer Science & Engineering, City, University of London, London EC1V 0HB, UK
*   Correspondence: huzhenhong@hrbeu.edu.cn; Tel.: +86-451-8256-8147

Received: 13 October 2019; Accepted: 3 December 2019; Published: 9 December 2019

Abstract: The smoothed particle hydrodynamics (SPH) method has been playing a more and more important role in violent flow simulations since it is easy to deal with the large deformation and breaking flows from its Lagrangian particle characteristics. In this paper, the incompressible SPH (ISPH) method was used to simulate the liquid sloshing in a 2D tank with water jet flows. The study compares the liquid sloshing under different water jet conditions to analyze the effects of the excitation frequency and the water jet on impact pressure. The results demonstrate that the water jet flows can significantly affect the impact pressures on the wall caused by violent sloshing. The main purpose of the paper is to test the ISPH ability for this study and some useful regulars that are obtained from different numerical cases and study the effect of their practical importance.

Keywords: ISPH; liquid sloshing; water jet flow; impact pressure; excitation frequency

---

1. Introduction

The phenomenon of violent sloshing appears widely in the field of Naval Architecture and Ocean Engineering, especially in liquid cargo carriers, such as the LNG (liquefied natural gas) carriers. During this process, the motion of liquid in partially filled tanks may cause large global and local loads on the tank walls when the frequency of sloshing is close to the natural frequency of the liquid tank. This consequence would be very serious in engineering practice, which may cause damage to the hull structure and even affect the stability of the carrier [1]. Therefore, sloshing is an old topic, but it still needs to be studied in depth.

At present, lots of studies on the violent sloshing flows have been carried out by the linear and nonlinear potential flow theory and the scaled model experiment. Faltinsen [2–4] used the incompressible potential flow theory to simulate liquid sloshing and obtained the formulas that have been widely used in the field of sloshing simulation. However, the method can be used to study sloshing tanks with relatively simple geometry and internal structure. In addition, Akyildiz et al. [5] investigated the pressure distribution on a rectangular tank during the process of sloshing by an experimental method. Sames et al. [6] studied sloshing in a rectangular tank with a baffle, and a cylindrical tank was also considered. Indeed, the experimental method can be applied to study the sloshing in the tank with more complex shapes, but it also requires high expenses for the site and facility. Hence, the numerical method has been getting more important in the simulation of liquid sloshing in recent years. The conventional numerical methods are carried out by using Euler grids. Wu et al. [7] simulated the sloshing waves in a 3D tank based on the finite element method (FEM). In the conventional grid-based methods, in order to track the moving free surface, some additional techniques, such as the Volume-of-Fluids, are used in the methods. The VOF uses the volume fraction of fluid in gird to define the free surface. However, the problems of numerical diffusion become serious when the surface cell becomes extremely complicated, such as in a liquid sloshing, which can easily fail to simulate because of the large deformation of grids. Recently, a kind of mesh-less method named smoothed particle hydrodynamics (SPH) has attracted quite a few researchers' attention [8]. It does not depend on any grids, and the computation is purely based on a group of discrete points that can move freely. So, it can capture the free surface flow conveniently, which is more suitable for treating the problems of large deformation of free-surfaces. Delorme and Colagrossi et al. [9] investigated impact pressure in the case of shallow water sloshing by the SPH method, compared the results with experimental ones, and then discussed the influence of viscosity and density re-initialization on the SPH results. Gotoh and Khayyer [10] simulated the violent sloshing flows using the incompressible SPH (ISPH) method and presented two schemes to enhance the accuracy of the simulation of impact pressures. Zheng and You [11] compared the effect of different baffle configurations on mitigating sloshing by the ISPH method. A great deal of research [12,13] shows that the ISPH method can improve the accuracy and stability of the calculation pressure, and the pressure field is smoother.

As a matter of fact, the marine environment is always very complex when sloshing happens. If an oil fire occurs, it would be a big disaster and hard to control, and the water jet flow outside would get into the tank to put out the oil tank fire, which would influence the impact effect of sloshing. Such a consequence may cause more serious damage to the hull structure, which would be very serious in engineering practice. With regard as the problem of jet flows, Hatton et al. [14,15] studied the trajectories of large water jets that are used in the design of fire-fighting systems, particularly those used in offshore situations, and evaluated the effects of flow-rate, pressure, and nozzle size during the process of the system design. Fischer et al. [16] used three different CFD codes, namely, the CHYMES multiphase flow model, the FEAT finite element code, and the Harwell-FLOw3D finite volume code, to simulate the problem of a laminar jet of fluid injected into a tank of fluid at rest and make a detailed comparison. Aristodemo et al. [17] studied the plane jets propagating into still fluid tanks and current flows by using the WCSPH method. Andreopoulos et al. [18] carried out an experiment on the flow generated by a plane with a buoyant jet discharging vertically into shallow water.

In this paper, the liquid sloshing with a water jet flow from the top of the tank will be studied by using the incompressible SPH (ISPH) method. Through the comparison of different situations, the sloshing effects and characteristics of the impact pressure are studied. The aim of this study is to summarize the influence of the water jet flow on sloshing, so as to give a reference for practical engineering.

2. ISPH Methodology

2.1. Governing Equations

The SPH model is based on the semi-Lagrangian form of the continuity equation and the momentum equation. In the ISPH method, the density of fluid is considered to be a constant, and thus, the governing equations are written as follows:

$$\nabla \cdot u = 0, \tag{1}$$

$$\frac{Du}{Dt} = -\frac{1}{\rho}\nabla P + g + v_0 \nabla^2 u, \tag{2}$$

where ρ is the density of fluid; u is the velocity of particle; t is the time; P is the particle pressure; g is the gravitational acceleration; v_0 is the kinematic viscosity; and ∇ is Hamilton operator, which is a vector operator.

2.2. Particle Approximation

The computational domain of the SPH method is composed of a group of discrete particles, and each particle is given corresponding physical information, such as density, volume, mass, velocity, and pressure. The physical information of each particle can be approximately obtained by the information carried by the surrounding particles, which is shown as follows.

$$f(\boldsymbol{r_i}) \;=\; \sum_{j=1}^{N} \frac{m_j}{\rho_j}\, f\,(\boldsymbol{r_j})\, W\,(\boldsymbol{r_{ij}}) \tag{3}$$

where $f(\boldsymbol{r})$ represents the physical information of particles, m is the mass of the particle, i and j are the center particle and neighbor particle, respectively. N is the number of neighbor particles. $W(\boldsymbol{r_{ij}})$ is the kernel function, which can reflect the different effects between different particles. In this paper, the cubic B-spline kernel proposed by Monaghan et al. [19] is used as follows:

$$W\,(\boldsymbol{r_{ij}},\, h) \;=\; \alpha_d \begin{cases} \frac{2}{3} - q^2 + \frac{1}{2}q^3, & 0 \le q < 1 \\ \frac{1}{6}(2-q)^3, & 1 \le q < 2 \\ 0, & 2 \le q \end{cases} \tag{4}$$

where h is the kernel smoothing length, r_{ij} is the distance between the i and j particle, α_d is a constant, and when the case is 2D, its value is $\frac{15}{7\pi h^2}$ and $q = \frac{r_{ij}}{h}$.

So the derivatives of $f(\boldsymbol{r})$ can be represented as:

$$\nabla f(\boldsymbol{r_i}) \;=\; \sum_{j=1}^{N} \frac{m_j}{\rho_j}\, f\,(\boldsymbol{r_j})\, \nabla_i W\,(\boldsymbol{r_{ij}}), \tag{5}$$

where ∇_i is the gradient, which is taken with respect to the particle i.

2.3. Poisson Pressure Equation

In the ISPH, a two-step projection method is used to solve the velocity and pressure field from the continuity equation and momentum equation [20]. The first step is the prediction of velocity without considering the pressure term. The second step is the correction step in which the pressure term is added through the pressure Poisson equation (PPE), then the PPE is obtained as follows:

$$\nabla^2 P^{t+\Delta t} \;=\; \frac{\rho \nabla \cdot \boldsymbol{u}^*}{\Delta t} \tag{6}$$

where $\boldsymbol{u}^*$ is the intermediate particle velocity at the first step.

Similarly, Shao and Lo [20] proposed a projection-based incompressible approach by imposing the density invariance on each particle, leading to the following PPE equation:

$$\nabla \cdot \left(\frac{1}{\rho^*} \nabla P^{t+\Delta t} \right) \;=\; \frac{\rho_0 - \rho^*}{\rho_0 \Delta t^2} \tag{7}$$

where ρ^* is the density at the intermediate time step, ρ_0 is the initial fluid density, and the combined PPE incorporates both the velocity-divergence-free condition and the zero-density-variation condition, which is obtained as:

$$\nabla^2 P^{t+\Delta t} \;=\; \alpha \frac{\rho_0 - \rho^*}{\Delta t^2} + (1-\alpha) \frac{\rho_0 \nabla \cdot \boldsymbol{u}^*}{\Delta t} \tag{8}$$

where α is a blending coefficient. If α is equal to 1, in Equation (8), the source term of PPE adopts the density variable effect, which may lead to substantial pressure noises and particle randomness caused by larger density changes. If α is equal to 0.0, the source term of PPE adopts the velocity divergence effects, which is smoother for source term distribution, but it will cause the pattern distribution of some particles. In order to make the source term more reliable, α is equal to 0.01 in this paper, according to lots of computational experience. And more advanced PPEs with the error-compensating source term (ECS) can follow the references [21,22].

2.4. Calculation of Spatial Derivatives

According to Equation (5), the spatial derivatives of pressure and velocity can be calculated as follows:

$$\nabla P_i = \rho_i \sum_{j=1}^{N} m_j \left(\frac{P_j}{\rho_j^2} + \frac{P_i}{\rho_i^2} \right) \nabla_i W\left(r_{ij}, h\right), \tag{9}$$

$$\nabla u_i = -\frac{1}{\rho_i} \sum_{j=1}^{N} m_j (u_i - u_j) \nabla_i W\left(r_{ij}, h\right). \tag{10}$$

Therefore, in the ISPH method, the viscous term adopts the following form:

$$\nabla(v_i \nabla \cdot u_i) = \sum_{j=1}^{N} 4m_j \left(\frac{v_i + v_j}{\rho_j + \rho_i} + \frac{u_{ij}\, r_{ij}}{r_{ij}^2 + \eta^2} \right) \nabla_i W\left(r_{ij}, h\right) \tag{11}$$

where η is a small parameter to avoid the singularity, and in this paper, the value is chosen to be 0.1 h. So, the PPE is discretized by combining the SPH gradient and divergence rules to obtain:

$$\nabla\left(\frac{1}{\rho^*}\nabla P\right) = \sum_{j=1}^{N} m_j \left(\frac{8}{(\rho_j + \rho_i)^2} \frac{P_{ij}\, r_{ij}}{r_{ij}^2 + \eta^2} \right) \nabla_i W\left(r_{ij}, h\right). \tag{12}$$

The treatment of free surface and solid boundary conditions follows the study of Zheng et al. [23].

2.5. Inlet Boundary Treatment

With regards to the case of a general closed boundary, sufficient particles are given at the initial time, and no particles are added during the middle steps. However, the model of sloshing with a water jet flow needs to add particles constantly at the top boundary. Hence, the condition of the top boundary should be treated as an inlet boundary to add the particles of the water jet. However, if the particles are added directly, there will be only a row of particles at the beginning, which would lead to big errors in the particle approximation.

In this paper, three rows of virtual particles are arranged at the top boundary, as shown in Figure 1a, which have the same velocity and physical information as the initial water jet particles, but they do not participate in the solving of PPE, they are just used in the particle approximation. The velocities of virtual particles remain constant, and when they have moved a distance of a particle size dx, as shown in Figure 1b, they will return to the initial position automatically, as shown in Figure 1c. So, the distance between virtual particles and the water particles is just enough to add a row of new water particles. Finally, the new row of water particles will be added, as shown in Figure 1d.

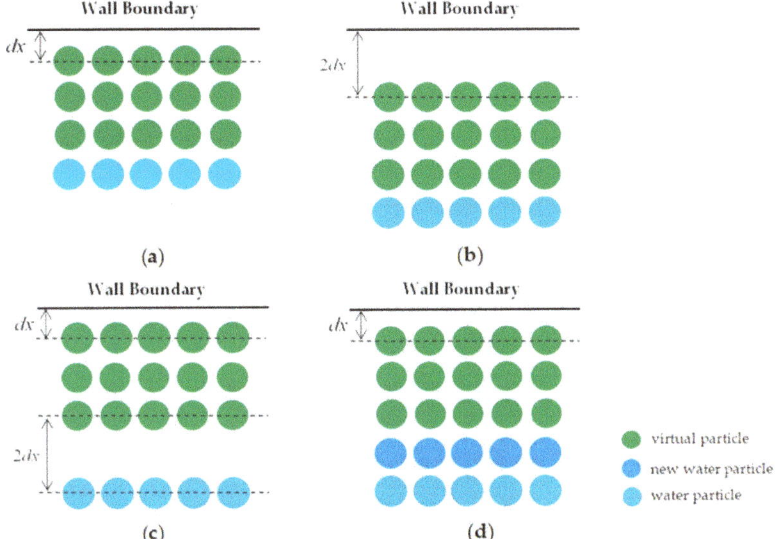

Figure 1. Different steps of the inlet boundary treatment: (**a**) Initial distribution of particles; (**b**) Particles has moved a distance of a particle size; (**c**) The virtual particles are return to the initial positions; (**d**) The new water particles are added.

3. Numerical Results and Validation

In this section, the numerical results of the sloshing tank and water jet calculated by the ISPH method are compared with the experimental data to validate the accuracy of the ISPH model.

3.1. Sloshing Tank Simulation and Validation

Figure 2 illustrates the rectangular sloshing tank that was used in the experiment of Liao and Hu [24], and the schematic view gives the geometrical dimensions of the sloshing tank; the length of the tank is 1.2 m, the height is 0.6 m, the initial depth of the liquid inside is 0.12 m, and the rotation center is located at the geometric center of the tank. M_1 and M_2 are two pressure sensors, which are used to record the pressure of the two points. In the experiment, the sloshing tank is imposed a rolling motion, and the amplitude and period of excitation are set as 10° and 1.85 s, respectively. The distribution of the particles in the ISPH computational model is uniform, with a particle spacing $dx = 0.002$ m. A constant time step of $dt = 0.0003$ s is used. The pressure data is extracted for 8 s after about twice the sloshing period T, which is relatively stable.

Figure 3 shows the snapshots of the sloshing motion of the ISPH model with the calculated pressure fields. Meanwhile, the corresponding experimental photographs are considered for the comparisons of the motion patterns. According to the comparisons, the ISPH model can get a good agreement on surface profile with the experimental results, and most of the features of the violent sloshing process have been captured by the ISPH model in a satisfactory manner. In addition, the ISPH method also provides a reliable regular distribution of pressure at the impact regions, and the pressure fields show a very stable pattern with little pressure noise.

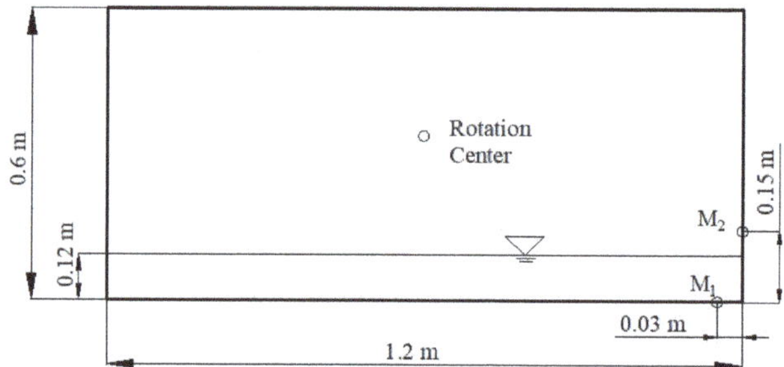

Figure 2. Schematic view of the sloshing tank following Liao and Hu [24].

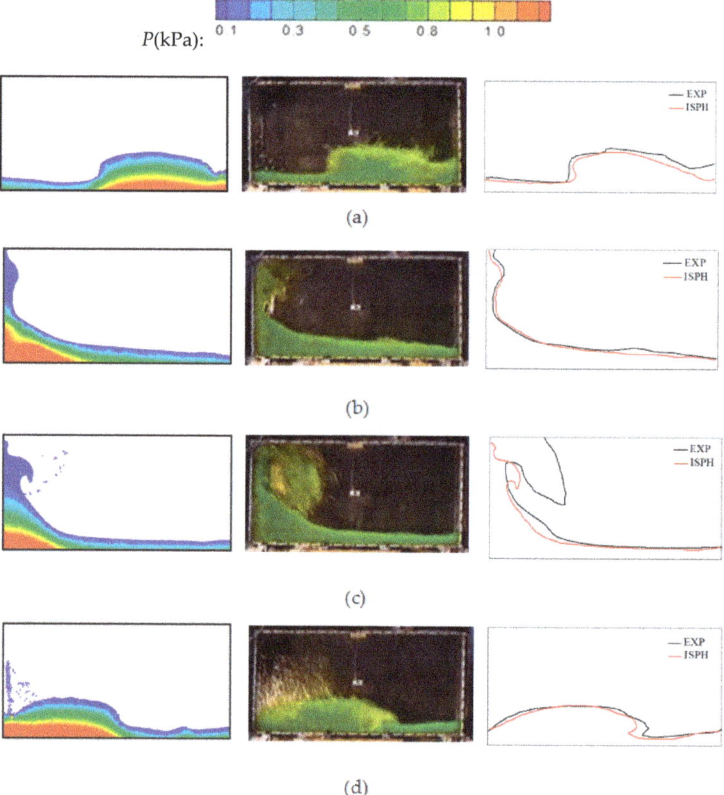

Figure 3. Sloshing snapshots computed by the ISPH (left) compared with the experimental photos (middle), and overlaid free surfaces (right): (**a**) 0.0*T*; (**b**) *T*/6; (**c**) *T*/3; (**d**) *T*/2.

Figure 4 presents the time histories of calculated pressure at points M_1 and M_2 by the ISPH method, which are compared with the experimental data. From the contrast, the pressure traces with both first and second peaks calculated by the ISPH method have made good accuracy with experimental results. So the comparisons between the ISPH results and experimental results demonstrate that the ISPH method can get accurate results for the maximum peak values and the phases of the pressure time histories.

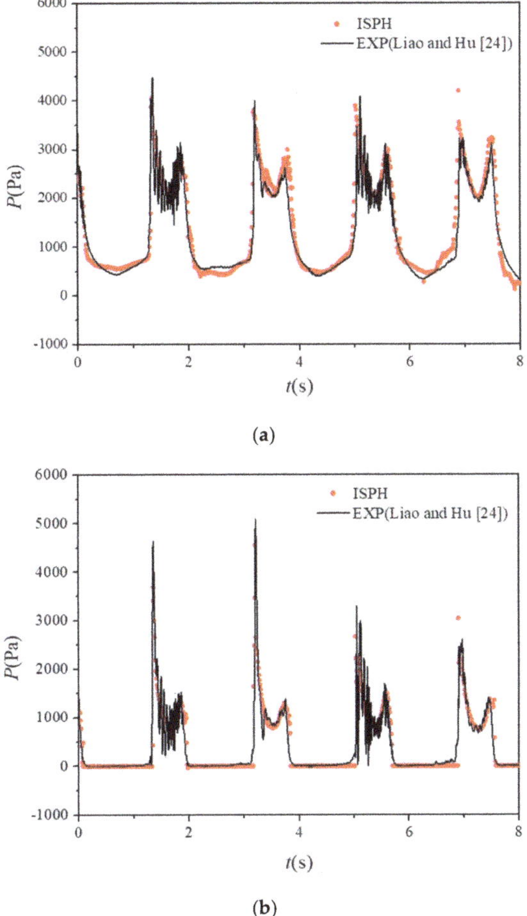

(a)

(b)

Figure 4. Comparisons of pressure time history between ISPH and experimental data of Liao and Hu [24] at: (**a**) M_1; (**b**) M_2.

3.2. Convergence Analysis of ISPH Model

In order to validate the convergence of the ISPH sloshing model, three different particle sizes have been tested, which are 0.002, 0.003, and 0.004 m, corresponding to the number of particles are 49,000, 16,000, and 9000, respectively. And the pressure at M_1 and M_2 calculated in three different particle sizes are compared with the experimental data, and the results are shown in Figure 5. It shows that the amplitude of non-physical pressure oscillations decreases when particle size is from 0.004 m to 0.002 m. Moreover, the enlarged parts manifest that the pressure traces are more smoothed and more closed to the experimental results. Furthermore, the quantitative comparisons in three different particle sizes are carried on M_1 and M_2, which obtained the values of the mean error in Equation (13), the results are shown in Table 1, more details can refer to Zheng et al. [23]. From the comparison of pressure results, the errors decrease with the particle number increasing, and the results demonstrate that the ISPH method has very good convergence and stability.

$$E_a = \frac{1}{N} \sum_{i=1}^{N} \left| \frac{\widetilde{P(t)} - P(t)}{P(t)} \right| \tag{13}$$

where *Ea* is the defined mean error; $\widetilde{P(t)}$ is the numerical results of the pressure; $P(t)$ is the experimental pressure; N is the number of sampling points.

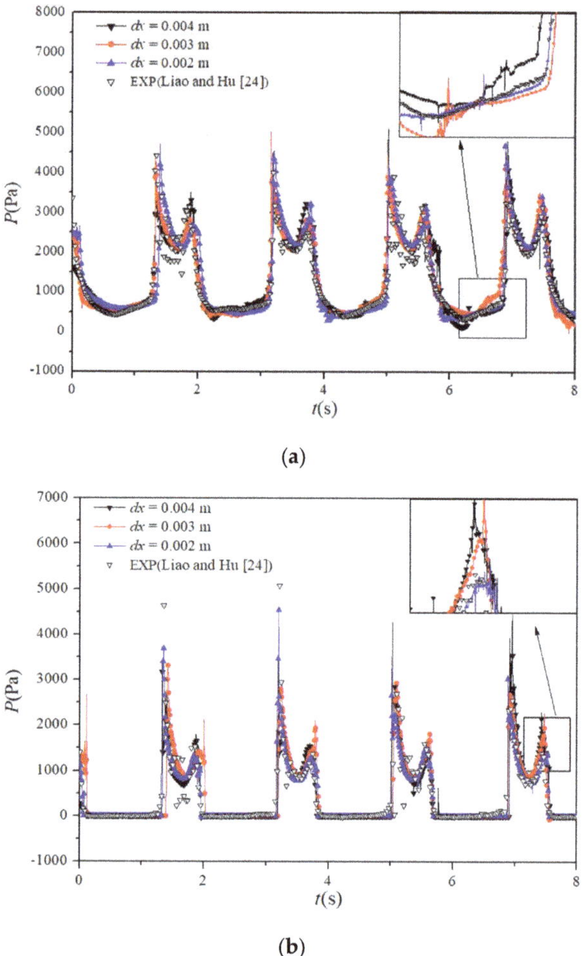

(**a**)

(**b**)

Figure 5. Comparisons of pressure time histories between incompressible smoothed particle hydrodynamics (ISPH) and experimental data of Liao and Hu [24] at (**a**) M_1 and (**b**) M_2.

Table 1. Numerical error of ISPH in three particle sizes.

Particle Size (m)	E_a (M_1)	E_a (M_2)
0.004	1.77%	1.35%
0.003	0.97%	0.93%
0.002	0.40%	0.36%

Figure 6 gives the convergence rate of the pressure results calculated by this ISPH method. It is shown that the numerical error E_a is closer to the second-order accuracy for the results of pressure calculated at M_1 and M_2. The ISPH method provides good performance for liquid tank sloshing; thus, it will be used for the study of the sloshing tank applications.

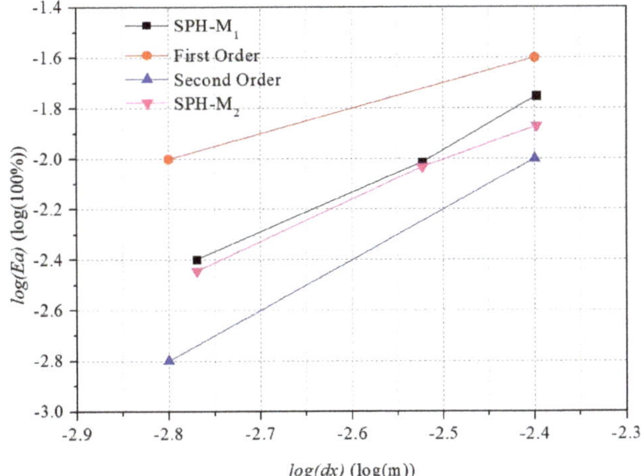

Figure 6. Convergence and error analysis of pressure calculated by the ISPH method.

3.3. Numerical Validation of the Water Jet Model

Figure 7 shows the schematic view of the water jet test case, which was used in the experiment of Kvicinsky et al. [25]. The diameter of the jet inlet is $D = 0.03$ m, and it is located at $H = 0.1$ m above a flat plate. The initial velocity of the water jet was $v = 19.81$ m/s. In the ISPH, a particle size $dx = 0.0015$ m was used, the time step $dt = 0.00001$ s, and the physical time of simulation was $T = 0.03$ s. The pressures at the plate were recorded.

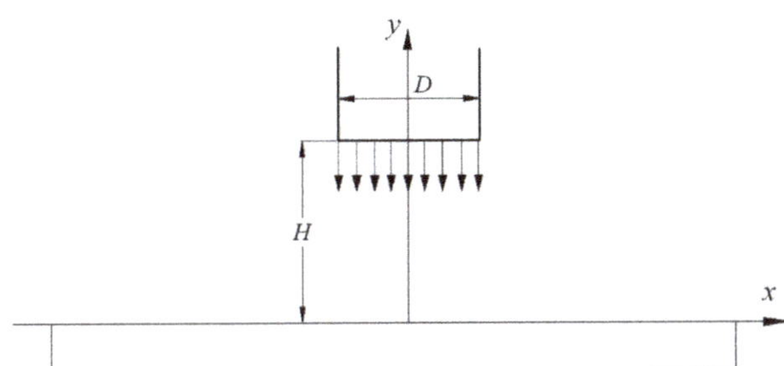

Figure 7. Schematic view of the water jet, following Kvicinsky et al. [25].

In the experiment, the pressure coefficient C_p on the flat plate was used as follows:

$$C_p = \frac{2P}{\rho v^2} \tag{14}$$

where P is the pressure at the flat plate, ρ is water density, and v is the initial velocity of the water jet. Figure 8 shows the snapshots of the numerical results. In order to get the stable pressure data, the pressures were extracted when the physical time is greater than 0.015 s. Figure 9 shows the comparison of pressure coefficient C_p among the ISPH results, the experimental data, and the CFD results, where x is the coordinate value on the flat plate. Figure 9 shows that the ISPH results have good agreement with both the experimental data and CFD results, which demonstrates that the ISPH method can have high accuracy on the impact pressure caused by the water jet flow.

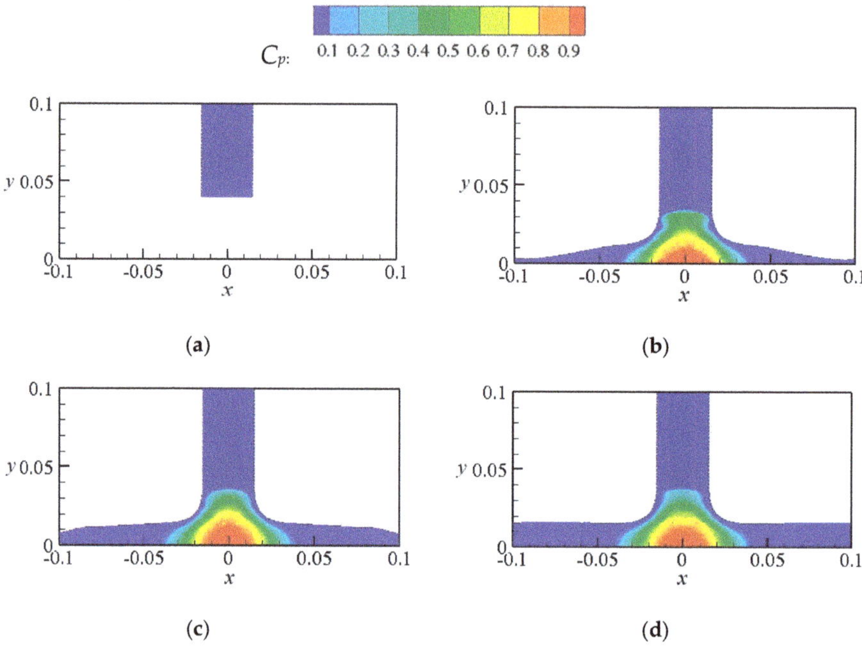

Figure 8. Pressure contour snapshots of the water jet case by ISPH: (**a**) *T*/10; (**b**) *T*/4; (**c**) *T*/3; (**d**) *T*.

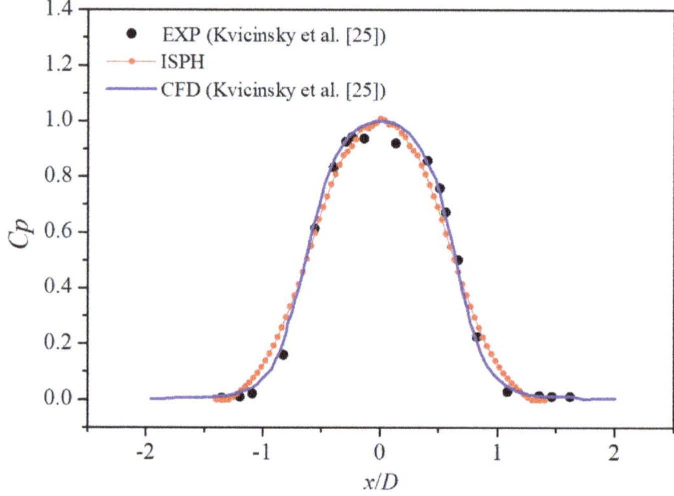

Figure 9. Comparisons of pressure coefficient C_p among the ISPH results, the CFD results, and the experimental data of Kvicinsky et al. [25].

3.4. Validation of Injected Water Jet Flow Model

Figure 10 gives the geometric dimensions of a rectangular tank. The height of the tank is $H = 0.5$ m, its width is $B = 0.3$ m, and the tank is full of water. $D_i = 0.08$ m, which is the inflow jet diameter in the top center of the tank, and $D_0 = 0.012$ m, which is the outflow diameter. The water jet is injected from D_i with the velocity $V = -0.45\,j$ m/s, and its direction is straight down.

Figure 10. Sketch of the jet injected at the top of a water tank.

Figure 11 shows the comparisons of the vertical component of the velocity field w between ISPH and the results of Aristodemo et al. [17] at three significant time instants. It can be seen from Figure 11 that the still water is accelerated progressively with the entry of the water jet flow, and the highest negative velocities are symmetrical by the jet centerline at the initial time. However, these negative velocities become non-symmetrical as time goes on. According to the results of the vertical velocity component by the ISPH method, it is in good agreement with the results of Aristodemo et al. [17].

To quantify the accuracy of the ISPH results, Figure 12 gives the comparisons of relative depth h/H of jet flow between the results of Aristodemo et al. and Fletcher et al. [16,17], where h is the depth of jet flow penetration. The results of ISPH have a high degree of coincidence with the ones of other numerical models. The ISPH method has good potentials for the injected water jet flow.

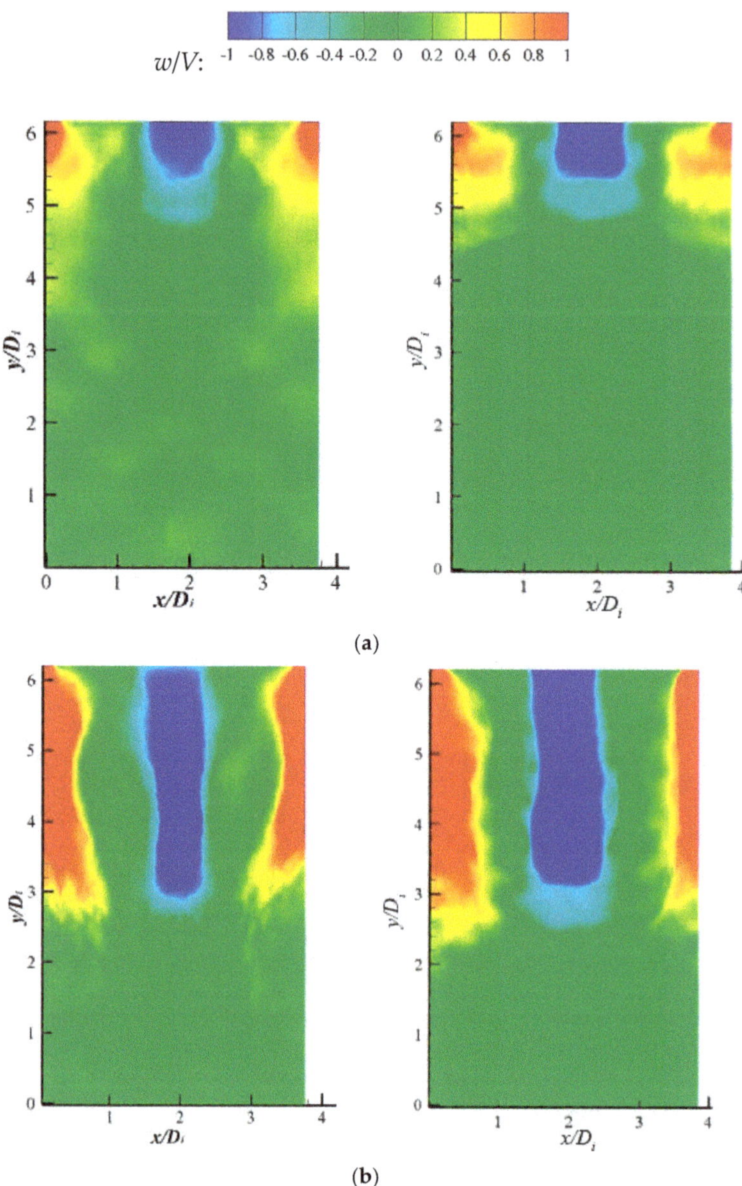

(a)

(b)

Figure 11. *Cont.*

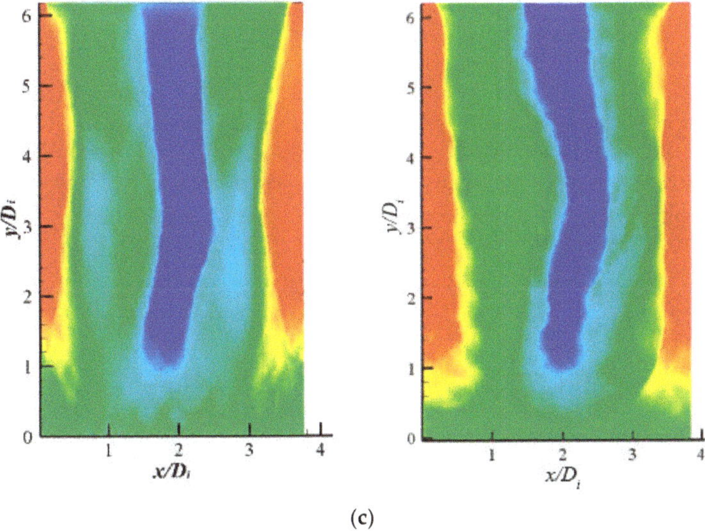

(c)

Figure 11. Comparisons of the vertical component of the velocity field between the results of the ISPH method (left) and the results of Aristodemo et al. [26] (right) at $t(V/D_i)$ = (**a**) 2.8; (**b**) 6.2; (**c**) 13.4.

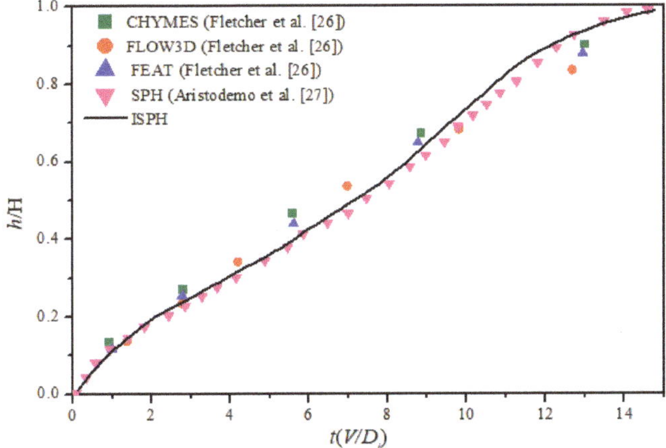

Figure 12. Comparison of the relative depth of jet flow penetration between ISPH and other numerical models.

4. Results and Analyses

The main scientific problem discussed in this paper is the effect of the water jet on sloshing. The calculation formula of the natural frequency of a liquid tank is shown as follows:

$$\Omega_0 = \sqrt{(g\pi/L)\tanh(\pi d/L)} \tag{15}$$

where g is the gravitational acceleration, L is the length of the tank, and d is the depth of water.

4.1. Sloshing Behaviors with Water Jet

In this section, the liquid sloshing in a rectangular tank with a water jet from the top of the tank is considered. The configuration of the sloshing tank is shown in Figure 13, and the geometrical dimensions of the sloshing tank are the same as the tank in Figure 2, but one difference is that there is water flow jetted from the center on the top of the tank. The tank experienced a rolling motion with an amplitude of 8° and an excitation period of 1.85 s. Four sensors, P_1–P_4, were placed, as shown in the figure, to monitor the pressure distributions. The initial water depth is 0.12 m, and the tank is rotated at the geometric center. The water jet enters the liquid tank through the center on top of the tank, and the initial velocity was imposed to 0.3 m/s.

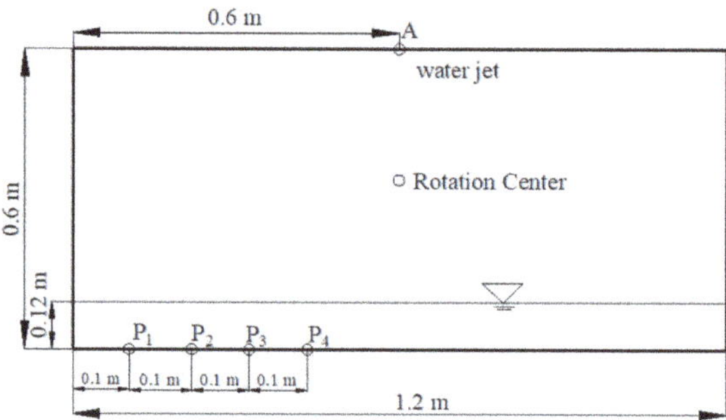

Figure 13. Schematic view of the sloshing tank with the water jet from the center on top.

The computed particle snapshots with pressure contours and streamlines for this case are shown in Figure 14. It shows the violent sloshing flow and water jet flow with strong interaction for a sloshing period T. The smooth and noise-free stable pressure fields indicate the robustness of the presented ISPH method, which is a good way to analyse the impact pressures. It also can be seen from the streamlines that the fluid is entrapped generating some recirculating counter-rotating cells in the violent sloshing process.

Then, the pressure histories at four different locations at the bottom are recorded and compared. Figure 15 gives the comparison of four pressure histories when the velocity is 0.3 m/s, it shows that the trends of four pressure histories are generally similar, the peak value of impact pressure gradually increases with the water jets into the tank, and the two-peak pressure patterns of four histories are remarkable. By way of contrast, the maximum pressure appears at the location of P_1, meanwhile the change range of impact pressure is also the largest. It can get higher pressure when the monitor location is closer to the side wall, the sloshing effect is also more violent in this area. Therefore, in the next cases, the impact pressure at P_1 will be studied and analyzed.

The liquid sloshing in a rectangular tank with a water jet of various velocities from the top of the tank is considered. The configuration of the sloshing tank is shown in Figure 13, and different initial velocities of the water jet flow are used in this case, which are 0.2, 0.3, 0.4, and 0.5 m/s, respectively. Figure 16 portrays the snapshots of the sloshing process with pressure fields at $t = 0.296$ s with different initial velocities. It can be seen that when the water jet flow gets into the free surface, the strong impact generates a large free surface deformation with two waves running out along the water jet flow, creating an open-air cavity. Then, the air cavity gradually closes, and the two free surfaces form a short-term bump. From the comparison, the open-air cavity becomes larger with the initial velocity increasing.

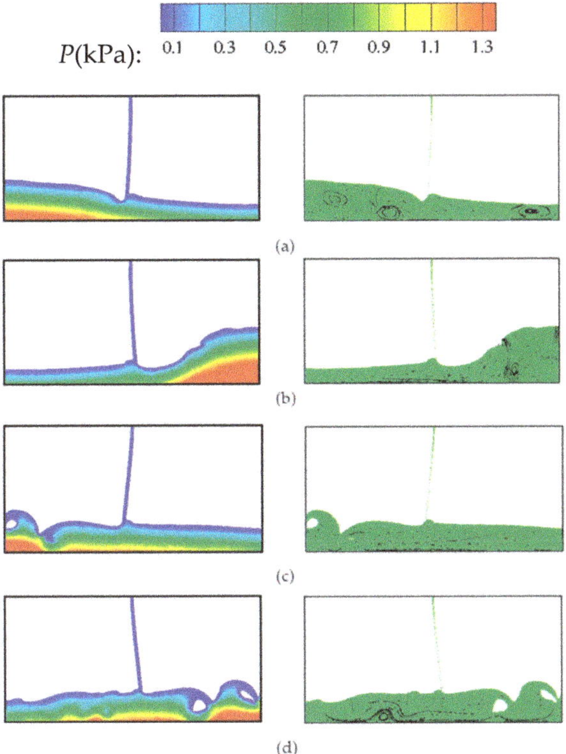

Figure 14. Particle snapshots with pressure contours (left) and streamlines (right) of sloshing flow interactions with water jet flow: (**a**) $T/4$; (**b**) $T/2$; (**c**) $3T/4$; (**d**) T.

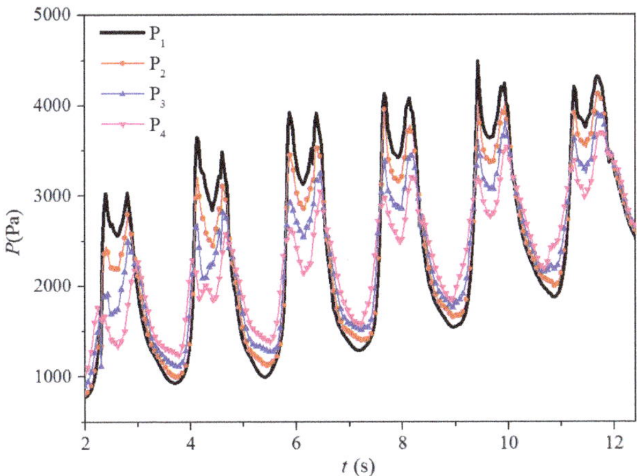

Figure 15. Comparison of pressure histories at four different locations at the bottom.

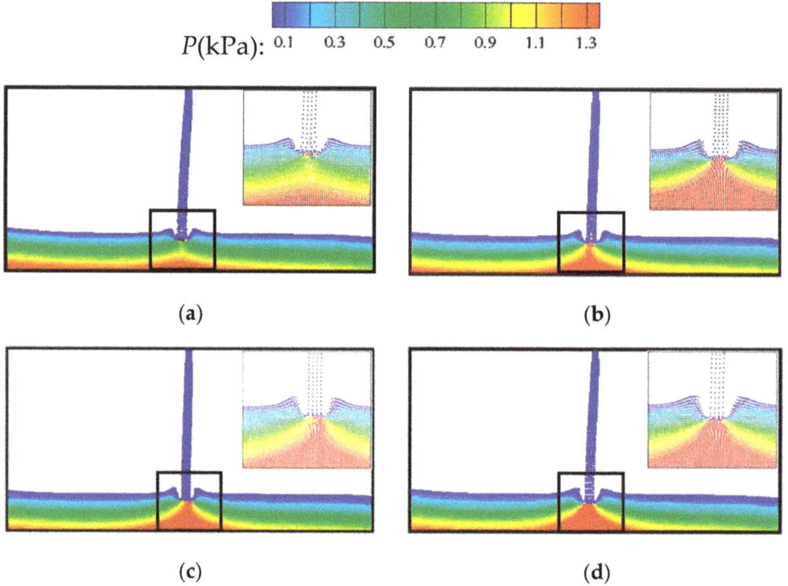

Figure 16. Comparison of sloshing patterns with different initial velocities at 0.296 s: (**a**) 0.2 m/s; (**b**) 0.3 m/s; (**c**) 0.4 m/s; and (**d**) 0.5 m

4.2. The Effects of the Water Jet Flow Position

Now, the liquid sloshing in a rectangular tank with a water jet flow from various positions at the top of the tank is compared. These correspond to the configurations shown in Figure 17, and there is also a harmonic rolling motion imposed on the geometry center of the tank with an amplitude 8° and a series of excitation frequencies, where P_1 is the point of pressure monitoring. A_1, A_2, and A_3 are three different locations for the water flow inlet, and d is 0.5, 0.3, and 0.1 m, respectively, which is the horizontal distance between the pressure probe and the water jet flow. The initial velocity of the water jet flow is 0.3 m/s.

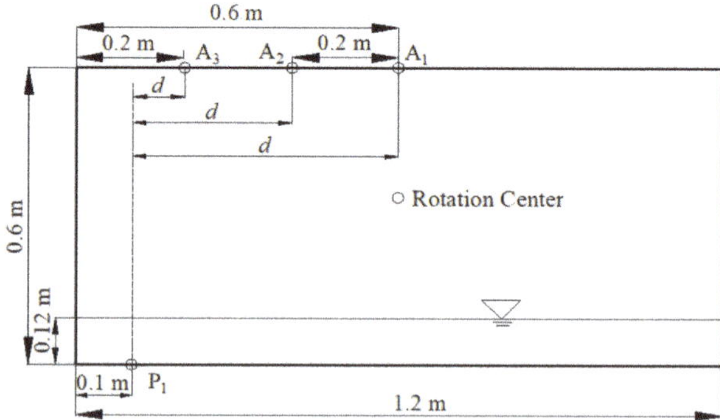

Figure 17. Schematic view of the sloshing tank with water jet flows from different positions.

Figure 18 shows the comparisons of free surface profiles at $t = 0.296$ s with the water jet from three positions at the top of the tank when the excitation period is 1.85 s. From the contrast of the three free surfaces, it can be seen that the deformations of the free surface generated by the impact of three water flows are in a basic agreement, which generates the air cavities with the same shapes and sizes.

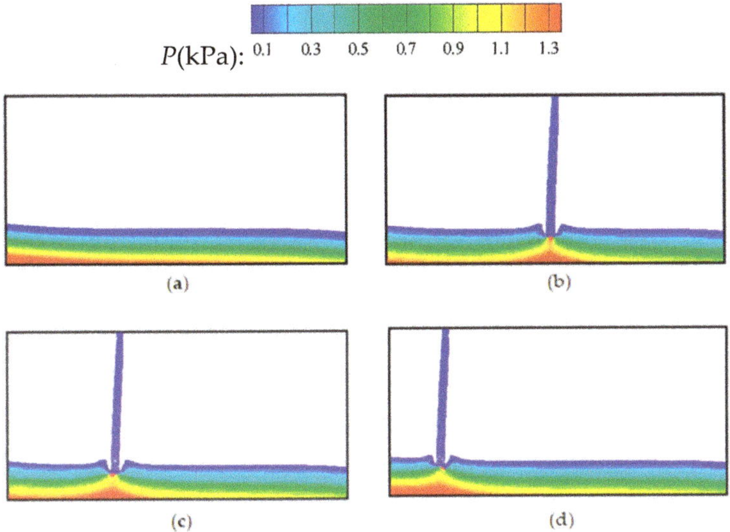

Figure 18. Comparison of free surface profiles at the instant $t = 0.296$ s with a water jet from three positions: (**a**) no water jet; (**b**) A_1; (**c**) A_2; (**d**) A_3.

Figure 19 gives the peak value, which is obtained from the impact pressure of P_1 at different excitation frequencies. It can be seen that the maximum pressure at P_1 decreases when a water flow enters the sloshing tank with an initial velocity, and the excitation frequency where the maximum pressure occurs increases. Also, they change accordingly with the position of the water flow. When the horizontal distance between the water flow and P_1 is closer, the effect is more obvious.

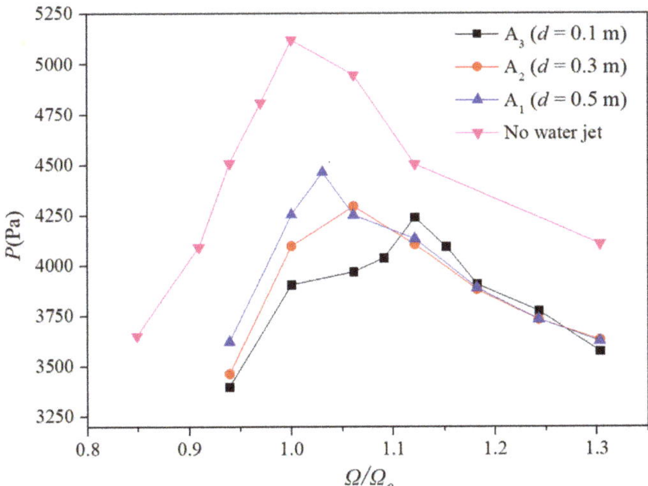

Figure 19. Comparisons of impact pressure with the water jet flow in different positions.

Through quantitative analysis, the maximum pressure is reduced by about 12.83% compared to the system with and without a water jet flow at A_1, and this difference reaches up to 17.22% when the water jet flow comes from A_3. It shows that the water jet flow can reduce the sloshing impact load on the tank wall. When the position of the water jet flow is closer to the pressure probe, the sloshing impact load becomes smaller. In addition, the water jet flow can also change the frequency where the maximum pressure occurs, so it is an effective way of avoiding the appearance of maximum pressure by using a water jet flow in an appropriate position.

4.3. The Effects of the Water Jet Flow Number

In this section, the effects of the water jet flow number are investigated based on the configuration shown in Figure 20. The same rectangular tank has a length of $L = 1.2$ m and a height of $h = 0.6$ m, which is partially filled up to an initial depth of 0.12 m. B_1, B_2, and B_3 are three different locations for the water jet flow inlet. Again, a harmonic rolling motion is imposed on the tank with the same amplitude previously used. The sensor P_1 is placed, as shown in the figure, to monitor the pressure variations. The same initial velocity of the water jet used in the previous section is also adopted here, which is 0.3 m/s.

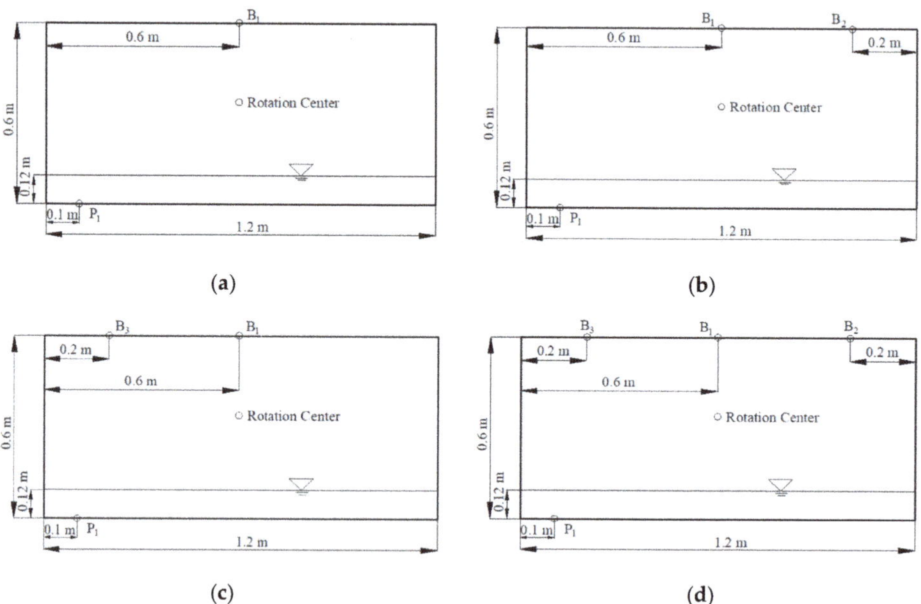

Figure 20. Schematic view of the sloshing tank with a water jet: (**a**) only one water flow; (**b**) two water flows at B_1 and B_2; (**c**) two water flows at B_1 and B_3; (**d**) three water flows at B_1, B_2, and B_3.

The particle snapshots of the pressure contour for the four cases at $t = 0.296$ s are shown in Figure 21, which illustrates the pattern of the free surface when the water jet flows enter the liquid. It shows that the water jet flow has played an important role in the pressure distribution of the sloshing process because it causes strong collisions between the water jet flow and the free surface below. It should be noted that there are some differences in the pressure distribution at different jet regions.

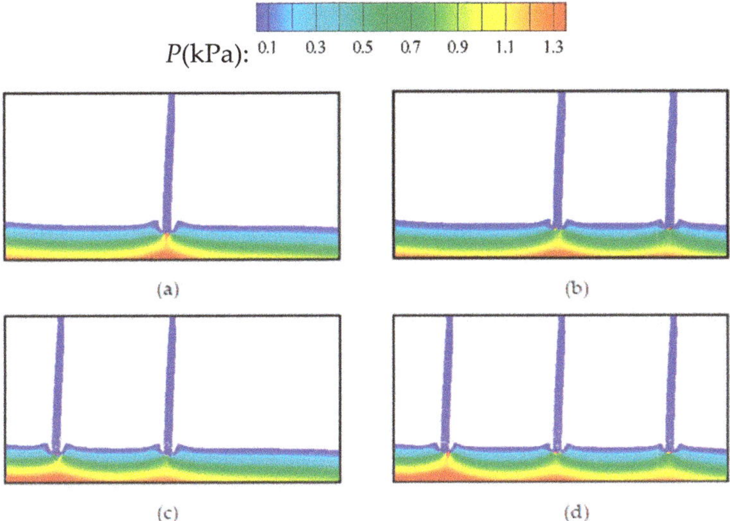

Figure 21. Comparison of free surface profiles at $t = 0.296$ s with a different number of water jets: (**a**) only one water flow; (**b**) two water flows at B_1 and B_2; (**c**) two water flows at B_1 and B_3; (**d**) three water flows at B_1, B_2, and B_3.

Figure 22 shows the peak value of impact pressure at P_1 obtained by ISPH at different excitation frequencies. The comparison includes the case of two water jet flows and only one water jet flow. In all the cases, the excitation frequency is a non-dimensional one, and its value is obtained between 0.9 and 1.5. All the cases correspond to the configurations shown in Figure 20a–c. By comparing the two water flows with only one water flow at B_1, it can be seen that the maximum pressure at P_1 can get a smaller value when another water flow is added at B_2 or B_3; meanwhile, the excitation frequency where the maximum pressure occurs also increases. Furthermore, the comparisons between different combinations of two water jet flows are shown in Figure 20b, c. The maximum pressure value of this case combined by B_1 and B_3 is the smallest. It is reduced by around 5.6% compared with only one water jet flow at B_3 and is reduced by around 2.7% compared with the case combining B_1 and B_2. The excitation frequency where the maximum pressure occurs moves right again when B_3 replaces B_2. In summary, it demonstrates that the maximum pressure at P_1 can be reduced when the other water jet flow is added, and it decreases more remarkably when the water flow is added closer to the pressure probe.

Figure 23 gives the peak value comparisons at P_1 for the case of two water jet flows and three water jet flows. The excitation frequency adopts a non-dimensional one, which is between 0.9 and 1.5. The results are obtained according to the configurations shown in Figure 20b–d. The contrast demonstrates that it can obviously be improved for reducing the maximum pressure when more water jet flow is added.

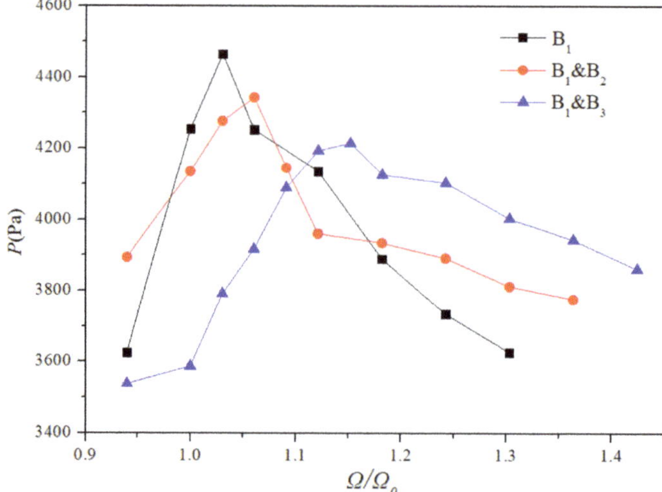

Figure 22. Comparisons of impact pressure with the single water jet flow and two water jet flows.

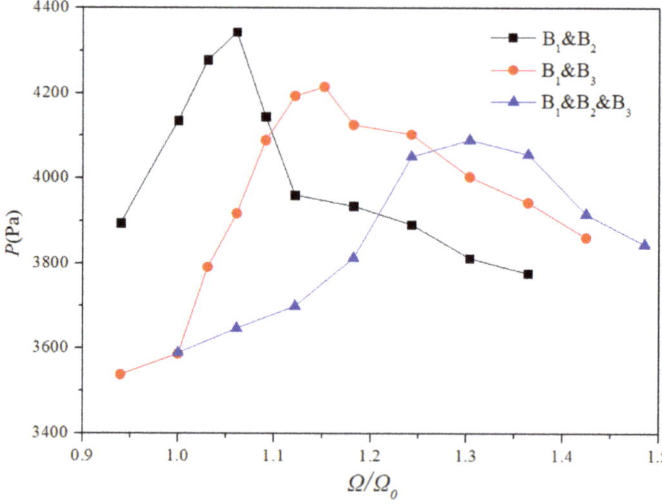

Figure 23. Comparisons of impact pressure with two water flows and three water flows.

5. Conclusions

In this paper, the incompressible SPH method was used to simulate liquid sloshing in a tank with different configurations of the water jet flow, including the water jet flow with different initial velocities, positions, and water jet flow numbers. ISPH shows good agreement in both the free surface profiles and the impact pressures with the experimental data and demonstrates its great potential in predicting violent sloshing flows. The main purpose of this paper was to study the practical importance of the effect of a water jet flow on liquid sloshing through follow-on model applications of different configurations.

The main conclusions of the paper lie in the following aspects. Firstly, adding a water jet flow at the top of the sloshing tank can reduce the maximum impact pressure effectively. Secondly, adding a water jet flow at the top of the sloshing tank can change the excitation frequency where the maximum pressure occurs. Thirdly, adding different numbers of water jet flows also can decrease the value of maximum impact pressure. Finally, when the horizontal distance between water flow and pressure probe is closer, the effect on maximum pressure and excitation frequency is more obvious.

Furthermore, it should be noted that there are some limitations in the present sloshing model as follows:

(1) The ISPH computations are based on a 2D model. According to previous experimental research, it seems there is not much difference between the 2D and 3D models, especially in the impact pressure and water surface;

(2) The compressibility of entrapped air also has effects on the violent sloshing process, and the maximum Mach number of all the particles in the violent sloshing process was smaller than 1%, which proves the ISPH model can be used;

(3) In the simulations of coastal and ocean engineering problems, the SPH method is mainly used for the impulsive impact on breaking waves, and the longer simulations are often carried out by traditional CFD methods;

(4) The presence of turbulence would produce fully three-dimensional flow structures in the breaking region at the tip of the wave crest [26]. However, this study focused on the macro liquid impact pressure on the tank walls and the general free surface deformation. Hence, a 2D model could also provide a reasonable simulation.

Author Contributions: H.J. and Y.Y. made the computations and data analysis; Z.H. made the data analysis and did the proofreading; X.Z. did the proofreading and editing; Q.M. guided the engineering project and provided the data; Y.Y. drafted the manuscript with others. All authors contributed to the work.

Funding: This research work was fundedby the National Natural Science Foundation of China (Nos. 51879051; 51739001; 51579056, and 51639004); Natural Science Foundation of Heilongjiang Province in China (E2018024); Foundational Research Funds for the Central Universities (Nos. HEUCF170104; HEUCDZ1202); Defense Pre Research Funds Program (No. 9140A14020712CB01158).

Acknowledgments: The fifth author also thanks the Chang Jiang Visiting Chair Professorship scheme of the Chinese Ministry of Education, hosted by HEU.

Conflicts of Interest: The authors declare no conflict of interest.

References

1. Ibrahim, R.A. *Liquid Sloshing Dynamics: Theory and Applications*; Cambridge University Press: Cambridge, UK, 2005.

2. Faltinsen, O.M. A nonlinear theory of sloshing in rectangular tanks. *J. Ship Res.* **1974**, *18*, 224–241.

3. Faltinsen, O.M. A numerical nonlinear method of sloshing in tank with two-dimensional flow. *J. Ship Res.* **1978**, *22*, 193–202.

4. Faltinsen, O.M. Multidimensional modal analysis of nonlinear sloshing in a rectangular tank with finite water depth. *J. Fluid Mech.* **2000**, *407*, 201–234. [CrossRef]

5. Akyildiz, H.; Ünal, E. Experimental investigation of pressure distribution on a rectangular tank due to the liquid sloshing. *Ocean Eng.* **2005**, *32*, 1503–1516. [CrossRef]

6. Sames, P.C.; Marcouly, D.; Schellin, E.T. Sloshing in rectangular and cylindrical tanks. *J. Ship Res.* **2002**, *46*, 186–200.

7. Wu, G.X.; Ma, Q.W.; Taylor, R.E. Numerical simulation of sloshing waves in a 3d tank based on a finite element method. *Appl. Ocean Res.* **1998**, *20*, 337–355. [CrossRef]

8. Liu, M.B.; Liu, G.R. Smoothed particle hydrodynamics (SPH): An overview and recent developments. *Arch. Comput. Methods Eng.* **2010**, *17*, 25–76. [CrossRef]

9. Delorme, L.; Colagrossi, A.; Souto-Iglesias, A.; Zamora-Rodriguez, R.; Botia-Vera, E. A set of canonical problems in sloshing, Part I: Pressure field in forced roll-comparison between experimental results and SPH. *Ocean Eng.* **2009**, *36*, 168–178. [CrossRef]

10. Gotoh, H.; Khayyer, A.; Ikari, H.; Arikawa, T.; Shimosako, K. On enhancement of Incompressible SPH method for simulation of violent sloshing flows. *Appl. Ocean Res.* **2014**, *46*, 104–115. [CrossRef]

11. Zheng, X.; You, Y.; Ma, Q.W.; Khayyer, A.; Shao, S.D. A Comparative Study on Violent Sloshing with Complex Baffles Using the ISPH Method. *Appl. Sci.* **2018**, *8*, 904. [CrossRef]

12. Khayyer, A.; Gotoh, H.; Shao, S.D. Enhanced predictions of wave impact pressure by improved incompressible SPH methods. *Appl. Ocean Res.* **2009**, *31*, 111–131. [CrossRef]

13. Zheng, X.; Ma, Q.W.; Duan, W.Y. Comparative study of different SPH schemes in simulating violent water wave impact flows. *China Ocean Eng.* **2014**, *28*, 791–806. [CrossRef]

14. Hatton, A.P.; Leech, C.M.; Osborne, M.J. Computer simulation of the trajectories of large water jets. *Int. J. Heat Fluid Flow* **1985**, *6*, 137–141. [CrossRef]

15. Hatton, A.P.; Osborne, M.J. The trajectories of large fire fighting jets. *Int. J. Heat Fluid Flow* **1979**, *1*, 37–41. [CrossRef]

16. Fletcher, D.F.; McCaughey, M.; Hall, R.W. Numerical simulation of a laminar jet flow: A comparison of three CFD models. *Comput. Phys. Commun.* **1993**, *78*, 113–120. [CrossRef]

17. Aristodemo, F.; Marrone, S.; Federico, I. SPH modeling of plane jets into water bodies through an inflow/outflow algorithm. *Ocean Eng.* **2015**, *105*, 160–175. [CrossRef]

18. Andreopoulos, J.; Praturi, A.; Rodi, W. Experiments on vertical plane buoyant jets in shallow water. *J. Fluid Mech.* **1986**, *168*, 305–336. [CrossRef]

19. Monaghan, J.J.; Lattanzio, J.C. A refined particle method for astrophysical problems. *Astron. Astrophys.* **1985**, *149*, 135–143.

20. Shao, S.D.; Lo, E.Y.M. Incompressible SPH method for simulating newtonian and non-newtonian flows with a free surface. *Adv. Water Resour.* **2003**, *26*, 787–800. [CrossRef]

21. Khayyer, A.; Gotoh, H. Enhancement of stability and accuracy of the Moving Particle Semi-Implicit Method. *J. Comput. Phys.* **2011**, *230*, 3093–3118. [CrossRef]

22. Khayyer, A.; Gotoh, H. *A Multiphase Compressible—Incompressible Particle Method for Water Slamming*; International Society of Offshore and Polar Engineers: Kona, HI, USA, 2015.

23. Zheng, X.; Ma, Q.W.; Duan, W.Y. Incompressible SPH method based on Rankine source solution for violent water wave simulation. *J. Comput. Phys.* **2014**, *276*, 291–314. [CrossRef]

24. Liao, K.P.; Hu, C.H. A coupled FDM-FEM method for free surface flow interaction with thin elastic plate. *J. Mar. Sci. Technol.* **2013**, *18*, 1–11. [CrossRef]

25. Kvicinsky, S.; Longatte, F.; Kueny, J.L.; Avellan, F. Free surface flows: Experimental validation of volume of fluid method in the plane wall case. In Proceedings of the 3rd ASME/JSME Conference, San Francisco, CA, USA, 18–23 July 1999.

26. Alberello, A.; Pakodzi, C.; Nelli, F.; Bitner-Gregersen, E.M.; Toffoli, A. Three dimensional velocity field underneath a breaking rogue wave. In Proceedings of the 36th International Conference on Ocean, Offshore and Arctic Engineering (ASME 2017), Trondheim, Norway, 25–30 June 2017.

 *water*

Article

Simulating the Overtopping Failure of Homogeneous Embankment by a Double-Point Two-Phase MPM

Yong-Sen Yang, Ting-Ting Yang, Liu-Chao Qiu * and Yu Han

College of Water Resources & Civil Engineering, China Agricultural University, Beijing 100083, China
* Correspondence: qiuliuchao@cau.edu.cn; Tel.: +86-10-6273-6423

Received: 25 May 2019; Accepted: 5 August 2019; Published: 8 August 2019

Abstract: Embankments are usually constructed along rivers as a defense structure against flooding. Overtopping failure can cause devastating and fatal consequences to life and property of surrounding areas. This motivates researchers to study the formation, propagation, and destructive consequences of such hazards in risk analysis of hydraulic engineering. This paper reports a numerical simulation of failure processes in homogeneous embankments due to flow overtopping. The employed numerical approach is based on a double-point two-phase material point method (MPM) considering water–soil interaction and seepage effects. The simulated results are compared to available laboratory experiments in the literature. It was shown that the proposed method can predict the overtopping failure process of embankments with good accuracy. Furthermore, the effects of the cohesion, internal fiction angle, initial porosity, and maximum porosity of soil on the embankment failure are investigated.

Keywords: embankments; overtopping failure; material point method; water–soil interactions; numerical simulation

---

1. Introduction

Embankments are frequently used along rivers against flooding events, the probability of which has increased due to increasing intensity extreme weather events in recent years [1]. Their overtopping failure may cause serious loss of life and property downstream. For example, the Banqiao dam-breaking flood due to Typhoon Nina in 1975 lead to about 62,000 deaths [2]. Therefore, a deep investigation of overtopping failure mechanisms and processes in embankments is important for the risk management and assessment. Among the available study methods, numerical modeling has distinctive advantages such as the relatively low cost and the flexible application to almost arbitrary site configurations. In the past decades, many numerical investigations of the overtopping failure in embankments have been performed. For example, Powledge et al. [3] studied the embankment erosion due to overflow, Tingsanchali and Chinnarasri [4] used a one dimensional model to study the overtopping failure in embankment. Chinnarasri et al. [5] investigated the overtopping-induced progressive damage in embankment. Leopardi et al. [6] reviewed the modeling of embankment overtopping, Pontillo et al. [7] employed a two-phase model to evaluate the dike erosion induced by overtopping, Volz et al. [8] modeled the overtopping failure of a non-cohesive embankment by the dual-mesh approach, Mizutani et al. [9] simulated the overtopping failure of embankment considering infiltration effects, Guan et al. [10] and Kakinuma and Shimizu [11] developed a 2D shallow-water model for embankment breach, Evangelista [12] simulated dike erosion induced by dam-breaking flows using a depth-integrated two-phase model, Larese et al. [2] modeled the overtopping and failure of rockfill dams using the particle finite element method (PFEM). Although many efforts have been made to numerically reproduce the main features of the process, the overtopping failure of embankments is still poorly understood. Moreover, most of the aforementioned numerical methods are mesh-based approaches and have difficulties in accurately modeling the time evolution of the embankment breach, since the real overtopping failure phenomenon is, in fact, an unsteady process involving large deformations, free-surface flows, moving boundary, and water–soil interactions. To deal with these challenges, several meshless methods, such as smoothed particle hydrodynamics (SPH) [13,14], moving particle semi-implicit method (MPS) [15], element-free Galerkin method (EFGM) [16], and material point method (MPM) [17,18], have been developed to model large deformation problems. For examples, Gotoh et al. [19,20] modeled embankment erosion due to overflow using MPS, Li et al. [21] simulated erosion of HPTRM levee based on SPH, Zhang et al. [22] modeled failures of dike due to water level-up and rainfall using SPH, Liu et al. [23] simulated piping erosion process of dike foundation by EFGM. Nikolic et al. [24] presented a discrete beam lattice model capable of simulating localized failure in a heterogeneous fluid-saturated poro-plastic solid. Zhao and Liang [25] applied MPM to model seepage flow through embankments, Martinelli et al. [26] modeled the failure of a sand dike due to seepage flow using two-phase MPM. In this work, the MPM was employed for modeling the overtopping failure of embankments because it has several advantages over other meshless methods such as EFGM, MPS, and SPH: The boundary condition realization is as easy as in the finite element method (FEM) for the use of background mesh in MPM [27]. Time-consuming neighbor particle searching is required in SPH, MPS and EFGM, but the MPM requires only the identification of particles relative to the background mesh. The MPM also avoids the tensile instability that is evident in the SPH [27].

The MPM was first proposed by Sulsky et al. [17] in 1994, which uses the Lagrangian particle and the Euler background mesh to describe large deformation problems. The particles move freely across the background mesh and carry all physical parameters such as stress, density, mass, velocity, and other historical variables. The MPM combines the advantages of the Euler method and the Lagrangian method to avoid the interference of convection terms and mesh distortion. It also can treat contact problems between different objects without any additional interface elements. In addition, the traditional soil constitutive laws such as Mohr–Coulomb, Drucker–Prager, and Cam Clay models can be easily implemented in MPM at the particle level [17,18]. Recent developments in the multiphase MPM formulations considering the soil–water interaction can be categorized into two main groups: Single-point approach and double-point approach. In the first approach, each material point (MP) possesses both the soil and water information. The water is described in the Eulerian framework while the soil skeleton is described in the Lagrangian framework. The single-point approach does not guarantee the mass conservation of water. In the second approach, two sets of MPs are used to represent water and solid maters. The double-point approach can guarantee both the solid and water mass conservation. Mackenzie-Helnwein et al. [28] used the double-point approach to model soil–water mixtures through a drag model for capturing the interphase interaction. Abe et al. [29] employed the double-point approach to model saturated soil without considering the relative acceleration between water and the soil skeleton. Bandara and Soga [30] developed a double-point two-phase MPM formulation based on the mixture theory considering the relative acceleration between water and the soil skeleton. This approach not only can be used for modeling rapid deformation problems but also guarantees both the soil skeleton and water mass conservation. It is therefore very suitable for modeling soil–water interaction problems such as currently considered embankment failure due to erosion.

The present work aims to numerically investigate the failure mechanism and failure process of homogeneous embankment due to overtopping flows based on a double-point two-phase MPM method.

2. Mathematical Formulation

The double-point two-phase MPM formulation [26,31] based on the mixture theory is used to model the water–soil interactions. In this approach, the soil skeleton and the pore liquid in the saturated soil are represented separately by two sets of material points: Solid material points and water material points (Figure 1). This paper focuses on the application of two-phase MPM to simulate embankment overtopping, only some of the basic equations solved are quoted for the convenience of readers in understanding the two-phase MPM formulation, and for detailed description of the algorithms readers are referred to the literature [26,31].

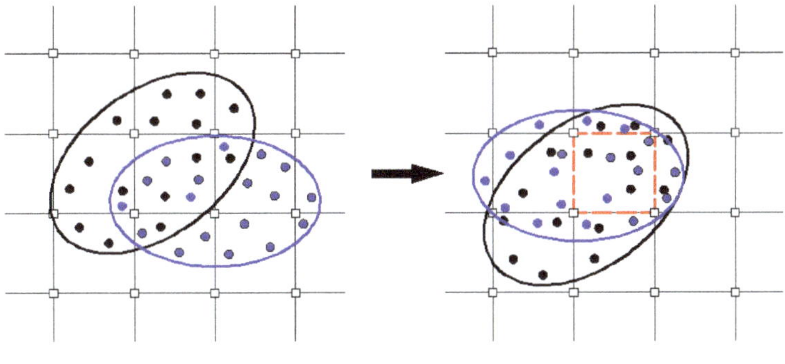

Reference configuration   Current configuration

- Solid skeleton material point
- Water material point
- Solid-water mixture
- Background mesh node
- Background mesh

Figure 1. The schematic model for the double-point two-phase material point method (MPM) [32].

The governing equations of the double-point two-phase MPM include the mass and momentum conservations of both solid and liquid phases with their respective constitutive equations. The mass conservation equations for the solid and liquid phase are respectively given by [31].

$$\frac{Dn_L}{Dt} = n_S \nabla \cdot \mathbf{v}_S \tag{1}$$

$$\frac{D\varepsilon_L}{Dt} = \frac{1}{n_L}[n_S \nabla \cdot \mathbf{v}_S + n_L \nabla \cdot \mathbf{v}_L + (\mathbf{v}_L - \mathbf{v}_S)\nabla n_L] \tag{2}$$

where n_S is the volumetric concentration ratio of solid, n_L is the volumetric concentration ratio of liquid. For saturated soils, n_L is equivalent to porosity of the soil skeleton and $n_L + n_S = 1$. n_L is a discontinuous function at the transition between free surface water and porous medium and it is characterized by two constant values: One in the free surface water and the value of the porosity n in the porous medium. Martinelli [33] proposed the concept of a transition zone in which an interpolated liquid concentration ratio is introduced which gives a smooth transition of n_L between the free surface water and the porous medium. $\mathbf{v}_S$ is the velocity vector of the solid phase, $\mathbf{v}_L$ is the velocity vector of the liquid phase. ε_L is the volumetric strain of the liquid. $\frac{D(\bullet)}{Dt}$ denotes the material time derivative. Equation (1) represents the variation of volumetric concentration ratio of the liquid phase (i.e., porosity), and Equation (2) is also known as the storage equation and represents the volumetric strain rate of the pore liquid.

The momentum conservation of the solid and liquid phase can be respectively expressed as [31]:

$$\bar{\rho}_S \mathbf{a}_S = \nabla \cdot \bar{\sigma}_S + \mathbf{f}_L^d + \bar{\rho}_S \mathbf{g} \tag{3}$$

$$\bar{\rho}_L \mathbf{a}_L = \nabla \cdot \bar{\sigma}_L - \mathbf{f}_L^d + \bar{\rho}_L \mathbf{g} \tag{4}$$

where $\bar{\rho}_S$ and $\bar{\rho}_L$ are respectively the partial densities of the solid and liquid, computed as the ratio of the mass of each constituent with respect to the reference volume. $\mathbf{g}$ is the gravity vector, $\mathbf{a}_L$ is the liquid phase acceleration, $\mathbf{a}_S$ is the solid phase acceleration. $\bar{\sigma}_S = \sigma' - n_S \sigma_L$ and $\bar{\sigma}_L = n_L \sigma_L$ are the partial stresses of solid and liquid phases respectively, σ' is the effective stress tensor, and σ_L is the liquid phase stress tensor. $\mathbf{f}_L^d$ is the drag force, which represents the water–soil interaction due to the velocity difference between the two phases, it can be calculated by [34]

$$\mathbf{f}_{\mathrm{L}}^{\mathrm{d}} = \frac{n_{\mathrm{L}}^2 \mu}{\kappa}(\mathbf{v}_{\mathrm{L}} - \mathbf{v}_{\mathrm{S}}) + \beta n_{\mathrm{L}}^3 \rho_{\mathrm{L}} |\mathbf{v}_{\mathrm{L}} - \mathbf{v}_{\mathrm{S}}|(\mathbf{v}_{\mathrm{L}} - \mathbf{v}_{\mathrm{S}}) + \sigma_{\mathrm{L}} \nabla n_{\mathrm{L}} \tag{5}$$

where μ is the liquid dynamic viscosity, $\kappa = \frac{D^2}{A} n_{\mathrm{L}}^3 / (1 - n_{\mathrm{L}})^2$ is the soil intrinsic permeability with D the average diameter of grains, $\beta = B / \sqrt{\kappa A n_{\mathrm{L}}^3}$ is the non-Darcy flow coefficient, the empirical constants A and B are respectively set to 150 and 1.75 [34].

In order to fully describe the behavior of saturated soils, the constitutive equations for both phases are required. Assuming the validity of Terzaghi's effective stress concept, the general form of stress–strain relationship for soil skeleton is given by

$$\frac{D\boldsymbol{\sigma}'}{Dt} = \mathbf{D}\frac{D\boldsymbol{\varepsilon}}{Dt} \tag{6}$$

where $\mathbf{D}$ is the tangent stiffness matrix, $\boldsymbol{\sigma}'$ is the effective stress vector, and $\boldsymbol{\varepsilon}$ is the total strain. For the liquid phase, the volumetric stresses is updated by

$$\frac{D\sigma_{\mathrm{Vol,\,L}}}{Dt} = K_{\mathrm{L}}\frac{D\varepsilon_{\mathrm{Vol,\,L}}}{Dt} \tag{7}$$

where K_{L} is the liquid bulk modulus, $\varepsilon_{\mathrm{Vol,\,L}}$ is the volumetric strain of the liquid. In addition, for pure liquid or fluidized mixture, the deviatoric part of the stress tensor is calculated by

$$\sigma_{\mathrm{dev,L}} = 2\mu\frac{D\varepsilon_{\mathrm{Vol,L}}}{Dt} \tag{8}$$

In the proposed double-point two-phase MPM, the saturated soils can be considered as a solid-like or liquid-like state according to the porosity. As shown in Figure 2a, in the case of low porosity, the solid skeleton grains are in contact and the soil behaves like a solid, its response thus can be modeled by a soil constitutive model. For a high-porosity soil, as shown in Figure 2b, the grains are not in contact and float together with the liquid phase and the soil response is modeled by the Navier–Stokes equation. In the present work, a pre-given maximum porosity n_{max} is used to distinguish the two aforementioned states. When the soil porosity is less than the maximum porosity ($n_{\mathrm{L}} = 1 - n_{\mathrm{S}} < n_{\mathrm{max}}$), the mean effective stress decreases as the porosity increases and the mean effective stress vanishes once the grains are not in contact. When the porosity is larger than n_{max}, fluidization occurs.

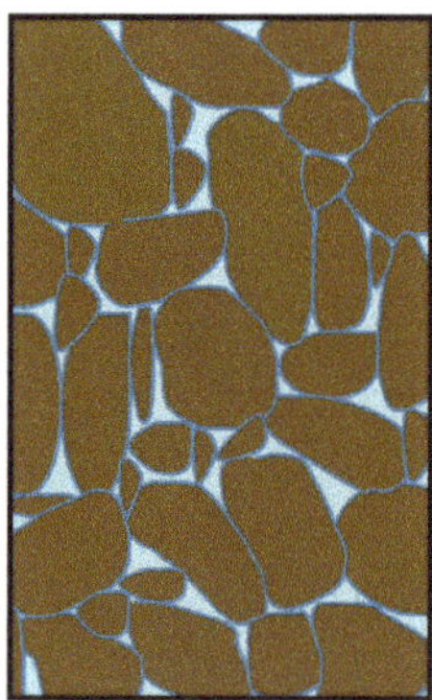 

(**a**) solid-like response (low porosity)   (**b**) liquid-like response (high porosity)

Figure 2. Schematic diagram for soil behavior [34].

3. Numerical Examples

In this section, the proposed double-point two-phase MPM is first validated by an example of flow through a porous block. Then a numerical investigation of failure process in a homogeneous embankment due to flow overtopping was performed and the effects of the cohesion, internal fiction angle, initial porosity, and maximum porosity of soils on the embankment failure are investigated. All calculations are conducted with the software Anura3D_v2016 (www.anura3d.com), which implements the double-point two-phase MPM formulation.

3.1. Flow through a Porous Block

In order to validate the capability of the current MPM algorithm to model the water–soil interaction, a numerical simulation of the flow through a porous block was performed and the calculated results were compared with the experimental data obtained by Liu et al. [35]. The experiment as sketched in Figure 3 was carried out in a transparent water tank with a size of 0.892 m × 0.44 m × 0.58 m, the porous block, which was made of gravel, was located at the center of the water tank (x = 0.30–0.59 m), and it was 0.29 m long, 0.44 m wide, and 0.37 m high. The water on the left side of the water tank was separated from the porous block by a movable gate with a thickness of 0.02 m, and the height was set to 0.35 m, and the porous block was confined in the initial region to ensure that the porous media was not allowed to move. The material parameters used in this simulation are listed in Table 1. The computational domain was divided into 4104 tetrahedral elements. Each mesh element in the porous block part contained four material points, while each element in the water contained 10 material points.

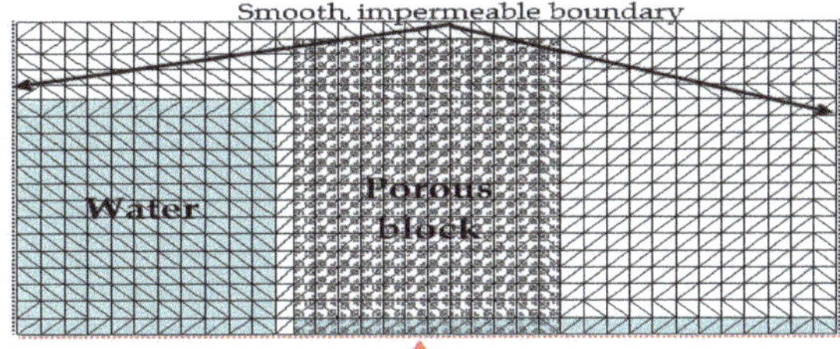

Figure 3. Geometry and discretization of the problem of flow through the porous block.

Table 1. The material parameters for flow through the porous block.

Material	Parameter	Value
Porous block	Density (kg/m^3)	2700
	Young modulus (kPa)	1000
	Poisson's ratio	0.3
	Initial porosity	0.49
	Mean diameter (mm)	15.9
Water	Density (kg/m^3)	1000
	Bulk modulus (kPa)	2.15×10^4
	Dynamic viscosity (kPa·s)	1×10^{-6}

The simulation process was divided into two stages, the first stage was the initialization of stress, also called the gravity loading process. At this stage, the lateral displacement of the water was constrained and the stress initialization was done by increasing the gravity until the solution converged to a quasi-static equilibrium, which was checked by evaluating the normalized kinetic energy and the normalized out-of-balance force falls below a predefined tolerance as given in [31]. In addition, a local damping coefficient of 0.75 was applied to accelerate the convergence to quasi-static equilibrium allowing a considerable reduction in the computational time. In the second stage, the lateral constraint of water was removed to simulate the instantaneous moving process of the gate, and then water flow impacted the porous block. Figure 4 shows some snapshots of the flow through the porous block at time of 0.35, 0.75, 1.15, and 1.95 s and the calculated free-surface profiles agree well with that of the experimental results [35], and comparisons of free surface profiles are shown in Figure 5. It is proven that the double-point MPM can simulate the process of flow through the porous medium considering the water–soil interaction.

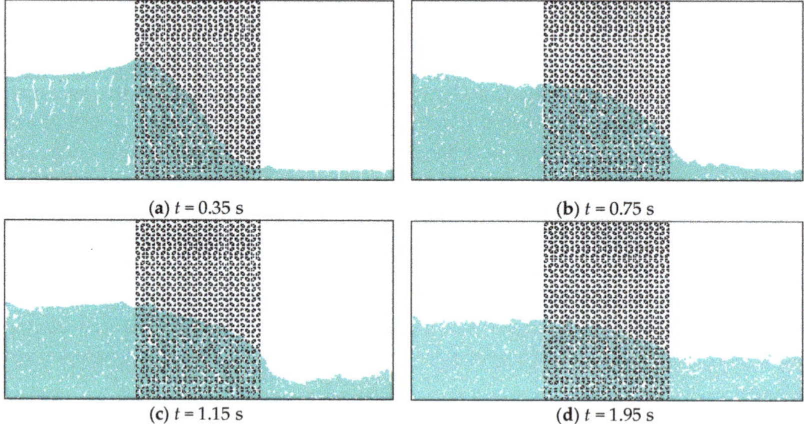

(a) *t* = 0.35 s   **(b)** *t* = 0.75 s

(c) *t* = 1.15 s   **(d)** *t* = 1.95 s

Figure 4. Some snapshots of the flow through the porous block.

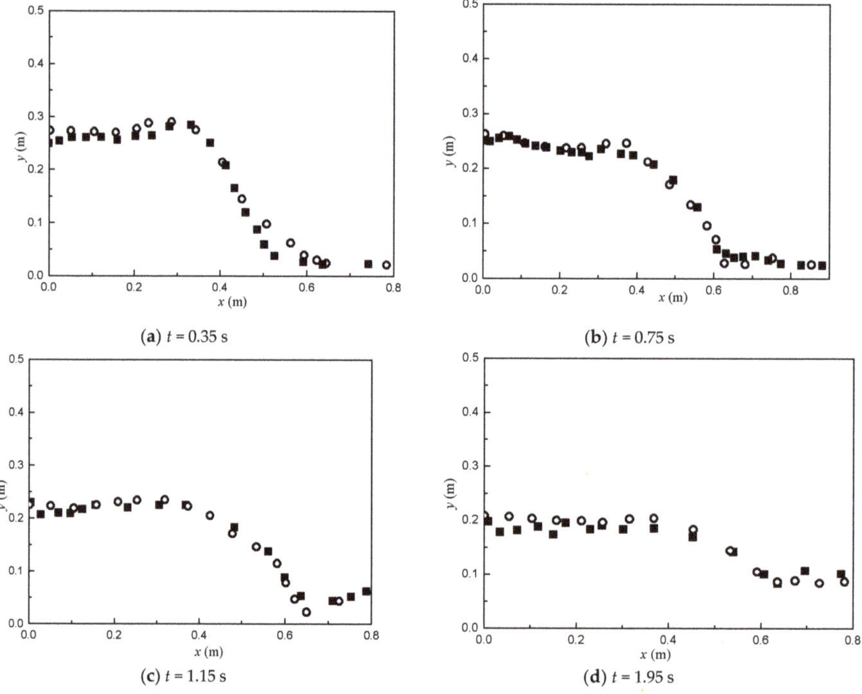

(a) *t* = 0.35 s   **(b)** *t* = 0.75 s

(c) *t* = 1.15 s   **(d)** *t* = 1.95 s

Figure 5. Comparison of free surface profile for flow through porous block: Simulation (■) and experiment (○).

3.2. Overtopping Failure of Homogeneous Embankments

For a better understanding of the embankments failure mechanism and process due to the overtopping flow, in this section, MPM modeling of the overtopping erosion of a homogeneous embankment considering free water surface flow, water–soil interaction, and seepage effects was performed and the simulated results were compared with the experimental data obtained by Evangelista et al. [12]. The effects of the cohesion, internal fiction angle, initial porosity, and maximum porosity of soil on the embankment failure were investigated. As sketched in Figure 6a, this experiment was carried out in a Perspex horizontal channel with a 0.40 m wide rectangular cross-section and transparent walls. A 3.00 m long tank with a moving gate was used to store water, in which H represents the water level. The gate was suddenly lifted to produce the dam-break flow, which propagates over the downstream channel bottom. The first segment of the rigid bed was 1.00 m long and had a mild slope (5%), the subsequent one was a 2.00 m long, 0.05 m thick horizontal segment, on which the embankment was built.

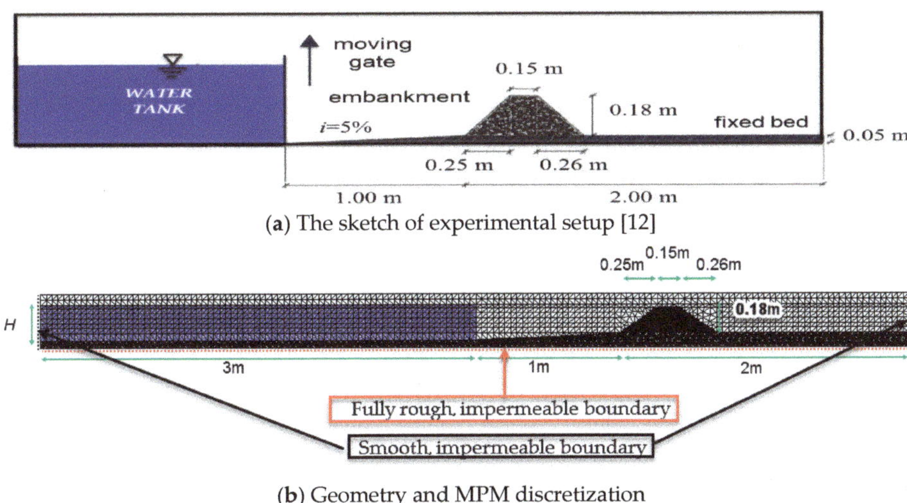

Figure 6. Experimental setup [12] and MPM model for the embankment overtopping problem.

According to Evangelista et al. [12], the friction at the lateral interface of the channel has no significant wall effects and the embankment undergoes an almost plane process because the flow overtops the entire embankment width. For these reasons, in our simulation, the overtopping failure process was simplified approximately as a two-dimensional problem. In addition, the gate opening was considered instantaneous following the work of Lauber and Hager [36]. The computational domain was divided into a set of tetrahedral elements with the size of 0.05 m. Each mesh element in the embankment part contained four material points, while each element in the water contained 10 material points (see Figure 6b). The water was modeled as a weak compressible fluid, and the Mohr–Coulomb model was used as the constitutive model of the soil. The material parameters used in this simulation are listed in Table 2. Three different cases with initial water level of H = 0.2, 0.25 and 0.395 m were conducted.

For the first case (H = 0.2 m), Figure 7 shows several snapshots of the simulated water surface and the embankment profiles along the channel at different times of 0.35, 0.8, 1.4, 2.0, and 3.0 s. Comparisons between the numerically simulated results and the experimental ones [12] are illustrated in Figure 8. Specifically, the embankment (black color) profiles and the water surface (blue color) at times 1.4 and 3.0 s, respectively, after the gate removal are plotted: Solid lines represent the numerical results and dotted lines indicate the experimental ones. In this case, the initial water level in the water tank was almost equal to the top elevation of embankment, thus the gate opening–induced dam break flow did not have enough energy to overtop the dike embankment, and therefore, it induced erosion only on the upstream slope of the embankment.

Table 2. The material parameters of the two phase materials.

Material	Parameter	Value
Sand	Density (kg/m^3)	2680
	Young modulus (kPa)	1000
	Poisson's ratio	0.3
	Internal friction angle ($^\circ$)	36
	Cohesion (kN)	0
	Initial porosity	0.40
	Maximum porosity	0.45
	Mean diameter (mm)	0.2
Water	Density (kg/m^3)	1000
	Bulk modulus (kPa)	2.15×10^4
	Dynamic viscosity (kPa·s)	1×10^{-6}

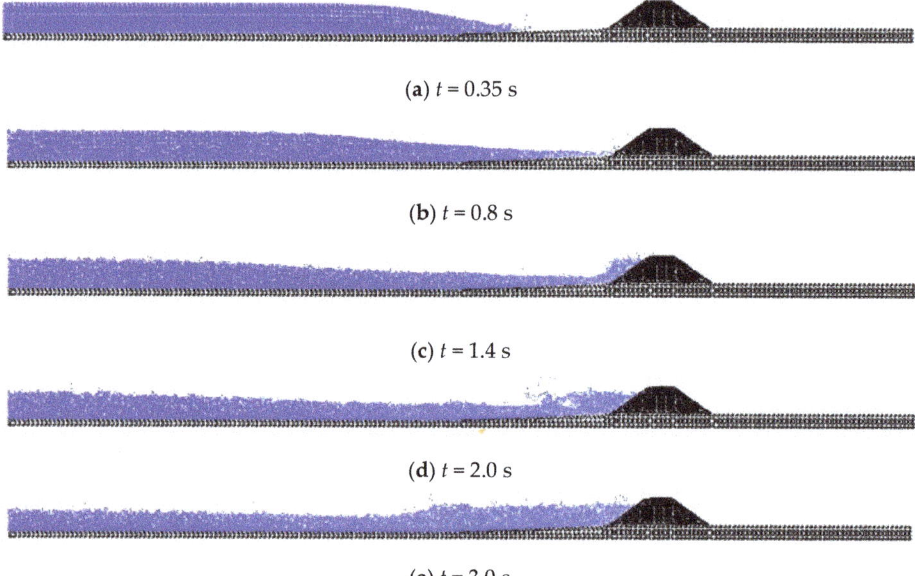

(**a**) $t = 0.35$ s

(**b**) $t = 0.8$ s

(**c**) $t = 1.4$ s

(**d**) $t = 2.0$ s

(**e**) $t = 3.0$ s

Figure 7. Snapshots of the simulated water surface and the embankment profiles along the channel ($H = 0.2$ m).

For the second case ($H = 0.25$ m), Figure 9 shows several snapshots of the simulated water surface and the embankment profiles along the channel at different times of 0.35, 0.8, 1.4, 2.0, and 3.0 s. Comparisons between the numerically simulated results and the experimental ones [12] are illustrated in Figure 10. Specifically, the embankment (black color) profiles and the water surface (blue color) at times 1.4 and 3.0 s, respectively, after the gate removal are plotted: Solid lines represent the numerical results and dotted lines indicate the experimental ones. In this case, the initial water level in tank was higher than the top elevation of embankment, thus the gate opening–induced dam breaking flow overtopped the embankment and eroded it along its entire profile. The upstream slope erosion was more prominent due to the higher potential energy of the dam breaking flow. This dam breaking flow also reached the crest of the embankment this time, progressively eroding it as it passed.

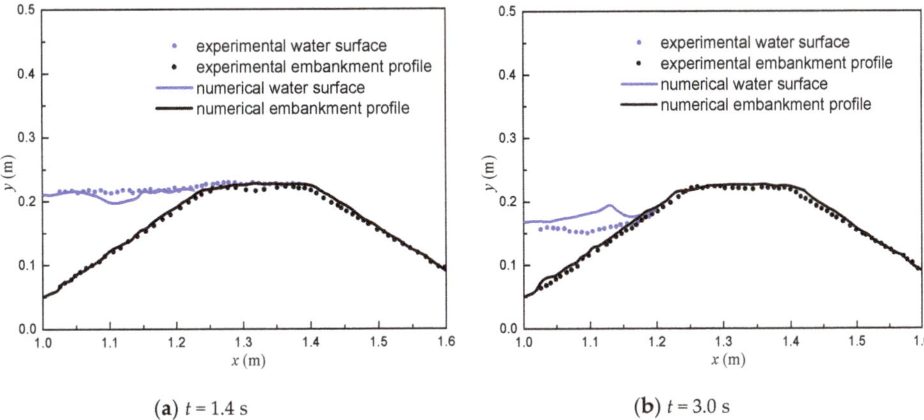

(**a**) $t = 1.4$ s (**b**) $t = 3.0$ s

Figure 8. Comparisons of the free water surface and embankment profiles between the simulation and experiment ($H = 0.2$ m).

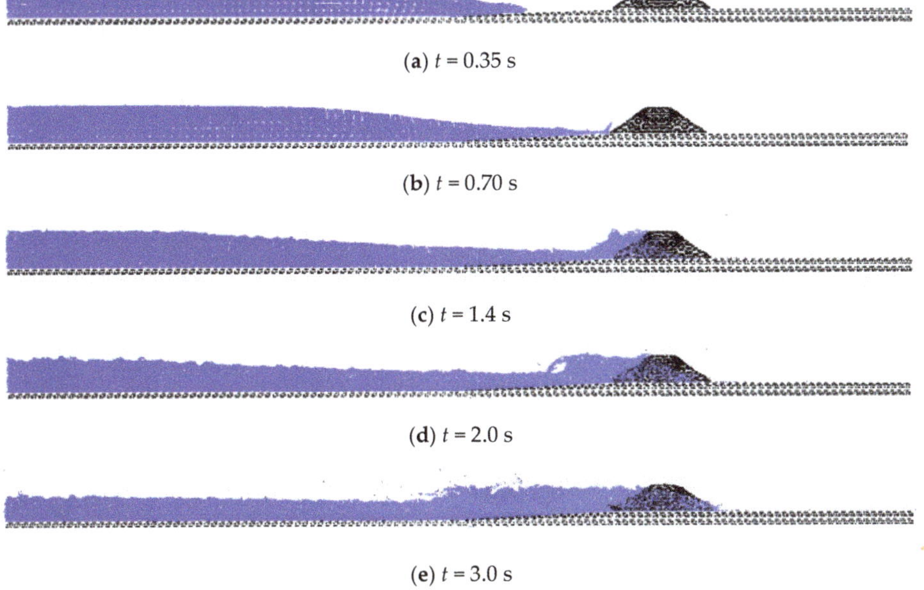

(**a**) $t = 0.35$ s

(**b**) $t = 0.70$ s

(**c**) $t = 1.4$ s

(**d**) $t = 2.0$ s

(**e**) $t = 3.0$ s

Figure 9. Snapshots of the simulated water surface and the embankment profiles along the channel ($H = 0.25$ m).

For the third case ($H = 0.395$ m), Figure 11 shows several snapshots of the simulated water surface and the embankment profiles along the channel at different times of 0.35, 0.8, 1.4, 2.0, and 3.0 s. Comparisons between the numerically simulated results and the experimental ones [12] are illustrated in Figure 12. Specifically, the embankment (black color) profiles and the water surface (blue color) at times 1.4 and 3.0 s, respectively, after the gate removal are plotted: Solid lines represent the numerical results and dotted lines indicate the experimental ones. In this case, the potential energy of the gate opening induced dam break flow was about two times that in the previous case, thus leading to more prominent erosion in the embankment. The erosion was quite homogeneous along the entire profile, and the embankment profile shape was rounded during the process. During the process, steep portions of the embankment slopes underwent a combination of surface erosion and sliding failures due to the loss of supporting material.

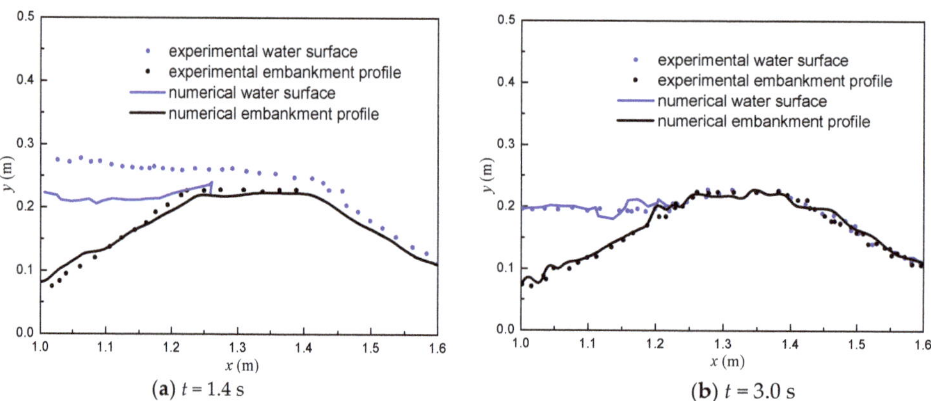

(**a**) $t = 1.4$ s

(**b**) $t = 3.0$ s

Figure 10. Comparisons of the free water surface and embankment profiles between the simulation and experiment ($H = 0.25$ m).

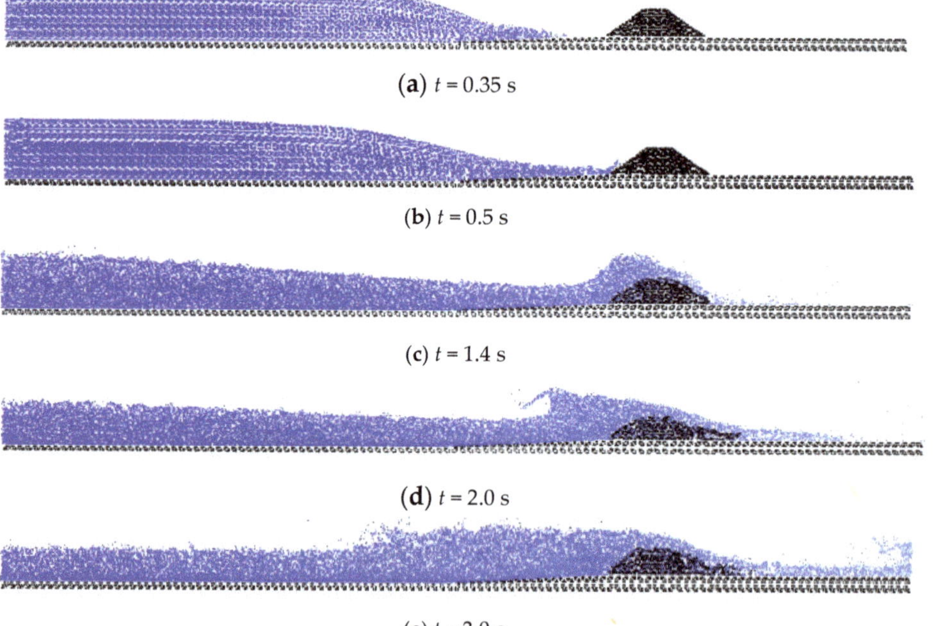

(**a**) $t = 0.35$ s

(**b**) $t = 0.5$ s

(**c**) $t = 1.4$ s

(**d**) $t = 2.0$ s

(**e**) $t = 3.0$ s

Figure 11. Snapshots of the simulated water surface and the embankment profiles along the channel ($H = 0.395$ m).

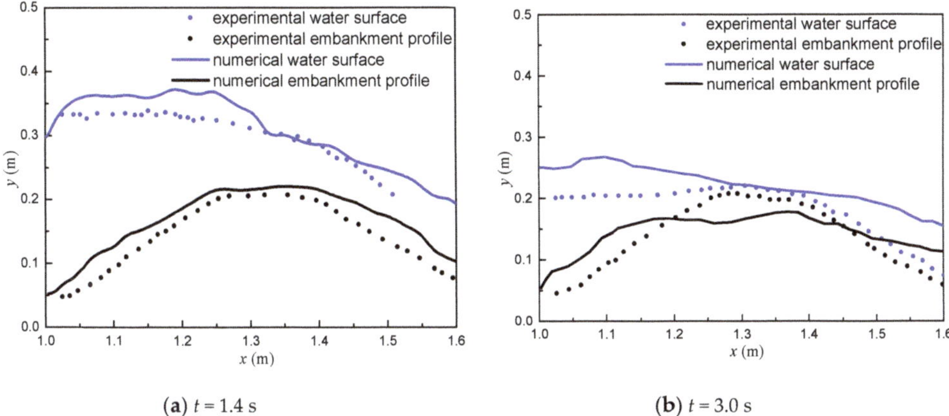

(**a**) $t = 1.4$ s   (**b**) $t = 3.0$ s

Figure 12. Comparisons of the free water surface and embankment profiles between the simulation and experiment ($H = 0.395$ m).

In general, the erosion process of the embankment was greatly affected by the upstream water level elevation. The higher the initial water level, the more serious was the rate of damage of the embankment. By analyzing the embankment erosion process at low, medium, and high water levels, it can be seen that when the water level is low or medium ($H = 0.2, 0.25$ m), the water flow coming to the embankment takes a long time, and the erosion is mainly concentrated on the upstream slope surface. When the water level is high ($H = 0.395$ m), the water flow has a larger energy and rises rapidly along the embankment and overtops the crest to the downstream. Under the erosion and seepage of the water flow, the entire embankment is damaged seriously. Overall, the numerical simulation can relatively reproduce the overtopping erosion phenomenon, which validates the capacity of the proposed double-point MPM to model the overtopping failure of the homogeneous embankments with considering free water surface flow, water–soil interaction, and seepage effects.

In the previous simulation, the homogeneous embankments were considered as non-cohesive. However, we must remember that many embankments were built from cohesive materials, which behave quite differently than non-cohesive materials. In order to study the effect of cohesion on the embankment erosion process, the previous third case ($H = 0.395$ m) was selected, the simulations were conducted with three cohesions of $c = 0$, 5, and 10 kPa, the other conditions remained unchanged. Simulated results are shown in Figure 13, the erosion rate of the upstream and downstream slopes was more serious when $c = 0$ kPa. However, when $c = 5$ and 10 kPa, the rate of embankment erosion was very small and almost negligible. It can be seen that the increase of soil cohesion can effectively improve the resistance. As the cohesion increases, the erosion caused by the water flow is weakened, but when increased to a certain extent, the effect is weakened.

The effect of the internal fiction angle on the embankment erosion process was also investigated. Three different internal fiction angles ($\varphi = 30°$, $36°$, $41°$) were used in the case of the initial water level $H = 0.395$ m, and the other parameters were the same as in the previous simulations. The calculated embankment profiles for different internal fiction angles at $t = 1.4$ s and $t = 3.0$ s are shown in Figure 14. Results show that the larger the internal friction angle is, the weaker the erosion will be. It is the internal friction angle that influences the contact between soil particles. For the small internal friction angle, it has weak resistance to overtopping flow, whereas a large friction angle can retard the flow.

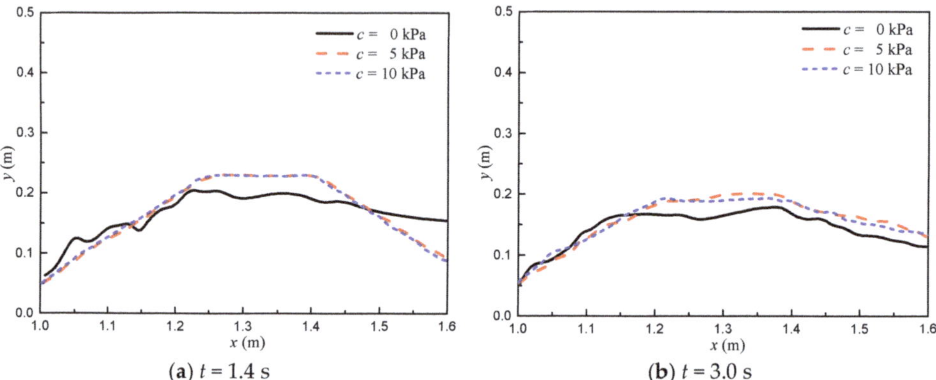

(**a**) t = 1.4 s   (**b**) t = 3.0 s

Figure 13. The effect of the cohesion on the embankment profiles (H = 0.395 m).

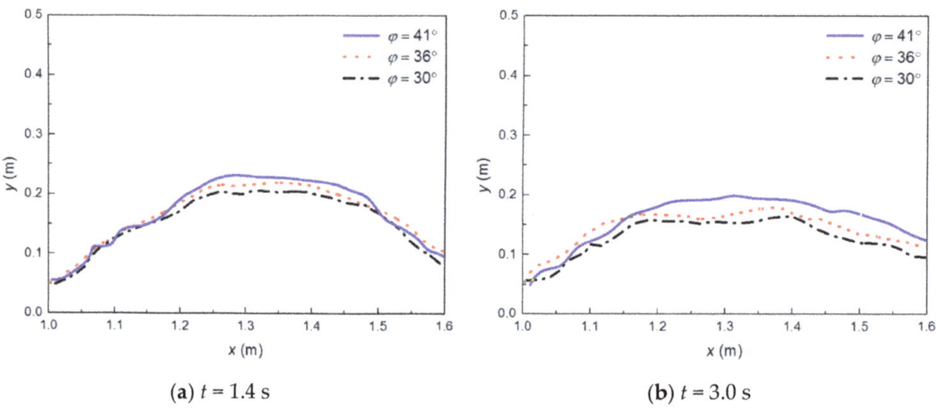

(**a**) t = 1.4 s   (**b**) t = 3.0 s

Figure 14. The effect of the internal fiction angle on the embankment profiles.

In order to investigate the effect of the initial porosity on the embankment erosion process, three different initial porosities (n = 0.2, 0.3, 0.4) were selected to model in the case of the initial water level H = 0.395 m, and the other parameters were also the same as in the previous simulations. The calculated embankment profiles for different porosities at t = 1.4 s and t = 3.0 s are shown in Figure 15. It can be seen that in the early stage (t = 1.4 s) of the embankment failure, the erosion was not very obvious, only surface erosion caused by the water flow occurred and there was no infiltration. However, the profiles of the embankments at t = 3.0 s were different. The larger the porosity of the soil, the more obvious the embankment erosion. It indicates that the water flows began to infiltrate into the embankment in the middle stage, and the embankment erosion was induced by the water flow and seepage, and the seepage failure mainly occurred in the middle and later stages. With the gradual infiltration of the water flow, the water flow infiltrated completely in the later stage, and the pores of the soil were filled with water, the soil became saturated. When the effective stress of the soil skeletons is 0, the soil may be liquefied and moved with the water flow. Hence, decreasing the porosity of the soil can slow down the infiltration rate and reduce the damage of the embankments.

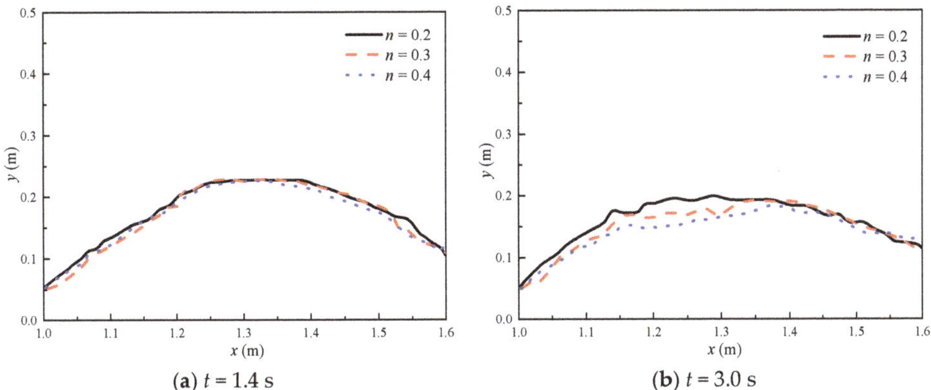

(a) $t = 1.4$ s   (b) $t = 3.0$ s

Figure 15. The effect of the initial porosity on the embankment profiles.

In the proposed double-point two-phase MPM, a pre-given maximum porosity n_{max} was used to distinguish the saturated soils as a solid-like or liquid-like state. When the porosity was larger than n_{max}, fluidization occurred. The grains are not in contact and float together with the liquid phase. Here, three cases of different porosities ($n_{max} = 0.45, 0.50, 0.55$) were conducted to study the effect of maximum porosity on the embankment erosion process, and the other conditions were the same as in the previous simulations. The calculated embankment profiles for different cases at $t = 1.4$ s and $t = 3.0$ s are shown in Figure 16. Results show that the maximum porosity plays an important role on the dam failure due to overtopping flow. The larger the maximum porosity is, the more difficult it is for soil particles to reach the fluidization state. Hence, it is very important to get the right maximum porosity in the simulations, and more accurate simulation results can only be obtained with the appropriate value.

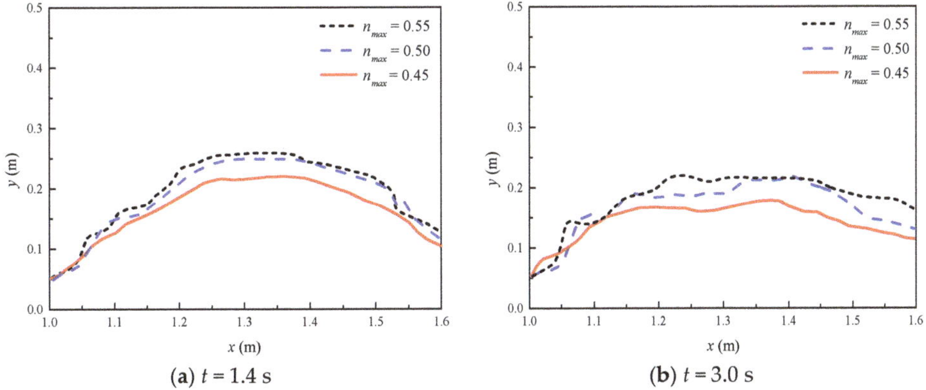

(a) $t = 1.4$ s   (b) $t = 3.0$ s

Figure 16. The effect of maximum porosity on the embankment profiles.

4. Conclusions

In this paper, a double-point two-phase material point method was employed to model problems involving free water surface flow, large deformations, and water–soil interaction. The proposed method was first validated by the example of the flow through a porous block problem. Then a numerical investigation of failure process in homogeneous embankments due to overtopping flow was performed. By comparing the water flow and embankments profiles with the experimental data, it was shown that the current double-point two-phase MPM can predict the overtopping flow pattern and the development of embankment erosion with good accuracy, and the higher the initial water level, the more obviously and faster the embankment is eroded. Furthermore, the effects of the cohesion, internal fiction angle, initial porosity, and maximum porosity of soil on the embankment failure were investigated. The results show that these parameters have important influence on the erosion process. With the increase of the cohesion, the resistance of the embankment can be improved to a certain extent, but the effect of the cohesion is gradually weakened. The larger the internal friction angle is, the weaker the erosion will be. In addition, the larger the initial porosity of the soil, the more obvious the embankment erosion. The larger the maximum porosity is, the more difficult it is for soil particles to reach the fluidization state. These investigated results indicate that the double-point two-phase MPM is capable of predicting and reproducing the failure process of embankments due to the overtopping flow. The proposed method is an alternative promising tool for investigating complex failure mechanism in water–soil interaction.

Author Contributions: Y.-S.Y. and T.-T.Y. conducted the numerical computations and data analysis; Y.-S.Y. and L.-C.Q. drafted the manuscript; Y.H. did the proof reading and editing. All authors contributed to the work.

Funding: This research was funded by the National Key R & D Program of China (No. 2018YFC0406604) and the National Natural Science Foundation of China (No. 11772351).

Acknowledgments: The authors would like to acknowledge Anura3D MPM Research Community for providing the external version of the software Anura3D_v2016.

References

1. Confalonieri, U.B.; Menne, R.; Akhtar, K.L.; Ebi, M.; Hauengue, R.S.K. *Climate Change 2007: Impacts, Adaptation and Vulnerability*; Contribution of Working Group II to the Fourth Assessment Report of the Intergovernmental Panel on Climate Change; Parry, M.L., Canziani, O.F., Palutikof, J.P., van der Linden, P.J., Hanson, C.E., Eds.; Cambridge University Press: Cambridge, UK, 2007.

2. Larese, A.; Rossi, R.; Oñate, E.; Toledo, M.Á.; Morán, R.; Campos, H. Numerical and experimental study of overtopping and failure of rockfill dams. *Int. J. Geomech.* **2015**, *15*, 1–23. [CrossRef]

3. Powledge, G.R.; Ralston, D.C.; Miller, P.; Chen, Y.H.; Clopper, P.E.; Temple, D.M. Mechanics of overflow erosion on embankments. II: Hydraulic and design considerations. *J. Hydraul. Eng.* **1989**, *115*, 1056–1075. [CrossRef]

4. Tingsanchali, T.; Chinnarasri, C. Numerical modeling of dam failure due to overtopping. *Hydrol. Sci. J.* **2001**, *46*, 113–130. [CrossRef]

5. Chinnarasri, C.; Tingsanchali, T.; Weesakul, S.; Wongwises, S. Flow patterns and damage of dike overtopping. *Int. J. Sediment Res.* **2003**, *18*, 301–309.

6. Leopardi, A.; Oliveri, E.; Greco, M. Numerical simulation of gradual earth-dam failure. In Proceedings of the 2nd International Conference New Trends in Water and Environmental Engineering for Safety and Life: Eco-Compatible Solutions for Aquatic Environments, Capri, Italy, 24–28 June 2002.

7. Pontillo, M.; Schmocker, L.; Greco, M.; Hager, W.H. 1D numerical evaluation of dike erosion due to overtopping. *J. Hydraul. Res.* **2010**, *48*, 573–582. [CrossRef]

8. Volz, C.; Rousselot, P.; Vetsch, D.; Faeh, R. Numerical modelling of non-cohesive embankment breach with the dual-mesh approach. *J. Hydraul. Res.* **2012**, *50*, 587–598. [CrossRef]

9. Mizutani, H.; Nakagawa, H.; Yoden, T.; Kawaike, K.; Zhang, H. Numerical modelling of river embankment failure due to overtopping flow considering infiltration effects. *J. Hydraul. Res.* **2013**, *51*, 681–695. [CrossRef]

10. Guan, M.; Wright, N.; Sleigh, P. 2D Process-Based Morphodynamic Model for Flooding by Noncohesive Dyke Breach. *J. Hydraul. Eng.* **2014**, *140*, 04014022. [CrossRef]

11. Kakinuma, T.; Shimizu, Y. Large-scale experiment and numerical modeling of a riverine levee breach. *J. Hydraul. Eng.* **2014**, *140*, 04014039. [CrossRef]

12. Evangelista, S. Experiments and Numerical simulations of dike erosion due to a wave impact. *Water* **2015**, *7*, 5831–5848. [CrossRef]

13. Gingold, R.A.; Monaghan, J.J. Smoothed particle hydrodynamics: Theory and applications to non-spherical stars. *Mon. Not. R. Astron. Soc.* **1977**, *181*, 375–389. [CrossRef]

14. Lucy, L. A numerical approach to testing the fission hypothesis. *Astron. J.* **1977**, *81*, 1013–1024. [CrossRef]

15. Koshizuka, S.; Oka, Y. Moving-particle semi-implicit method for fragmentation of incompressible fluid. *Nucl. Sci. Eng.* **1996**, *123*, 421–434. [CrossRef]

16. Belytschko, T.; Lu, Y.Y.; Gu, L. Element free Galerkin methods. *Int. J. Numer. Meth. Eng.* **1994**, *37*, 229–256. [CrossRef]

17. Sulsky, D.; Chen, Z.; Schreyer, H.L. A particle method for history-dependent materials. *Comput. Methods Appl. Mech. Eng.* **1994**, *118*, 176–196. [CrossRef]

18. Sulsky, D.; Zhou, S.J.; Schreyer, H.L. Application of a particle-in-cell method to solid mechanics. *Comput. Phys. Commun.* **1995**, *87*, 236–252. [CrossRef]

19. Gotoh, H.; Hyashi, M.; Oda, K.; Sakai, T. Gridless analysis of slope failure of embankment by overflow. *Ann. J. Hydraul. Eng. JSCE* **2002**, *46*, 439–444. [CrossRef]

20. Gotoh, H.; Ikari, H.; Tanioka, H.; Yamamoto, K. Numerical simulation of river-embankment erosion due to overflow by particle method. *Proc. Hydraul. Eng.* **2008**, *52*, 979–984. [CrossRef]

21. Li, L.; Rao, X.; Amini, F.; Tang, H.W. SPH Modeling of Hydraulics and Erosion of HPTRM Levee. *J. Adv. Res. Ocean Eng.* **2015**. [CrossRef]

22. Zhang, W.; Maeda, K.; Saito, H.; Li, Z.; Huang, Y. Numerical analysis on seepage failures of dike due to water level-up and rainfall using a water–soil-coupled smoothed particle hydrodynamics model. *Acta Geotech.* **2016**, *11*, 1401–1418. [CrossRef]

23. Liu, C.J.; Ning, B.H.; Ding, L.Q.; Zhang, S.F. Simulation of piping erosion process of dike foundation with element free method. *Water Resour. Power* **2012**, *30*, 58–61.

24. Nikolic, M.; Ibrahimbegovic, A.; Miscevic, P. Discrete element model for the analysis of fluid-saturated fracturedporo-plastic medium based on sharp crack representation with embedded strong discontinuities. *Comput. Methods Appl. Mech. Eng.* **2016**, *298*, 407–427. [CrossRef]

25. Zhao, X.Y.; Liang, D.F. MPM modelling of seepage flow through embankments. In Proceedings of the Twenty-Sixth International Ocean and Polar Engineering Conference, Rhodes, Greece, 26 June–1 July 2016.

26. Martinelli, M.; Rohe, A.; Soga, K. Modeling dike failure using the Material Point Method. *Procedia Eng.* **2017**, *175*, 341–348. [CrossRef]

27. Ma, S.; Zhang, X.; Qiu, X.M. Comparison study of MPM and SPH in modeling hypervelocity impact problems. *Int. J. Impact Eng.* **2009**, *36*, 272–282. [CrossRef]

28. Mackenzie-Helnwein, P.; Arduino, P.; Shin, W.; Moore, J.A.; Miller, G.R. Modeling strategies for multiphase drag interactions using the material point method. *Int. J. Numer. Methods Eng.* **2010**, *83*, 295–322. [CrossRef]

29. Abe, K.; Soga, K.; Bandara, S. Material Point Method for Coupled Hydromechanical Problems. *J. Geotech. Geoenviron. Eng.* **2013**, *140*, 04013033. [CrossRef]

30. Bandara, S.; Soga, K. Coupling of soil deformation and pore fluid flow using material point method. *Comput. Geotech.* **2015**, *63*, 199–214. [CrossRef]

31. Fern, J.; Rohe, A.; Soga, K.; Alonso, E. *The Material Point Method for Geotechnical Engineering: A Practical Guide*; CRC Press, Taylor & Francis Group: Boca Raton, FL, USA, 2019.

32. Soga, K.; Alonso, E.; Yerro, A.; Kumar, K.; Bandara, S. Trends in large-deformation analysis of landslide mass movements with particular emphasis on the material point method. *Géotechnique* **2015**, *66*, 1–26.

33. Martinelli, M. *Soil-Water Interaction with Material Point Method*; Technical Report; Deltares: Delft, The Netherlands, 2016.

34. Ceccato, F.; Yerro, A.; Martinelli, M. Modelling Soil-Water Interaction with the Material Point Method. Evaluation of single-point and double-point formulations. In Proceedings of the NUMGE, Porto, Portugal, 25–29 June 2018.

35. Liu, L.-F.; Lin, P.Z.; Chuang, K.-A.; Sakakiyama, T. Numerical modelling of wave interaction with porous structures. *J. Waterw. Port Coast. Ocean. Eng. ASCE* **1999**, *125*, 322–330. [CrossRef]

36. Lauber, G.; Hager, W.H. Experiments to dambreak wave: Horizontal channel. *J. Hydraul. Res.* **1998**, *36*, 291–307. [CrossRef]

Article

Local Scour of Armor Layer Processes around the Circular Pier in Non-Uniform Gravel Bed

Manish Pandey [1], Su-Chin Chen [2,*], P. K. Sharma [3], C. S. P. Ojha [3] and V. Kumar [4]

1   Department of Soil and Water Conservation, National Chung Hsing University, Taichung City 40227, Taiwan
2   Innovation and Development Centre of Sustainable Agriculture (IDCSA), Department of Soil and Water Conservation, National Chung Hsing University, Taichung City 40227, Taiwan
3   Department of Civil Engineering, Indian Institute of Technology Roorkee, Roorkee 247667, India
4   Civil Engineering Department, National Institute of Technology Hamirpur (NIT Hamirpur), Hamirpur 177005, India
*   Correspondence: scchen@nchu.edu.tw

Received: 27 June 2019; Accepted: 8 July 2019; Published: 10 July 2019

Abstract: Flume experiments have been carried out under clear water scour conditions to analyze the maximum equilibrium scour depth and scour processes in armored streambeds. A total of 85 experiments have been carried out using different diameters of circular piers and non-uniform gravels. A graphical approach for dimensionless scour depth in equilibrium condition around the circular pier in armored streambeds has been developed. As per this curve, the maximum dimensionless scour depth variation with dimensionless armor particle size depends on the densimetric particle Froude number ($F_{rd_{50}}$), and the decreasing rate of dimensionless scour depth decreases with the value of $F_{rd_{50}}$.

Keywords: pier scour; non-uniform sediment; armor layer; equilibrium scour depth processes; clear water scour condition

1. Introduction

Scour around bridge elements like abutments and piers have been recognized as the main cause of bridge collapse [1]. Hence, for the economical and safe design of bridge elements, it is necessary to estimate precise calculations of scour depths. Several studies are available on this subject, mostly dealing with uniform non-cohesive sediment beds [2–9]. For refraining bridges with pier hazards, it is necessary to propose a pier scour study based on many factors, classified as structural parameters of pier, bed sediment characteristics and hydraulic conditions prevailing at the site. So far, those objectives have not been met.

Characteristically, riverbeds in upper reaches deal with non-uniform coarser sediments such as non-uniform gravels, sand-gravel mixtures, sand-boulder-gravel mixtures etc. However, present study only focuses on non-uniform gravels. It is hard to accurately predict scour depths using uniform sediment bed approaches [10]. Sediment transport difficulties involving non-uniform gravel beds could not be simply explained critically. This is because the physical process of sediment transport phenomenon has not yet been ideally defined. Jueyi et al. [11] stated that a protecting layer of coarser particles formed around the pier is identified as the armor layer. In other words, the formation of an armor layer in a streambed stops the scouring process; however, some finer particles from the stream can be transported downstream through the armored bed [12]. An armor layer plays a significant role in river evolution.

When the incipient condition of uniform sediment in a streambed exceeds, the particles start to move. However, the entrainment development of a single particle in a non-uniform sediment bed is complex. The resistance to single particle motion is a function of shape, size and relative density of the particle. Schmidt and Gintz [13] stated that Platy shaped particles have a lesser possibility to be entrained than more compacted sediments. In addition, resistance to particle movement is also affected by the exposure and hiding effects in non-uniform sediments [13].

The development of scour processes in non-uniform streambeds stops because of the development of a steady armor coating, due to the removal of fine sediment particles by the flowing water from the channel bed material (parent bed material) near the pier. Usually, the value of geometric standard deviation of the stratified layer particles is less than parent bed material and these particles are larger in median diameter than parent bed particles [14]. It was observed experimentally by Jueyi et al. [11] that the removal of finer sediment tends to occur earlier because it entails less energy to move. The development of armor layer by flushing fine particles from the scoured hole, and gradual exposure of coarser particles does not allow the further removal of sediment from the scoured hole. This phenomenon is also known as the end condition for scour in non-uniform sediments, and the scoured zone does not undergo further changes with time. The condition of end scour and the formation of the armor layer is dependent on the approach flow parameters, sediment properties, channel bed geometry and pier shape.

The armor layer at equilibrium scour condition on streambed plays a significant role in pier scour. In the literature, only a few studies are available related to scour processes and the behavior of different parameters around bridge piers in armored streambeds [11,15–17]. Most recently, Jueyi et al. [11] derived a maximum dimensionless scour depth equation in the presence of the armoring layer. Guo [15] and Kim et al. [16] conducted pier scour study in non-uniform sediment and proposed maximum scour depth relationships. However, they did not consider the armor layer in their study. Jueyi et al. [11] conducted an experimental study for abutment scour in non-uniform gravel bed and proposed an empirical relationship to describe the maximum depth of scour at equilibrium state.

Presently, only a few studies are available to describe the scour processes around bridge piers in non-uniform streambeds [10,14–16]. In this study, we investigated the scour processes and behavior of different parameters on maximum scour depth in armored streambed. All experiments have been carried out in a laboratory flume under the clear water scour condition using different sizes of non-uniform gravels and flow conditions.

2. Experimental Setup and Procedure

In this study, a total of 85 tests were carried out in a 20.0 m long and 1.0 m wide rectangular flume, as shown in Figure 1. Four different diameters (*b*) of circular bridge piers, i.e., 6.6 cm, 8.4 cm, 11.5 cm and 13.5 cm were used for present experiments. All piers were made of hollow cast iron pipes.

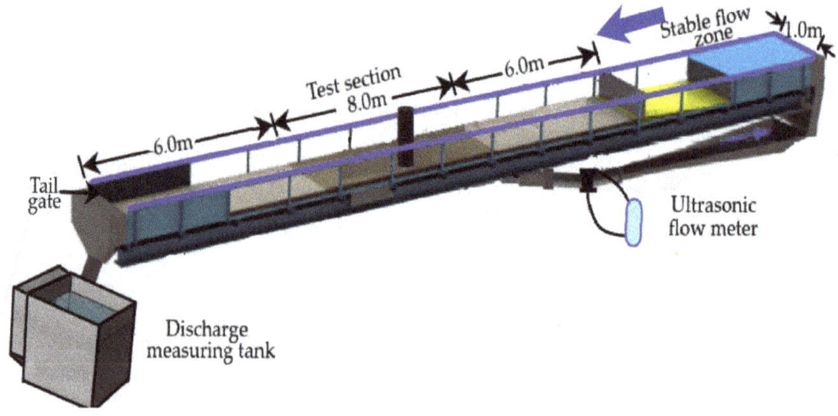

Figure 1. Schematic view of the experimental setup.

All experiments have been completed under different flow conditions and sediment properties like time-averaged velocities (U), approach flow depths (y), pier diameters (b), median diameter of parent streambed particles (d_{50}) and geometric standard deviation (σ). Test limits are mentioned in Table 1 for cases (a–h).

Table 1. Experimental data of non-uniform gravel bed.

	Experimental Conditions						**Outcomes**	
Case	b (cm)	d_{50} (mm)	σ	U (m/s)	U/U_c	y (cm)	h_s (cm)	d_{50a} (mm)
(a)	6.6–11.5	3.26	1.57	0.44–0.54	0.78–0.74	10.5–13.0	4.1–12.2	4.25–4.95
(b)	6.6–13.5	3.82	1.66	0.45–0.55	0.76–0.93	11.0–13.0	3.1–11.2	6.15–6.82
(c)	6.6–13.5	4.38	1.88	0.53–0.61	0.85–0.98	10.0–14.0	3.0–14.2	8.8–9.61
(d)	6.6–13.5	4.94	1.56	0.46–0.61	0.71–0.94	10.0–14.0	2.6–13.5	5.81–6.36
(e)	6.6–13.5	7.5	1.59	0.69–0.73	0.92–0.97	13.0	2.9–10.1	8.38–9.56
(f)	6.6–13.5	8.6	1.63	0.70–0.74	0.90–0.95	13.0	1.8–9.7	12.03–13.73
(g)	11.5–13.5	9.1	1.51	0.72–0.76	0.89–0.94	13.0	6.7–9.1	14.6–15.11
(h)	6.6–13.5	10.7	1.71	0.74–0.79	0.84–0.90	12.5	2.2–8.1	13.53–13.7

The working section starts from 6.0 m of the flume entrance, as shown in Figure 1. The working section (8.0 m long and 0.35 m deep) was completely filled with eight different sizes of non-uniform gravels ($\sigma > 1.4$) having 3.26, 3.82, 4.38, 4.94, 7.5, 8.6, 9.1 and 10.7 mm median diameter (d_{50}) and 1.57, 1.66, 1.88, 1.56, 1.59, 1.63, 1.71 and 1.51 geometric standard deviation of parent particle size distribution, respectively. The discharge was measured with an ultrasonic flow meter, which was provided at inlet pier of the flume. Approach flow depth was adjusted using a tailgate, which was located at the downstream end of the flume. A wave regulator or straighter was facilitated at the flume entrance to produce a uniform or near uniform flow condition in the experimental flume. In the present study, the maximum scour depth (h_s) was measured with a Vernier point gauge. All tests were prepared for 24 hours. However, experimentally it was observed that the equilibrium scour stage was reached within 14 hours. After the equilibrium time of scour,, i.e., 14 hours, the scour depth at different points near the pier was the same at every 30 minutes' interval. Before the start of each experiment, the working section of the flume was made perfectly level with respect to flume bed and covered with a thin Perspex sheet. Once preset flow conditions were achieved, the Perspex sheet was separated very carefully to avoid undesirable scour around the pier. Figure 2 shows a definition sketch of pier scour in armored bed for case (b), followed by Table 1.

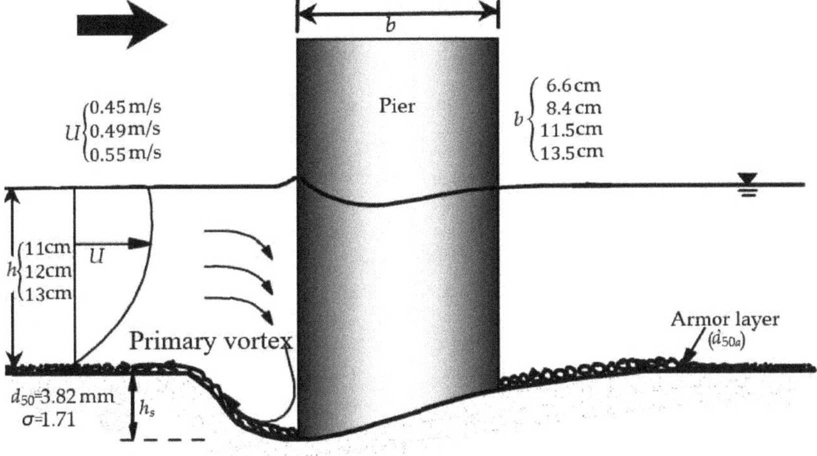

Figure 2. Scour around circular pier in armored streambed for case (b), refer to Table 1.

After equilibrium scour state, the maximum scour depth (h_s) was measured with a Vernier point gauge at upstream nose of the pier. We also collected the armor layer sediment for further particle size distribution and used the median diameter of armor layer (d_{50a}) for scour depth analysis.

Critical velocity U_c, was computed by Melville and Sutherland [18] and varied with present flume experiments. Shield's approach was obtained for calculating the critical shear velocity (u_{*c}).

3. Scour Processes and Armor Layer Formation

The interaction between gravel particles and the flow is complex, and this a significant information for designing a pier in gravel-bed streams. Armoring is a common phenomenon in rivers and occurs when selective entrainment of finer particles leave the stream-bed with a coarser gravel size than that of the underlying bed. Experimentally, it was stated by various researchers that when shear velocity (u_*) exceeds the critical shear velocity (u_{*c}) of streambed particles then incipient motion condition develops in the stream, and then further increment in shear velocity develops scour processes near the bridge elements [6,14]. Melville and Sutherland [18] stated that the scour near the bridge elements occurs when approach shear velocity is almost 0.5 times of u_{*c} of armor bed particles. At the critical condition, the selective entrainment of gravels leaves the higher, more stable gravels on the stream-bed. Further increase in flow, the entrainment frequency increases and an armor layer of coarser gravel particles starts to form. The scour hole starts forming due to the formation of primary-vortex and drops into the scour hole as the scour hole volume increases with time [16]. Hydrodynamic forces are induced on particles and eliminate the particles around the pier. The drag and lift forces decrease with a rise in scour depth. In actuality, the depletion of the particles near the bridge elements (pier and abutments) occur layer-by-layer [14]. The scour processes end in non-uniform sediment beds with the development of a stable armor layer [16]. At equilibrium scour condition, this armor layer does not permit the further removal sediment from the scoured hole. An armor layer holds partially of cluster creations, which are more constant than specific particles. The cluster creations have a noteworthy influence on pier scour, they can offer more stable bed after equilibrium scour state. After flume experimental study, it has been found that the stable armor layer usually has uniform sediment, with lesser geometric standard deviation in comparison to parent streambed material and depends on the strength and quantity of coarser particles [14].

The strength of the scour hole depends on movement of coarser particles [11]. In this study, it has been noted by us that coarser particles are more stable inside the scour hole because of their larger mass (more energy is required to remove from scour hole). At the same phase, finer gravels are protected by coarser gravels. Slowly, this process results in the development of clusters of different sizes of gravels. For a definite expanse, these clusters show the same characteristics [11]. Kothyari et al. [14] stated that the formation of armor layer may be noticed as a nonstop process of developments and collapses of clusters.

4. Results and Discussion

The stability of armor layer around the pier is influenced by parent particle size, median diameter of armor particles and approach flow properties. The maximum scour depth (h_s) variations at equilibrium condition with time-averaged flow velocity (U) for different diameters of piers along with different sizes of gravels is shown in Figure 3a–h. It can be noted that coarser gravels show less variation of maximum scour depth with approach flow velocity, while medium gravels show higher variation. It can also be noted from Figure 3, the maximum scour depth in armored streambeds increases with approach flow velocity and that for any sizes of streambed particles, different sizes of particles of armored streambeds illustrate a different maximum scour depth. It has been found that the maximum scour depth variation is higher for fine gravel beds ($d_{50} = 3.26 - 4.94$ mm) and lower for coarser gravel bed ($d_{50} = 7.5 - 10.7$ mm). For particular streambed sediment, the maximum scour depth increases with increase in pier diameter and approach velocity.

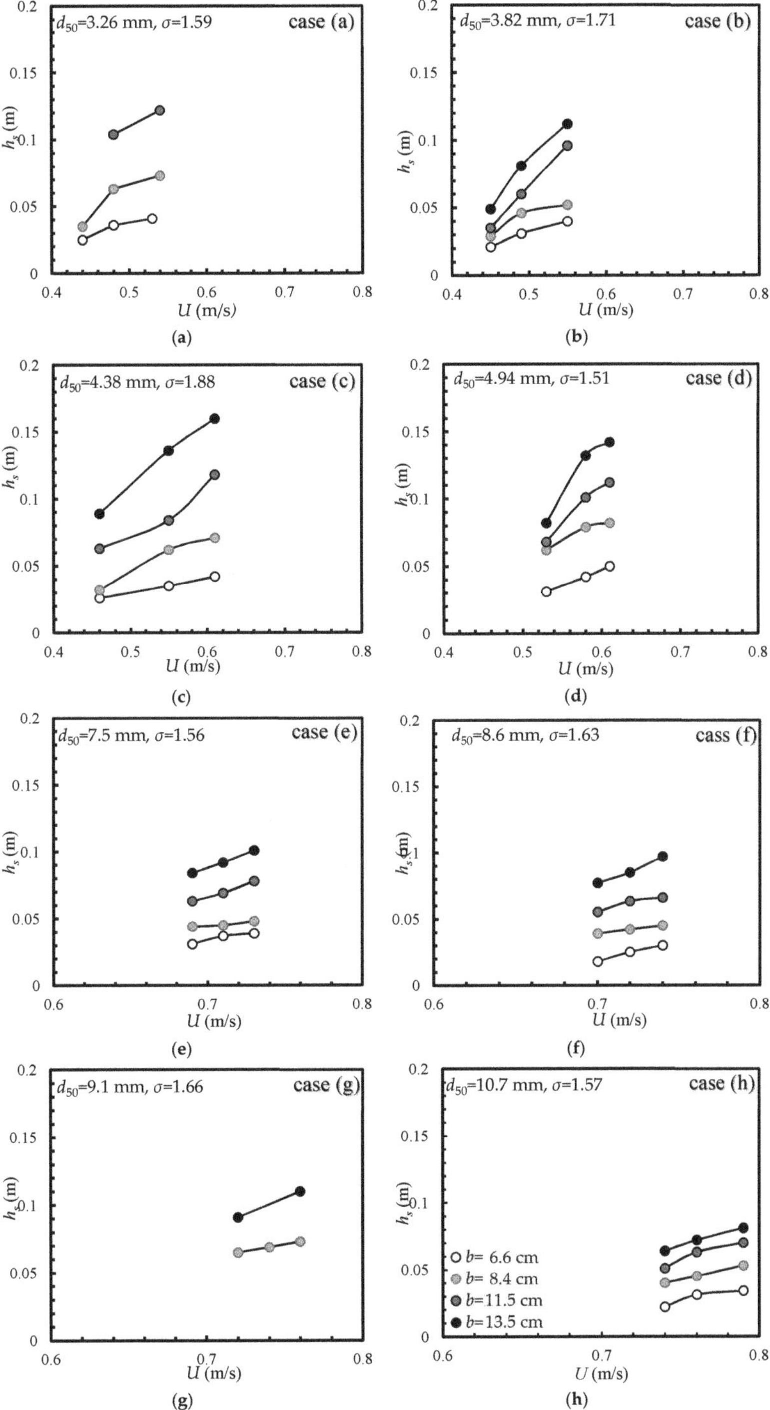

Figure 3. Maximum scour depth variation with approach flow velocity.

We observed that the median diameter of armoring layer (d_{50a}) at equilibrium scour condition depends on σ and increases with σ, as shown in Figure 4. Armoring ratio (d_{50a}/d_{50}) defines the degree of armoring and mainly depends on the flow properties [19]. For $\sigma = 1.63 - 1.88$, the armoring ratio varied from 1.7–2.5, while for the lesser values of σ, i.e., $\sigma = 1.5 - 1.6$, the armoring ratio varied from 1.25–1.7, as can be seen in Figure 4. Wu et al. [17] stated that the size of equilibrium scour-hole illustrates a strong dependency with parent particle size distribution. Kothyari et al. [14] concluded that the maximum scour depth variation in non-uniform and non-cohesive sediment mixture is a function of geometric standard deviation of parent particle size distribution. Similar results were noted by us in the present study and the value of armoring ratio increases with σ. However, the increment in dimensionless scour depth with armor ratio is very gradual up to $h_s/R_L = 0.5$, where R_L is the reference length and defined as $(b^2y)^{1/3}$ by Oliveto and Hager [20,21] and Kothyari et al. [14]. After $h_s/R_L = 0.5$, there is no change in dimensionless scour depth with armor ratio, as can be seen in Figure 4. It indicates that the dimensionless scour depths have -less dependency on degree of armoring ratio, but maximum scour depths have shown noteworthy changes with pier diameter and time-averaged velocity, as can be seen in Figure 3.

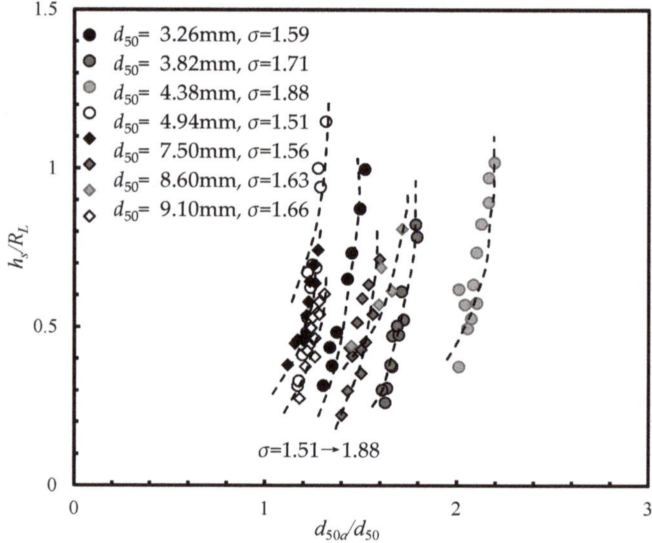

Figure 4. Dimensionless scour depth variation with armor ratio.

Figure 5 shows the dimensionless scour depth (h_s/R_L) variation with y/d_{50} ratio, where y is approach flow depth. For a particular sediment property, the maximum scour depth increases with approach flow depth (y). Figure 5 clearly shows that the maximum scour depth variation with approach flow depth is higher for $d_{50} = 3.26 - 4.94$ mm (fine gravel) and lowest for $d_{50} = 7.5 - 10.7$ mm (coarse gravel). Moreover, the finer streambed particles ($d_{50} = 3.26 - 4.94$ mm) result in a higher dimensionless scour depth variation than that of coarser streambed particles ($d_{50} = 7.5 - 10.7$ mm).

Figure 6 shows the dimensionless scour depth (h_s/R_L) variation with d_{50a}/R_L, for different ranges of critical velocity ratio (U/U_c), and critical velocity ratio is known as the ratio of the approach mean velocity to the critical velocity of sediment. The range of critical velocity ratio is from 0.70 to 0.98. It can be identified that the dimensionless scour depth variation with d_{50a}/R_L is highest for maximum range of critical velocity ratio, i.e., $U/U_c \approx 0.95 - 0.98$. Hence, maximum number of experiments were carried out for $U/U_c \approx 0.90 - 0.95$ and it was noted that for all ranges of critical velocities, h_s increases with decrease of d_{50a}. It has been observed that the lowest variation of dimensionless scour depth with d_{50a}/R_L for $U/U_c \approx 0.70 - 0.80$, as can be seen in Figure 6. Figure 6 clearly illustrates that the dimensionless scour depth variation increases with U/U_c. It can also be noted that the time to reach near equilibrium depths for low values of U/U_c is larger than the higher values of U/U_c.

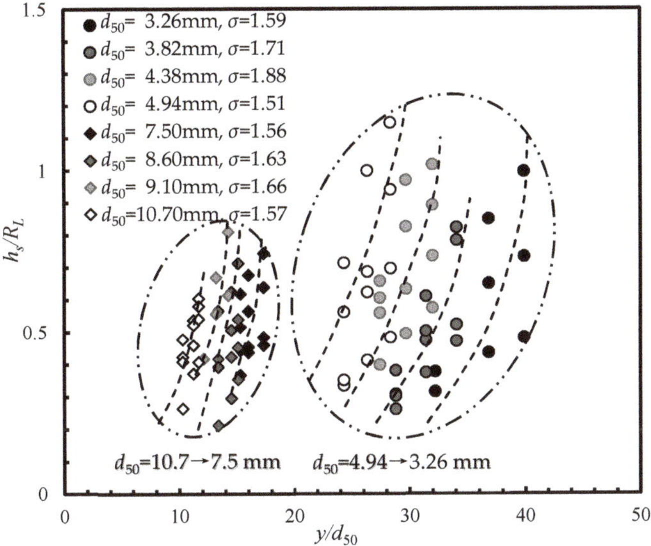

Figure 5. Dimensionless scour depth variation with y/d_{50}.

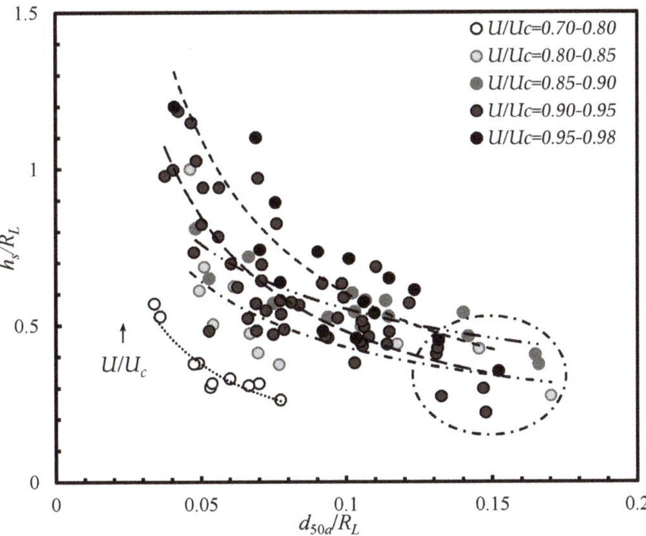

Figure 6. Dimensionless scour depth (h_s/R_L) variation with d_{50a}/R_L.

Implications for Design

In the past studies, several investigators have derived numerous dimensionless scour depth relationships [2,14,22–28]. Present experimental results show that scour depth is not only a function of flow parameters, but also depends on particle size and geometric standard deviation of parent bed material and armor layer particle size. Pier scour experimental data collected under armor streambeds are shown in Table 1, and were used to illustrate the dimensionless scour depth variation with parent densimetric particle Froude number, $F_{rd_{50}} = U/[(S - 1)gd_{50}]^{0.5}$ and ratio of armor layer particle size – reference length (d_{50a}/R_L), as shown in Figure 7. Where S is specific gravity of gravel equal to 2.65. It can be noted that for a particular range of $F_{rd_{50}}$, the dimensionless scour depth decreases with an increase in d_{50a}/R_L. The reducing frequency of dimensionless scour depth was very rapid for the higher value of densimetric Froude number, as can be seen in Figure 7. It can be noted that the decreasing rate of dimensionless scour depth variation with d_{50a}/R_L is lowest for $F_{rd_{50}} = 1.6$–1.7. Meanwhile, Figure 5 shows the dimensionless scour depth gradually increasing with increase in armor ratio, for a particular sediment property. Figures 6 and 7clearly illustrate that the dimensionless scour depth variation with d_{50a}/R_L depends on U/U_c and $F_{rd_{50}}$, respectively. During the experiments, it was observed that the armor layer around the pier is washed out for higher value of U/U_c due to the higher critical velocity of armor particles than the parent particles. For an armor-layered streambed, maximum scour depth increases with the decrease of parent streambed particle size and the increase of pier diameter and approach velocity, as shown in Figures 3 and 5. Therefore, under fixed flow conditions and parent streambed properties, the maximum scour depth in equilibrium condition is larger in finer parent particle streambeds.

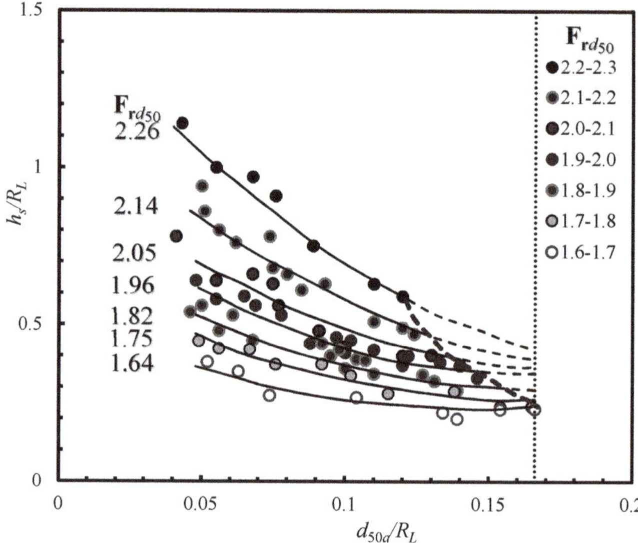

Figure 7. Dimensionless scour depth variation with d_{50a}/R_L and F_{rd50}.

5. Conclusions

To date, very few studies are available for local scour around pier founded in armored gravel-bed streams. In the present study, a total of 85 experiments were carried out under different armor streambeds, along with different pier diameters and flow conditions. Using these different parameters, the maximum scour depth in armor streambeds was analyzed. A new graphical approach was proposed for the maximum scour depth computation.

It was observed that the scour hole around the pier occurs through the formation of an armor layer and scour hole has been reached within equilibrium scour state, when the armor layer particles cover the scour hole and no further movement allows from parent bed material. We found that the maximum scour depth is not only a function of flow parameters, but also depends on particle size and geometric standard deviation of parent bed material and armor layer particle size. The dimensionless scour depth variation with d_{50a}/R_L depends on the parent densimetric particle Froude number, and the decreasing rate of dimensionless scour depth decreases with $F_{rd_{50}}$. We found that the influence of armor ratio on dimensionless scour depth is less, up to $h_s/R_L = 0.5$, however, after $h_s/R_L = 0.5$, change in dimensionless scour depth was negligible. The maximum scour depth around the pier increases with increase in pier diameter, approach velocity, critical velocity ratio and decreases with the size of parent streambed particles. The maximum scour depth decreases with an increase of armor layer particle size for a particular bed material having the same parent bed material. This research indicates the necessity for additional non-uniform coarser bed scour study as it relates to fluvial hydraulics.

Author Contributions: M.P., P.K.S. and C.S.P.O. conducted a series of experiments; M.P., V.K. and S.-C.C. analyzed data, applied methodology and validated this experimental study; S.-C.C., P.K.S., V.K., M.P. and C.S.P.O. reviewed and edited; funding acquisition by S.-C.C.

Funding: This research was funded by the Ministry of Education under the Higher Education Sprout Project.

Acknowledgments: The authors kindly acknowledge the staff of hydraulic laboratory, I.I.T. Roorkee and students of department of soil and water conservation, National Chung Hsing University, Taichung, Taiwan.

Conflicts of Interest: The authors declare no conflict of interest.

Abbreviations

b	Pier diameter
d_{50a}	Median diameter of armor particles
d_{50}	Median diameter of parent streambed particle size
d_{16}	Particle size at 16% finer
d_{84}	Particle size at 84% finer
F_{rd50}	Densimetric particle Froude number of parent bed $U/[(S-1)gd_{50}]^{0.5}$
g	Gravitational acceleration
h_s	Maximum scour depth at equilibrium scour stage
R	Hydraulic radius
R_L	Reference length $(b^2 y)^{1/3}$
S	Relative density
S_0	Channel bed slope
y	Approach flow depth
U	Approach mean velocity
U_c	Critical velocity of parent sediment
u_*	Shear velocity $(gRS_0)^{0.5}$
u_{*c}	Critical shear velocity
σ	Standard deviation of particle size distribution $\{(d_{84}/d_{16})^{1/2}\}$

References

1. Raudkivi, A.J.; Ettema, R. Clear-water scour at cylindrical piers. *J. Hydraul. Eng.* **1983**, *109*, 338–350. [CrossRef]
2. Chen, S.C.; Tfwala, S.; Wu, T.Y.; Chan, H.C.; Chou, H.T. A hooked-collar for bridge piers protection: Flow fields and scour. *Water* **2018**, *10*, 1251. [CrossRef]
3. Sheppard, D.M.; Melville, B.; Demir, H. Evaluation of existing equations for local scour at bridge piers. *J. Hydraul. Eng.* **2013**, *140*, 14–23. [CrossRef]
4. Deng, L.; Wang, W.; Yu, Y. State-of-the-art review on the causes and mechanisms of bridge collapse. *J. Perform. Constr. Facil.* **2015**, *30*, 04015005. [CrossRef]
5. Qi, M.; Li, J.; Chen, Q. Comparison of existing equations for local scour at bridge piers: Parameter influence and validation. *Nat. Hazards* **2016**, *82*, 2089–2105. [CrossRef]
6. Qi, M.; Li, J.; Chen, Q. Applicability analysis of pier-scour equations in the field: Error analysis by rationalizing measurement data. *J. Hydraul. Eng.* **2018**, *144*, 04018050. [CrossRef]
7. Wang, C.; Yu, X.; Liang, F. A review of bridge scour: Mechanism, estimation, monitoring and countermeasures. *Nat. Hazards* **2017**, *87*, 1881–1906. [CrossRef]

8.   Pandey, M.; Sharma, P.K.; Ahmad, Z.; Karna, N. Maximum scour depth around bridge pier in gravel bed streams. *Nat. Hazards* **2018**, *91*, 819–836. [CrossRef]

9.   Yang, Y.; Qi, M.; Li, J.; Ma, X. Evolution of Hydrodynamic Characteristics with Scour Hole Developing around a Pile Group. *Water* **2018**, *10*, 1632. [CrossRef]

10.   Dey, S.; Raikar, R.V. Clear-water scour at piers in sand beds with an armor layer of gravels. *J. Hydraul. Eng.* **2007**, *133*, 703–711. [CrossRef]

11.   Jueyi, S.U.I.; Afzalimehr, H.; Samani, A.K.; Maherani, M. Clear-water scour around semi-elliptical abutments with armored beds. *Int. J. Sediment Res.* **2010**, *25*, 233–245. [CrossRef]

12.   Raikar, R.V.; Dey, S. Clear-water scour at bridge piers in fine and medium gravel beds. *Can. J. Civ. Eng.* **2005**, *32*, 775–781. [CrossRef]

13.   Schmidt, K.H.; Gintz, D. Results of bedload tracer experiments in a mountain river. *River Geomorphol.* **1995**, *2*, 145–158.

14.   Kothyari, U.C.; Hager, W.H.; Oliveto, G. Generalized approach for clear-water scour at bridge foundation elements. *J. Hydraul. Eng.* **2007**, *133*, 1229–1240. [CrossRef]

15.   Guo, J. Pier scour in clear water for sediment mixtures. *J. Hydraul. Res.* **2012**, *50*, 18–27. [CrossRef]

16.   Kim, I.; Fard, M.Y.; Chattopadhyay, A. Investigation of a bridge pier scour prediction model for safe design and inspection. *J. Bridge Eng.* **2014**, *20*, 04014088. [CrossRef]

17.   Wu, P.; Hirshfield, F.; Sui, J. Armour layer analysis of local scour around bridge abutments under ice cover. *River Res. Appl.* **2015**, *31*, 736–746. [CrossRef]

18.   Melville, B.W.; Sutherland, A.J. Design method for local scour at bridge piers. *J. Hydraul. Eng.* **1988**, *114*, 1210–1226. [CrossRef]

19.   Wilcock, P.R.; DeTemple, B.T. Persistence of armor layers in gravel-bed streams. *Geophys. Res. Lett.* **2005**, *32*. [CrossRef]

20.   Oliveto, G.; Hager, W.H. Temporal evolution of clear-water pier and abutment scour. *J. Hydraul. Eng.* **2002**, *128*, 811–820. [CrossRef]

21.   Oliveto, G.; Hager, W.H. Further results to time-dependent local scour at bridge elements. *J. Hydraul. Eng.* **2005**, *131*, 97–105. [CrossRef]

22.   Melville, B.W.; Coleman, S.E. *Bridge Scour*; Water Resources Publication: Highlands Ranch, CO, USA, 2000.

23.   Bozkuş, Z.; Özalp, M.C.; Dinçer, A.E. Effect of pier inclination angle on local scour depth around bridge pier groups. *Arab. J. Sci. Eng.* **2018**, *43*, 5413–5421. [CrossRef]

24.   Brandimarte, L.; Paron, P.; Di Baldassarre, G. Bridge pier scour: A review of processes, measurements and estimates. *Environ. Eng. Manag. J.* **2012**, *11*, 975–989. [CrossRef]

25.   Pu, J.; Wei, J.; Huang, Y. Velocity distribution and 3D turbulence characteristic analysis for flow over water-worked rough bed. *Water* **2017**, *9*, 668. [CrossRef]

26.   Azamathulla, H.M.; Ghani, A.A.; Zakaria, N.A.; Guven, A. Genetic programming to predict bridge pier scour. *J. Hydraul. Eng.* **2010**, *136*, 165–169. [CrossRef]

27.   Pandey, M.; Sharma, P.K.; Ahmad, Z.; Singh, U.K. Evaluation of existing equations for temporal scour depth around circular bridge piers. *Environ. Fluid Mech.* **2017**, *17*, 981–995. [CrossRef]

28.   Khan, M.; Azamathulla, H.M.; Tufail, M. Gene-expression programming to predict pier scour depth using laboratory data. *J. Hydroinform.* **2012**, *14*, 628–645. [CrossRef]

Article

Simulated Flow Velocity Structure in Meandering Channels: Stratification and Inertia Effects Caused by Suspended Sediment

Fei Yang [1], Xuejun Shao [1], Xudong Fu [1,*] and Ehsan Kazemi [2]

[1] State Key Laboratory of Hydroscience and Engineering, Tsinghua University, Beijing 100084, China; yangfeihaoyun@163.com (F.Y.); shaoxj@tsinghua.edu.cn (X.S.)
[2] Department of Civil and Structural Engineering, University of Sheffield, Sheffield S1 3JD, UK; e.kazemi@sheffield.ac.uk
* Correspondence: xdfu@tsinghua.edu.cn; Tel.: +86-10-6279-7071

Received: 28 April 2019; Accepted: 11 June 2019; Published: 15 June 2019

Abstract: In this study, the coupled effects of sediment inertia and stratification on the pattern of secondary currents in bend-flows are evaluated using a 3D numerical model. The sediment inertia effect, as well as the stratification effect induced by the non-uniform distribution of suspended sediment, is accounted for by adopting the hydrodynamic equations without the Boussinesq approximation. The 3D model is validated by existing laboratory experimental results. Simulation results of a simplified meandering channel indicate that sediment stratification effect enhances the intensity of secondary flow via reducing eddy viscosity, while sediment inertia effect suppresses it. The integrated effects result in an increase and a reduction in the secondary flow, respectively, at lower and higher concentrations (near-bed volumetric concentrations of 0.015 and 0.1 are, respectively, considered in this study). This suggests that the dominance of the suspended sediment effect depends on the sediment concentration profile. With the increase of concentration under a specific sediment size, the secondary flow rises to reach a maximum, and then decreases. Moreover, as the sediment concentration increases, an exponentially decaying rate has been found for the secondary flow. It is concluded that in the numerical simulation of flow in meandering channels, when concentration is high, the variable-density hydrodynamic equations without the Boussinesq approximation should be considered.

Keywords: stratification effect; inertia effect; secondary flow; meandering; sediment laden flows

---

1. Introduction

In fluvial process studies, suspension has always been known to have a fundamental role in defining the preferential bend migration, since the stratification brought by suspended load modifies the turbulence, flow structure and sediment entrainment. Early works relate sediment stratification effect on flow velocity to concentration profiles. Vanoni [1] and Einstern and Chien [2] classically documented the sediment stratification effect on the velocity profile in turbulent open channel flows. Vanoni showed that the von Karman constant κ reduces at high sediment concentration. This was based on the hypothesis that the suspended sediment dampens turbulence and increases velocity gradient near the wall. Later, many laboratory investigations [3–5] confirmed this finding. Castro-Orgaz et al. [6] analytically indicated that the reduction of κ in the high concentration condition is a physically coherent approximation. Suspension stratification effect modulates the turbulence intensity which in turn reduces the vertical sediment diffusivity. A few attempts referred the suspension effect to heat or salinity induced stratification effect [7]. By analogy with thermally stratified boundary layers, the Flux Richardson number or the gradient Richardson number for sediment-laden flow is defined as a measure of stratification effect or turbulence damping [8,9] and the Monin-Obukhov length scale is introduced to establish damping functions for the eddy viscosity [10]. Herrmann and Madsen [11] obtained the optimal parameters for damping eddy diffusivity from observations. Nezu and Azuma [12] presented simultaneous measurements of both sediment and fluid in dilute flows. They pointed out that flow turbulence intensity is slightly dampened far from the bed while it rises in the near-wall region owing to the velocity difference between fluid and sediment. The ratio of concentration to slope is a primary indicator of stratification effect (which is greater in low slope rivers) according to Wright and Parker [13]. For high suspension concentration, Lamb et al. [9] experimentally observed that the Richardson number critical value of ~1/4 for completely damped turbulence [7] is not applicable where the major source of suspension is transported turbulence; and Muste et al. [14] provided the priority of two-phase over traditional mixed-flow perspective on sediment laden flows from image velocimetry experiments.

Although sediment laden flows have been widely studied by using various numerical models, most studies concern low suspension concentration using the advection-diffusion equation. The fundamental assumptions of these models limit their applicability. The advection-diffusion theory regulates the upward transfer of suspension due to turbulent diffusion with the downward settlement. Winterverp [15,16] numerically explained the sediment stratification effect by a buoyancy destruction term in the standard k-ε turbulence model at low concentration. Later, concerning hindered settling, van Maren et al. [17] extended this explanation to hyperconcentrated flows. Byun and Wang [18] coupled the sediment transport model and the 3D tidal hydrodynamics model to examine the reduction of turbulence and bottom shear stress due to the sediment stratification effect. Amoudry and Souza [19] performed 3D simulations and showed that sediment stratification effect leads to non-negligible modifications for bed evolution. Cantero et al. [20] exploited Direct Numerical Simulation (DNS)s of sediment laden flows under the dilute flow assumption and Boussinesq approximation. For higher settling velocity, the vertical Reynolds fluxes could be completely suppressed locally and relaminarization occurs [20,21]. For the condition of complete turbulence suppression, the critical product of the Richardson number and sediment settling velocity has a logarithmic dependence on the Reynolds number [22]. Dutta et al. [23] further investigated the sediment stratification effect on the vertical sediment diffusivity by DNSs. Dallali and Armenio [24] used Large Eddy Simulation (LES) to investigate suspension transport, supporting that the buoyancy effect on flow momentum is not neglectable for small particles. Alternatively, two-phase flow theory addresses water and sediment phases separately [25]. Recently, several models based the two-phase flow theory have been developed to include small scale mechanics [26–28]. The drift velocity shows that the inertia of fluid, particle turbulence and collisions effect contribute to sediment dispersion [26].

The Boussinesq approximation [29] is the basis for mathematical formulation of various flow problems. According to Kundu and Cohen [30], after subtracting a static reference state, i.e., ∇p_0 ($= \rho_0 \vec{g}$, where ρ_0 and p_0 = reference mixture density and pressure, and $\vec{g}$ = acceleration of gravity), the momentum equation can be expressed as

$$\left(1 + \frac{\rho'}{\rho_0}\right)\frac{d\vec{u}}{dt} = -\frac{1}{\rho_0}\nabla p' + \frac{\rho'}{\rho_0}\vec{g} + \nu\nabla^2\vec{u} \tag{1}$$

where t = time; p' and ρ' = local deviations of pressure and mixture density from static reference state; $\vec{u}$ = flow velocity vector; and ν = kinematic viscosity.

Both the inertia and buoyancy terms (the term on the left-hand side and the second term on the right-hand side of the above equation, respectively) contain the density ratio. With the Boussinesq approximation, the sediment stratification effect is shown as the buoyancy term while the inertia effect (the density ratio in the inertia term) is neglected. One then has

$$\frac{d\vec{u}}{dt} = -\frac{1}{\rho_0}\nabla p' + \frac{\rho'}{\rho_0}\vec{g} + \nu\nabla^2\vec{u} \tag{2}$$

Sediment concentration near the water surface is always much lower than that near the bed for the condition of heavy particles. A large vertical density gradient which has obvious inertia effect on flow can be a result of high bulk sediment concentration. The vertical concentration gradient is often very large with significant inertia variations. Therefore, the applicability of the Boussinesq approximation needs to be assessed in such a case. To the best of our knowledge, there have been no reported documents on the numerical comparison of sediment laden open-channel flows with and without considering the Boussinesq approximation.

Because the role of fluid inertia is highly important in the acceleration/deceleration processes of fluid flow, a 3D modelling method is adopted to thoroughly analyze the inertia effects of suspended sediment under complicated flow boundary conditions. The 3D model is substantially different from traditional hydrodynamic models under Boussinesq approximation which only consider the stratification effect of suspensions as the buoyance term. In the following sections, first, model setup is provided; second, a few cases are selected for model verification; third, numerical experiments with and without Boussinesq approximation are presented for comparison; and finally, discussions and suggestions with regards to the sediment inertia effect are given.

2. Model Setup

2.1. Hydrodynamic Model

The governing hydrodynamic equations are 3D shallow water equations based on the hydrostatic pressure. The mass and momentum equations are written as follows:

$$\frac{\partial u}{\partial x} + \frac{\partial v}{\partial y} + \frac{\partial w}{\partial z} = 0 \tag{3}$$

$$\rho\frac{du}{dt} = -\frac{\partial p}{\partial x} + \frac{\partial}{\partial z}\left(\rho v_z \frac{\partial u}{\partial z}\right) + \frac{\partial}{\partial x}\left(\rho v_h \frac{\partial u}{\partial x}\right) + \frac{\partial}{\partial y}\left(\rho v_h \frac{\partial u}{\partial y}\right) \tag{4}$$

$$\rho\frac{dv}{dt} = -\frac{\partial p}{\partial y} + \frac{\partial}{\partial z}\left(\rho v_z \frac{\partial v}{\partial z}\right) + \frac{\partial}{\partial x}\left(\rho v_h \frac{\partial v}{\partial x}\right) + \frac{\partial}{\partial y}\left(\rho v_h \frac{\partial v}{\partial y}\right) \tag{5}$$

$$-\rho g + \frac{\partial p}{\partial z} = 0 \tag{6}$$

where x, y and z = Cartesian coordinates; u, v and w = components of velocity in the x, y and z directions; p = pressure; ρ = density of mixture; and v_h and v_z = eddy viscosity in the horizontal (x and y) and vertical (z) directions. The density of the mixture is determined by

$$\rho = \rho_w + (\rho_s - \rho_w)c, \tag{7}$$

where c = volumetric concentration of sediment; ρ_w = water density; and ρ_s = sediment density.

Similar to layer-integrated models [31,32], the vertical velocity is calculated from the integrated continuity equation as follows

$$w = w_{z_b} - \frac{\partial}{\partial x}\int_{z_b}^{z} u\,dz - \frac{\partial}{\partial y}\int_{z_b}^{z} v\,dz \tag{8}$$

where z_b = bed elevation; and w_{z_b} is the vertical velocity at the bed. At the water surface,

$$\frac{\partial H}{\partial t} = w_{z_b} - \frac{\partial}{\partial x}\int_{z_b}^{H} u\,dz - \frac{\partial}{\partial y}\int_{z_b}^{H} v\,dz \tag{9}$$

where H = water surface elevation.

The pressure in the horizontal momentum equations is given by integration of vertical momentum equation as

$$p = p_a + g \int_z^H \rho \, dz \tag{10}$$

where p_a = pressure at the water surface which is set to zero in the present simulations. The pressure gradient is obtained as follows

$$\frac{\partial p}{\partial y} = g \rho_w \frac{\partial H}{\partial y} + g \frac{\partial}{\partial y} \int_z^H (\rho - \rho_w) dz \tag{11}$$

Eddy diffusivity and viscosity are obtained from a two-equation k-ω turbulence closure model [33]. Complete extinction of local turbulence is unsustainable according to Cantero et al. [22]. To prevent the turbulence to be totally damped by the sediment stratification effect, a prescribed upper threshold 0.2 is set on the flux Richardson number. As the Reynolds-Averaged Navier–Stokes (RANS) model is incapable of modeling flow under conditions of strong stratification [34], an empirical distribution [25] for viscosity is reserved here as follows.

$$v_h = v_z = \begin{cases} \kappa U_* z (h - z)/h & y < 0.5h \\ \frac{1}{4}\kappa U_* h & y \geq 0.5h \end{cases} \tag{12}$$

where h = water depth; and U_* = friction velocity.

2.2. Sediment Transport Model

Sediment transport is represented by the advection-diffusion equation,

$$\frac{dc}{dt} = -\frac{\partial c w_f}{\partial y} + \frac{\partial}{\partial z}\left(\varepsilon_z \frac{\partial c}{\partial z}\right) + \frac{\partial}{\partial x}\left(\varepsilon_h \frac{\partial c}{\partial x}\right) + \frac{\partial}{\partial y}\left(\varepsilon_h \frac{\partial c}{\partial y}\right) \tag{13}$$

where w_f = sediment settling velocity; and ε_h and ε_z = eddy diffusivity in the horizontal (x and y) and vertical (z) directions.

2.3. Boundary Condition

There are two thin layers of δ at the water surface and the bed boundaries, thus the thickness of the calculation domain is $h - 2\delta$. Vertical boundary conditions are implemented at these two interfaces. Normal gradients of horizontal velocities as well as the net vertical flux of sediment are zero at the upper interface. At the bottom one, we enforce the bottom frictional stress for momentum equations as

$$\rho v_z \left(\frac{\partial u}{\partial z}, \frac{\partial v}{\partial z}\right)_{z_b} = \rho C_D \sqrt{u_b^2 + v_b^2}(u_b, v_b) \tag{14}$$

where u_b and v_b = near-bed components of horizontal velocity; and C_D = drag coefficient to be calibrated. Besides, the near-bed net flux of sediment is

$$\left(\varepsilon_z \frac{\partial c}{\partial z} - c w_f\right)_{z_b} = q_{su} - c_b w_f \tag{15}$$

where q_{su} = sediment entrainment rate; and c_b = near-bed volumetric concentration.

Flow condition at the inlet is given by flow discharge. In other words, the flow velocity profile at the inlet can be estimated through the transport equation for turbulent components and momentum equations with the assumption that flow is locally uniform at this boundary. Similarly, the suspended sediment concentration profile at the inlet is determined by assuming local equilibrium in the inlet sediment entrainment. At the outlet, the water level is set to prescribed values as the flow is assumed to be subcritical.

2.4. Domain Discretization

For discretization of the computational domain, a Spectral Element Method (SEM) is used in the vertical direction (due to the importance of the vertical variations in the present work) and the finite difference method is employed in the horizontal directions. Besides, for ease of implementation of the boundary conditions, vertical domain is discretized into three layers (due to the existence of two thin layers adjacent to water surface and bed) with unstretched z-coordinates. First, vertical space is divided into a number of sublayers with evenly thickness of h_0. The top/bottom sublayer is combined with its adjacent sublayer, producing a layer with a thickness between h_0 and $2h_0$. Then, all the rest inner sublayers are combined to form an inner layer with its thickness being multiple of h_0. Therefore, there will be three vertical layers throughout the whole calculation domain. All these three layers are variable according to water surface and bed changes during calculation process. Thickness thresholds z_{min} and z_{max} are imposed over top and bottom layers. When the thickness of the top/bottom layer is out of range ($>z_{max}$ or $<z_{min}$), after the calculation of water surface/bed elevation at a certain time step, the layer interface will move a distance of h_0 inward or outward to ensure the thickness stays in the range. Let $h_0 = (z_{max} - z_{min})/2$, after adjustment, its thickness will be close to $(z_{max} - z_{min})/2$. Then, there is no need to readjust the thickness for small surface/bed changes. Every thickness adjustment is accompanied by local redistribution of physical quantities.

This kind of vertical discretization allows us to conveniently employ the finite dimensional space/polynomial of relatively higher degree for inner layer and lower degree for top/bottom layers. For the horizontal diffusion/viscosity calculations at the inner layer, if the horizontally adjacent inner layers are consistent, the calculations will be straight-forward as in the z-coordinates method, otherwise, additional interpolations need to be carried out separately for each calculation. For the calculations at the top/bottom layer, the horizontally adjacent layers are always inconsistent, and interpolations are required.

Horizontal domain is discretized using curvilinear coordinates into staggered grids. The governing equations need to be converted accordingly for implementation. Here, we present only the framework in the Cartesian coordinates for simplicity.

2.5. Solution of Generalized Equation by SEM

The horizontal momentum, sediment and turbulence transport equations can be generalized as

$$\begin{cases} \frac{d\phi}{dt} = \frac{\partial}{\partial z}\left(w_f\phi + \frac{v_t}{\sigma}\frac{\partial\phi}{\partial z}\right) + f, z \in (z_b, H) \\ \frac{v_t}{\sigma}\frac{\partial\phi}{\partial z}\Big|_H = g_1, \ \frac{v_t}{\sigma}\frac{\partial\phi}{\partial z}\Big|_{z_b} = g_2 \end{cases} \tag{16}$$

where ϕ = the physical quantity to be solved; w_f is set only for sediment transport; σ = the Prandtl–Schmidt number for ϕ; and f = other terms. The top and bottom boundaries are chosen according to the specific equation to be solved. At each vertical layer i with thickness $L_i = z_{i+1} - z_i$, generalized equation is conducted under local vertical coordinate $\zeta = 2 (z - z_i)/L_i - 1$. We set an approximation such that ϕ belongs to the finite dimensional space of degree N. Choosing the Legendre polynomials as the local basis functions, the approximate solution is defined by

$$\phi = \sum_{k=0}^{N} \phi_k h_k(\zeta) \ \text{with,} \ h_k(\zeta) = \frac{1}{N(N+1)L_N(\zeta_k)}\frac{\left(1-\zeta^2\right)L'_N(\zeta)}{\zeta-\zeta_k} \tag{17}$$

where ϕ_k and $h_k(\zeta)$ = the coefficients and the Lagrange form of Legendre polynomials, respectively; interpolation point ζ_k = Legendre–Gauss–Lobatto point with its weight ω_k; and $L_N(\zeta)$ and $L'_N(\zeta)$ = the

Legendre polynomials and their derivatives. Generalized equation is multiplied by the test function and integrated over ζ (between [−1, 1]) and time step. We choose the local basis function as the test function. Then, the weak formulation after including boundary condition, can be derived as

$$\frac{L_i}{2}\sum_{j=0}^{N}\phi_j{}^{n+1}B_{jk} + \frac{2}{L_i}\Delta t\sum_{m=0}^{N}v_m\sum_{j=0}^{N}\phi_j{}^{n+1}D_{jk,m} + \Delta t\sum_{j=0}^{N}w_f\phi_j{}^{n+1}C_{jk}$$
$$= \frac{L_i}{2}\sum_{j=0}^{N}\phi_j{}^{n}B_{jk} + \frac{L_i}{2}\Delta t\sum_{j=0}^{N}f_j{}^{n}B_{jk}, \quad k = 0,\cdots,N. \tag{18}$$

The integral coefficients are given as

$$B_{jk} = \begin{cases}\omega_j & j=k \\ 0 & j\neq k\end{cases}, \; C_{j,k} = \frac{dh_k(\zeta_j)}{d\zeta}\omega_j, \; D_{kj,m} = \frac{dh_k(\zeta_m)}{d\zeta}\frac{dh_j(\zeta_m)}{d\zeta}\omega_m \tag{19}$$

where ω_j = the weighting function of Gaussian quadrature for ζ_j. The depth integrated continuity and horizontal momentum equations are solved in the coupled way.

Depth integrated continuity equation is discretized with a semi-implicit approach in curvilinear coordinates as

$$\begin{aligned}H^{n+1} &= z_b + h^{n+1} \\ &= z_b + h^n - J\Delta t\Big(\frac{\partial}{\partial\xi}\frac{h^n\bar{u}^n}{J} + \frac{\partial}{\partial\eta}\frac{h^n\bar{v}^n}{J}\Big)(1-\theta) - J\Delta t\Big(\frac{\partial}{\partial\xi}\frac{h^n\bar{u}^{n+1}}{J} + \frac{\partial}{\partial\eta}\frac{h^n\bar{v}^{n+1}}{J}\Big)\theta\end{aligned} \tag{20}$$

where ξ and η = horizontal curvilinear coordinates; $\bar{u}$ and $\bar{v}$ = depth-averaged values of u and v; θ = the implicitness factor for temporal discretization; and J = Jacobi determinant. Substituting the discretized continuity equation into the horizontal momentum equations,

$$\begin{aligned}&\frac{L_i}{2}\sum_{m=0}^{N}u_m{}^{n+1}B_{mk} + \frac{2}{L_i}\Delta t\sum_{p=0}^{N}v_p\sum_{m=0}^{N}u_m{}^{n+1}D_{mk,p} - \Delta t\Big(\frac{2}{L_i}\frac{v}{\sigma}\frac{\partial u}{\partial\zeta}\Big)\Big|_{\Gamma_2} \\ &-\Delta t^2 g\frac{L_i}{2}\beta_1\theta^2\sum_{m=0}^{N}B_{mk}\frac{\partial}{\partial\xi}\Big(J\frac{\partial}{\partial\xi}\frac{h^n\bar{u}^{n+1}}{J} + J\frac{\partial}{\partial\eta}\frac{h^n\bar{v}^{n+1}}{J}\Big) = \frac{L_i}{2}\sum_{m=0}^{N}u_{m*}{}^{n}B_{mk} + \frac{L_i}{2}\Delta t\sum_{m=0}^{N}f_m{}^{n}B_{mk} \\ &-\Delta tg\frac{L_i}{2}\beta_1\sum_{m=0}^{N}B_{mk}\frac{\partial}{\partial\xi}\Big[-\Delta t\theta(1-\theta)J\Big(\frac{\partial}{\partial\xi}\frac{h^n\bar{u}^n}{J} + \frac{\partial}{\partial\eta}\frac{h^n\bar{v}^n}{J}\Big) + H^n\Big], \quad k = 0,\cdots,N.\end{aligned} \tag{21}$$

where $\beta_1 = \xi_x^2 + \xi_y^2$; $v = v_h$ for simplification; σ = Schmidt number; and $u_{m*}{}^{n}$ = the backtracked value of u_m at time step n. Momentum equation for v has a similar form. Underlined terms are zero at the outlet for imposing water level condition for subcritical flows. In the momentum and sediment transport equations, advection terms are incorporated in the total derivatives. By backtracking characteristic lines accurately under Eulerian–Lagrangian context [35], the backtracked value at the foot of the characteristic lines at time step n can be approximated by interpolation. The final set of momentum equations can be solved by Conjugate Gradient. Equation (21) is semi-implicit with respect to v. Hence, v is firstly taken as an explicit quantity in the horizontal momentum for u, and then one more iteration is applied after the solution of the horizontal momentum equations. Once the horizontal velocities are known, water surface level and vertical velocity can be computed using the depth integrated continuity equation.

3. Model Verification and Results

The developed 3D numerical model is employed to simulate flow and suspended sediment transport in straight and meandering channels, and validated by comparing the measured and predicted distributions of flow hydrodynamics and sediment transport characteristics. To examine the effect of suspended sediment on the flow structure, a meandering channel flow with considerable suspended sediment is considered. As there was no experimental data available to us for suspended sediment motion in meandering channels, we firstly validated the model for the cases with suspended sediment at equilibrium and non-equilibrium conditions in straight channels with comparing the model results with a set of existing experimental data (Section 3.1); then applied the model for a meandering channel (but without presence of suspended sediment, i.e., only clear water) and validated the accuracy of the model against existing experiments (Section 3.2); and finally, numerically simulated flow and sediment motion in a meandering channel with investigating the effects of suspended sediment on the velocity structure (Section 3.3).

3.1. Uniform and Steady Open Channel Flows

A data set of laboratory experiment carried out by Coleman [36] was used to verify the model for flows with suspended sediment at equilibrium condition. The experiment was carried out in a 15 m long and 0.356 m wide recirculating flume with the slope of ~0.002. The measurements were taken at 12 m downstream of the inlet. The flow discharge and depth were ~0.064 m^3/s and ~0.169 m, respectively. The velocity and sediment concentration were measured with a Pitot-static tube. Forty runs were performed with sediments of three different grain diameters and different concentrations. Runs 9, 20, 25, 31, and 40 are selected for comparisons with numerical results. Diameter of the sand grains were about 0.105, 0.210, and 0.420 mm, respectively, in runs {9 and 20}, {25 and 31}, and {40}. The Schmidt number for sediment vertical diffusivity is closely related to sediment diameter. Hence, we use the Schmidt numbers of 0.7, 1.2, and 2, respectively, for those runs.

Figure 1 presents the estimated velocity and sediment concentration profiles in comparison with the experimental ones. A good agreement between them is observed except in the upper part of the profiles, i.e., near the water surface. One reason could be that the effects of lateral walls on the flow are neglected in the present simulations, while those effects could be relatively high in the present case due to the low aspect ratio (width to depth) of the flume.

The model performance is tested for non-equilibrium condition, too. The non-equilibrium suspended sediment laboratory experiment of Wang and Ribberink [37] is considered. This experiment was carried out in a 30 m × 0.5 m × 0.5 m straight flume with a slope of 0.00097. The flow was 0.216 m deep and the discharge was 0.0601 m^3/s. The flume was composed of a 10 m inflow section with a rigid bed, a 16 m test section with a perforated bed and a 4 m outflow section. Sediment was released at upstream section through the water surface. Nearly uniform sediment was used with settling velocity of $w_f = 0.007$ m/s. The velocity was measured with a micropropellor and the sediment concentration was estimated by using the siphon method. The sediment profile was calculated at the inflow section as in an equilibrium state under given near-bed condition of $q_{su} = 1.19 \times 10^{-6}$ m/s. In the test section, q_{su} was set to zero. According to Wang and Ribberink [37], fluctuations were observed in the measured profile of the sediment concentration.

The numerical result is presented in Figure 2 in comparison with the experimental data. The comparison implies that the model performance is quite good for the non-equilibrium condition, too. The model is thus validated for the flow with suspended sediment at equilibrium and non-equilibrium conditions in straight channels.

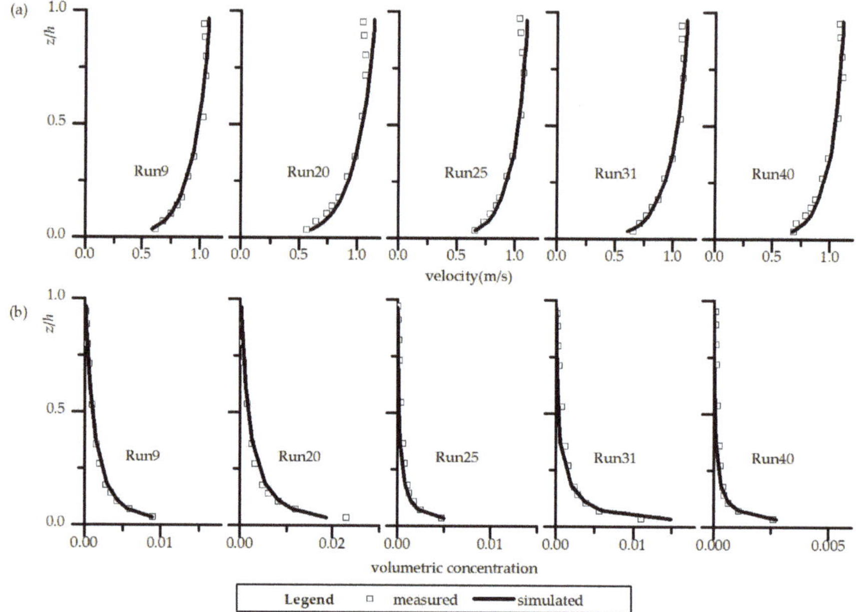

Figure 1. Equilibrium condition: estimated profiles of velocity (**a**) and sediment concentration (**b**) in comparison with the experimental data.

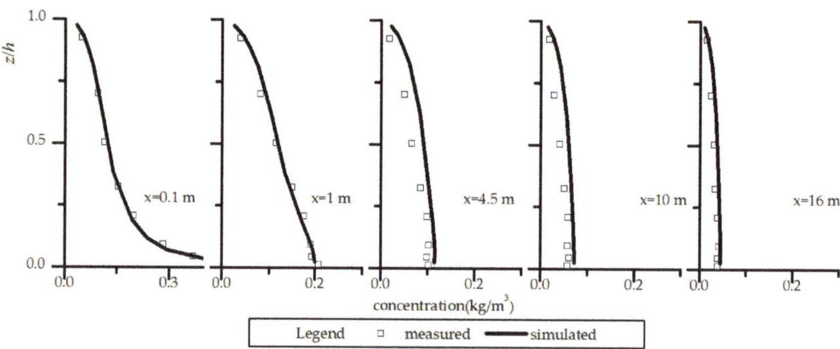

Figure 2. Non-equilibrium condition: calculated sediment concentration profiles in comparison with the experimental data.

3.2. Steady Flows in Meandering Channels

The sharp bend experiment of Blanckaert [38] is used for model evaluation in curved bend flows. The experiment was carried out in a zero-slope flume consisted of a 9 m straight inflow, a 193° curved bend and a 5 m straight outflow section (Figure 3). The curvature radius of the bend was R = 1.7 m. The width of the channel was B = 1.3 m, and the lateral walls were vertical and hydraulically smooth. The velocity was measured by an acoustic Doppler velocity profiler. For the present simulation, we selected one steady case with the water depth of 0.159 m and discharge of 0.089 m^3/s.

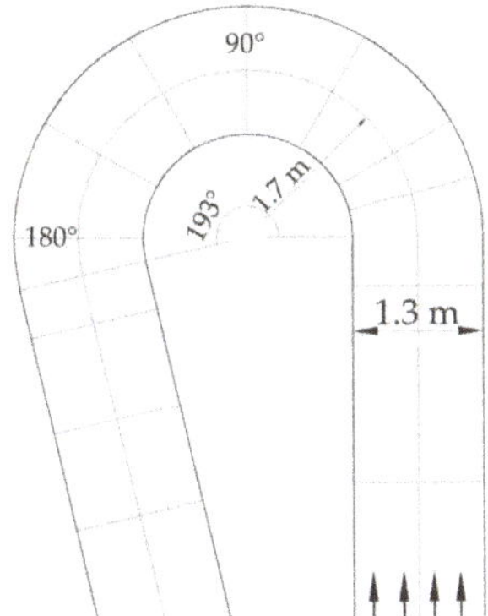

Figure 3. Plan view of Blanckaert's laboratory flume.

The calculation was performed on a horizontal staggered grid with 130 × 20 cells. Calculation domain contains the curved bend and two 2.6 m length sections of inlet and outlet. The time step was set to 0.1 s. Inlet depth was identical to the measured one after adjusting the drag coefficient. Friction resistance is not taken into account for the side walls.

Figure 4a shows the distribution of the measured depth-averaged streamwise velocity. The short flow route leads to an acceleration close to the inner bank which is strengthened by transversal pressure gradient. At the same time, the streamline curvature induced a cross-sectional secondary flow, which shifts the larger velocity core toward the outer bank. These two mechanisms together generate the gradual outward shift of the larger velocity core at the channel bend. In the present calculation, magnitude of vertical viscosity is set to half of the horizontal one. Figure 4b shows the result of the numerical model. Outward shifting process of larger velocity core is closely consistent with the experiment. Figure 5 presents the streamwise velocity profiles in comparison with the experiments at two cross-sections shown on Figure 3, i.e., cross-sections 90° and 180°. Although the present model is hydrostatic with isotropic turbulence model, only some small local differences with the experiment are observed near the side walls (Figure 5). Figure 6 presents the estimated transverse velocity profiles at the same cross-sections in comparison with the experimental data. It shows that the model is also capable of predicting transverse velocity with good accuracy.

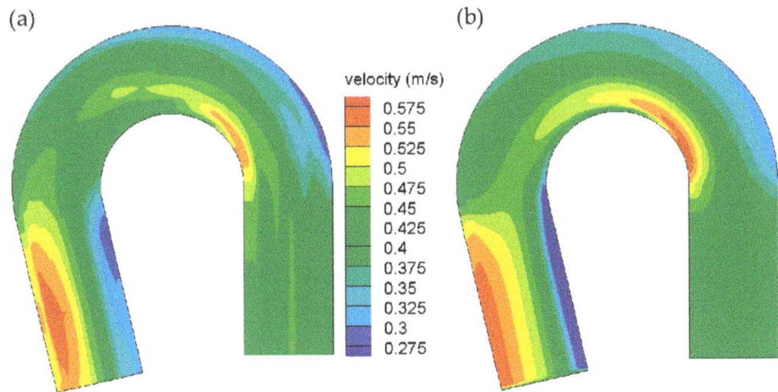

Figure 4. Distribution of the depth-averaged streamwise velocity: (**a**) Experiment; (**b**) modelling.

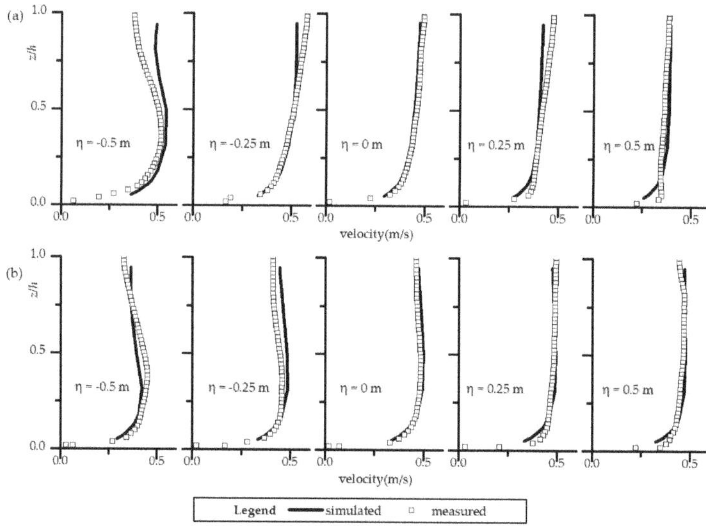

Figure 5. Calculated streamwise velocity profiles in comparison with the measurements: (**a**) Cross-section 90°; (**b**) cross-section 180°. η is the position in the lateral direction from the central line towards the outer bank.

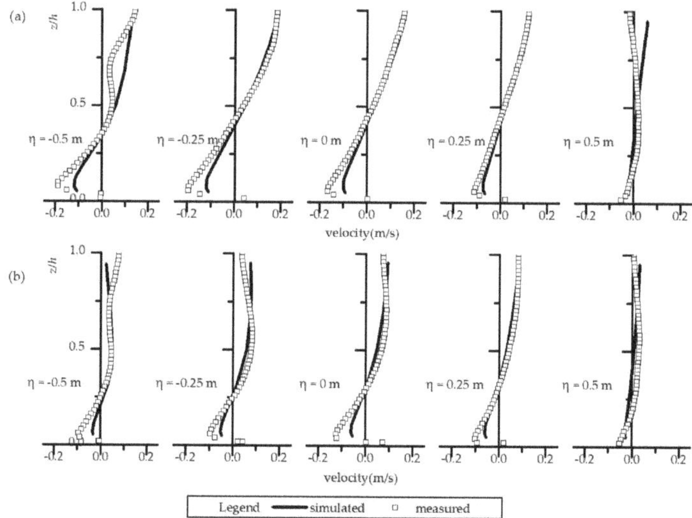

Figure 6. Calculated transverse velocity profiles in comparison with the measurements: (**a**) Cross-section 90°; (**b**) cross-section 180°. η is the position in the lateral direction from the central line towards the outer bank.

3.3. Effects of Suspended Sediment on Velocity Structure

Flows in curved open channels modulate the velocity components both in the streamwise and transverse directions as shown previously. Inertia of flow plays an important role in the acceleration and deceleration processes. In the dilute sediment laden flows, the mixture density can be approximately treated as a homogeneous quantity, thus the effect of sediment inertia can be neglected. Since sediment has a greater density than water, if concentration increases, how the sediment inertia affects the laden flow is still unclear. To investigate this issue, here, we carry out a numerical study on the inertia effect in a meandering channel. Channel geometry and flow discharge in Section 3.2 are adopted. The parameters employed in the simulation are reported in Table 1. The simulations are carried out both with and without considering the Boussinesq approximation, with investigating the effect of near-bed volumetric concentration c_b. Runs BAL and N0L are those with and without the Boussinesq approximation for c_b of 0.015; and runs BAH and N0H are also with and without applying the Boussinesq approximation but for a higher concentration, i.e., c_b of 0.1. Run CL, which is the clear water case presented in Section 3.2, is also included for comparison.

Table 1. Conditions of numerical simulations.

Run	BAL	N0L	CL	BAH	N0H
D (mm)	0.05	0.05	-	0.05	0.05
Boussinesq approximation	Yes	No	-	Yes	No
c_b	0.015	0.015	0	0.10	0.10
$\rho_s/\rho_w - 1$	1.65	1.65	-	1.65	1.65

According to the results of the simulations, all the runs have similar water surface elevation distributions with different transversal slope. Here, we investigate the effect of suspended sediment on the flow structure. Figure 7 presents the results associated with the cross-section 90° for the lower concentration case ($c_b = 0.015$), i.e., the runs BAL and N0L, in comparison with the clear water (CL) case. As can be seen, the streamwise velocity near the inner bank is smaller in the BAL and N0L cases compared to the CL case, while it is larger near the outer bank for BAL and N0L runs. That means the secondary flow is stronger in the BAL and N0L runs with respect to the CL case. Besides, it is clear that the suspended sediment has negligible inertia effect on the flow structure in present condition, since the velocity and sediment profiles almost overlap in runs BAL and N0L.

Existence of sediment with low concentration reduces the eddy viscosity so that the vertical gradient of the streamwise velocity is quite large close to the bed. In transversal direction, shear stress is balanced with outward centrifugal force and inward transversal pressure gradient. We can suppose that the direct impact of eddy viscosity on the shear stress is clearer than its indirect impact on the centrifugal force and pressure gradient. Lower eddy viscosity should cause larger transversal velocity gradient under nearly constant shear stress condition. Therefore, the transversal velocity will be larger, which leads to a stronger secondary flow.

Figure 8 shows the results for the higher concentration case (BAH and N0H) in comparison with the CL case. The secondary flow in the runs BAH and N0H seems weaker than the CL case. Similar to the lower concentration case, the eddy viscosity is reduced because of the sediment stratification effect, but the secondary flow is not strengthened in this case. This indicates that the sediment stratification effect on the suppression of the turbulence is not the key factor. As mentioned above, in the cases BAH and N0H, density is taken as a variable quantity in Equation (11). The buoyant force from the sediment stratification effect in the transversal momentum equation controls the secondary flow behavior. For example, consider the profile at centerline ($\eta = 0$) in Figure 8. The sediment concentration is larger close to the inner bank and transversal gradient of sediment concentration is very large. Transversal buoyant force direction is outward which is in opposite direction to the transversal surface gradient and the total transversal pressure gradient is smaller. The centrifugal force and the resultant transversal velocity are smaller than the clear water case. Run BAH and N0H have almost the same transversal surface gradient which has decisive role on the total transversal pressure gradient in the upper depth. Run N0H has smaller density in the upper depth compared to BAH. Therefore, in run N0H, the transversal velocity is larger at the upper part of the depth; and accordingly, the transversal surface gradient becomes larger, thus the transversal velocity in the lower depth increases in the reverse direction. Therefore, the secondary flow in run N0H is stronger than that in run BAH. This means the effect of sediment inertia on the flow momentum is significant while the Boussinesq approximation cannot disclose the flow information at higher concentration in meandering channels.

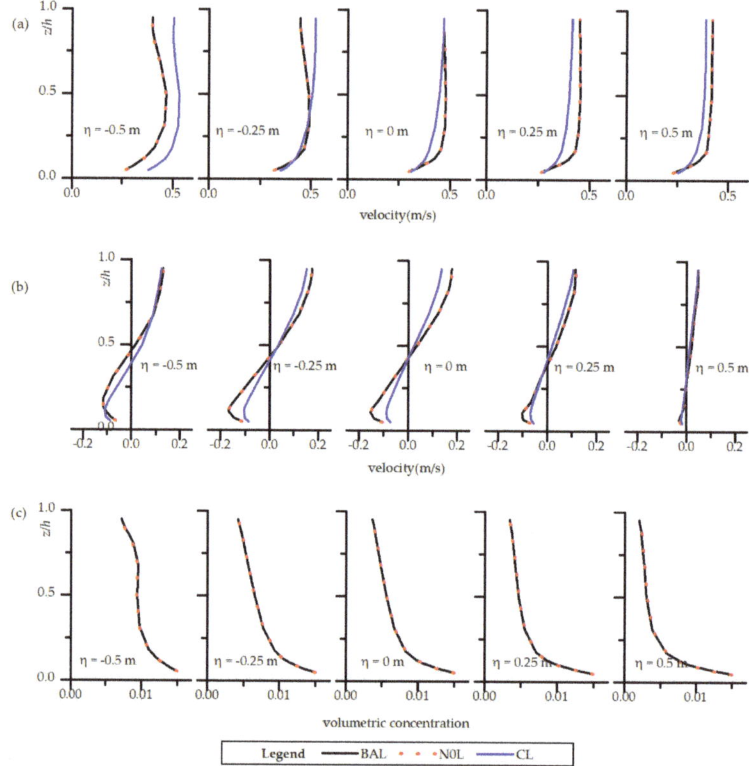

Figure 7. Velocity and sediment concentration profiles at the cross-section 90° for lower concentration case: (**a**) Streamwise velocity; (**b**) transversal velocity; (**c**) sediment volumetric concentration. (velocity unit: m/s).

Runs N0H and BAH have higher concentration where sediment inertia effect is equally important as stratification effect. To illustrate the integrated effect on the secondary flow with different concentrations and sediment sizes, the sensitivity analysis on the size and concentration of sediment is discussed here. Two normalized quantities $\langle f_n^2 \rangle$ and $\langle f_s f_n \rangle$ were adapted as characteristic parameters which represent the secondary flow strength and advecting flow momentum [39]. The sediment size is varied from 0.01 to 0.1 mm with depth averaged volumetric concentration from 0.01 to 0.2 at the inlet. The sediment buoyancy effect on turbulence is excluded here as the simple profile presented in Equation (12) for eddy viscosity was employed in this analysis. The results are summarized in Figure 9. The secondary flow strength decreases as the concentration increases. The finer sediment has a nearly uniform profile and only slightly affects the secondary flow strength. For 0.05- and 0.10-mm sediments, the non-uniformity of concentration profile at low concentration is much higher, vertical density gradient is large and the secondary flow strength declines efficiently to $\frac{1}{4}$ with concentration larger than 0.075. At higher concentration, the group hinder settling velocity is much smaller and the profile approaches to uniformity. Thus, the decrease retards and an exponentially decaying rate is found for the secondary flow here.

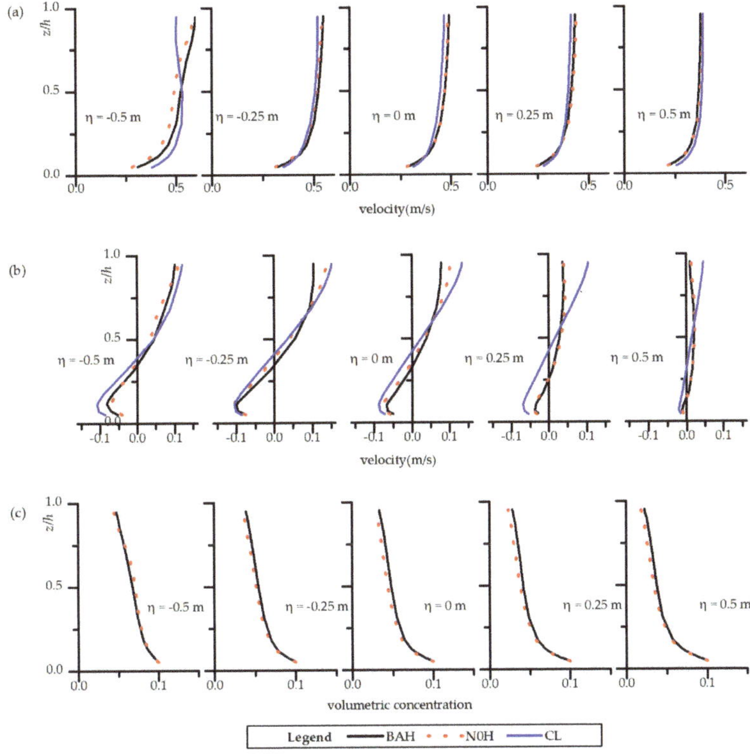

Figure 8. Velocity and sediment concentration profiles at the cross-section 180° for higher concentration case: (**a**) Streamwise velocity; (**b**) transversal velocity; (**c**) sediment volumetric concentration. (velocity unit: m/s).

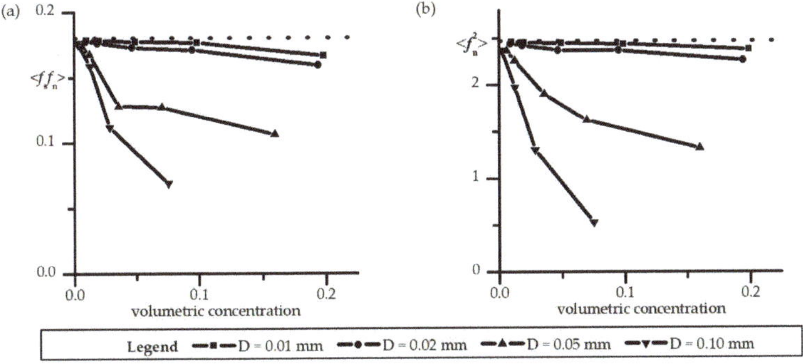

Figure 9. $<f_n^2>$ and $<f_s f_n>$ variation with suspended sediment at the center of cross-section 90°. The dot line represents the clear water (CL) case.

4. Conclusions

This paper presents 3D numerical simulations of sediment laden flows in open channel meanders concerning suspension inertia effect on the flow structure. The 3D governing equations without the Boussinesq approximation were used here and both the inertia and buoyant effects were considered. The velocity and concentration calculations performed in bend channels with comparison between the cases with and without Boussinesq approximation. Sediment size and specific density of sediment were studied to demonstrate the integrated effect. Although the additional turbulence transport produced by the presence of sediment could not be simulated by the present model due to the limitations with the employed turbulence closure model, the modifications in the vertical profiles of concentration and velocity due to the sediment were still estimated successfully, similar to the previous studies.

The Boussinesq approximation that neglects all other density effects except buoyancy for dilute flows results in obvious discrepancy in flow structure and sediment transport for high sediment concentration in meandering channels. Both the inertia and stratification effects of sediment play important roles in river meandering processes. The stratification effect promotes secondary flow at lower concentration by reducing eddy viscosity and suppresses secondary flow at higher concentration by the buoyance force acting on transversal pressure. The inertia effect suppresses secondary flow which is important at higher concentration. Depending on the sediment concentration and diameter, the integrated effect results in an increase and a decrease of the secondary flow at lower and higher concentrations, respectively. This suggests that the dominance of the sediment effect depends on the concentration. With the increase of concentration under a specific sediment size, secondary flow increases to reach a maximum and then declines. Moreover, as the sediment concentration rises, an exponentially decaying rate has been found for the secondary flow subject to inertia effect (Figure 9). It is concluded that numerical simulations without considering the Boussinesq approximation is necessary when concentration is high in meandering channels.

The present qualitative predictions based on the 3D numerical simulations provide a new perspective on sediment laden flow. A quantitative analysis of the inertia and stratification performance in meandering channels provides a comprehensive understanding of the water-sediment interactions. The lower Yellow River carries a heavy sediment load with bed material mainly composed of fine sand and silt. The main stream channel migrates very fast during the occurrence of hyperconcentrated flows while the meandering mechanics is still an open question. Extensive research efforts, including numerical modelling, have been continuously carried out on meandering mechanisms. Highly efficient 2D models are still the main tools for river engineering and perform quite satisfactorily in straight or mild meandering rivers [40]. This study would give some cues to 2D and 3D modelling of fluvial process for sediment laden flows.

Author Contributions: Funding acquisition, writing—review, project administration and supervision, X.S. X.F. and E.K.; methodology, software, validation, and writing—original draft preparation, F.Y. and X.F.

Funding: This research was funded by National Key R&D Program of China (Grant No. 2018YFC0407402) and National Natural Science Foundation of China (NSFC, Grant Nos. 51525901 and 91747207).

References

1. Vanoni, V.A. Transportation of Suspended Sediment by Water. *Trans. Am. Soc. Civ. Eng.* **1946**, *111*, 67–102.
2. Einstein, H.A.; Chien, N. *Effects of Heavy Sediment Concentration near the Bed on Velocity and Sediment Distribution*; University of California, Institute of Engineering Research: Oakland, CA, USA, 1955.
3. Smith, J.D.; McLean, S.R. Spatially averaged flow over a wavy surface. *J. Geophys. Res.* **1977**, *82*, 1735–1746. [CrossRef]
4. Lyn, D.A. A Similarity Approach to Turbulent Sediment-Laden Flows in Open Channels. *J. Fluid Mech.* **1988**, *193*, 1–26. [CrossRef]
5. Valiani, A. An open question regarding shear flow with suspended sediments. *Meccanica* **1988**, *23*, 36–43. [CrossRef]
6. Castro-Orgaz, O.; Giráldez, J.V.; Mateos, L.; Dey, S. Is the von Kármán constant affected by sediment suspension? *J. Geophys. Res. Earth Surf.* **2012**, *117*. [CrossRef]
7. Turner, J.S. *Buoyancy Effects in Fluids*; Cambridge University Press: New York, NY, USA, 1973.

8. Styles, R.; Glenn, S.M. Modeling stratified wave and current bottom boundary layers on the continental shelf. *J. Geophys. Res. Oceans* **2000**, *105*, 24119–24139. [CrossRef]

9. Lamb, M.P.; D'Asaro, E.; Parsons, J.D. Turbulent structure of high-density suspensions formed under waves. *J. Geophys. Res. Oceans* **2004**, 109. [CrossRef]

10. Taylor, P.A.; Dyer, K.R. Theoretical models of flow near the bed and their implications for sediment transport. *Sea* **1977**, *6*, 579–601.

11. Herrmann, M.J.; Madsen, O.S. Effect of stratification due to suspended sand on velocity and concentration distribution in unidirectional flows. *J. Geophys. Res.* **2007**, 112. [CrossRef]

12. Nezu, I.; Azuma, R. Turbulence Characteristics and Interaction between Particles and Fluid in Particle-Laden Open Channel Flows. *J. Hydraul. Eng.* **2004**, *130*, 988–1001. [CrossRef]

13. Wright, S.; Parker, G. Density Stratification Effects in Sand-Bed Rivers. *J. Hydraul. Eng.* **2004**, *130*, 783–795. [CrossRef]

14. Muste, M.; Yu, K.; Fujita, I.; Ettema, R. Two-phase versus mixed-flow perspective on suspended sediment transport in turbulent channel flows. *Water Resour. Res.* **2005**, 41. [CrossRef]

15. Winterwerp, J.C. Stratification effects by fine suspended sediment at low, medium, and very high concentrations. *J Geophys. Res. Oceans* **2006**, 111. [CrossRef]

16. Winterwerp, J.C. Stratification effects by cohesive and noncohesive sediment. *J Geophys. Res. Oceans* **2001**, *106*, 22559–22574. [CrossRef]

17. van Maren, D.S.; Winterwerp, J.C.; Wu, B.S.; Zhou, J.J. Modelling hyperconcentrated flow in the Yellow River. *Earth Surf. Process. Landf.* **2009**, *34*, 596–612. [CrossRef]

18. Byun, D.S.; Wang, X.H. The effect of sediment stratification on tidal dynamics and sediment transport patterns. *J. Geophys. Res. Oceans* **2005**, 110. [CrossRef]

19. Amoudry, L.O.; Souza, A.J. Impact of sediment-induced stratification and turbulence closures on sediment transport and morphological modelling. *Cont. Shelf Res.* **2011**, *31*, 912–928. [CrossRef]

20. Cantero, M.I.; Balachandar, S.; Parker, G. Direct numerical simulation of stratification effects in a sediment-laden turbulent channel flow. *J. Turbul.* **2009**, 10. [CrossRef]

21. Ozdemir, C.E.; Hsu, T.J.; Balachandar, S. A numerical investigation of fine particle laden flow in an oscillatory channel: The role of particle-induced density stratification. *J. Fluid Mech.* **2010**, *665*, 1–45. [CrossRef]

22. Cantero, M.I.; Shringarpure, M.; Balachandar, S. Towards a universal criteria for turbulence suppression in dilute turbidity currents with non-cohesive sediments. *Geophys. Res. Lett.* **2012**, 39. [CrossRef]

23. Dutta, S.; Cantero, M.I.; Garcia, M.H. Effect of self-stratification on sediment diffusivity in channel flows and boundary layers: A study using direct numerical simulations. *Earth Surf. Dyn.* **2014**, *2*, 419–431. [CrossRef]

24. Dallali, M.; Armenio, V. Large eddy simulation of two-way coupling sediment transport. *Adv. Water Resour.* **2015**, *81*, 33–44. [CrossRef]

25. Cao, Z.; Wei, L.; Xie, J. Sediment-laden flow in open channels from two-phase flow viewpoint. *J. Hydraul. Eng.* **1995**, *121*, 725–735. [CrossRef]

26. Zhong, D.; Zhang, L.; Wu, B.; Wang, Y. Velocity profile of turbulent sediment-laden flows in open-channels. *Int. J. Sediment Res.* **2015**, *30*, 285–296. [CrossRef]

27. Liang, L.; Yu, X.; Bombardelli, F. A general mixture model for sediment laden flows. *Adv. Water Resour.* **2017**, *107*, 108–125. [CrossRef]

28. Chauchat, J.; Cheng, Z.; Nagel, T.; Bonamy, C.; Hsu, T.-J. SedFoam-2.0: A 3D two-phase flow numerical model for sediment transport. *Geosci. Model Dev. Discuss.* **2017**, *10*, 1–42. [CrossRef]

29. Zeytounian, R.K. Joseph Boussinesq and his approximation: A contemporary view. *Comptes Rendus Mécanique* **2003**, *331*, 575–586. [CrossRef]

30. Kundu, P.K.; Cohen, I.M. *Fluid Mechanics*; Elsevier: Amsterdam, The Netherlands, 2008; p. 297.

31. Lin, B.; Roger, A.F. Numerical modelling of three dimensional suspended sediment for estuarine and coastal waters. *J. Hydraul. Res.* **1996**, *34*, 435–456. [CrossRef]

32. Wu, W. *Computational River Dynamics*; CRC Press: Boca Raton, FL, USA, 2007; p. 297.

33. Warner, J.C.; Sherwood, C.R.; Arango, H.G.; Richard, P.S. Performance of four turbulence closure models implemented using a generic length scale method. *Ocean Model.* **2005**, *8*, 81–113. [CrossRef]

34. Yeh, T.; Cantero, M.; Cantelli, A.; Pirmez, C.; Parker, G. Turbidity current with a roof: Success and failure of RANS modeling for turbidity currents under strongly stratified conditions. *J. Geophys. Res. Earth Surf.* **2013**, *118*, 1975–1998. [CrossRef]

35. Zhang, Y.; Baptista, A.M.; Myers, E.P. A cross-scale model for 3D baroclinic circulation in estuary–plume–shelf systems: I. Formulation and skill assessment. *Cont. Shelf Res.* **2004**, *24*, 2187–2214. [CrossRef]

36. Coleman, N.L. Effects of suspended sediment on the open-channel velocity distribution. *Water Resour. Res.* **1986**, *22*, 1377–1384. [CrossRef]

37. Wang, Z.B.; Ribberink, J.S. The validity of a depth-integrated model for suspended sediment transport. *J. Hydraul. Res.* **1986**, *24*, 53–67. [CrossRef]

38. Blanckaert, K. Saturation of curvature-induced secondary flow, energy losses, and turbulence in sharp open-channel bends: Laboratory experiments, analysis, and modeling. *J. Geophys. Res. Earth Surf.* **2009**, *114*. [CrossRef]

39. Blanckaert, K.; de Vriend, H.J. Nonlinear modeling of mean flow redistribution in curved open channels. *Water Resour. Res.* **2003**, *39*. [CrossRef]

40. Pu, J.H.; Hussain, K.; Shao, S.; Huang, Y. Shallow sediment transport flow computation using time-varying sediment adaptation length. *Int. J. Sediment Res.* **2014**, *29*, 171–183. [CrossRef]

 *water* 

Review

A Review on Hydrodynamics of Free Surface Flows in Emergent Vegetated Channels

Soumen Maji [1], Prashanth Reddy Hanmaiahgari [2,*], Ram Balachandar [3], Jaan H. Pu [4,*], Ana M. Ricardo [5] and Rui M.L. Ferreira [6]

[1]   Department of Civil Engineering, Central Institute of Technology Kokrajhar, Kokrajhar 783370, Assam, India; s.maji@cit.ac.in

[2]   Department of Civil Engineering, Indian Institute of Technology Kharagpur, Kharagpur 721302, India

[3]   Department of Civil and Environmental Engineering, University of Windsor, Windsor, ON N9B 3P4, Canada; rambala@uwindsor.ca

[4]   School of Engineering, Faculty of Engineering and Informatics, University of Bradford, Bradford BD7 1DP, UK

[5]   CERIS—Civil Engineering Research and Innovation for Sustainability, Av. Rovisco Pais, 1049-003 Lisboa, Portugal; ana.ricardo@tecnico.ulisboa.pt

[6]   CERIS and Department of Civil Engineering, Architecture and Georesources, Instituto Superior Tecnico, Av. Rovisco Pais, 1049-001 Lisboa, Portugal; ruimferreira@ist.utl.pt

*   Correspondence: hpr@civil.iitkgp.ac.in (P.R.H.); j.h.pu1@bradford.ac.uk (J.H.P.); Tel.: +91-3222-282420 (P.R.H.); +44-1274-234556 (J.H.P.)

Received: 27 February 2020; Accepted: 16 April 2020; Published: 24 April 2020

Abstract: This review paper addresses the structure of the mean flow and key turbulence quantities in free-surface flows with emergent vegetation. Emergent vegetation in open channel flow affects turbulence, flow patterns, flow resistance, sediment transport, and morphological changes. The last 15 years have witnessed significant advances in field, laboratory, and numerical investigations of turbulent flows within reaches of different types of emergent vegetation, such as rigid stems, flexible stems, with foliage or without foliage, and combinations of these. The influence of stem diameter, volume fraction, frontal area of stems, staggered and non-staggered arrangements of stems, and arrangement of stems in patches on mean flow and turbulence has been quantified in different research contexts using different instrumentation and numerical strategies. In this paper, a summary of key findings on emergent vegetation flows is offered, with particular emphasis on: (1) vertical structure of flow field, (2) velocity distribution, 2nd order moments, and distribution of turbulent kinetic energy (TKE) in horizontal plane, (3) horizontal structures which includes wake and shear flows and, (4) drag effect of emergent vegetation on the flow. It can be concluded that the drag coefficient of an emergent vegetation patch is proportional to the solid volume fraction and average drag of an individual vegetation stem is a linear function of the stem Reynolds number. The distribution of TKE in a horizontal plane demonstrates that the production of TKE is mostly associated with vortex shedding from individual stems. Production and dissipation of TKE are not in equilibrium, resulting in strong fluxes of TKE directed outward the near wake of each stem. In addition to Kelvin–Helmholtz and von Kármán vortices, the ejections and sweeps have profound influence on sediment dynamics in the emergent vegetated flows.

Keywords: turbulence; emergent vegetation; flexible vegetation; rigid vegetation; coherent structures; shear layer

---

1. Introduction

Vegetation is ubiquitous in rivers, estuaries, lake shores and some coastal areas [1]. It can be of different types including submerged, floating and emergent, whose incidence in a channel is schematized in Figure 1. Plants having varied stiffness, density, flexibility and height influence the water flow in different ways [2,3]. Its spatial distribution is seldom uniform [4]. Vegetation is frequently arranged in patches [5,6]. The presence of vegetation in water bodies has major impacts on flow hydrodynamics, including mean flow and variables describing turbulence. Certain important features like flow rate, changes in the bed and sediment carrying capacity of the stream are influenced by riparian plantation and its interactions with hydrodynamics.

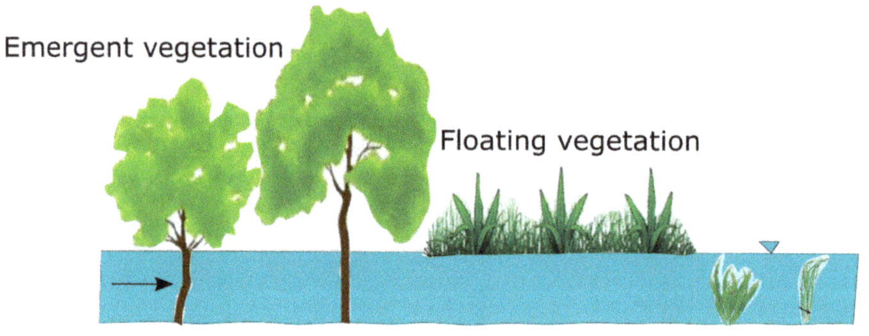

Figure 1. Different types of vegetation in open channel flow.

Parameters such as increased length, density and height of vegetation result in increased roughness, increased water levels and decreased velocities. In some cases vegetation offers protection against bank erosion, tsunamis and high waves. Experimental results show that the presence of vegetation leads to the generation of turbulence and additional drag forces. It is well known that the vertical distribution of flow velocity depends on the density of vegetation. The increase in stem density decreases the flow velocity through the interior of vegetation and increases the velocity in the regions without vegetation. Lichtenstein [7] classified aquatic plants into four types: algae, floating-leaved plants, submerged plants and emerged plants as shown in Figure 2a.

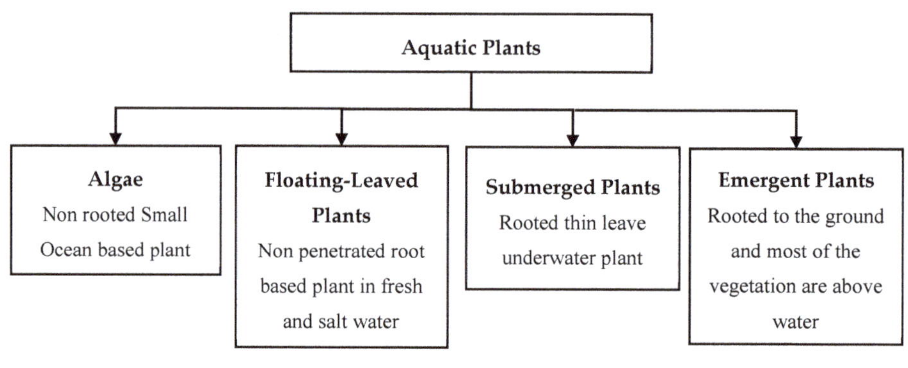

(a)

Figure 2. *Cont.*

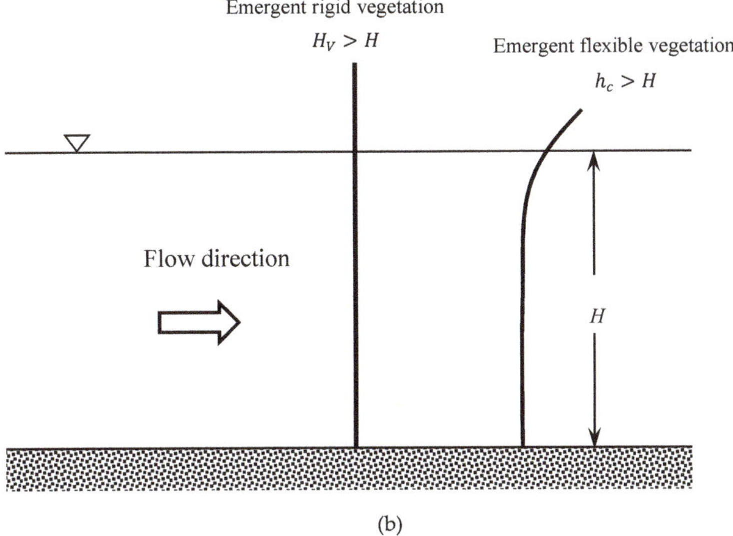

Figure 2. (a) Classification of aquatic plants; (b) emergent vegetation types based on plant stiffness.

Vegetation may be broadly divided into rigid and flexible, as shown in Figure 2b. The effect of flexible vegetation is quite different from that of rigid vegetation. The rigid vegetation causes the flow to produce separation, wakes and eddies which in turn dissipate the flow energy [8]. Figure 2 shows definition of emergent, rigid and flexible vegetations, in which, H_V = height of rigid vegetation, h_c = height of bent vegetation and H = water depth.

Given the complexity of emergent vegetated flows with ecological, geomorphologic and hydrodynamic effects, a great deal of research work including sophisticated field, laboratory and numerical studies have been undertaken to understand it. Significant advances have been achieved in the field of emergent vegetation hydrodynamics with respect to horizontal and vertical structures, [3,4,8–13], drag and frictional characteristics, [9,10,14–19]. Furthermore, with increasing computer power and improvements in numerical modelling, impressive advances were made in the study of the origin and development of coherent structures in the emergent vegetated flows [11,12]. These numerous investigations require a consolidated review to help researchers working on the hydrodynamics of emergent vegetated flows for deeper understanding. Therefore, the aim and scope of the paper is to make a synthesis of the state of the art relevant research works on the hydrodynamics of emergent vegetated flows, with emphasis on mean flow, turbulence and drag. In line with these objectives, present review highlights important research contributions of emergent vegetation hydrodynamics including complex interactions between turbulence and vortex structures.

The review of research findings on open channel flows with emergent vegetation is further divided into: (i) vertical structure of flow field (ii) velocity distribution, 2nd order moments, and distribution of turbulent kinetic energy (TKE) in the horizontal plane, (iii) horizontal features which include wake and shear flows and (iv) drag effect of vegetation on the flow. In addition, future directions are deliberated for researchers working on the emergent vegetated flows, and finally conclusions of the paper are presented at the end.

2. Hydrodynamics of Emergent Vegetated Flows

Vegetation extending from the bed all the way beyond the free surface is called emergent vegetation. Good examples of emergent vegetation include jute plantations, mangroves, and often partially submerged trunks of palm trees in flood water. Studying emergent vegetation is sought after because it modifies velocity vertical profiles and turbulence intensities. Commonly, emergent vegetation consists of relatively stiff stems, which are usually circular in cross section and without leaves below the water surface [3]. Therefore, emergent rigid cylindrical stems have been extensively employed in research works as a proxy for natural stems.

In unbounded open channels, the mean flow through groups of emergent vegetation is decreased. The emergent canopy has a significant role on coherent structures and accompanying mass and momentum fluxes at the boundary between emergent vegetation and open channel. In particular, emergent canopy patch density in the open channel flow affects mass and momentum transfer, roughness, sedimentation, velocity of flow, bed shear, turbulence quantities, biological processes and aquatic life within the canopy. Additionally, characteristics of flow through emergent vegetation consist of significant velocity gradients (in transverse, stream wise and vertical directions) and drag discontinuity at the interface resulting in the formation of a shear layer between the vegetation stems and the flow exterior of the vegetation. The shear layer in turn generates large coherent vertical structures due to Kelvin–Helmholtz instability. These coherent structures grow downstream due to creation of vortex-street, vortex pairing and eventually dissipate into the flow.

2.1. Vertical Structure

The vertical structure of flows within the stems of emergent vegetation is divided in three layers: (i) the reach close to the bottom, where the flow is highly 3D due to interaction with the bed; (ii) the layer close to the free surface, which is affected by the oscillations of the free surface; and (iii) the intermediate layer, sufficiently away from the bed and from the free surface, where the flow is controlled by the vertical stems and the flow properties are approximately constant in the vertical direction [9–11,20] as shown in Figure 3.

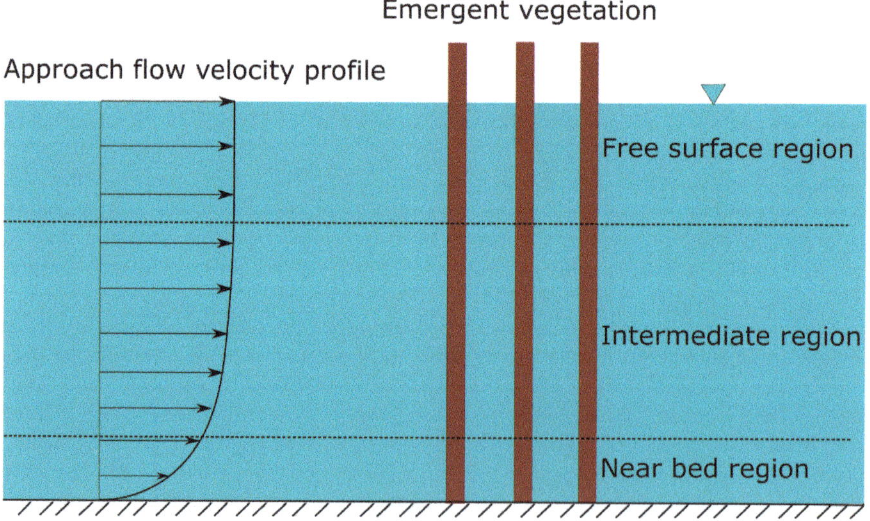

Figure 3. Vertical structure of flow in emergent vegetation.

Stoesser et al. [11] numerically studied three types of spacing between cylinders on the flow field. Very small vertical velocities are observed in the large spacing cases except just behind the cylinder where relatively large values of vertical velocity indicates considerable upward movement of fluid observed just behind the cylinder. However, upward and downward movement of fluid is observed in the case of low spacing between the cylinders. Contours of the time-averaged streamwise velocity and streamlines at about half depth for the three different vegetation densities are investigated.

Flow separation occurs at approximately 95° and a relatively large recirculation region comprising two counteracting vortices that are about the size of the cylinder diameter are observed in low and medium vegetation densities. In high vegetation density, flow separates considerably later and the recirculation region behind the cylinder is much smaller.

Ricardo et al. [3] addressed the influence of the longitudinal variation of the stem areal number density on the spatial distribution of the turbulent flow variables at the inter-stem scale. They concluded that Reynolds stresses are not sensitive to local spatial gradients of the stem distribution, being determined by the local number of stems per unit area. However, form-induced stresses depend on the local number of stems and its spatial distribution. Ricardo et al. [4] identified a spatial memory of the flow manifested by the upstream generated complexity of the flow that subsists further downstream and is combined with the complexity locally generated, resulting in the increased spatial variability relative to that of a flow with a uniform stem distribution.

Chang and Constantinescu [12] numerically investigated the flow and turbulence structure interior and exterior of a circular array of emergent rigid vegetation patch by varying the solid volume fractions and relative cylinder diameter arrays. These studies were performed without considering a different bedforms effect, which could be a crucial factor in flows through natural bedforms as studied by Pu et al. [21]. The flow was found to be highly three dimensional in the array when solid volume fraction (SVF) was more than 0.1 and this effect was relatively small when the magnitude of SVF was less than 0.025. Von Kármán vortex streets were developed for cylinder Reynolds number greater than 10,000. Furthermore, Chang and Constantinescu [12] found that at sufficiently high SVFs such as 0.2, von Kármán wake cylinder interactions become important resulting in the formation of von Kármán vortex street. In low SVFs such as 0.023, wake and cylinder interactions were weak and flow advects through the cylinders without a vortex street. However, shear layers are observed irrespective of formation of von Kármán vortex streets. For larger values of SVF, cylinders in the downstream imped the formation of strong vortices.

Liu and Shan [22] used an analytical model to predict longitudinal profile of depth averaged streamwise velocity for simulated emergent cylindrical vegetation and used an experimental setup to validate the proposed model for short and long arrays of vegetation. Tong et al. [23] investigated the effect of emergent rigid vegetation in a y shaped confluence channel and concluded that non submerged vegetation is responsible for changing the flow structure. Due to the vegetation, middle of the non-vegetated area in the vicinity of tributaries showed existence of high velocity region, and the disappearance of the secondary flow. The flow in non-vegetated areas was more intense and turbulent kinetic energy of the non-vegetated area was lesser than the vegetated area.

2.2. Mean Flow and Distribution of Turbulent Kinetic Energy (TKE) in the Horizontal Plan

Ricardo et al. [8] carried out a laboratory work simulating rigid and emergent cylinders, with longitudinally varying density where 2D instantaneous velocity maps were measured with a Particle Image Velocimetry (PIV) aiming at characterizing the mean flow field at the inter-cylinder and patch scales. Figure 4a,b exemplifies a horizontal map of longitudinal velocity and out-of-plane vorticity map for a patch of 1200 cylinders/m^2 in a 3.5 m-long reach populated with rigid cylinders and with longitudinally varying density of cylinders.

The mean flow field within vegetation covered areas is heterogeneous at large scales, as reach or patch scales, with zones of low longitudinal velocity at the vegetation elements wake alternated by high longitudinal velocity zones between those elements [8,24]. These low/high velocity patterns are observed independently of the density and distribution of the vegetation elements.

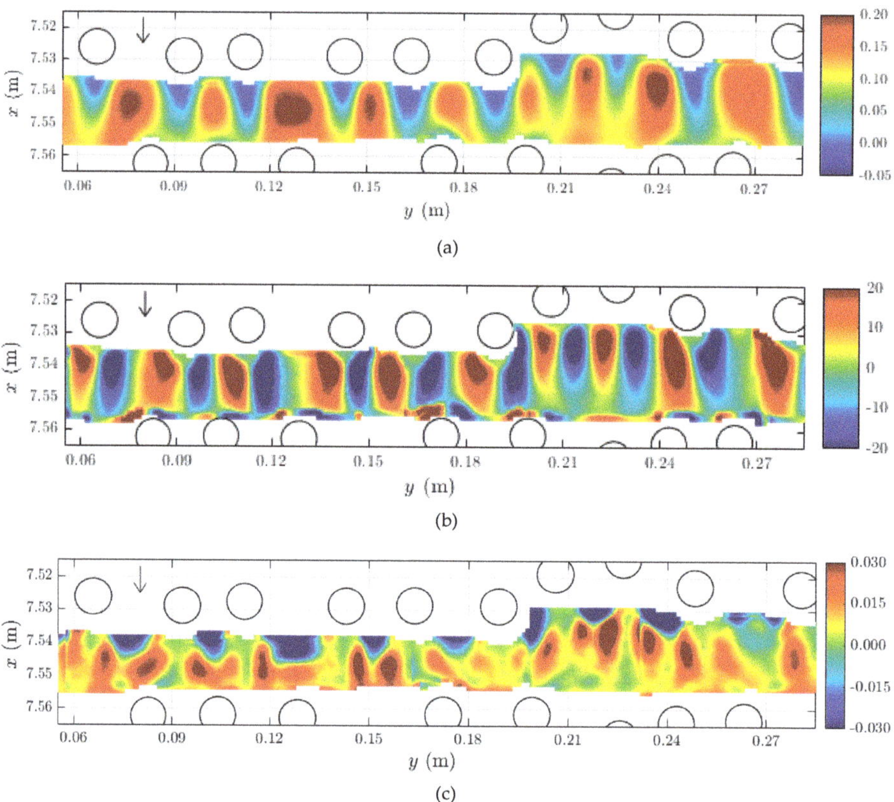

Figure 4. (**a**) Time-averaged longitudinal velocity map (m/s); (**b**) out-of-plane vorticity map (m^{-1}); (**c**) rate of production of TKE (m^2 s^{-3}), at longitudinal position P8 and at 3.1 cm above the bed corresponding to 60% of the flow depth. (Source: Ricardo et al., 2014).

Relative to the vorticity distribution, a repeating pattern of paired vortexes caused by the unsteady separation of the flow on the cylinders is observed. These quasi-symmetric high vorticity patterns behind the cylinders identify von Kármán vortex streets. Comparing vorticity maps for the different density of cylinders, the conclusion is that the cylinders induce a regular structure of vortex patterns independently space between cylinders. The main difference is that the space necessary to fully develop the vortex pattern is strongly reduced at smaller distances between cylinders. At higher cylinder densities, the vortex pattern is forced to compress due to the proximity of the next cylinder, while at positions with lower cylinder density cases the vortex pattern has space to develop a vortex street.

Ricardo et al. [8] performed laboratory experiments on rigid emergent vegetation aimed at the characterization of the key terms of the TKE equation. Three dimensional laser doppler anemometry (LDA) and two dimensional particle image velocimetry (PIV) measurements have been employed. They characterized the spatial distribution of turbulent production, convective rate of change of TKE and turbulent diffusion and the dissipation rate of TKE was also computed. They observed that the production of TKE is mostly associated with vortex shedding from individual stems (Figure 4c). The magnitude of the rate of production is higher in the wake region and negative production of TKE was identified between close cylinders, associated with strong local accelerations in the flow field. Turbulent transport is particularly important along the von Kármán vortex street and convective rate of change of TKE and pressure diffusion are most relevant in the vicinity of the cylinders. The rate of dissipation was found to increase with the stem areal number density. Ricardo et al. [8] observed that the rates of production and dissipation are not in equilibrium in the inter-stem space, revealing important interactions of turbulence with mean flow and pressure field and turbulent transport of TKE. The cumulative effect of convection and turbulent transport of TKE is the generation of background turbulence, i.e., non-locally generated turbulence.

Ricardo et al. [4] investigated the impact of the background turbulence caused by randomly placed cylinders on the vortex shedding regime. The strong background turbulence, generated by the randomly placed cylinders in the array, causes a premature vortex detachment increasing the shedding frequency. It was concluded that the background turbulence acts on the faster loss of vortex coherence as they travel downstream. Furthermore, Ricardo et al. [4] observed that the impact of the background turbulence on decay of the longitudinal vorticity flux is small, the latter is mainly caused by the vorticity cancellation imposed by the local distribution of neighbouring cylinders.

2.3. Horizontal Structure

Density, shape and arrangement of the stems in the emergent vegetation patch have a dominant role on the mean flow and turbulence structures in the interior and exterior of the patch in horizontal direction.

The mean flow is largely dominated by the wakes which tend to reduce the flow between the bodies and this effect is larger than the inviscid kinematic effect of the bodies blocking the flow. The difference between Lagrangian and Eularian velocities with different void ratios is one of the aspects of horizontal structure. At the upstream of the array of cylinders, the flow is irrotational and slows down due to the blocking effect of the entire patch and the drag force from the cylinders. The flow bleeds through the vegetation with low velocity and high velocity flow takes through the region without vegetation. Shear layers form on either sides of the vegetation patch as shown in Figure 5. At what distances upstream and downstream of the flow is influenced by the array of cylinders is a function of vegetation density, width of the patch, Reynolds and Froude numbers.

Stoesser et al. [11] carried out large eddy simulations (LES) of flow through the canopies with solid volume fractions 0.016, 0.063 and 0.251. In their analysis, the streamwise turbulence intensity profiles were significantly different from unobstructed channel flow not only in shape but also in magnitude. Higher streamwise turbulence intensities were observed in high vegetation density cases. Similarly, normalized vertical turbulence intensities have higher magnitudes for higher density vegetation. The turbulence in the high vegetation density case is rather generated by high streamwise velocity gradients, while von Kármán vortex shedding occurs in low density cases [11].

Nicolle and Eames [25] studied the effects of the patch of emergent and rigid cylinders on the flow field in a uniform air flow using direct numerical simulations (DNS). Nicolle and Eames [25] found that the von Kármán vortex-street behind the circular cylinder array is similar to that of flow past a single cylinder. The most obvious limitation of their study is the implementation of 2D model to study the 3D flow field where additional drag is caused by vortex stretching.

Nepf [26] concluded that emergent vegetation develops two different types of turbulent scales i.e., canopy scale turbulence and stem scale turbulence in which canopy scale vortices control the high level of turbulent diffusion in the exchange zone. In general, the turbulence in the interior of emergent vegetation is less than that of the outside the vegetation in free surface flows. However, the decrease in turbulence due to decreased mean velocity near the bed is too small as compared to the additional turbulence generated by stem vortices. White and Nepf [13] and White and Nepf [27] considered horizontal flow structure and studied the momentum transfer at the interface between the flow inside the emergent vegetation and the region outside the vegetation. Shear layer is created due to the drag discontinuity at the interface and coherent vortices are generated in the shear layer due to Kelvin–Helmholtz instability. The exchange of fluxes between the open channel flow through the vegetation and the region outside the vegetation is dominated by the energetic shear layer vortices. Vortices in the shear layer are found to be continuously growing in the downstream direction predominantly through vortex pairing. In addition a quasi-coherent vortical structure with two times the spacing of the shear layer vortices is generated by the interaction of paired shear layer vortices.

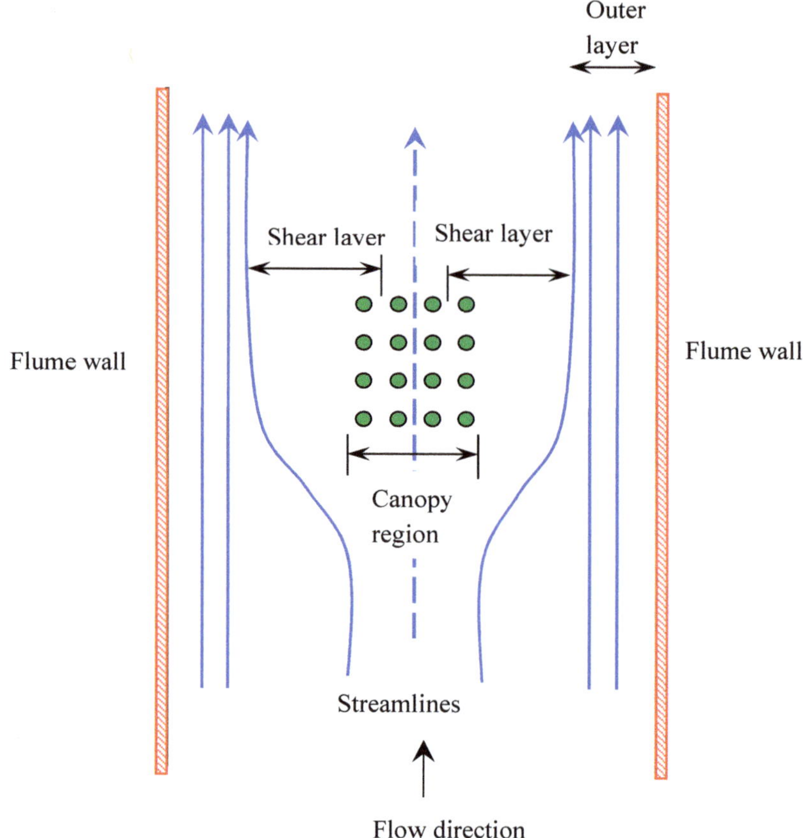

Figure 5. Horizontal structure of the flow through a vegetation patch.

Meftah and Mossa [28] studied channel partially obstructed by emergent vegetation with an emphasis to understand the effect of contraction ratio. They observed the formation of a shear layer immediately next to the interface, followed by an adjacent free stream region of full velocity flow. In addition, the transversal flow velocity profile at the obstructed and unobstructed interface was analyzed. The experimental results demonstrated that the contraction ratio significantly affects the flow hydrodynamic structure. Finally, a general modified log-law was proposed to describe the typical transversal profile of the mean velocity and its application was in good agreement with analytical solutions.

Naot et al. [29] used computational fluid dynamics (CFD) with the $k - \epsilon$ model to simulate hydrodynamic behaviour of compound channel with rigid and emergent vegetation on the flood plain, they found that flow behaviour depends on shading factor and reference length of the vegetation. A two dimensional LES model was used by Ikeda et al. [30] to get a relation between the square of flow velocity and the waving motion of a flexible emergent (where plant height is equal to water depth) plant. The interesting observation in this study is the motion of the plant increases the frequency of vortices whereas increased strength of vortices increases the motion of plants.

Stoesser et al. [11], and Kim and Stoesser [31] concluded that von Kármán-type vortices, Kelvin–Helmholtz instability and vortex shedding are clearly visible in sparse vegetation. As the vegetation density increases, the influence of vortices that are shed from upstream cylinders affects more irregular shedding behavior of downstream cylinders. Increase in vegetation patch density increases the vorticity magnitude which is accompanied by the linear increase of vertical vorticity. Furthermore, the frequency of vortex shedding increases as the patch density increase.

Chang and Constantinescu [12] numerically found that coherent structures formed in the separated shear layer (SSL) are larger with SVF = 0.2 as compared to SVF = 1.0. For low SVF values, the separated shear layer (SSL) was found to be longer and eddies in the SSL have more energy as compared to that of the SSL in high SVF patch. Furthermore, the behaviour of eddies in SSL of low SVF patch resembles that of a mixing layer. Von Kármán Street was formed only in flows with SVF values higher than 0.2 and in which vortical structures were impeded and, therefore, not producing the regular wake vortices. For SVF values less than 0.05, shedding of wake vortices behind a cylinder was strongly disturbed by the vortices shedding from other cylinders in the vicinity. Chang and Constantinescu [12] concluded that for SVF = 0.023, shedding behind an individual cylinder in the patch is similar to that of an isolated cylinder in the flow.

Anjum and Tanaka [32] used discontinuous and vertically doubled layered patches and simulated it numerically for varying vegetable density and patches. Large-scale turbulence followed by saw-tooth distribution within the patches and low turbulence in the non-vegetation regions was observed. Maximum of 13% turbulence intensity for dense vegetation arrangements and maximum of 9% turbulence intensity for sparse vegetation was observed. They concluded the presence of non-uniform flow through discontinuous and double layered vegetation patches. Yamasaki et al. [33] studied the hydrodynamics of flow-vegetation interactions using a computational fluid dynamics model.

2.4. Drag and Frictional Characteristics

In open channel flows without vegetation, streamwise velocity distribution along the depth is a function of Reynolds shear stress, whereas velocity vertical profile in the vegetated channel is a function of frictional drag imposed by the vegetation since the vegetation drag is far greater than the frictional drag at the bed in high-density vegetated flows [10,34]. The drag force exerted by the flow on emergent vegetation can be estimated based on the available knowledge of flow around a circular cylinder.

Many studies are available in the literature to predict or measure frictional resistance in emergent vegetation conditions [9–11,14–16,35,36]. To compute the flow resistance in a vegetated channel, the drag caused by a single stem and the addition of individual drags of all the stems in that canopy, the blockage effect of the canopy in addition to the already existing bed roughness must be considered. Fathi-Maghadam and Kouwen [16] were among the first researchers who studied emergent flexible vegetation effect on the drag coefficient and in-turn on Manning's roughness coefficient. Fathi-Maghadam and Kouwen [16] experimentally analysed the effect of proposed non-dimensional variables on C_d and concluded that (i) Manning's roughness proportionally increases with an increase in the square root of flow depth and increase in density of vegetation, and (ii) Manning's roughness inversely proportional to the average flow velocity and vegetation flexibility.

In the riverine environment, however, canopies occur in finite patches and make a random angle with the direction of flow; for example, patches of mangroves in Purna River, India and Daintree River Queensland, Australia as shown in Figure 6. In these scenarios, flow is divided into flow passing through interior and exterior of the vegetation. According to Mitul et al. [18], mangroves are the densest canopies in coastal water with stem diameters between 4 cm to 9 cm and their solid volume fraction is around 0.55.

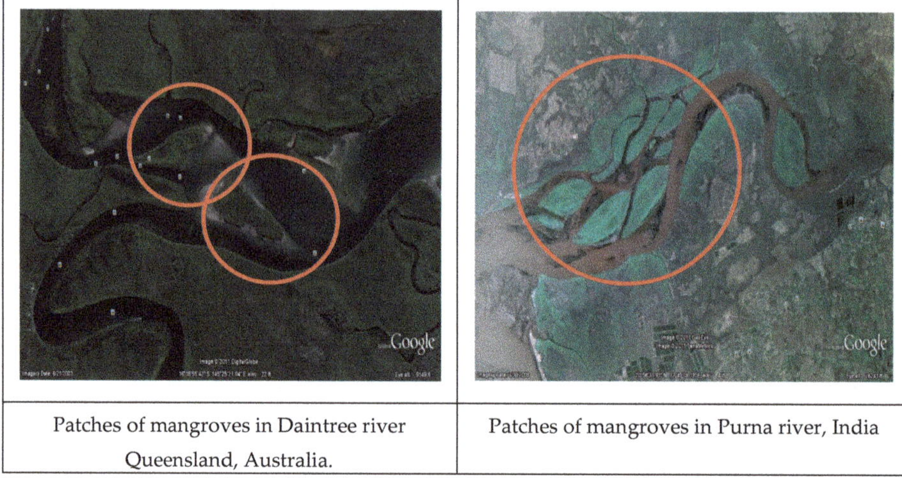

Patches of mangroves in Daintree river Queensland, Australia.	Patches of mangroves in Purna river, India

Figure 6. Patches of vegetation (highlighted in circles) present in the river systems (Background image is sourced from Google Earth).

Maji et.al. [37] investigated the turbulence characteristics in open channel flow with an emergent, sparse and rigid vegetation patch. They observed that the nature of time-averaged velocities along streamwise, lateral and vertical directions are not consistent upto half the length of the patch from its leading edge; however, velocity profile becomes uniform after that length and their results are similar to various un-obstructed uniform and non-uniform flows studied by Pu et al. [38]. Inside the patch, the magnitude of vertical normal stress is small as compared to the streamwise and lateral normal stresses and Reynolds shear stress decreases along the centreline of the patch in the downstream. Furthermore, flow velocity decreases within the vegetation, which can lead to deposition of sediment which accelerates with the time and acts like a catalyst to the larger morphological changes in the river [37,39].

Heidari et al. [40] studied the drag created by a slender emergent cylinder in open channel flow and proposed that wake stability parameter significantly affects the drag on the cylinder away from the bed. Kothyari et al. [41] developed a drag model applicable to emergent plants in which drag coefficient decreases as the stem Reynolds number increases. Luna et al. [42] defined a term called 'global flow resistance' similar to Chezy's coefficient, however the global flow resistance is a combination of bed shear stress and vegetation drag on the flow. Luna et al. [42] assessed various global flow resistance models for emergent vegetation and found that Baptist [43] model accomplishes the bestfit. Takemura and Tanaka [44] investigated the fluid structures in a colony type of emergent vegetation and coefficient of drag was experimentally measured. Their results restated that vortex patterns, strouhal number and drag coefficient of colony are function of L/D where L is cross stream spacing between neighboring cylinders and D is cylinder diameter. Takemura and Tanaka [44] termed a von Kármán vortex-street behind an individual cylinder as primitive von Kármán vortex street (PKV) and its instability governs large scale vortical structures.

Tanino and Nepf [9] examined the canopy drag as a function of $R_{ep} = (\langle \overline{U} \rangle D/v$, where $\langle \overline{U} \rangle$ = double averaged pore velocity, v is kinematic viscosity and D is stem diameter) which is in the range of 25–685 and solid volume fraction ranging from 0.091 to 0.35. They found that the C_d value of a canopy patch is inversely proportional to R_{ep} and proportional to solid volume fraction. Average drag per unit cylinder length non-dimensionalized by the product of $\overline{U}$ and dynamic viscosity (μ) is found to be a linear function of R_{ep}. This linear relation was further investigated by Ferreira et al. [10], who proposed that the increase of the solid fraction generates a momentum sink observed as an increase in the gradients of dispersive stresses. This may justify the increase of the drag coefficient with the increase of solid fraction at intermediate values of the latter. Liu and Zeng [45], however, suggest that, in general, the drag coefficient decreases with the solid fraction. Another important observation is that the drag of an individual cylinder is not a function of solid volume fraction. Chen et al. [17] studied the wake region behind a circular vegetation patch and found that for low flow blockages sediment deposition occurs, whereas for high flow blockages erosion takes place in the wake region.

Musleh and Cruise [36] combined the effects of flow depth, velocity, stem diameter, lateral spacing and longitudinal spacing into a non-dimensional variable called relative density ratio (ratio of total surface area of the vegetation to bed cross-sectional area) that is linearly related to Darcy's friction factor and mannings coefficient. Musleh and Cruise [36] explained that for rigid vegetation, flow resistance of vegetation non-linearly increases with increasing flow depth especially when vegetation density is high, whereas flow resistance found to be nonlinearly decreasing with increasing mean flow velocity for flexible emergent vegetation flows. In addition, they found that the arrangement of rigid emergent cylinders plays an important role on the flow resistance and friction factor linearly varies with the relative density ratio. Musleh and Cruise [36] found that lateral spacing and stem diameter of cylinders have a more significant effect on the flow resistance as compared to the streamwise spacing of cylinders for a given stem density. The important observation in this experimental study is the drag effect of stem diameter is more than the transverse and streamwise spacings of cylinders. Furthermore, Musleh and Cruise [36] found that lateral spacing between the rods significantly affects the drag on rods. Wu et al. [19] experimentally found that the drag coefficient of emergent rigid vegetation increases with the increasing flow Reynolds number.

Chen et al. [14] used a tapered emergent flexible vegetation plant model to derive an analytical solution for flow resistance, bending behaviour and potential rupture locations. In this method, power law was used to determine the velocity distribution which is responsible for drag force on the rigid as well as flexible stems. Analytically computed deflections were validated with experimentally measured deflections. Finally, Chen et al. [14] concluded that the location of rupture depends on the tapering angle and the base diameter of the stem. Their results were consistent with earlier research that drag is inversely proportional to the average flow velocity.

Stoesser et al. [11] found that value of C_d is linearly proportional to canopy density. In highly dense canopies, the downstream cylinders experience a drag force which is slightly smaller than for an isolated cylinder, but the lift force is much larger than the isolated cylinder because downstream cylinders are sitting in the wakes of two upstream cylinders [25]. In sparse canopies (SVF $\ll$ 1.0) upstream cylinders experience a drag force similar to that of an isolated cylinder but the lift force, while still relatively small is determined by the position of the cylinders in downstream. Cylinders located in the wake of other cylinders experience larger drag and lift forces as a result of the accelerated flow induced between the upstream cylinders. For intermediate void fractions, the force on the individual cylinders is steady and the lift force acting on the whole patch is negligible. The aggregate drag force on the canopy is dominated by the contributions from the upstream rows of cylinders, which experience a drag force comparable to that of a single cylinder. For high void fractions, the lift forces are significant and are mainly induced by the wake of the whole array. The effect of increasing void fraction leads to a slight increase in the scatter of the forces on the upstream cylinders and significantly increases the magnitude of the wake forces on the downstream cylinders.

Stoesser et al. [11] also studied the forces on cylindrical stems in a canopy and derived the following conclusions. Pressure drag, friction drag components are high and bed shear is low on the single cylinder in high patch density as compared to low patch density. In general the drag forces, pressure drag constitutes about 90% and remaining forces are only 10% (frictional drag is around 7% and bed shear is around 3%). Bed shear stress decreases with increasing patch density. The friction drag is almost constant regardless of vegetation density. Chang and Constantinescu [12] found that for numerical simulations, drag coefficient of a cylinder in the array is a very important parameter.

They have also added that combined drag parameter for porous cylinders increases as frontal area per unit volume of patch increases.

Rooijen et al. [46] used an isolated emergent cylinder to assess canopy drag and obtained good agreement in drag computation by their proposed analytical model. In the case of a submerged cylinder, the authors suggested that their results are consistent with isolated cylinder theory. Wang et al. [47] used an experimental model to investigate the drag coefficient due to rainfall over vegetation. Experiments were conducted with varying stem density and applied water to obtain a link between friction slope and local kinetic energy head induced by steady non uniform flow. Spatial variation of drag coefficient exhibited a monotonic reduction during rain or non-monotonic hump formation when rain was absent.

Shan et al. [48] examined the effect on the change of drag force and velocity inside mangrove forest model for different vegetation stem arrangements. They found that the force at root is most dominant, randomly distributed branches inside a tree may experience more or less same force, and however, drag force increases with tree density. Razmi et al. [49] used a field scale large eddy simulation model by using distributed-drag approach and velocity-dependent models to study the effect of canopy reconfiguration on canopy posture and canopy drag. This model removed the flaws in the previous research of not capturing vertical spread in peak Reynold stress that is associated with the movement of the canopy interface.

2.5. Vegetation in Bank Protection and Sediment Transport

The effect of riparian vegetation on river morphology is well documented [50,51]. The role of vegetation density on improved riverbank stability is an emerging topic and important for further research [52]. River bankside vegetation can effectively improve bank strength and stability by mechanical, hydrologic, and hydraulic effects [53–55] and hence it may be considered essential to plant vegetation at erosion-prone locations on the riverbank [56]. Mechanical method comprised of root reinforcement along with flow resistance increment [53,57–59]. Hydrologic method protects the river bank from current induced erosion by reduction of soil moisture and pore water pressure, and increasing the root-induced compaction [60]. By contrast, the hydraulic method causes increase in hydraulic roughness and toe erosion reduction [61–63]. Interconnection between soil and root by apparent cohesion can increase bank stability [64]. But plant root can also negatively affect stability if extra weight of the vegetation increases downward force causing bank failure [58,65]. In addition, vegetal cover can reduce the danger of river course change and braid formation, which helps in maintaining river morphology [59,66–70]. Yu et al. [71] used the bank stability and toe erosion model (BSTEM) to find the effects of different root conditions on channel bank strength and verified that roots are good at reinforcing the unconsolidated banks and erosion control, hence proving the effectiveness of riverbank vegetation.

Yang and Nepf [72] proposed that near bed turbulent kinetic energy is better than bed shear stress for indicating bed load transport in vegetated flows. They proposed a turbulent kinetic energy based Einstein–Brown equation and satisfactorily compared the bed load transport with experimental results. Armanini and Cavedon [73] used a probabilistic/deterministic model to estimate the flow field changes induced by vegetation and associated bed load transport by refining their previous bed load model [74] and recalibrating [75] parameters. Their model shows good resemblance with experimental observations.

3. Further Research Prospects

Most of the research studies available are based on using rigid cylinders as vegetation stems, but real vegetation has a different shape, roughness and bending stiffness, and therefore laboratory experiments may not substitute the field investigations and this could be the focus in the future. In fact, a few researchers are working towards this achievement. Plant morphological studies were conducted with real vegetation for uniformly vegetated flow [2,76–79]. But there is a lack of dedicated study for partly vegetated flow with real vegetation. Caroppi et al. [80] are carrying out pioneering research by using naturally grown plants in an open channel for the study of reconfiguration and mixing effect of flow. In the real channels, aquatic seeds grow into aquatic plants and further experimental studies are required to study the transient flow field during aquatic plants growing stage. Similarly, capabilities of numerical simulations must be extended to field plants and scenarios. The feedback between flow field and sediment dynamics in the interior and exterior of the vegetation is beginning to be recognized as vital. Such research now provides an outstanding background for beginning to address areas of greater complexity that will enable a better understanding of these important natural features. In this context, studies conducted by [81–83] on threshold magnitude of velocity fluctuations and duration of velocity fluctuations for particle entrainment from the bed are useful for developing new sediment transport theories based on the energy approach in emergent vegetated flows. It is recognized that sweep events play a dominant role in dislodging bed particles. Flow conditions and the solid volume fraction of emergent vegetation affect the individual contributions of sweep and ejection coherent structures in the interior and wake regions. Therefore, the direct relation between the sediment transport and instantaneous coherent flow structures in vegetated flows are to be probed further. The effect of patch size of aquatic plants has not yet been fully investigated and elaborated clearly. As per the authors' knowledge, the effect of changing the size of the patch in hydrodynamics and sediment transport through vegetation was studied only by [84] and gave an idea of the minimum size of a patch for triggering the flow interaction but it required the study of detailed turbulence characteristics for further investigation.

4. Conclusions

The additional drag created by vegetation resulting in a decrease in discharge in irrigation channels across the world has prompted researchers to characterize and quantify the effect of emergent vegetation on the flow field. Furthermore, the advent of accurate invasive and non-invasive velocity-measuring devices has accelerated the research in this area since early 2000. It was well accepted that the presence of emergent vegetation in the rivers and open channel flows has a significant role on mean velocity, turbulence levels, coherent structures and the accompanying mass and momentum fluxes at the boundary between emergent vegetation and an open channel. Decreased flow velocity within the vegetation leads to the deposition of sediment which accelerates with the time and acts like a catalyst to the larger morphological changes in the river. Another important and consistent research finding is that universal logarithmic law is not applicable for spatially and time averaged velocity vertical profile within the flow region of emergent vegetation.

Experimental evidence shows that the magnitude of vertical normal stress is small as compared to the streamwise and lateral normal stresses within the vegetation patch. The streamwise and vertical turbulence intensities, vorticity and vortex shedding have higher magnitudes for higher solid volume fraction of the vegetation. Investigation of spatial distribution of the turbulent flow variables at the inter-stem scale demonstrated that Reynolds stresses are not sensitive to local spatial gradients of the stem distribution, whereas form-induced stresses depend on the local number of stems and its spatial distribution. The production of TKE is mostly associated with vortex shedding from individual stems and therefore the magnitude of the rate of TKE production is higher in the wake region. Another important and consistent research finding is that the rate of TKE production and dissipation is found to be increasing with the increasing stem areal density. Few researchers studied the momentum transfer at the interface between the flow region inside and outside the emergent vegetation and observed the formation of coherent vortices in the shear layer due to Kelvin–Helmholtz instability. Researchers found the motion of the flexible emergent plant increases the frequency of vortices whereas the increased strength of vortices increases the motion of plants.

The coefficient of drag of the emergent vegetation patch was found to be inversely proportional to the cylinder Reynolds number, flow velocity and vegetation flexibility. However, the coefficient of drag was found to be proportional to the solid volume fraction, flow depth, and bed slope. It was also established that for low flow blockages, Kelvin–Helmholtz vortices are dominant, whereas for high flow blockages in which the vegetation patch acts like a bluff body von Kármán vortices are dominant. It was established that stem diameter of cylinders has a more significant effect on the flow resistance as compared to the streamwise and lateral spacing of cylinders for a given stem density.

Author Contributions: S.M.: writing—original draft preparation, funding acquisition; P.R.H.: writing—review and editing, project administration; R.B.: writing—review and editing, data curation; J.H.P.: writing—review and editing, supervision; A.M.R.: writing—review and editing; R.M.L.F.: writing—review and editing. All authors contributed to the work. All authors have read and agreed to the published version of the manuscript.

Funding: This research received no external funding.

Conflicts of Interest: The authors declare no conflict of interest.

References

1. Yager, E.M.; Schmeeckle, M.W. The influence of vegetation on turbulence and bed load transport. *J. Geophys. Res. Earth Surf.* **2013**, *118*, 1585–1601. [CrossRef]

2. Järvelä, J. Flow resistance of flexible and stiff vegetation: A flume study with natural plants. *J. Hydrol.* **2002**, *269*, 44–54. [CrossRef]

3. O'Hare, M.T. Aquatic vegetation—A primer for hydrodynamic specialists. *J. Hydraul. Res.* **2015**, *53*, 687–698. [CrossRef]

4. Ricardo, A.M.; Franca, M.J.; Ferreira, R.M.L. Turbulent flows within random arrays of rigid and emergent cylinders with varying distribution. *J. Hydraul. Eng.* **2016**, *142*, 4016022. [CrossRef]

5. Schoelynck, J.; De Groote, T.; Bal, K.; Vandenbruwaene, W.; Meire, P.; Temmerman, S. Self-organised patchiness and scale-dependent bio-geomorphic feedbacks in aquatic river vegetation. *Ecography* **2012**, *35*, 760–768. [CrossRef]

6. Meire, D.W.S.A.; Kondziolka, J.M. Nepf, H.M. Interaction between neighboring vegetation patches: Impact on flow and deposition. *Water Resour. Res.* **2014**, *50*, 3809–3825. [CrossRef]

7. Lichtenstein, D. Types of Aquatic Plants Sciencing.com. Available online: https://sciencing.com/types-of-aquatic-plants-12003789.html (accessed on 9 January 2020).

8. Ricardo, A.M.; Koll, K.; Franca, M.J.; Schleiss, A.J.; Ferreira, R.M.L. The terms of turbulent kinetic energy budget within random arrays of emergent cylinders. *Water Resour. Res.* **2014**, *50*, 4131–4148. [CrossRef]

9. Tanino, Y.; Nepf, H.M. Laboratory investigation of mean drag in a random array of rigid, emergent cylinders. *J. Hydraul. Eng.* **2008**, *134*, 34–41. [CrossRef]

10. Ferreira, R.M.L.; Ricardo, A.M.; Franca, M.J. Discussion of "Laboratory Investigation of Mean Drag in a Random Array of Rigid, Emergent Cylinders" by Yukie Tanino and Heidi M. Nepf. *J. Hydraul. Eng.* **2009**, *135*, 690–693. [CrossRef]

11. Stoesser, T.; Kim, S.J.; Diplas, P. Turbulent flow through idealized emergent vegetation. *J. Hydraul. Eng.* **2010**, *136*, 1003–1017. [CrossRef]

12. Chang, K.; Constantinescu, G. Numerical investigation of flow and turbulence structure through and around a circular array of rigid cylinders. *J. Fluid Mech.* **2015**, *776*, 161–199. [CrossRef]

13. White, B.L.; Nepf, H.M. Shear instability and coherent structures in shallow flow adjacent to a porous layer. *J. Fluid Mech.* **2007**, *593*, 1–32. [CrossRef]

14. Chen, L.; Stone, M.C.; Acharya, K.; Steinhaus, K.A. Mechanical analysis for emergent vegetation in flowing fluids. *J. Hydraul. Res.* **2011**, *49*, 766–774. [CrossRef]

15. Tinoco, R.O.; Cowen, E.A. The direct and indirect measurement of boundary stress and drag on individual and complex arrays of elements. *Exp. Fluid.* **2013**, *54*, 1–16. [CrossRef]

16. Fathi-Maghadam, M.; Kouwen, N.; Nonsubmerged, N. Vegetative Roughness on Floodplains. *J. Hydraul. Eng.* **1997**, *123*, 51–57. [CrossRef]

17. Chen, Z.; Ortiz, A.; Zong, L.; Nepf, H.M. The wake structure behind a porous obstruction and its implications for deposition near a finite patch of emergent vegetation. *Water Resour. Res.* **2012**, *48*, W09517. [CrossRef]

18. Mitul, L.; Rominger, J.; Nepf, H. Interaction between flow, transport and vegetation spatial structure. *Environ. Fluid Mech.* **2008**, *8*, 423–439. [CrossRef]

19. Wu, F.-C.; Shen, H.W.; Chou, Y.-J. Variation of roughness coefficients for unsubmerged and submerged vegetation. *J. Hydraul. Eng.* **1999**, *125*, 934–942. [CrossRef]

20. Liu, D.; Diplas, P.; Fairbanks, J.D.; Hodges, C.C. An experimental study of flow through rigid vegetation. *J. Geophys. Res.* **2008**, *113*, 1–16. [CrossRef]

21. Pu, J.H.; Wei, J.; Huang, Y. Velocity Distribution and 3D Turbulence Characteristic Analysis for Flow over Water-Worked Rough Bed. *Water* **2017**, *9*, 668. [CrossRef]

22. Liu, C.; Shan, Y. Analytical model for predicting the longitudinal profiles of velocities in a channel with a model vegetation patch. *J. Hydrol.* **2019**, *576*, 561–574. [CrossRef]

23. Tong, X.; Liu, X.; Yang, T.; Hua, Z.; Wang, Z.; Liu, J.; Li, R. Hydraulic Features of Flow through Local Non-Submerged Rigid Vegetation in the Y-Shaped Confluence Channel. *Water* **2019**, *11*, 146. [CrossRef]

24. Heidari, M. Wake Characteristics of Single and Tandem Emergent Cylinders in Shallow Open Channel Flow. Ph.D. Thesis, University of Windsor, Windsor, ON, Canada, 2016.

25. Nicolle, A.; Eames, I. Numerical study of flow through and around a circular array of cylinders. *J. Fluid Mech.* **2011**, *679*, 1–31. [CrossRef]

26. Nepf, H.M. Flow and transport in regions with aquatic vegetation. *Annu. Rev. Fluid Mech.* **2012**, *44*, 123–142. [CrossRef]

27. White, B.L.; Nepf, H.M. A vortex-based model of velocity and shear stress in a partially vegetated shallow channel. *Water Resour. Res.* **2008**, *44*, WR005651. [CrossRef]

28. Meftah, M.B.; Mossa, M. Partially obstructed channel: Contraction ratio effect on the flow hydrodynamic structure and prediction of the transversal mean velocity profile. *J. Hydrol.* **2016**, *542*, 87–100. [CrossRef]

29. Naot, D.; Nezu, I.; Nakagawa, H. Hydrodynamic Behavior of Partly Vegetated Open Channels. *J. Hydraul. Eng.* **1996**, *122*, 625–633. [CrossRef]

30. Ikeda, S.; Yamada, T.; Toda, Y. Numerical study on turbulent flow and honami in and above flexible plant canopy. *Int. J. Heat Fluid Flow* **2001**, *22*, 252–258. [CrossRef]

31. Kim, S.J.; Stoesser, T. Closure modelling and direct simulation of vegetation drag in flow through emergent vegetation. *Water Resour. Res.* **2011**, *47*, W10511. [CrossRef]

32. Anjum, N.; Tanaka, N. Study on the flow structure around discontinued vertically layered vegetation in an open channel. *J. Hydrodyn.* **2019**. [CrossRef]

33. Yamasaki, T.N.; de Lima, P.H.; Silva, D.F.; Cristiane, G.D.A.; Janzen, J.G.; Johannes, G.; Nepf, H.M. From patch to channel scale: The evolution of emergent vegetation in a channel. *Adv. Water Res.* **2019**, *129*, 131–145. [CrossRef]

34. Pu, J.H.; Hussain, A.; Guo, Y.; Vardakastanis, N.; Hanmaiahgari, R.; Lam, D. Submerged Flexible Vegetation Impact toward Open Channel Flow Velocity Distribution: An Analytical Modelling Study on Drag and Friction. *Water Sci. Eng.* **2019**, *12*, 121–128. [CrossRef]

35. Nepf, H.M. Drag, Turbulence, and diffusion in flow through emergent vegetation. *Water Resour. Res.* **1999**, *35*, 479–489. [CrossRef]

36. Musleh, F.; Cruise, J. Functional Relationships of Resistance in Wide Flood Plains with Rigid Unsubmerged Vegetation. *J. Hydraul. Eng.* **2006**, *132*, 163–171. [CrossRef]

37. Maji, S.; Pal, D.; Hanmaiahgari, P.R.; Pu, J.H. Phenomenological Features of Turbulent Hydrodynamics in Sparsely Vegetated Open Channel Flow. *J. Appl. Fluid Mech.* **2016**, *9*, 2865–2875. [CrossRef]

38. Pu, J.H.; Tait, S.; Guo, Y.; Huang, Y.; Hanmaiahgari, R. Dominant Features in Three-Dimensional Turbulence Structure: Comparison of Non-Uniform Accelerating and Decelerating Flows. *Environ. Fluid Mech.* **2018**, *18*, 395–416. [CrossRef]

39. Maji, S.; Pal, D.; Hanmaiahgari, R.; Gupta, U. Hydrodynamics and turbulence in emergent and sparsely vegetated open channel flow. *Environ. Fluid Mech.* **2017**, *17*, 853–877. [CrossRef]

40. Heidari, M.; Balachandar, R.; Roussinova, V.; Barron, R.M. Characteristics of flow past a slender, emergent cylinder in shallow open channels. *Phys. Fluid.* **2017**, *29*, 065111. [CrossRef]

41. Kothyari, U.C.; Hashimoto, H.; Hayashi, K. Effect of tall vegetation on sediment transport by channel flows. *J. Hydraul. Res.* **2009**, *47*, 700–710. [CrossRef]

42. Vargas-Luna, A.; Crosato, A.; Uijttewaal, W.S.J. Effects of vegetation on flow and sediment transport: Comparative analyses and validation of predicting models. *Earth. Surf. Proc. Landf.* **2015**, *40*, 157–176. [CrossRef]

43. Baptist, M.J. Modelling Floodplain Biogeomorphology. Ph.D. Thesis, Delft University of Technology, Delft, The Netherlands, 2005.

44. Takemura, T.; Tanaka, N. Flow structures and drag characteristics of a colony-type emergent roughness model mounted on a flat plate in uniform flow. *Fluid Dyn. Res.* **2007**, *39*, 694–710. [CrossRef]

45. Liu, X.; Zeng, Y. Drag coefficient for rigid vegetation in subcritical open-channel flow. *Environ. Fluid Mech.* **2017**, *17*, 1035–1050. [CrossRef]

46. Van Rooijen, A.; Lowe, R.; Ghisalberti, M.; Conde-Frias, M.; Tan, L. Predicting Current-Induced Drag in Emergent and Submerged Aquatic Vegetation Canopies. *Front. Mar. Sci.* **2018**, *5*, 449. [CrossRef]

47. Wang, W.J.; Huaia, W.X.; Thompson, S.; Peng, W.Q.; Katul, G.G. Drag coefficient estimation using flume experiments in shallow non-uniform water flow within emergent vegetation during rainfall. *Ecol. Indic.* **2018**, *92*, 367–378. [CrossRef]

48. Shan, Y.; Liu, C.; Nepf, H. Comparison of drag and velocity in model mangrove forests with random and in-line tree distributions. *J. Hydrol.* **2019**, *568*, 735–746. [CrossRef]

49. Razmi, A.; Chamecki, M.; Nepf, H.M. Efficient numerical representation of the impacts of flexible plant reconfiguration on canopy posture and hydrodynamic drag. *J. Hydraul. Res.* **2019**. [CrossRef]

50. Perucca, E.; Camporeale, C.; Ridolfi, L. Significance of the riparian vegetation dynamics on meandering river morphodynamics. *Water Resour. Res.* **2007**, *43*, W03430. [CrossRef]

51. Motta, D.; Langendoen, E.J.; Abad, J.D.; García, M.H. Modification of meander migration by bank failures. *J. Geophys. Res. Earth Surf.* **2014**, *119*, 1026–1042. [CrossRef]

52. Liu, D.; Valyrakis, M.; Williams, R. Flow Hydrodynamics across Open Channel Flows with Riparian Zones: Implications for Riverbank Stability. *Water* **2017**, *9*, 720. [CrossRef]

53. Thorne, C.R. Effects of vegetation on riverbank erosion and stability. In *Vegetation and Erosion*; Thorne, J.B., Ed.; John Wiley and Sons: Chichester, UK, 1990; pp. 125–144.

54. Simon, A.; Collison, A.J.C. Quantifying the mechanical and hydrologic effects of riparian vegetation on streambank stability. *Earth. Surf. Proc. Landf.* **2010**, *27*, 527–546. [CrossRef]

55. Hopkinson, L.; Wynn, T. Vegetation impacts on near bank flow. *Ecohydrology* **2009**, *2*, 404–418. [CrossRef]

56. Baets, S.D.; Poesen, J.; Reubensm, B.; Wemans, J.; Baerdemaeker, D.; Muys, B. Root tensile strength and root distribution of typical Mediterranean plant species and their contribution to soil shear strength. *Plant Soil* **2008**, *305*, 207–226. [CrossRef]

57. Abernethy, B.; Rutherfurd, I.D. The effect of riparian tree roots on the mass-stability of riverbanks. *Earth. Surf. Proc. Landf. J. Br. Geomorphol. Res. Group* **2000**, *25*, 921–937. [CrossRef]

58. Pollen, N. Temporal and spatial variability in root reinforcement of streambanks: Accounting for soil shear strength and moisture. *Catena* **2007**, *69*, 197–205. [CrossRef]

59. Tal, M.; Paola, C. Effects of vegetation on channel morphodynamics: Results and insights from laboratory experiments. *Earth. Surf. Proc. Landf.* **2010**, *35*, 1014–1028. [CrossRef]

60. Krzeminska, D.; Kerkhof, T.; Skaalsveen, K.; Stolte, J. Effect of riparian vegetation on stream bank stability in small agricultural catchments. *Catena* **2019**, *172*, 87–96. [CrossRef]

61. Hupp, C.R.; Osterkamp, W.R. Riparian vegetation and fluvial geomorphic processes. *Geomorphology* **1996**, *14*, 277–295. [CrossRef]

62. Tooth, S.; Nanson, G.C. The role of vegetation in the formation of anabranching channels in an ephemeral river, Northern Plains, arid central Australia. *Hydrol. Process.* **2000**, *14*, 3099–3117. [CrossRef]

63. Micheli, E.R.; Kirchner, J.W.; Larsen, E.W. Quantifying the effect of riparian forest versus agricultural vegetation on river meander migration rates, central Sacramento river, California, USA. *River Res. Appl.* **2010**, *20*, 537–548. [CrossRef]

64. Hickin, E.J. Vegetation and river channel dynamics. *Can. Geogr.* **1984**, *28*, 111–126. [CrossRef]

65. Pollen, N.; Simon, A.; Collision, A.J.C. Advances in assessing the mechanical and hydrologic effect of riparian vegetation on streambank stability. *Riparian Veg. Fluv. Geomorphol.* **2004**, *8*, 125–139.

66. Graf, W.L. *Fluvial Processes in Dryland Rivers*; Springer: Berlin, Germany, 1988.

67. Birken, A.S.; Cooper, D.J. Processes of Tamarix invasion and floodplain development along the lower Green River Utah. *Ecol. Appl.* **2006**, *16*, 1103–1120. [CrossRef]

68. Braudrick, C.A.; Dietrich, W.E.; Leverich, G.T.; Sklar, L.S. Experimental evidence for the conditions necessary to sustain meandering in coarse bedded rivers. *Proc. Natl. Acad. Sci. USA* **2009**, *106*, 16936–16941. [CrossRef] [PubMed]

69. Camporeale, C.; Perucca, E.; Ridolfi, L.; Gurnell, A.M. Modeling the interactions between river morphodynamics and riparian vegetation. *Rev. Geophys.* **2013**, *51*, 379–414. [CrossRef]

70. Gran, K.; Wartman, E.D. Co-evolution of riparian vegetation and channel dynamics in an aggrading braided river system, Mount Pinatubo. Philippines. *Earth Surf. Proc. Landf.* **2015**, *40*, 1101–1115. [CrossRef]

71. Yu, M.-H.; Wei, H.-Y.; Wu, S.-B. Experimental study on the bank erosion and interaction with near-bank bed evolution due to fluvial hydraulic force. *Int. J. Sediment Res.* **2015**, *30*, 81–89. [CrossRef]

72. Yang, J.Q.; Nepf, H.M. A turbulence-based bed-load transport model for bare and vegetated channels. *Geophys. Res. Lett.* **2018**, *45*, 10428–10436. [CrossRef]

73. Armanini, A.; Cavedon, V. Bed-load through emergent vegetation. *Adv. Water Resour.* **2019**, *129*, 250–259. [CrossRef]

74. Armanini, A.; Cavedon, V.; Righetti, M. A probabilistic/deterministic approach for the prediction of the sediment transport rate. *Adv. Water Res.* **2015**, *81*, 10–18. [CrossRef]

75. Einstein, H.A. *The Bed-Load Function for Sediment Transportation in Open Channel Flows*; Technical Report No. 1026; U.S. Department of Agriculture: Washington, DC, USA, 1950.

76. Rowiński, M.; Kubrak, J. A mixing-length model for predicting vertical velocity distribution in flows through emergent vegetation. *Hydrol. Sci. J.* **2002**, *47*, 893–904. [CrossRef]

77. Rubol, S.; Ling, B.; Battiato, I. Universal scaling-law for flow resistance over canopies with complex morphology. *Sci. Rep.* **2018**, *8*, 4430. [CrossRef] [PubMed]

78. Siniscalchi, F.; Nikora, V.I.; Aberle, J. Plant patch hydrodynamics in streams: Mean flow, turbulence, and drag forces. *Water Resour. Res.* **2012**, *48*, 1–14. [CrossRef]

79. Västilä, K.; Järvelä, J. Modeling the flow resistance of woody vegetation using physically based properties of the foliage and stem. *Water Resour. Res.* **2014**, *50*, 229–245. [CrossRef]

80. Caroppi, G.; Västilä, K.; Järvelä, J.; Rowiński, M.; Giugni, M. Turbulence at water-vegetation interface in open channel flow: Experiments with natural-like plants. *Adv. Water Resour.* **2019**, *127*, 180–191. [CrossRef]

81. Valyrakis, M.; Diplas, P.; Dancey, C.L. Entrainment of coarse particles in turbulent flows: An energy approach. *J. Geophys. Res. Earth Surf.* **2013**, *118*, 42–53. [CrossRef]

82. Valyrakis, M.; Diplas, C.L.; Dancey, K.; Greer, K.; Celik, A.O. Role of instantaneous force magnitude and duration on particle entrainment. *J. Geophys. Res.* **2010**, *115*. [CrossRef]

83. Diplas, P.; Dancey, C.L.; Celik, A.O.; Valyrakis, M.; Greer, K.; Akar, T. The role of impulse on the initiation of particle movement under turbulent flow conditions. *Science* **2008**, *322*, 717–720. [CrossRef]

84. Licci, S.; Nepf, H.M.; Delolme, C.; Marmonier, P.; Bouma, T.J.; Puijalon, S. The role of patch size in ecosystem engineering capacity: A case study of aquatic vegetation. *Aquat. Sci.* **2019**, *81*. [CrossRef]

Article

Evaluation of Ecosystem Services in the Dongting Lake Wetland

Li Ma [1], Ruoxiu Sun [1], Ehsan Kazemi [2], Danbo Pang [3], Yi Zhang [4], Qixiang Sun [5], Jinxing Zhou [1,*] and Kebin Zhang [1,*]

1   School of Soil and Water Conservation, Beijing Forestry University, Beijing 100083, China; mmyy6363@163.com (L.M.); sunruoxiu0331@126.com (R.S.)

2   Department of Civil and Structural Engineering, University of Sheffield, Sheffield S1 3JD, UK; e.kazemi@sheffield.ac.uk

3   Breeding Base for State Key Laboratory of Land Degradation and Ecological Restoration in Northwest China, Ningxia University, Yinchuan 750021, China; pang89028@163.com

4   WHO Collaborating Center for Tropical Diseases, National Center for International Research on Tropical Diseases, Shanghai 200025, China; zhang1972003@163.com

5   Institute of Forestry, Chinese Academy of Forestry, Beijing 100091, China; sunqixiang@163.net

*   Correspondence: zjx9277@126.com (J.Z.); ctccd@126.com (K.Z.)

Received: 8 October 2019; Accepted: 29 November 2019; Published: 5 December 2019

Abstract: The Aeronautical Reconnaissance Coverage Geographic Information System (ArcGIS) 10.2 and Integrated Valuation of Ecosystem Services and Trade-offs (InVEST) model are used to comprehensively evaluate ecosystem services in the Dongting Lake Wetland, focusing on water yield, soil conservation, carbon storage, and snail control and *schistosomiasis* prevention. The spatial and temporal variations of these services, as well as their variations between different land use types in a period of 10 years from 2005 to 2015, are investigated, and the value of such services is then estimated and analyzed. The results of this study show various temporal and spatial trends in the ecosystem services, such as (1) the overall increase of all these services during the study period (although significant in some services, such as *schistosomiasis* patient reduction, by 86.8%; and, very slight in some others such as soil conservation, only by 0.02%); (2) different orders of the services values that are based on different land use types; and, (3) the temporal changes in the proportion of the values of different ecosystem services with respect to the total services value. Besides, it is concluded that the evaluation of ecosystem services of a certain wetland is heavily dependent on the characteristics of the area where the wetland is located, and the assessment indicators and methods should be selected based on such characteristics through the analysis of the results and a comparison with the findings of literature.

Keywords: InVEST model; wetland; ecosystem service assessment; value analysis; *schistosomiasis* prevention

---

1. Introduction

Ecosystem service is the natural environmental condition and utility that the ecosystem and ecological process form and maintain [1,2]. The millennium ecosystem assessment (MEA) has found that 60% of global ecosystem services have deteriorated, with human activity being one of the main causes of this adverse change [3,4]. Wetlands, as important habitats of humans, only cover 1% of the earth's surface, but they provide habitat for about 20% of all species worldwide [5,6]. Wetland ecosystem services are crucial for the survival of humans and a range of other animals, by supplying freshwater and aquatic products, flood control, climate regulation, biodiversity protection, etc. [7]. It has been estimated that lake wetland ecosystem services provide 23.2% of the total value of global ecosystem services [8].

Numerous scholars have assessed wetland ecosystems at different geographical locations and scales. For instance, Great Lakes Restoration Initiative Task Force systematically assessed the ecosystem services of the Great Lakes of North America in 2010, and showed that a number of management and engineering measures improved these services [9]. Wong et al. (2016) tested a new approach to evaluate water storage and local climate regulation of a green infrastructure project on the Yong ding River in Beijing, China, showing that the lakes and wetlands have increased local evapotranspiration on the Yongding Corridor by approximately 0.71 mm per day [10]. In another study, Bikash et al. (2015) pointed out that the economic values of the selective ecosystem services of the Koshi Tappu Wildlife Reserve, Nepal, equaled $16 million per year [11]. Around the world, several countries, including Canada and China, have evaluated their wetland ecosystem services, but no universal evaluation index system of lake wetland ecosystem services has been established, and there has also been no evaluation of the effects of wetland ecosystem services on human health [12–17].

There is a lack of regional validation of parameters in the use of evaluation models, with only a few evaluations on cultural or support services in wetland ecosystem services, which makes the selection of indicators one-sided. In this sense, it is necessary to further unify the methods, indicators, and other aspects of ecosystem services of different scales and landscape types. At present, numerous studies are available regarding the quantitative evaluation, model simulation, and scenario analysis of ecosystem services, while using different types of ArcGIS (Aeronautical Reconnaissance Coverage Geographic Information System)-based support tools to fully integrate geography, ecology, economics, etc. The number and types of models for ecosystem services assessment have dramatically increased over the past decade. IMAGE (Integrated Assessment of Global Environmental Change), SAORES (Spatial Assessment and Optimization Tool for Regional Ecosystem Services), ARIES (Artificial Intelligence for Ecosystem Services), Costing Nature, and LUCI (Land Utilization and Capability Indicator) are some examples. For instance, ARIES is a model built via an expert knowledge method, which might result in the one-sidedness of evaluation results due to experts' knowledge reserve and subjective factors. The models IMAGE, Costing Nature, and LUCI are based on a global data set for ecosystem service evaluation [18,19]. The SAORES model evaluates ecosystem services in the context of ecological restoration and the management of the Loess Plateau in China, ignoring the accuracy verification and uncertainty analysis of service evaluation [20]. The InVEST (Integrated Valuation of Ecosystem Services and Trade-offs) model inputs the relevant data of the study area and reflects the impacts of ecosystem structure and function on service flow and value through the simplification of biophysical processes [21]. The multi-scale data input and the result output have made the InVEST model a comprehensive ecosystem service evaluation and balance model that is widely used in 20 countries and regions, for example in the assessment of freshwater ecosystem services in the Tualatin and Yamhill basins, the Chubute River valley in southern Argentina, the Beijing mountain forest, etc. [22–25].

With the rapid development of social economy and the expansion of global population, the demand for land is constantly increasing in order to survive and obtain economic benefits. Consequently, wetland resources are under unprecedented pressure and are decreasing at the global level [26,27]. The Dongting Lake Wetland is the largest lake that leads to the Yangtze River and the second largest freshwater lake in China. It provides favorable conditions for the breeding of snails and it is a severe epidemic area of *schistosomiasis*, seriously endangering the health of the local people. *Schistosomiasis* is a parasitic disease that affects both humans and animals and, globally, it is the second largest tropical parasitic disease after malaria [28,29]. It is endemic in about 78 countries in the tropical and subtropical regions of the world [30]. More than 95% of the *Schistosomiasis* in China are distributed in the Yangtze River basin tidal land, which seriously endangers wetland ecosystems. *Oncomelania hupensis*, which is a tropical freshwater snail, is the only intermediate host of *Schistosoma japonicum* and it can live in moist, shaded, and alternate land and water wetland environments. Thus, snail eradication is an effective measure for blocking the transmission of *schistosomiasis* [31,32]. At present, there are various technical systems for snail control, including drug snail control, agricultural engineering measures, and water conservancy projects. The impermanence of snail control effectiveness, the higher costs on capital, and the labor-intensity are substantial disadvantages between water conservancy projects and ecological environmental protection. The 65th World Health Assembly has passed Resolution WHA65.21, which proposes eliminating *schistosomiasis*, a neglected tropical disease, in low-transmission areas of the world. Previous research has shown that the *Schistosomiasis* Prevented Forestry Project is an important measure for the comprehensive control of *schistosomiasis* in China, with the advantages of long-term effects and prevention [33–37]. The project is based on the construction of forests for snail control and *schistosomiasis* prevention, which aims to change the environment for snail breeding, inhibit the growth and development of snails, isolate the source of infection, and control the epidemic of *schistosomiasis* [38].

Recently, China has built a total of 5189 km^2 of snail control and *schistosomiasis* prevention forests, decreasing the snail density and infection rate of human by 89.8% and 51.0%, respectively [39]. The construction of these forests, which also provide a variety of other ecosystem services, such as carbon fixation, water and soil conservation, and biodiversity protection, has reached 1640 km^2, in 2015, along the Dongting Lake Wetland beach. However, such ecosystem services, especially snail control and *schistosomiasis* prevention, have been paid less attention in previous studies on the Dongting Lake Wetland area (e.g., Mao et al., 2007, where some ecosystem services, such as agricultural and fishery productions, tourism and leisure, climate regulation, and flood storage were investigated [40]). On the other hand, most studies regarding the snail control and *schistosomiasis* prevention in the literature have mainly focused on this issue in terms of mechanism and construction technology [39,41]. This study focuses on the ecosystem services in the Dongting Lake area that resulted from the construction of disease prevention forests, including snail control and *schistosomiasis* prevention, water yield, soil conservation, and carbon storage; and, evaluates them in a period of 10 years between 2005 and 2015. For this purpose, ArcGIS 10.2 and InVEST models (with the aid of market value and shadow engineering methods for service value estimations) are applied to (1) study the temporal and spatial variations of these ecosystem services in the study area; (2) explore their variations between different land use types; and, (3) estimate their value in 2005, 2010, and 2015.

2. Study Area and Data

2.1. Study Area

Dongting Lake is the second largest freshwater lake in China, which covers an area of 18,780 km^2 and combines Xiangjiang, Zijiang, Yuanjiang, Lijiang, and other water-flooded lakes in the Yangtze River floods during the flood season. It is located between N 27°39′~30°25′ and E 111°19′~113°34′, with an average elevation of 33.5 m above sea level. Average annual temperature in the lake is 17.0 °C, with an average annual precipitation of 1100–1400 mm. The soil pH ranges between 7 and 8.4. From June to August, the area is generally flooded; the difference between the highest water level in the wet season and the lowest one in the dry season in winter can reach 8–10 m, which makes the Dongting Lake Wetland a *schistosomiasis* epidemic area [42]. Figure 1 shows the geographic position of this area, which includes 17 counties and cities, such as Yueyang City, and Xiangyin, Huarong, and Li Counties in Hunan Province, covering an area of 3.17 × 10^6 km^2, accounting for 16.5% of the area of the province.

By the end of 2010, the area of snail control and *schistosomiasis* prevention forests totaled 701.86 km^2, while, in 2015, this area amounted to 1334.66 km^2, which greatly inhibited snail numbers and decreased the number of *schistosomiasis* patients. The proportion of arable land in the Dongting Lake Wetland accounts for more than 45%, followed by woodland and water area. The proportions of residential, unutilized, and grassland areas are comparatively small and below 5%. From 2005 to 2015, arable land, woodland, grassland, and water area decreased at different degrees, i.e., by 1% (313.90 km^2), 0.35% (111.09 km^2), 0.15% (45.99 km^2), and 0.72% (228.25 km^2), respectively. In contrast, the proportions of residential and unutilized lands continued to increase by 1.34% (424.26 km^2) and 0.88% (279.60 km^2), respectively. Table 1 presents the land cover types and areas from 2005 to 2015.

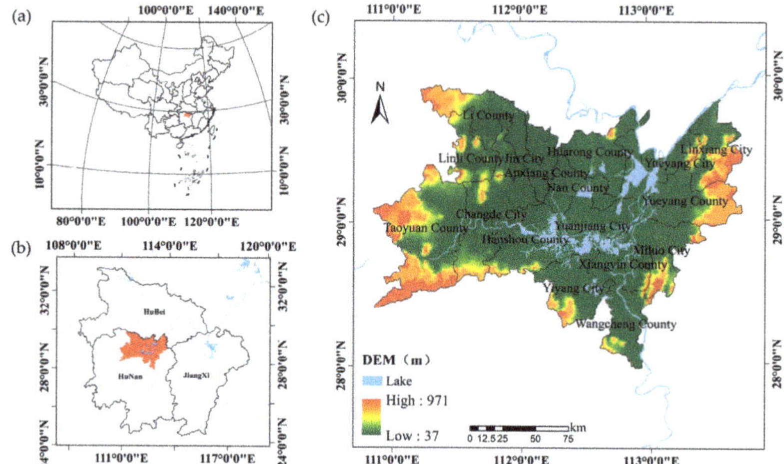

Figure 1. Geographic location of the study region (**a**) in China; (**b**) in the vicinity of Hubei, Hunan and Jiangxi provinces; and (**c**) elevation map of the study area including Dongting lake.

Table 1. Land cover types of the Dongting Lake Wetland between 2005 and 2015 ($\times 10^2$ km^2).

Land Use Type	Area			
	2005	2010	2015	Change Area
Arable land	147.69	146.02	144.55	−3.14
Woodland	102.29	101.93	101.18	−1.11
Grassland	3.34	2.93	2.88	−0.46
Waters	48.27	48.32	45.99	−2.28
Residential land	8.99	11.01	13.24	4.24
Unutilized land	6.85	7.25	9.64	2.80

2.2. Data Sources and Processing

The land use maps (at a spatial resolution of 30 × 30 m) of the Dongting Lake Wetland from 2000 to 2015 were downloaded from the Resource and Environment Data Cloud platform [43]. Precipitation data of the Wetland during this period were acquired from the National Meteorological Information Center [44]. ArcGIS 10.2 conducted Kriging interpolation in the study area and the surrounding areas on precipitation data from 59 meteorological and hydrological stations and, subsequently, grid maps with a resolution of 30 m were generated for the related years. The Digital Elevation Model (DEM) data were generated in ArcGIS 10.2 by means of interpolation, cutting, and filling digital elevation data of Global Digital Elevation Model Version 2 (GDEM V2), with a high resolution of 30 m, which was sourced from Geospatial Data Cloud [45]. Soil data, including soil texture, topsoil organic carbon, and fractions of topsoil sand, silt, and clay were obtained from the Soil Map of China based on the Harmonized World Soil Database (version 1.1, with a 1:1 million map scale), which was created by the Cold and Arid Regions Sciences Data Center in 2009 [46], and ArcGIS 10.2 was used to obtain the soil grid maps with a resolution of 30 m. The carbon density values for each land use were collected from Li et al. (2003) and Wang et al. (2010) [47,48] based on the principle of similarity within the natural environment of the Dongting Lake Wetland.

3. Methods

The data were firstly processed by using the ArcGIS 10.2 to develop an input for the analysis of the ecosystem services in this section, as discussed in the previous section. The InVEST model is then used, as described in Sections 3.1–3.3, to estimate water yield (the difference between precipitation and actual evapotranspiration), soil conservation, and carbon storage. Moreover, the Raster Calculator of ArcGIS 10.2 is applied to evaluate the snail control and *schistosomiasis* prevention, as presented in Section 3.4. Besides, the market value and shadow engineering methods were employed to calculate the values of the services.

3.1. Water Yield Model

The InVEST water yield module is an estimation method that is based on the water balance method, where the water yield of a grid unit refers to precipitation minus actual evapotranspiration, and the generated water yield data are the total amount and depth of water yield [49]. The module formula is as follows:

$$Y_{xj} = \left(1 - \frac{AET_{xj}}{P_x}\right) \cdot P_x \tag{1}$$

where Y_{xj} and AET_{xj} are the annual water yield and the actual evapotranspiration of the land cover type j in Grid x in mm; and, P_x is the precipitation of Grid x in mm. The annual potential evaporation for the 87 meteorological stations in the Dongting Lake Wetland was estimated by the Penman–Monteith equation [50]. Subsequently, Kriging spatial interpolation was used to obtain a map of the annual average potential evapotranspiration of the Wetland. Besides, the market value method was employed to calculate the value of the water yield service.

3.2. Soil Conservation

The soil retention module, which was based on the Revised Universal Soil Loss Equation (RUSLE), was used to estimate the amount of soil erosion within the region. The formulas of the module are as follows:

$$A = R \times K \times LS \times C \times P \tag{2}$$

$$RKLS = R \times K \times LS \tag{3}$$

$$SD = RKLS - A \tag{4}$$

where $RKLS$ is the total potential soil loss (t) per pixel in the original land cover without the application of factors C or P from the RUSLE (i.e., equivalent to the soil loss for bare soil); A is the estimated average soil loss (t) per pixel in the original land cover (t·km^{-2}·yr^{-1}); SD is the amount of soil retention (t); and, R is the rainfall erosivity (MJ·mm/(km^2·h·yr)), which is calculated while using the formula of Wischmeier according to the average monthly and annual precipitations in the Dongting Lake Wetland [51]; K is the soil erodibility (t·km^2·h/(km^2·MJ·mm)) calculated while using the formula of Williams based on relevant soil data [52]; LS is the slope length gradient factor; C is the cover management factor obtained by the FAO [53]; and, P is the support practice factor that is available from the USDA Agriculture Handbook [54]. The value of soil conservation was estimated while using shadow engineering method.

3.3. Carbon Storage

The carbon module uses land use maps and the carbon density (i.e., above ground biomass, belowground biomass, soil, and dead biomass) of each land use type to estimate the amount of carbon that is currently stored in a landscape, as follows [55]:

$$C_t = C_{above} + C_{below} + C_{dead} + C_{soil} \tag{5}$$

where C_t is the total carbon storage (t); and, C_{above}, C_{below}, C_{soil}, and C_{dead} are the carbon storage (t) in aboveground biomass, belowground biomass, soil, and dead biomass, respectively. The value of carbon storage was calculated while using the market value method.

3.4. Snail Control and Schistosomiasis Prevention

The purpose of constructing snail control and *schistosomiasis* prevention forests has been to reduce the density and area of snails and effectively prevent and control the spread of *schistosomiasis*. Therefore, in this study, the snail-control function is evaluated by the reduction of the snail area and the reduction in the number of *schistosomiasis* patients. The Raster Calculator of ArcGIS 10.2 is used to analyze the quality and value of snail control and disease prevention, while using the following equations:

$$\Delta_A = A_{2010} - A_{2015} \tag{6}$$

$$\Delta_N = N_{2010} - N_{2015} \tag{7}$$

where Δ_A and Δ_N denote the reduced areas that are colonized by the snail and the reduction number of *schistosomiasis* patients, respectively; A_{2010} and N_{2010} are the snail area (10^4 m^2) and the number of schistosomiasis patients in 2010; and, A_{2015} and N_{2015} are those quantities in 2015.

The market value and the shadow engineering methods, and then were used to calculate the value of snail control and *schistosomiasis* prevention. The value of snail and disease prevention function (C) is computed as the following:

$$C = C_{Snail} \cdot \Delta_A + C_{Patient} \cdot \Delta_N \tag{8}$$

where C_{Snail} is the cost of snail control (in CNY/m^2) and $C_{Patient}$ is the annual treatment cost per patient. Table 2 summarizes the sources used for the collection of parameters.

Table 2. Parameter collection methods.

The Parameter Name	Content or Price	The Data Source
Water market transaction price	4.51 CNY/t (0.64 USD/t)	Circular on issues relating to the standards for the collection of water resources charges of the People's Republic of China
The cost of digging up a unit of soil	63.44 CNY/m^3 (9.01 USD/m^3)	The water conservancy construction project budget quota of the ministry of water resources of the People's Republic of China
Carbon price	48.52 CNY/t (6.89 USD/t)	A Report by Forest Trends' Ecosystem Market
The price of medicine to kill snails	0.27 CNY/m^2 (0.04 USD/m^2)	Chinese journal of schistosomiasis control [56]
Per capita annual cost of treatment	3230.7 CNY/yr (458.89 USD/yr)	National Bureau of Statistics
The area of the snail control and *schistosomiasis* prevention forest	—	State forestry administration
The snail area and the number of *schistosomiasis* patients	—	National Bureau of Statistics

4. Results and Analysis

In this section, the results of the present study on the variations of different ecosystem services in the Dongting Lake Wetland and an analysis of their values are presented.

4.1. Temporal and Spatial Variations of the Ecosystem Services

Figure 2 shows the results of the study on the temporal and spatial variations of the water yield, soil conservation, and carbon storage; and, Figures 3 and 4 present those variations in the reduced snail area and the number of schistosomiasis patients, in the Dongting Lake Wetland, from 2005 to 2015.

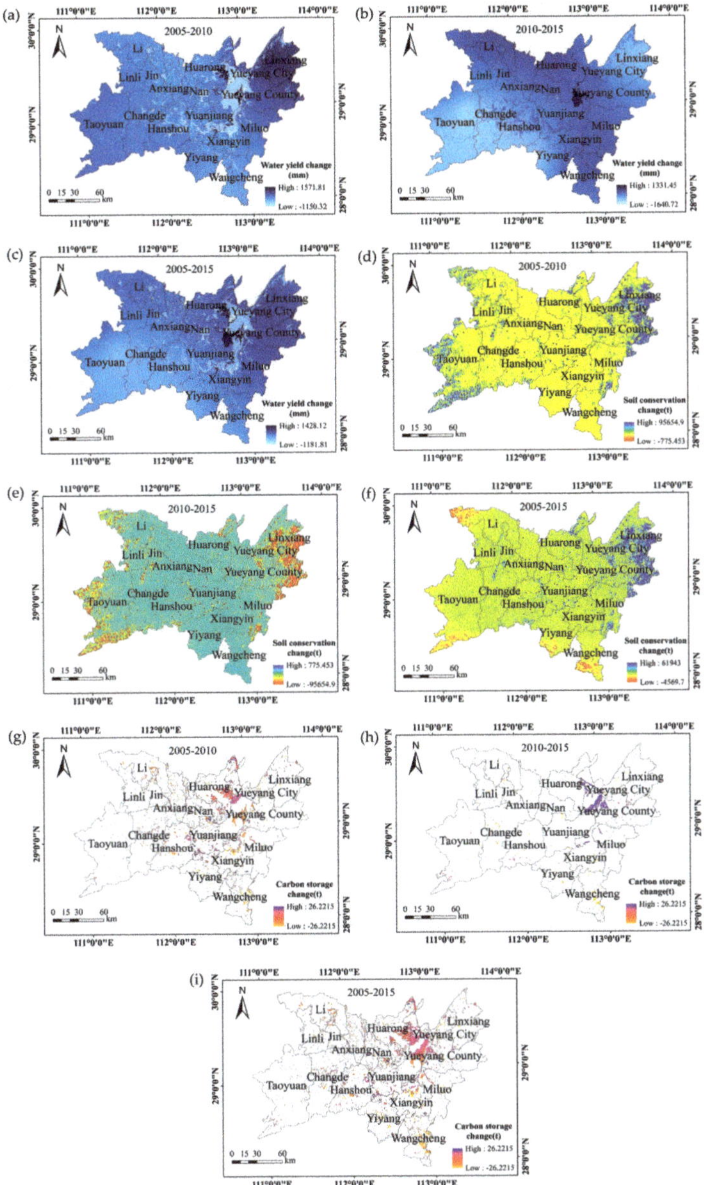

Figure 2. Distribution of changes in the ecosystem services in the Dongting Lake Wetland from 2005 to 2015.

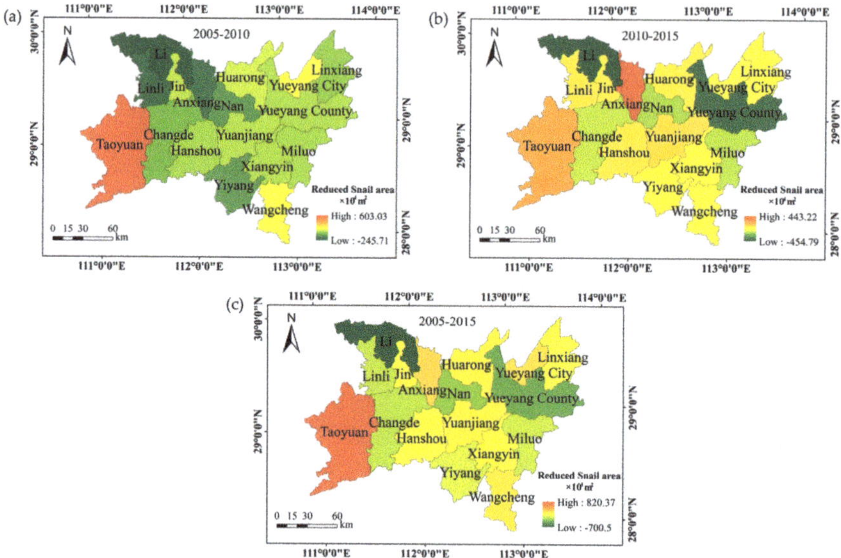

Figure 3. Temporal and spatial variations of the reduced snail area from (**a**) 2005 to 2010; (**b**) 2010 to 2015; and (**c**) 2005 to 2015.

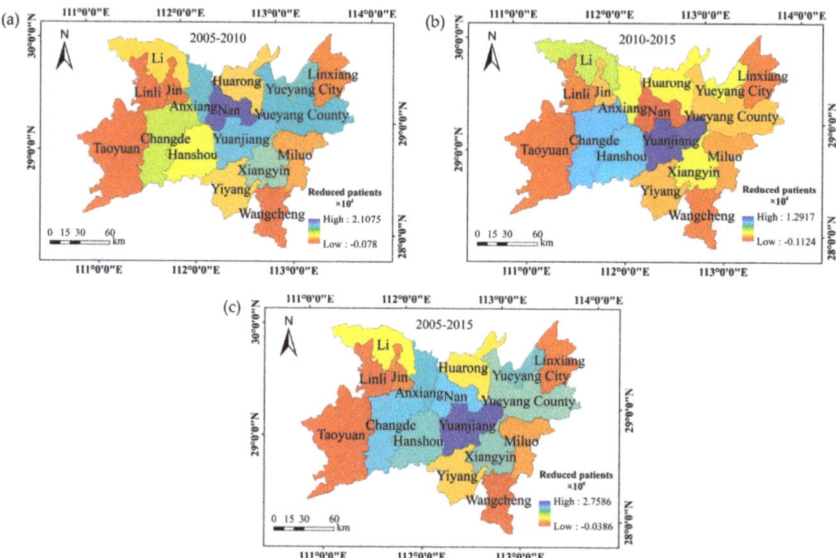

Figure 4. Temporal and spatial variations of the reduction in the number of *schistosomiasis* patients from (**a**) 2005 to 2010; (**b**) 2010 to 2015; and (**c**) 2005 to 2015.

4.1.1. Variations of Water Yield

Land use and climatic changes both influence water yield. In 2005, 2010, and 2015, water yield in the Dongting Lake Wetland was 236.91×10^8 m^3, 339.61×10^8 m^3, and 302.80×10^8 m^3, respectively. This shows that it increased from 2005 to 2010 by 43.3% and then decreased 10.8% between 2010 and 2015. The reason for these variations was that the amount of rainfall was higher than evapotranspiration between 2005 and 2010, while the latter became larger from 2010 to 2015. However, the net change from 2005 to 2015 was an increase of 65.89×10^8 m^3 (equivalent to 27.81%). Spatially, water yield during this period decreased in the southwest (Taoyuan and Hanshou Counties), with the largest reduction of 1313.0 m^3/km^2, while it increased in the northeast (Yueyang County, and Linxiang and Yueyang Cities), where the largest rise was 1586.8 m^3/km^2.

4.1.2. Variations of Soil Conservation

In 2005, soil conservation was equivalent to 1257.23×10^4 t, while, in 2010 and 2015, it accounted for 1931.06×10^4 t and 1257.44×10^4 t, respectively, i.e., a total increase of 0.21×10^4 t equivalent to a growth of 0.02% from 2005 to 2015. However, the change was not uniform across the area, as in the northwest (Li County) and southwest (Taoyuan County) a decrease was observed with the largest value of 4569.7 t; while, in the northeast (Linxiang City, Yueyang County, and Yueyang City), the amount of conserved soil became larger (with a maximum change of 61,943 t). See Figure 2 for details.

A variety of factors, such as climate, topography, land use, and human activities, influence soil conservation. Among these, precipitation and rainfall erosivity, with temporal variations that were similar to the soil conservation (i.e., increase from 2005 to 2010 and then decrease from 2010 to 2015), were the most significant causes of the large variations in the amount of the conserved soil between 2005 and 2015.

4.1.3. Variations of Carbon Storage

In 2005, 2010, and 2015, the total carbon storage was 3.57×10^8 t, 3.63×10^8 t, and 3.56×10^8 t, respectively. This shows a total decrease of 0.01×10^8 t (i.e., 0.28%) from 2005 to 2015. Spatially, during this period, the carbon storage mainly changed in the northwest (Li County), northeast (Yueyang County), and southeast (Wangcheng County), with a maximum reduction of 26.2215 t; while, in the areas in the middle (Huarong county and Yuanjiang City), it increased with a maximum change of 26.2215 t (Figure 2). Temporal and spatial changes in the carbon storage were both associated with the changes in the land use. For example, higher variations in the carbon storage were observed in the regions with higher changes of land use.

4.1.4. Variation of Snail Control and Schistosomiasis Prevention

The construction of snail control and *schistosomiasis* prevention forests in the Dongting Lake Wetland has considerably changed the environmental conditions for snail breeding. Figure 3 shows the temporal and spatial reductions of snail areas in the study area between 2005 and 2015. In 2005, snails colonized $176,042.5 \times 10^4$ m^2 of the area. Subsequently, from 2005 to 2015, 10 counties experienced a reduction, with the total amount of 326.01×10^4 m^2 (i.e., −0.19%), while an increase of snail area was observed in seven other counties. These variations were as follows. In Taoyuan, Anxiang, Wangcheng, Huarong, Hanshou, and Xiangyin Counties, as well as Yueyang, Yuanjiang, Jin, and Linxiang Cities, there was a reduction in the snail area with a maximum of 820.37×10^4 m^2 in Taoyuan County; while, in Nan, Yueyang, and Li Counties and Yiyang, Miluo, Linli, and Changde Cities, the area that was colonized by snails became greater with a maximum of 700.5×10^4 m^2 in Li City. The increase in the snail area in the latter, in spite of the presence of the constructed snail control and *schistosomiasis* prevention forests, was that these areas are located along the *Oncomelania hupensis* snail-ridden regions outside embankments of Yangtze river and Dongting Lake, where the infectious sources are various, large in number, and difficult to manage, thus a high risk of schistosomiasis still exists [57].

Except in Linli County, where the number of *schistosomiasis* patients increased from 2005 to 2015 (386 people), this number decreased in all other counties and cities of the study area. In 2005, the total number of *schistosomiasis* patients in the Dongting Lake Wetland area was 210,651, which then reduced by 182,829 (86.8%) from 2005 to 2015. The largest reduction was observed in Yuanjiang City with 27,586 patients, followed by Nan County, while the smallest decrease was found in Wangcheng County with 154 patients. The increase in the number of patients in Linli County was mainly due to the fact that, in this county, there is a large number of cattle and sheep, which increases the human and animal infection rate [58]. Figure 4 shows these temporal and spatial variations in the study area between 2005 and 2015.

4.2. Variations in Ecosystem Services between Different Land Use Types

Different types of land use showed different changes in the water yield. The amount of water yield per km^2 in unutilized, residential, arable, forest, and grassland areas had a growth of 63.83%, 32.86%, 28.19%, 23.25%, and 25.79%, respectively (Figure 5a). Moreover, it can be seen from this figure that the order of water yield per km^2 in different land use types was as follows: woodland > arable land > grassland > residential land > unutilized land > water area.

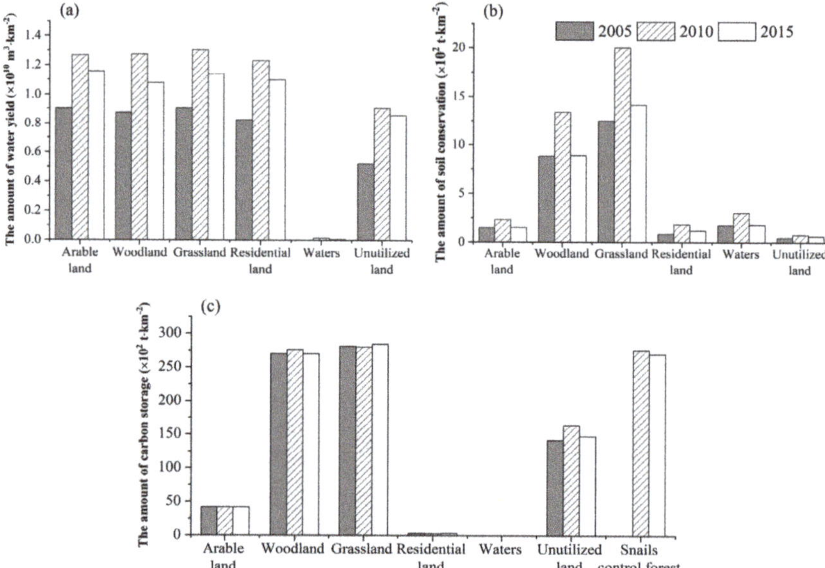

Figure 5. Ecosystem services per km^2 of different land cover types in the Dongting Lake Wetland from 2005 to 2015: the amounts of (**a**) water yield; (**b**) soil conservation; and (**c**) carbon storage.

The change in the total amount of conserved soil was different every five years. As shown in Section 4.1.2, it increased by 53.60% in the first five years of the study period, while it decreased by 34.88% in the second 5 years. Soil conservation per km^2 of different land use types from 2005 to 2015 is shown in Figure 5b. Among them, the residential, unutilized and grassland areas experienced an increase of soil conservation per km^2 by 38.74%, 37.40%, and 12.76%, respectively, while the others had a smaller increase. In summary, the soil conservation per km^2 in different land use types between 2005 and 2015 followed this order: grassland > woodland > water area > arable land > residential land > unutilized land.

As shown in Section 4.1.3, from 2005 to 2010, the total amount of carbon storage increased by 1.68%, and then a decrease of 1.93% was observed from 2010 to 2015. Figure 5c presents the changes in the carbon storage capacity per km^2 of different land use types for every five years during the study period. As can be seen, there has been an increase in unutilized land and grassland by 3.94% and 1.14%, respectively; and, a decrease of 4.87%, 0.15%, and 0.03%, respectively, in the residential, arable, and woodland areas. According to the study of Chen et al., (2012) [59] on the carbon storage in Dongting Lake area from 1979 to 2004, soil organic carbon was largely lost when grassland was converted into woodland due to the establishment of forests in the Wetland area. According to the present results, the order of the carbon storage was as follows: grassland > woodland > unutilized land > residential land > arable land.

4.3. Analysis of the Value of Ecosystem Services

Figure 6 presents the results of the analysis that was carried out on the value of the ecosystem services in the Dongting Lake Wetland while using the market value and shadow engineering methods. The value of these services was estimated to be 1249.84×10^8 CNY in 2005, of which water yield, carbon storage, soil conservation, and snail control and *schistosomiasis* prevention accounted for 85.49%, 13.87%, 0.64%, and 0.00%, respectively. In 2010, this value increased to 1723.85×10^8 CNY, with the rise in the value of water yield, soil conservation and snail control, and *schistosomiasis* prevention by 3.36%, 0.07%, and 0.23%, respectively, and a reduction of 3.67% in carbon storage. In 2015, the overall ecosystem value reduced to 1552×10^8 CNY. This was associated with decreases of 0.88% and 0.20% in the water yield and soil conservation, respectively, and growth of 0.93% and 0.15% in the values of carbon storage and snail control and *schistosomiasis* prevention.

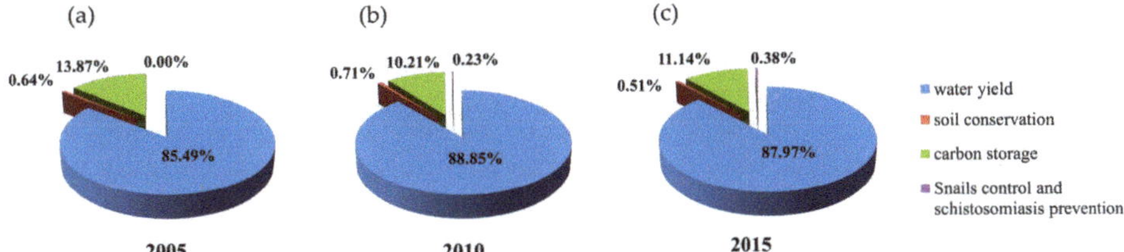

Figure 6. Values of different ecosystem services of the Dongting Lake Wetland in (**a**) 2005; (**b**) 2010; and (**c**) 2015.

Besides, it was observed that the values of different land use types per km^2 followed the order of snail control and *schistosomiasis* prevention forest > forest (non-snail control and *schistosomiasis* prevention) > arable land > residential land > unutilized land > water area. This can be seen in Figure 7, where the values per km^2 of different land cover types are presented for years 2005, 2010, and 2015.

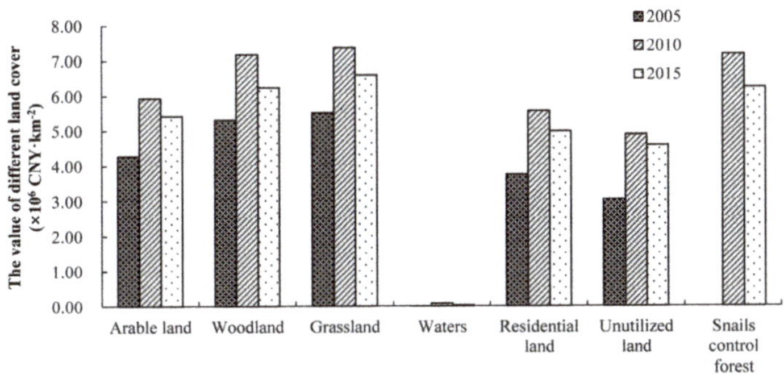

Figure 7. Values per km^2 of different land cover types.

In 2010, the value of the ecosystem service of snail control and *schistosomiasis* prevention forests in the Dongting Lake Wetland area was 54.33×10^8 CNY, which corresponded to the water yield, soil conservation, carbon storage, and snail control, and *schistosomiasis* prevention by 74.25%, 1.10%, 17.31%, and 7.34%, respectively. In 2015, this value reached 89.11×10^8 CNY. Only the carbon storage value increased from 2010 to 2015 (by 2.33%), but the value of the other factors reduced, as seen in Figure 8.

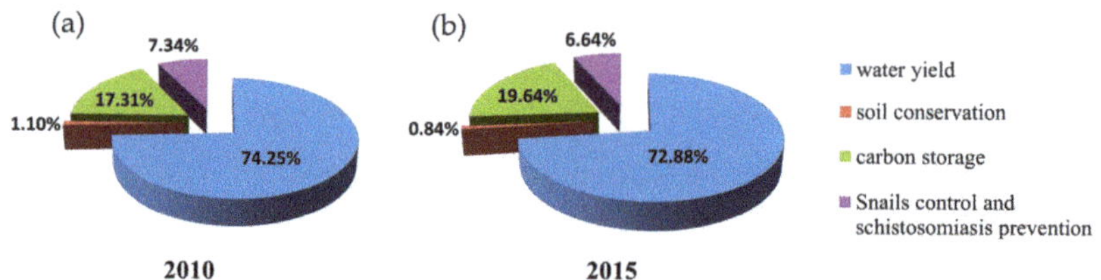

Figure 8. Values of different ecosystem services of snail control and *schistosomiasis* prevention forests in the Dongting Lake Wetland area in (**a**) 2010; and (**b**) 2015.

5. Discussion

5.1. Factors Affecting the Ecosystem Services

Previous studies have found that land use change, precipitation, evapotranspiration, and rainfall erosivity are the main mechanisms that affect ecosystem services [60]. We have carried out a correlation analysis of the ecosystem services with evapotranspiration, precipitation, soil erodibility, and rainfall erosivity while using Statistical Product and Service Solutions (SPSS) to investigate the effects of such factors in the present study. Table 3 presents the result, where it can be seen that, from 2005 to 2015, the ecosystem services show different degrees of correlation with these factors. Specifically, all of the factors significantly positively influenced water yield, soil conservation, and carbon storage; and, the rainfall erosivity positively influenced the reduced snail area at a moderate degree.

Table 3. Correlation analysis of ecosystem services and various influencing factors during 2005–2015.

Ecosystem Services	Influencing Factors			
	Evapotranspiration	Precipitation	Soil Erodibility	Rainfall Erosivity
Water yield	0.931 **	0.969 **	0.895 **	0.954 **
Soil conservation	0.803 **	0.840 **	0.841 **	0.847 **
Carbon storage	0.835 **	0.828 **	0.851 **	0.811 **
Reduced snail area	0.089	0.322	0.296	0.347 **
Reduced patient	−0.008	−0.05	−0.023	−0.073

Asterisks indicate significant differences at ** $p < 0.01$.

5.2. Selection of Appropriate Evaluation Indexes and Methods

Wetlands provide a variety of ecosystem services. The evaluation indexes of wetland ecosystem services are different due to different characteristics of wetlands. For example, Schallenberg et al. (2013) conducted ecosystem services evaluation on eight lakes in New Zealand, with an area of over 100 km^2 including water supply, fish supply, water purification, flood control, lakeside protection, climate regulation, and recreation, and discovered that they were in the trend of degradation [61]. Zhang et al. (2014) calculated the ecological service value of lakes and wetlands in China from three aspects of biodiversity, water quality, and economic indices. The results showed that the values of lake and marsh wetlands ecosystem services accounted for 54.64% of the total value of China's natural grassland ecosystem services and 30.34% of the total value of forest ecosystem services, respectively [62]. Sun et al. (2017) compared the different results of food security and biodiversity between China's Poyang Lake and Bangladesh's Tanguar Haor wetlands and showed that the trends in food security and biodiversity are declining [63]. Wondie (2018) selected nine wetlands in the region of the Lake Tana (Ethiopia) and estimated the value of wetland ecosystem services in terms of supply, regulation, culture, and support services. They showed that the wetland utilization planning, investment intervention, public land ownership, and other policy decision-making supports are poor, thus resulting in a major challenge to the sustainable use of wetland resources [64]. Mao et al. (2007) estimated the value of the ecological services in the Dongting Lake Wetland to be 411.10×10^8 CNY using the market value, shadow engineering, carbon tax, and alternative expenses methods, but without considering the effect of snail control and *schistosomiasis* prevention forests [40]. We observed that in 2015, this value was 1552×10^8 CNY, and the snail control and *schistosomiasis* prevention forests had the highest ecosystem service value among the different land use types. This is mainly due to the fact that such forests provide various benefits, such as water yield, soil conservation, and carbon storage. However, more importantly, they control and prevent *schistosomiasis*, which thereby reduces the cost of snail eradication and *schistosomiasis* treatment and decreases the income loss of *schistosomiasis* patients.

In summary, different assessment indicators are adopted for wetlands in different areas, and quite different results are often obtained. It can be concluded that there is no consensus on a singular best approach for the assessment of wetland ecosystem services, but the selection of appropriate assessment indicators and methods for a certain wetland is heavily dependent on the characteristics of the area where the wetland is located.

5.3. Limitation of the Present Study

Our study has some limitations, as follows. (1) This study only evaluated water yield, soil conservation, carbon storage, and snail control and *schistosomiasis* prevention functions in the evaluation of the ecosystem services of snail control and *schistosomiasis* prevention forests. However, other ecosystem services, such as the provision of wood and forest products, climate regulation, and air purification, which may have significant effects, were neglected. (2) The dynamic impact of afforestation years on the ecosystem services, such as the difference between an old-growth forest and a newly planted forest, was disregarded.

6. Conclusions

In this study, water yield, soil conservation, carbon storage, and snail control and *schistosomiasis* prevention were evaluated in the Dongting Lake Wetland area from 2005 to 2015. Their temporal and spatial changes were studied in detail, where different trends and variations were observed. Different land use types also showed various changes in these services. Subsequently, the value of the ecosystem services was estimated, and a total increase of 24.18% was observed over the 10-year period, where the water yield showed the highest value, which accounted for more than 70% of the total value of the four evaluated ecosystem services; while, the value of the snail control and *schistosomiasis* prevention was about 5% of it. Moreover, through a correlation analysis, it was shown that evapotranspiration, precipitation, soil erodibility, and rainfall erosivity significantly positively influenced some of the ecosystem services in the study area.

There is no singular best approach for the assessment of ecosystem services of wetlands, but the appropriate methods should be selected based on the characteristics of the area where the wetland is located. The potential improvements in the selection of evaluation index system and methods for a similar case to our study include considering the dynamic impact of afforestation years on the ecosystem services and taking the effect of other ecosystem services, such as the provision of wood and forest products, climate regulation, and air purification into account.

Author Contributions: Conceptualization: L.M., R.S., D.P.; methodology, software, validation and writing—original draft preparation: L.M.; writing—review, editing and revision: L.M., E.K., Y.Z., D.P., J.Z., K.Z.; funding acquisition: J.Z.; project administration: J.Z.; supervision: Q.S.

Funding: The National Science and Technology Support Program of China (grant No. 2015BAD07B07), and the National Natural Science Foundation of China (NSFC) (grant No. 41071334) supported this research.

Acknowledgments: We would like to thank Yilin Zhuang for her support and suggestions in the preparation of the manuscript.

Conflicts of Interest: Authors declare there is no conflict of interest.

References

1. Ouyang, Z.Y.; Wang, X.K.; Miao, H. A primary study on Chinese terrestrial ecosystem services and their ecological-economic values. *Acta Ecol. Sin.* **1999**, *19*, 607–613. (In Chinese)
2. Costanza, R.; D'Arge, R.; De Groot, R. The value of the world's ecosystem services and natural capital. *World Environ.* **1997**, *387*, 253–260. (In Chinese) [CrossRef]
3. Millennium Ecosystem Assessment (MEA). *Ecosystems and Human Well-Being: Biodiversity Synthesis*; Island Press: Washington, DC, USA, 2019.
4. Li, W.H. *Contemporary Ecological Study of China-Volume of Ecosystem Management*; Science Press: Beijing, China, 2013.
5. Wang, H.Q.; Zhang, C.H.; Zou, J.H. The game analysis of the protection of wetland resources. *Ecol. Econ.* **2011**, *235*, 174–178. (In Chinese)
6. Dugan, P. *Wetlands in Danger: A World Conservation Atlas*; Oxford University Press: New York, NY, USA, 1993.
7. Rudolf, S.G.; Wilson, M.A.; Boumans, R.J. A typology for the classification, description and valuation of ecosystem functions, goods and services. *Ecol. Econ.* **2002**, *41*, 393–408.
8. Costanza, R.; de Groot, R.; Sutton, P.; Ploeg, S.; Anderson, S.J.; Kubiszewski, I.; Farber, S.; Turner, R.K. Changes in the global value of ecosystem services. *Glob. Environ. Chang.* **2014**, *26*, 152–158. [CrossRef]
9. US Environmental Protection Agency. Great Lakes Restoration Initiative Task Force. In *Great Lakes Restoration Initiative Action Plan II Measures Reporting Plan*; 2017. Available online: https://www.glri.us/sites/default/files/glri-measures-reporting-plan-20170926-156pp.pdf (accessed on 12 October 2018).
10. Wong, C.P.; Jiang, B.; Bohn, T.J. Lake and wetland ecosystem services measuring water storage and local climate regulation. *Water Resour. Res.* **2017**, *53*, 3197–3223. [CrossRef]
11. Bikash, S.; Golam, R.; Nakul, C.; Sharma, B.; Rasul, G.; Chettri, N. The economic value of wetland ecosystem services: Evidence from the Koshi Tappu Wildlife Reserve, Nepal. *Ecosyst. Serv.* **2015**, *12*, 84–93.
12. Simon, G.S.; John, B.; Kate, S. Farmer perceptions of wetlands and waterbodies: Using social metrics as an alternative to ecosystem service valuation. *Ecol. Econ.* **2016**, *126*, 58–69.
13. Dearing, J.A.; Yang, X.; Dong, X. Extending the timescale and range of ecosystem services through paleoenvironmental analyses, exemplified in the lower Yangtze basin. *Proc. Natl. Acad. Sci. USA* **2012**, *109*, E1111–E1120. [CrossRef]
14. Li, C.; Zheng, H.; Li, S. Impacts of conservation and human development policy across stakeholders and Scales. *Proc. Natl. Acad. Sci. USA* **2015**, *112*, 7396–7401. [CrossRef]
15. Li, X.; Yu, X.; Jiang, L.; Li, W.; Liu, Y.; Hou, X. How important are the wetlands in the middle-lower Yangtze River region: An ecosystem service valuation approach. *Ecosyst. Serv.* **2014**, *10*, 54–60. [CrossRef]
16. Xu, M.; Dong, X.; Yang, X. Using palaeolimnological data and historical records to assess long-term dynamics of ecosystem services in typical Yangtze shallow lakes (China). *Sci. Total Environ.* **2017**, *584*, 791–802. [CrossRef] [PubMed]
17. Feng, J.G.; Ding, L.B.; Wang, J.S. Case-based evaluation of forest ecosystem service function in China. *Chin. J. Appl. Ecol.* **2016**, *27*, 1375–1382. (In Chinese)
18. Li, T.; Lü, Y.H. A review on the progress of modeling techniques in ecosystem services. *Acta Ecol. Sin.* **2018**, *38*, 5287–5296.
19. Silvestri, S.; Kershaw, F. *Framing the Flow: Innovative Approaches to Understand, Protect and Value Ecosystem Services Across Linked Habitats*; Environmental Policy Collection; UNEP World Conservation Monitoring Centre: Cambridge, UK, 2010; pp. 26–33.
20. Hu, H.T.; Fu, B.J.; Lü, Y.H.; Zheng, Z.M. SAORES: A spatially explicit assessment and optimization tool for regional ecosystem services. *Landsc. Ecol.* **2015**, *30*, 547–560. [CrossRef]
21. Sharp, R.; Tallis, H.T.; Ricketts, T.; Guerry, A.D.; Wood, S.A.; Chaplin-Kramer, R.; Nelson, E.; Ennaanay, D.; Wolny, S.; Olwero, N.; et al. *InVEST Version 3.2.0 User's Guide. The Natural Capital Project. The Nature Conservancy, and World Wildlife Fund*; Stanford University: Stanford, CA, USA; University of Minnesota: Minnesota, MN, USA, 2016.
22. Dai, E.F.; Wang, X.L.; Zhu, J.J.; Zhao, D.S. Methods, tools and research framework of ecosystem service trade-offs. *Geogr. Res.* **2016**, *35*, 1005–1016. (In Chinese)
23. Hoyer, R.; Chang, H. Assessment of freshwater ecosystem services in the Tualatin and Yamhill basins under climate change and urbanization. *Appl. Geogr.* **2014**, *53*, 402–416. [CrossRef]

24. Pessacg, N.; Flaherty, S.; Brandizi, L. Getting water right: A case study in water yield modelling based on precipitation data. *Sci. Total Environ.* **2015**, *537*, 225–234. [CrossRef]

25. Yu, X.X.; Zhou, B.; Lü, X.Z.; Yang, Z.G. Evaluation of water conservation function in mountain forest areas of Beijing based on InVEST model. *Sci. Silvae Sin.* **2012**, *48*, 1–5. (In Chinese)

26. Wilson, M.A.; Carpenter, S.R. Economic Valuation of Freshwater Ecosystem Services in the United States: 1971–1997. *Ecol. Appl.* **1999**, *9*, 772.

27. William, J.M.; James, G.G. *Wetlands*, 5th ed.; John Wiley & Sons Inc.: Hoboken, NJ, USA, 2015.

28. Wang, Y. Schistosomiasis epidemic in elation analysis between its prevention and control in China. *Chin. Prim. Health Care* **2010**, *24*, 69–72. (In Chinese)

29. Zhou, X.N.; Bergquist, R.; Leonardo, L. Schistosomiasis japonica control and research needs. *Adv. Parasitol.* **2010**, *72*, 145–178. [PubMed]

30. Steinmann, P.; Keiser, J.; Bos, R. Schistosomiasis and water resources development: Systematic review, meta-analysis, and estimates of people at risk. *Lancet Infect. Dis.* **2006**, *6*, 411–425. [CrossRef]

31. Ma, L.; Yang, X. Ecosystem services of forests for snail control and schistosomiasis prevention in the Yangtze River Basin. *J. Zhejiang A F Univ.* **2019**, *36*, 133–140. (In Chinese)

32. WHO. Resolution WHA 65.21. Elimination of schistosomiasis. In Proceedings of the Sixty-fifth World Health Assembly, Geneva, Switzerland, 21–26 May 2012.

33. Zhang, L.J.; Shen, D.W. The evolution of schistosomiasis control strategy in China. *J. Hubei Univ. Sci. Technol.* **2013**, *27*, 457–460. (In Chinese)

34. Yang, X.; Zhang, Y.; Sun, Q.X. SWOT analysis on snail control measures applied in the national schistosomiasis control programme in the People's Republic of China. *Infect. Dis. Poverty* **2019**, *8*, 13. [CrossRef]

35. Liu, Z.C.; He, H.B. Recent advances in control techniques on Oncomelania hupensis in lake and marshland regions. *Chin. J. Vector Biol. Control* **2015**, *26*, 537–540. (In Chinese)

36. Guan, W.; Hong, Q.B. Research progress of control techniques on Oncomelania hupensis. *Chin. J. Schist. Control* **2017**, *29*, 246–251. (In Chinese)

37. Shi, X.D. Research progress of ecological snail extermination technology. *J. Trop. Dis. Parasitol.* **2018**, *16*, 117–121. (In Chinese)

38. Peng, Z.H.; Jiang, Z.H. *China's New Type of Forest Studies on the Snails Control and Schistosomiasis Prevention Forest*; China Forestry Press: Beijing, China, 1995. (In Chinese)

39. Yang, X.; Sun, Q.X. Discussion on strategy of development of forestry schistosomiasis control programs of China in new period. *China. J. Schist. Control* **2018**, *30*, 472.

40. Mao, D.H.; Wu, F.; Li, J.B. Evaluation on Ecosystem Service Value of Dongting Lake Wetland and Ecological Restoration Counter measures. *Wetl. Sci. Manag.* **2007**, *5*, 39–44. (In Chinese)

41. Tang, Y.X.; Wu, M. Effects of Lakeside Plantations on Schistosomiasisi Control. *Wetl. Sci. Manag.* **2006**, *2*, 8–13.

42. Peng, Z.H. Mechanism of Forestry Schistosomiasis-Control Project on Snail Control and Schistosomiasis Prevention. *Wetl. Sci. Manag.* **2003**, *9*, 8–10. (In Chinese)

43. Resource and Environment Data Cloud. Available online: http://www.resdc.cn (accessed on 2 July 2018).

44. National Meteorological Information Center. Available online: http://data.cma.cn (accessed on 10 July 2018).

45. Geospatial Data Cloud. Available online: http://www.gscloud.cn (accessed on 2 July 2018).

46. Cold and Arid Regions Science Data Center at Lanzhou. Available online: http://westdc.westgis.ac.cn (accessed on 15 September 2018).

47. Li, K.R.; Wang, S.Q.; Cao, M.K. Carbon storage of vegetation and soil in China. *Sci. China* **2003**, *33*, 72–80. (In Chinese) [CrossRef]

48. Wang, Y.R.; Zhou, J.X.; Zhou, Z.X.; Sun, Q.X. Effects of Different Land Use Patterns under Converting Polders back into Wetlands on Soil Nutrient Pools in the Dongting Lake Region. *Resour. Environ. Yangtze Basin* **2010**, *19*, 634–639. (In Chinese)

49. Tallis, H.T.; Ricketts, T.; Guerry, A.D.; Wood, S.A.; Sharp, R.; Nelson, E.; Ennaanay, D.; Wolny, S.; Olwero, N.; Vigerstol, K.; et al. *InVEST 2.2.1 User's Guide*; The Natural Capital Project: Stanford, CA, USA, 2011.

50. Allen, G.; Pereira, L.S.; Raes, D. *Crop Evapotranspiration: Guidelines for Computing Crop Water Requirements*; FAO: Rome, Italy, 1998.

51. Wischmeier, W.H.; Smith, D.D. Rainfall energy and its relationship to soil loss. *Trans. Am. Geophys. Union.* **1958**, *39*, 285–291. [CrossRef]

52. Williams, J.R.; Arnold, J.G. A system of erosion—Sediment yield models. *Soil Technol.* **1997**, *11*, 1143–1155. [CrossRef]

53. Roose, E. *Land Husbandry: Components and Strategy*; FAO Soils Bulletin: Rome, Italy, 1996.

54. Renard, K.G.; Foster, G.R.; Weesies, G.A. *Predicting Soil Erosion by Water: A Guide to Conservation Planning with the Revised Universal Soil Loss Equation (RUSLE)*; USDA Agriculture Handbook: Washington, DC, USA, 1997; Volume 703, pp. 1–367.

55. Jia, F.F. InVEST Model Based Ecosystem Services Evaluation with Case Study on Ganjiang River Basin. Ph.D. Thesis, China University of Geosciences, Beijing, China, 2014. (In Chinese).

56. Tao, H.Y.; Xia, A.; Zhao, Y.M. Effect and cost-benefit of Oncomelania snail control by plowing and palnting in Jiaobei Beach of Zhengjiang City. *Chin. J. Schist. Control* **2012**, *24*, 576–578. (In Chinese)

57. Hu, B.J.; Zhao, Z.Y.; Xia, M. Study on spatial—Temporal characteristics of Schistosoma japonicum infections among human in Hunan Province, 2004–2011. *Chin. J. Schist. Control* **2017**, *29*, 406–411.

58. Wang, Y.M.; Wang, X.Y.; Su, P.C. Terrospective investigation on the schistosomiasis status in Linli County. *J. Trop. Dis. Parasitol.* **2016**, *14*, 204–207. (In Chinese)

59. Chen, W.W.; Zhang, J.X. Change rules of soil carbon storage in Dongting lake region within 20 years and its influencing factors. *J. Hunan Inst. Eng.* **2012**, *22*, 88–94. (In Chinese)

60. Wang, F.; Zhang, S.L.; Hou, H.P.; Yang, Y.J.; Gong, Y.L. Assessing the Changes of Ecosystem Services in the Nansi Lake Wetland, China. *Water* **2019**, *11*, 788. [CrossRef]

61. Schallenberg, M.; Winton, D.; Verburg, P. Ecosystem services of lakes. In *Ecosystem Services in New Zealand: Conditions and Trends*; Manaaki Whenua Press: Lincoln, New Zealand, 2013; pp. 203–225.

62. Zhang, Y.; Zhou, D.; Niu, Z. Valuation of lake and marsh wetlands ecosystem services in China. *Chin. Geogr. Sci.* **2014**, *24*, 269–278. [CrossRef]

63. Sun, C.; Zhen, L.; Miah, M.G. Comparison of the ecosystem services provided by China's Poyang Lake wetland and Bangladesh's Tanguar Haor wetland. *Ecosyst. Serv.* **2017**, *26*, 411–421. [CrossRef]

64. Wondie, A. Ecological conditions and ecosystem services of wetlands in the Lake Tana Area, Ethiopia. *Ecohydrol. Hydrobiol.* **2018**, *18*, 231–244. [CrossRef]

Article

Study on Multi-Scale Coupled Ecological Dispatching Model Based on the Decomposition-Coordination Principle

Tao Zhou *, Zengchuan Dong, Wenzhuo Wang, Rensheng Shi, Xiaoqi Gao and Zhihong Huang

Hydrology and Water Resources College, Hohai University, Nanjing 210098, China
* Correspondence: 160401010016@hhu.edu.cn

Received: 9 May 2019; Accepted: 9 July 2019; Published: 12 July 2019

Abstract: Studies on environmental flow have developed into a flow management strategy that includes flow magnitude, duration, frequency, and timing from a flat line minimum flow requirement. Furthermore, it has been suggested that the degree of hydrologic alteration be employed as an evaluation method of river ecological health. However, few studies have used it as an objective function of the deterministic reservoir optimal dispatching model. In this work, a multi-scale coupled ecological dispatching model was built, based on the decomposition-coordination principle, and considers multi-scale features of ecological water demand. It is composed of both small-scale model and large-scale model components. The small-scale model uses a daily scale and is formulated to minimize the degree of hydrologic alteration. The large-scale model uses a monthly scale and is formulated to minimize the uneven distribution of water resources. In order to avoid dimensionality, the decomposition coordination algorithm is utilized for the coordination among subsystems; and the adaptive genetic algorithm (AGA) is utilized for the solution of subsystems. The entire model—which is in effect a large, complex system—was divided into several subsystems by time and space. The subsystems, which include large-scale and small-scale subsystems, were correlated by coordinating variables. The lower reaches of the Yellow River were selected as the study area. The calculation results show that the degree of hydrologic alteration of small-scale ecological flow regimes and the daily stream flow can be obtained by the model. Furthermore, the model demonstrates the impact of considering the degree of hydrologic alteration on the reliability of water supply. Thus, we conclude that the operation rules extracted from the calculation results of the model contain more serviceable information than that provided by other models thus far. However, model optimization results were compared with results from the POF approach and current scheduling. The comparison shows that further reduction in hydrologic alteration is possible and there are still inherent limitations within the model that need to be resolved.

Keywords: ecological operation; multi-scale; decomposition-coordination; hydrologic alterations

1. Introduction

Reservoir ecological operation has become increasingly more significant and has recently drawn a lot of attention from research scientists in numerous scientific disciplines. By 2005, there were >45,000 reservoir dams higher than 15 m throughout the world, which are collectively capable of storing ~15% of annual global runoff [1]. While these reservoirs make a sizeable contribution to developing the economy, they are also the source of numerous ecological problems. The ultimate goal of ecological operation is to integrate ecological factors into standard operation objectives, such that ecological protection is given the same consideration as increasing profit margins. Essentially, the target of ecological operation is to mediate the conflict between society's demand and the river ecosystem's needs [2]. As early as 1971, Schlueter put forward that reservoir management should maintain river diversity [3]. Following that pronouncement, a number of river ecological water demand calculation methods were developed—i.e., the Tennant Method, Wetted Perimeter Method, and IFIM Method, etc. [4]. Since then, increasing discharge from reservoirs to reserve enough ecological flow has become a common method of ecological operation [5–7].

Progressively more in-depth research on the riverine ecosystems has widened our understanding of the river system itself and the ecosystem it supplies. Thus, river ecological water demand no longer exclusively concerns the reserved water downstream. Many scholars have reached a broad consensus that riverine ecosystems adapt to the natural river flow pattern [8], and every river has its own characteristic flow pattern and corresponding biological community [9]. Thus, any flow pattern change—such as flood runoff, flow velocity, arrival time, and flood pulse frequency, etc., can cause a series of ecological consequences. As a result of these insights, ecological water requirements have been transformed from merely considering reserved flow [10,11] to considering the full extent of natural flow to which riverine biological communities must adapt [11–13]. Based on the above consensus, Richter et al., 1997 suggested implementing the Range of Variability Approach (RVA), which uses the Indicator of Hydrologic Alteration (IHA) to quantify how much the reservoir operation alters the riverine ecosystem. Homa et al. [11] presented the concept of ecological deficit and proposed minimizing ecological deficit as an objective of reservoir operation. Richter [14] offered a definition of sustainable water management called "Sustainability Boundary Approach (SBA)". Francisco et al. used SBA to assess and manage three major river basins in Spain with good results [15,16]. Typical reservoir operating rule curves (RORCs) [17] were improved by using RVA as evaluation indicators [18,19].

While it is now recognized that a comprehensive evaluation of ecological disruption requires an investigation over the entire river system, the water need in river ecosystems must also be considered on multiple scales. The minimum ecological discharge is mainly reflected on a monthly scale, but pulse flow and ecological flow velocity are reflected on smaller scales. As such, studies that are restricted to any single scale are inherently missing important ecological information. That being said, the large scales used for current research on reservoir ecological operation are too high-level to describe ecosystem demand with more than an overview, as flow pattern and flood pulse, etc., are too specific to be considered.

Given that the ecological water demand is present on multiple scales, any ecological dispatching model needs to include various scales. Steinschneider et al. [20] developed a large-scale optimization model to evaluate the contribution of reservoir management measures to ecological benefits; and Tsai et al. [21] proposed a new hybrid method to quantify river ecosystem demand based on artificial intelligence. Clearly, the importance of a large-scale optimization model and small-scale characteristic flow regimes has been recognized, but few studies couple them together in a feasible and practical way. Although the multi-scale dispatching model covers more ecological information, the number of model decision variables dramatically increases accordingly. For example, the number of decision variables in this study is nearly 3000. Increasing the number of model decision variables and the non-linear forms of hydrologic alteration makes the overall system model difficult to solve and even harder to understand. Furthermore, current studies usually employ larger step sizes (such as ten-day period) and take scheduling rules parameters as decision variables [17], but they fail to acquire all the characteristics of daily flows. Finally, the large-scale dispatching model is unable to be learned by machine-learning algorithms. In order to overcome these challenges, a hierarchical control method of the decomposition-coordination principle was applied in this study. The large system decomposition coordination theory is a new branch of science that has developed rapidly since the 1970s. In its most fundamental form, the theory proposes decomposing a complicated system into several independent subsystems and handling the correlation between subsystems by coordinators. In 1984, Adigüzel [22] decomposed a complicated system into several subsystems according to the objective function based on the decomposition-coordination principle. Currently, the decomposition-coordination principle is one of the most commonly used methods for solving complex systems. Over time, as the decomposition-coordination principle was more frequently applied, a multi-layer hierarchical control system was established. In the multi-layer hierarchical control system, the superior subsystem result is not only the boundary but is also harmonized with the result of the subordinate subsystem.

As stated above, researchers acknowledge the multi-scale features of ecological water demand [23,24]; yet few studies have been performed on the multi-scale coupled ecological dispatching model. Thus, in order to obtain a more reasonable deterministic optimization model, a multi-scale coupled ecological dispatching model based on the decomposition-coordination principle was built in this study. The scheduling model consists of large-scale and small-scale models, which use the month and day, respectively, as the calculation scale. The large-scale model was implemented for solving problems associated with water supply and river ecological base flow; while the small-scale model was used for solving problems related to extremely low flows, flood pulses, and ecological hydraulic parameters. Due to the introduction of RVA, disturbances are kept within reasonable bounds, which is beneficial to the river ecosystems. Furthermore, the simulative daily flows can be applied to the training of ecological regulations by a machine learning algorithm, which is important to the implicit stochastic optimization. Investigations that take scheduling rules parameters as decision variables or use larger step sizes cannot obtain valid daily flow data. The multi-scale coupled model also provides a framework for the formulation of multi-scale scheduling rules. This model can help decision-makers to determine whether there is still room for improvement in ecological scheduling and look for ways to improve.

2. Materials and Methods

2.1. Generation and Evaluation of Characteristic Flow Regimes

The importance of natural flow regime to the riverine ecosystems has been widely accepted. Richter et al. [4,25] put forward IHA and RVA indicators to quantify the degree of flow regime alteration. Consequently, the effect of reservoir ecological operation can be quantified. Mathews and Richter [26] suggested that flows can be divided into five categories: small floods (flows ≥ the overbank flow, but <10 year flood), high-flow pulses (flows < the overbank flow, but > seasonal base flow), low flows (base flows in different months), extreme low flows (flows ≤ 95th percentile flow) and large floods (flows ≥ 10 year flood). Yin et al. [19] combined small floods and large floods into one flood category. Furthermore, to consider additional physical mechanisms, we take into account the hydraulic parameters that do not belong to the IHA in this study.

In order to build a comprehensive multi-scale coupled ecological dispatching model, the flow regime alteration level must be quantified, and the approaches for generating the characteristic flow regime must also be explained. In this study, all environmental flow components were accounted for by considering four kinds of characteristic flow regimes: base flows, extreme low flows, flood pulse with high flow pulses, and ecological hydraulic parameters. The generation mechanisms for these characteristic flow regimes are based on their ecological significance.

2.1.1. Base Flows

Ecological functions of base flows include maintaining suitable water quality, sustaining hydraulic habitats for aquatic organisms, and supporting hyporheic organisms [19]. Thus, maintaining base flow in a river is very important for sustaining the riverine ecosystem. Based on the ecological functions of base flow, the full spectrum of base flow magnitudes was selected as evaluation indicators in the multi-scale coupled ecological dispatching model.

Depending on the dates in question, a method (e.g., Tennant or 7Q10 methods, etc.) was selected and implemented to calculate the base flows. The base flows were then used as constraint conditions in the large-scale scheduling model, which has an account step of 1 month. Merging the large- and small-scale factors guaranteed that monthly discharge ≥ base flows.

With respect to the model's objective functions, RVA was used to quantify the degree of deviation between the discharged flow's base flow and the natural flow regime. In this study, the daily flow from 1958 to 1977 were used as the reference series—as the values closely reflect natural runoff; 2007–2014 was selected as the scheduling period. In order to avoid ecological flow flattening, the full spectrum of base flows within the reference period was divided into three target ranges using the 25th and 75th percentiles as parameter values. The parameters in every target range were preserved. The corresponding target ranges were integrated into the objective functions and mathematically expressed as follows:

$$D_i = \sum_{La=1,2,3} \left| \frac{N_i(La) - N_e(La)}{N_e(La)} \right| \tag{1}$$

where $La = 1, 2, 3$ represents the high, middle and low target ranges, respectively; D_i refers to the degree of change of the i-th index; $N_i(La)$ refers to the number of years in which the i-th index is observed in the corresponding La target ranges; $N_i(La)$ refers to the number of years in which the i-th index is expected to be observed in the corresponding La target ranges which were calculated according to the reference series.

As the number of parameters from the simulation period that are observed in the target ranges increases, the degree of hydrologic alteration decreases in parallel.

2.1.2. Extreme Low Flows

Extreme low flows concentrate prey species for predators, provide habitat for terrestrial animals, and purge exotic species. At the same time, the extremely low flows can reduce the connectivity of rivers and confine the movement of certain aquatic organisms.

For ecological reasons, the magnitude, frequency, duration, and timing of extremely low flows were taken into consideration in the small-scale scheduling model. As with base flows, the full spectrum of the parameters (the magnitude, frequency, duration, and timing) was grouped into three target ranges using the 25th and 75th percentiles as parameter values. The mathematical expression is identical to that shown in Formula (1). Just as with the base flows, the more parameters from the simulation period that are observed in target ranges, the lower the degree of hydrologic alteration.

2.1.3. Flood Pulses and High Flow Pulses

Junk [27] introduced the Flood Pulse Concept (FPC). Flood pulse is the primary driving force behind biological survival, productivity, and interaction of the river-flood plain system. The flood pulse's activity has the following ecological significance: (1) It promotes energy exchange and material circulation between aquatic species and terrestrial species; (2) it improves the dynamic connectivity of river systems, which provide shelter and habitat for different species; and (3) it transmits abundant and intense information to species, which is important for spawning, hatching, growth, refuge, and other migration activities.

Flood pulse and high flow pulse generation mechanisms are similar; they differ in threshold values. Using flood pulse as an example, in order to represent its ecological purposes, the magnitude, duration, timing, and frequency are considered in the small-scale scheduling model. Thus, the flood pulse generation mechanism is as follows:

When the control section flow exceeds a certain limit, a flood pulse incident is triggered.

$$Q_T > T \tag{2}$$

where T is a pre-set limit value that depends on the stage-discharge relationship and the control section shape; and Q_T is the control section flow.

When the control section flow decreases below a certain limit, the flood pulse incident ends. In the small-scale scheduling model, a flood pulse is induced by artificial floods due to the release flow from reservoirs. Thus, reservoirs need to control the release to compensate for the flood pulse as follows.

$$Q_P = Q_R - Q_T \qquad (3)$$

where Q_P represents the released flow from reservoirs; Q_R represents the control section target flow; and Q_T represents the control section flow.

Since flood pulse incidents usually last for a few days or less, conventional reservoir scheduling models with 1 month account steps, cannot incorporate the requirement. Thus, the magnitude, frequency, timing, and duration of all flood pulse incidents during the ecological critical period were evaluated in the small-scale scheduling model. The mathematical expression is the same as that shown in Formula (1).

2.1.4. Ecological Hydraulic Parameters

Using hydraulic methods to calculate ecological water demand is both prominent and important. Hydraulic methods are based on hydrodynamic parameters as well as specific biological criteria. Conventional hydraulic methods include the wet perimeter method and R2-CROSS [7,28–30]. Liu et al. [31] introduced the concept of ecological flow velocity, which combines hydrodynamic parameters and specific ecological targets.

Ecological flow velocity differs from the above-mentioned characteristic flow regimes in that the ecological hydraulic parameters reference standard is based on both the natural flow regime and species requirements. Because ecological flow velocity reflects the ecological target requirement, it is expected to vary within a certain range. In addition, specific water depths—which will differ depending on the local channel morphology and the local ecology—are essential for maintaining the basic ecological function of the ecosystem. In order to fully incorporate the biological requirements and hydrodynamic parameters, this study accounts for the appropriate hydraulic parameters, such as the depth and flow velocity of specific species in different growth cycles. The selected species primarily represent those that dominate the riverine ecosystem. For example, the growth cycle of *Gymnocypris eckloni herzenstein*—a keystone species found upstream of the Yellow River—was divided into three stages: spawn, brood, and growth. Liu [32] determined that maintaining a larger flow velocity during the spawning period stimulates spawning. In contrast, slow to moderate flow velocity is optimal for the brood or growth stage. In terms of depth, Xu [33] established that fish require a minimum water depth more than three times of their length to survive (See Table 1).

Table 1. Gymnocypris eckloni herzenstein.

	Length/cm	**Suitable Flow Velocity/(m/s)**	**Critical Flow Velocity/(m/s)**
Gymnocypris	20–25	0.3–0.8	1.00
eckloni herzenstein	25–35	0.3–0.8	1.10

Table 1 shows ecological parameters of the *Gymnocypris eckloni herzenstein*. This data is from the Advanced Aquatic Biology [32].

Flow was related to the hydraulic parameters via an open channel uniform flow theory. The conventional reservoir scheduling models were unable to meet the requirement because the hydraulic parameters are sensitive to flow rate. Therefore, the small-scale scheduling model uses a day scale as the account step. The ecological hydraulic parameters evaluation function is the same as that shown in Formula (1). However, in this case, the target range is based on the natural flow and species requirements. The more hydraulic parameters, from within the corresponding natural flow values that are observed in the suitable ranges, the lower the degree of hydrologic alteration that will be obtained.

The descriptive parameters of the four discussed characteristic flow regimes are summarized in Table 2:

Table 2. Characteristic flow regimes.

Characteristic Flow Regimes	Descriptive Parameters	Hierarchy of Model
Base Flows	Magnitude of base flows	Large-scale
Extreme Low Flows	Magnitude, frequency, duration, and timing of extremely low flows	Small-scale
Flood Pulses and High Flow Pulses	The magnitude, frequency, duration, and timing of flood and high flow pulses	Small-scale
Ecological Hydraulic Parameters	The hydraulic parameters for specific species in different growth cycles	Small-scale

Table 2 shows descriptive parameters of characteristic flow regimes and which model will handle them.

In this study, the characteristic flow regimes' degree of alteration was averaged and used as the degree of hydrologic alteration.

2.2. Construction of the Multi-Scale Coupled Ecological Dispatching Model

Two key problems with the multi-scale coupled model include: (1) increased decision variables and (2) coordination between the different scale models. In order to resolve these issues, the decomposition-coordination principle and multiple hierarchical control system, which is based on the theory of large-scale systems, were applied to this work.

2.2.1. The Space and Time Decomposition-Coordination Method

To resolve the issue of increased decision variables due to the introduction of small-scale scheduling models, the whole basin scheduling model for the entire scheduling period was divided into numerous subsystems, each of which includes part basin and part scheduling period. Because the whole basin scheduling model is regarded as a large complex system, the connection between the overall system and subsystems is reflected by the subsystems' objective functions, which include Lagrange multipliers. In this way, the large system optimization problem is transformed into several optimization problems of the subsystem objective functions. This method greatly reduces decision variables while retaining the model's optimal conditions [34]. With respect to space, the large-scale system is usually divided by reservoir control basins. In terms of time, the entire scheduling period is divided into small periods according to the model requirements. In general, subsystems coordinate with each other via subsystem hydraulic connections and cooperative work between subsystems—i.e., the interconnection constraints. The subsystem is mathematically expressed as follows:

$$L = \sum_{i=1}^{n} L_i = \sum_{i=1}^{n} f_i(x_i) + \lambda \left[\sum_{i=1}^{n} g_i(x_i) - b \right] + \mu \left[\sum_{i=1}^{n} h_i(x_i) - d \right]$$
$$\sum_{i=1}^{n} g_i(x_i) = b \tag{4}$$
$$\sum_{i=1}^{n} h_i(x_i) \geq d$$

where L represents the large system Lagrange function; L_i represents the Lagrange function of the i-th subsystem; n represents the number of subsystems; x_i represents the decision vector of the i-th subsystem (in this study, it represents water released from reservoirs); $f_i(x_i)$ represents the evaluation function of the i-th subsystem; $\sum_{i=1}^{n} g_i(x_i) = b$ is the equality constraint between subsystems; $\sum_{i=1}^{n} h_i(x_i) \geq d$ is the inequality constraint between subsystems; λ is the equality Lagrange multiplier constraint; and μ is the inequality Kuhn-Tacker multiplier constraint.

As shown in Equation (4), the equality constraint and inequality constraint are frequently used for constraining water balance and constraining cooperative work, respectively.

In this work, the subsystems' optimal solutions were solved using an adaptive genetic algorithm (AGA). The coordination between subsystems was implemented by means of coordinating variable iterations (Lagrange multiplier or Kuhn-Tacker multiplier) within the coordination level. The mixing method, which includes the interaction balance method and interaction prediction method, was employed for solving coordinating variables.

2.2.2. Coordination between Multi-Scale Subsystems

It is clear that the subsystems' hydraulic connections and cooperative work between subsystems are often regarded as interconnection constraints between subsystems of the same scale. The multi-scale coupled ecological dispatching model accounts for the coordination between different scale subsystems. At first, the water balance constraint is the only interconnection constraint in the multiple hierarchical control system:

$$\sum_{t=t_0}^{t_f} x_s(t) = x_i \tag{5}$$

where t_0 and t_f are the start and finish time of subordinate subsystems, respectively (smaller scale); x_s represents the subordinate subsystem decision variable; and x_i is the result of superior subsystem (larger scale).

As shown in Equation (5), the sum of the discharge flows in the small-scale model is equal to the discharge flow in the large-scale model for the corresponding period. Clearly, the water balance constraint is an equality constraint. Therefore, a new Lagrange multiplier, λ_s, was introduced to regulate the water balance constraint. Usually, the Lagrange equation L is used as the whole basin scheduling model objective function. Thus, the mathematical expression of the multiple hierarchical control system is as follows:

$$L = \sum_{i=1}^{n} \left\{ f_i(x_i) + \sum_{t=t_0}^{t_f} f_s(x_s) + \lambda_s \left[\sum_{t=t_0}^{t_f} x_s(t) - x_i \right] \right\} + \lambda \left[\sum_{i=1}^{n} g_i(x_i) - b \right] + \mu \left[\sum_{i=1}^{n} h_i(x_i) - d \right]$$

$$\sum_{i=1}^{n} g_i(x_i) = b$$

$$\sum_{i=1}^{n} h_i(x_i) \geq d \tag{6}$$

$$\sum_{t=t_0}^{t_f} x_s(t) = x_i$$

where λ_s is the Lagrange multiplier that regulates the water balance constraint between subsystems of different scales; $f_s(x_s)$ represents the subordinate subsystem's (smaller scale) evaluation function belonging to the i-th subsystem. For our purposes, b represents the vector of the initial water amount in the reservoir at each time; d represents the vector of common water supply mission between reservoirs. The other symbols are the same as those defined in Equations (4) and (5).

As demonstrated in the above formulas, subsystems of different scales have different evaluation functions. However, solving a hierarchical control model with different objective functions for different layers is very difficult. As such, in this work, large-scale subsystem objective functions were converted to the subordinate subsystem's constraint conditions. In this way, objective functions of the large-scale subsystems were implemented by the subordinate subsystems. Unlike common multi-scale models that use results of the superior subsystem merely for subordinate subsystem boundary conditions, multi-scale subsystems—which are based on the theory of large-scale systems—consider coordination between different scale subsystems via an iteration of the coordination variables (Lagrange multiplier or Kuhn-Tacker multiplier).

The implication of coordination variables is deduced by the following interaction prediction method:

$$\text{Let}: \frac{\partial L}{\partial \lambda_s} = \sum_{t=t_0}^{t_f} x_s(t) - x_i = 0 \tag{7}$$

In order to satisfy Karush–Kuhn–Tucker conditions [35]:

$$\frac{\partial L}{\partial x_i} = \frac{\partial f_i}{\partial x_i} - \lambda_s + \lambda \frac{\partial g_i}{\partial x_i} + \mu \frac{\partial h_i}{\partial x_i} = 0 \tag{8}$$

$$\frac{\partial L}{\partial [\sum\limits_{t=t_0}^{t_f} x_s(t)]} = \frac{\partial [\sum\limits_{t=t_0}^{t_f} f_s(x_s)]}{\partial [\sum\limits_{t=t_0}^{t_f} x_s(t)]} + \lambda_s = 0 \tag{9}$$

Thus:

$$\lambda_s = \frac{\partial f_i}{\partial x_i} + \lambda \frac{\partial g_i}{\partial x_i} + \mu \frac{\partial h_i}{\partial x_i} = -\frac{\partial [\sum\limits_{t=t_0}^{t_f} f_s(x_s)]}{\partial x_i} \tag{10}$$

As shown in the above formula, the large system global optimal solution is obtained exclusively at the moment when the derivative of the subordinate subsystem to the decision variable and the derivative of the superior subsystem to the decision variable are equal.

2.2.3. Construction of the Subsystems

After a coordination level was assigned for the coordination variables, the subsystem was converted into a relatively simple problem. Hence, AGA was used to mathematically solve the subsystem. The subsystem's objective function is written as follows:

$$L_i = f_i(x_i) - \alpha \lambda_s x_i + \lambda_i g_i(x_i) + \mu_i h_i(x_i) \tag{11}$$

where L_i represents the Lagrange function of the i-th subsystem; $f_i(x_i)$ represents the evaluation function of the i-th subsystem; α is a 0-1 variable, which is used as a subordinate subsystem structural discrimination coefficient, 1 and 0 denote whether the i-th subsystem has a subordinate subsystem or not, respectively; λ_i and μ_i refer to the corresponding coordination variable components of the i-th subsystem; and the other symbols are the same as those defined in Equations (4) and (5).

The mathematical expression of $f_i(x_i)$ is as follows:

$$f_i(x_i) = (1-\alpha)\left(\sum_{t=1}^{T} \frac{W - x_i}{W} + \sum_{j=1}^{M} D_j\right) + \alpha \sum_{j=1}^{N} D_j \tag{12}$$

where W refers to the subsystem water demand; x_i refers to the subsystem water supply; D_j refers to the degree of change of the j-th index; M and N represent the number of superior subsystem and subordinate subsystem indicators, respectively; and the other symbols are the same as those defined in Equations (4) and (5).

2.3. A Solution of the Multi-Scale Coupled Ecological Dispatching Model

The gradient method was used to mathematically resolve the coordination level. The coordination variables were all initialized to zero; and the subsystem's calculation results were passed to the coordination level. The iterative formulas are as follows:

$$\sum_{i=1}^{n} g_i(x_i) = b \tag{13}$$

$$\sum_{t=t_0}^{t_f} x_s(t) = x_i \tag{14}$$

$$\Lambda_s{}^{j+1} = -\frac{\partial[\sum_{t=t_0}^{t_f} f_s(x_s)]}{\partial x_i} \tag{15}$$

$$\mu_i{}^{j+1} = \mu_i + [h_i(x_i) - d] * R \tag{16}$$

$$\lambda_i{}^{j+1} = \frac{\partial f_{i+1}}{\partial x_i} - \Lambda_{s_{i+1}} + \lambda_{i+1} + \mu_{i+1} \tag{17}$$

where the symbol superscripts represent the number of iterations; unmarked symbols represent symbols of the j-th iteration; symbol subscripts represent the number of subsystems; R refers to the account step of the coordination variables; $\frac{\partial f_{i+1}}{\partial x_i}$ represents the derivative of the objective function of the i + 1-th subsystem to initial water of this subsystem (equal to surplus water of the last subsystem); $\frac{\partial[\sum_{t=t_0}^{t_f} f_s(x_s)]}{\partial x_i}$ represents the derivative of the objective function of the inferior subsystem to the superior subsystem output; and the other symbols are the same as those defined in Equations (4)and (5).

After the calculation, the coordination variables were passed to the subsystems until the convergence condition was met. The coordination variables' convergence conditions are expressed as:

$$\left|\Lambda_s{}^{j+1} - \Lambda_s{}^{j}\right| \leq \varepsilon_{\Lambda_s} \tag{18}$$

$$\left|\mu_i{}^{j+1} - \mu_i{}^{j}\right| \leq \varepsilon_{\mu_i} \tag{19}$$

$$\left|\lambda_i{}^{j+1} - \lambda_i{}^{j}\right| \leq \varepsilon_{\lambda_i} \tag{20}$$

where $\varepsilon_{\Lambda_s}, \varepsilon_{\mu_i}, \varepsilon_{\lambda_i}$ refer to the coordination variables' convergence thresholds.

Unlike the traditional targets, the derivative of the ecological objective function to initial water is difficult to calculate. Thus, in this work, a differential response model between the ecological objective function and initial water of a small-scale subsystem was established. The results were calculated in advance for coordination level.

2.4. Description of the Study Area

The lower reaches of the Yellow River (Figure 1) refer to the Yellow River stream segment below Peach Blossom Valley. At this location, the river length is 786 km and the drainage area covers 23,000 km^2. Furthermore, the stream gradient is small and the river's ecological problems are serious. The main water conservancy projects within this area are the Xiaolangdi reservoir and the Sanmenxia reservoir; which effectively store 5.1 billion cubic meters (bcm) and 0.462 bcm of water, respectively. Annual consumptive water use in the study area is 7.5 bcm. During flood season, there are water resources to ensure the implementation of ecological operation.

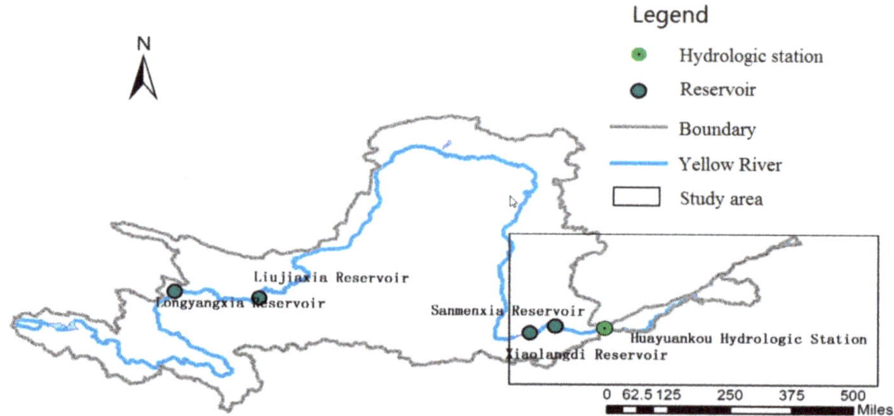

Figure 1. Locations of the downstream region of the Yellow River.

The ecological problems in the lower reaches of the Yellow River differ depending on the time of year. (1) During the major flood period—from July to September—the main ecological problem is the disappearance of the flood and high flow pulses, resulting in a flat, non-fluctuating flow; (2) during the drought period—from November to March—the river suffers a water shortage and fails to meet the minimum instreaming ecological flow requirements; (3) during medium season, the primary problem is habitat change in the riverine biological communities.

Because the characteristic flow regimes are relatively complete in the major flood period, for this study, the major flood period was selected as the ecological critical period that was assessed using the small-scale scheduling model. Water demand data used in this work came from the Comprehensive Planning of the Yellow River Basin [36], which is a book compiled by the Yellow River Conservancy Commission (YRCC) of the ministry of water resources of China; and the daily runoff data was taken from the Hydrological Yearbook of the People's Republic of China [37].

The multi-scale coupled ecological dispatching model shown in Figure 2 was established based on the characteristics of the study area described above.

In Figure 2, x refers to the discharge from the reservoirs; λ_1 and λ_2 are the Lagrange multipliers that regulate the water balance constraint between the Xiaolangdi and Sanmenxia reservoir superior subsystems, respectively, at different times; λ_3 and λ_4 are the Lagrange multipliers that regulate the water balance constraint between the superior subsystems and subordinate subsystems, respectively, of the Xiaolangdi and Sanmenxia reservoirs; and μ is the Lagrange multiplier that regulates the water balance constraint between the superior subsystems during the same period.

The objective function of the model is as follows:

(1)   Minimum water shortage rate:

$$Minf_{S-D} = \sum_{t=1}^{T} \sum_{m=1}^{M} \frac{D_{(m,t)} - S_{D(m,t)}}{D_{(m,t)}} \tag{21}$$

where f_{S-D} refers to the objective function of the water deficient ratio; m refers to the serial number of the reservoir; M refers to the sum of reservoirs; t refers to the scheduling period number; T refers to scheduling period length; $D_{(m,t)}$ refers to water demand of reservoir m at the t-th period; and $S_{D(m,t)}$ refers to the water supply of $D_{(m,t)}$;

(2) The minimum degree of hydrologic alteration is expressed as:

$$Min \sum_{i=1}^{I} D_i = \sum_{L=1,2,3} \left| \frac{N_i(La) - N_e(La)}{N_e(La)} \right| \tag{22}$$

where I represents the total number of indicators; La = 1, 2, 3 represents the high, middle and low target ranges, respectively; D_i refers to the degree of change of the i-th index; $N_i(La)$ refers to the number of years in which the i-th index is observed in the corresponding target ranges L; $N_e(La)$ refers to the number of years in which the i-th index is expected to be observed in the corresponding target ranges L.

The objective function (Lagrange function) of the overall system model is obtained by substituting Formula (21) and Formula (22) into Formula (6).

Constraint conditions of the model include water level constraint, flow constraint, water balance constraint, and flow balance constraint.

(1) Water level constraint:

$$Z_{min}(m,t) \le Z(m,t) \le Z_{max}(m,t) \tag{23}$$

(2) Flow constraint:

$$Q_{Omin}(m,t) \le Q_O(m,t) \le Q_{Omax}(m,t) \tag{24}$$

(3) Water balance constraint:

$$V(m,t+1) = V(m,t) + Q_I(m,t) - Q_O(m,t) \tag{25}$$

(4) Flow balance constraint:

$$Q_I(m+1,t) = Q_O(m,t) + q(m,t) \tag{26}$$

(5) Water supply reliability constraint:

$$\begin{aligned} reliability_{agricultural} &\ge 0.75 \\ reliability_{urban} &\ge 0.95 \end{aligned} \tag{27}$$

where $Z(m,t)$ refers to the water level of reservoir m at period t; $Z_{max}(m,t), Z_{min}(m,t)$ are the top and bottom limitations of the water level of reservoir m, respectively at period t; $Q_O(m,t)$ refers to the discharged flow of reservoir m at period t; $Q_{Omax}(m,t), Q_{Omin}(m,t)$ are the top and bottom limitations of the discharged flow of reservoir m at period t, respectively; $V(m,t)$ refers to the storage of reservoir m at period t; $Q_I(m,t)$ refers to the inflow of reservoir m at period t; $q(m,t)$ refers to the local inflow of reservoir m at period t; and $reliability_{agricultural}$ and $reliability_{urban}$ refer to the reliability of agricultural and urban water supply, respectively (urban water supply includes the industrial water supply and domestic water supply). The reliability data comes from the Comprehensive Planning of the Yellow River Basin.

Among the constraint conditions, water balance constraint between scheduling periods and flow balance constraint were implemented by coordination variables.

In this study, the daily flow from 1958 to 1977 was used as the reference series and the daily flow from 2007 to 2014 was selected as the scheduling period. The reservoir inflow, domestic water demand, agricultural water demand, and ecological water demand, etc., are required inputs of the multi-scale coupled ecological dispatching model.

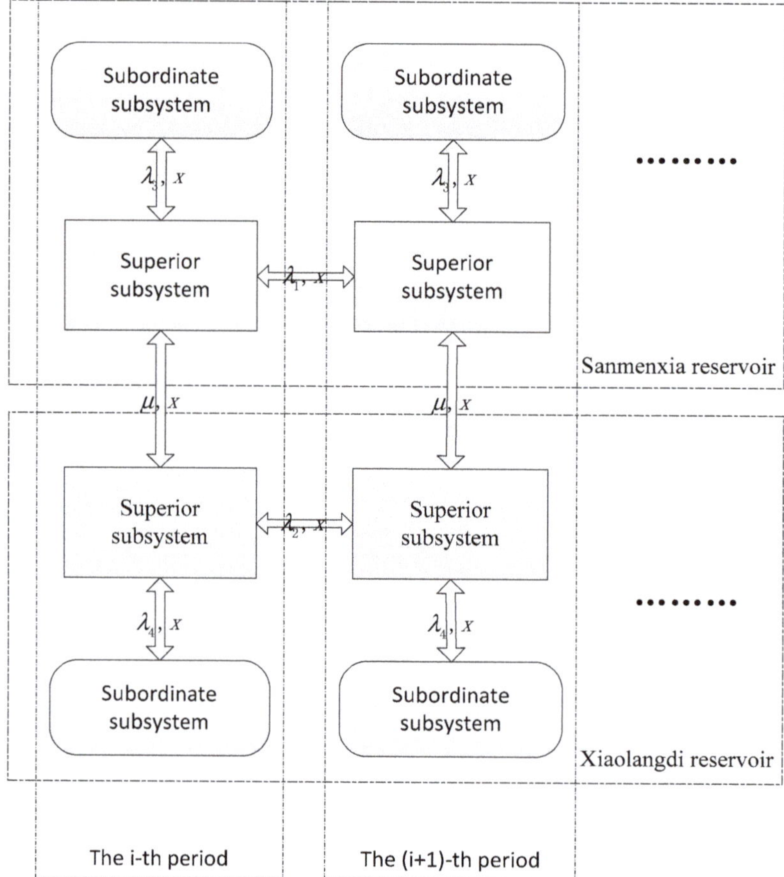

Figure 2. Information transfer structure diagram of the model.

3. Results

For comparison purposes, the hydrologic alterations caused by the operation of the current reservoir in the middle and lower Yellow River were assessed using the RVA method. The daily flow from 1958 to 1977 was used as the pre-impact series, and 2007–2014 were selected as the post-impact series. The degree to which the RVA target range is not attained is a measure of hydrologic alteration and is calculated as follows:

$$(|\text{Observed frequency} - \text{Expected frequency}|)/\text{Expected frequency} \tag{28}$$

The RVA uses 33 hydrological parameters to evaluate the hydrologic alterations [4], which are categorized into five groups: magnitude of monthly water conditions, magnitude and duration of annual extreme conditions, timing of annual extreme water conditions, frequency and duration of high and low pulses, and rate and frequency of water condition changes. As shown in Figure 3, the integrated hydrologic alteration reaches 86%. In terms of indicators, most change was observed in the magnitude of monthly water conditions and the magnitude and duration of annual extreme conditions. Among those, the change in magnitude of monthly water conditions during the flood season is the largest. In terms of RVA categories, the high RVA categories reduced the most.

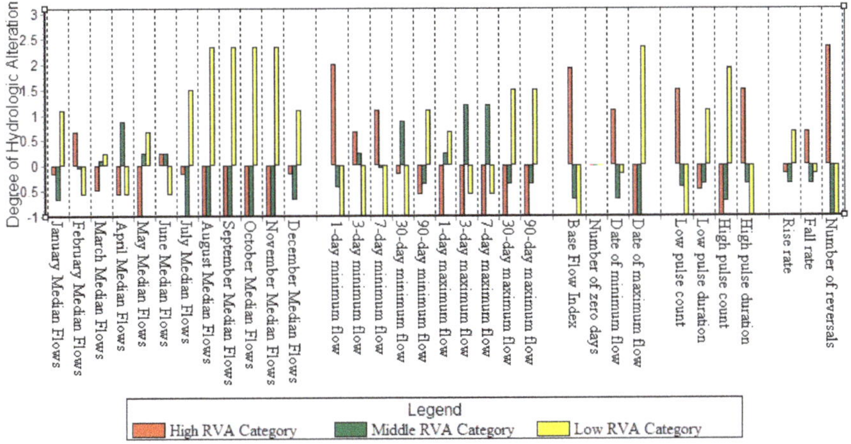

Figure 3. Hydrologic alteration of current scheduling.

The multi-scale coupled ecological dispatching model was established to generate small-scale ecological flow regimes and avoid dimension disaster. In this study, subsystems were solved by AGA and coordinated via the coordination variables. As such, the degree of hydrologic alteration can be used as an evaluation function of the reservoir scheduling model, which is unattainable for the conventional models. Thus, the multi-scale coupled ecological dispatching model provides more information to reservoir managers. Compliments of decomposition-coordination algorithms and AGA, large-scale model and small-scale model results can efficiently converge to the optimal solutions within 200 iterations as shown in Figure 4.

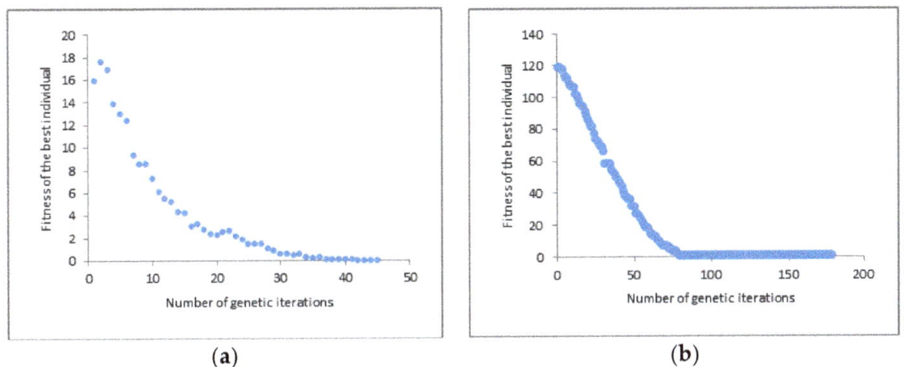

Figure 4. (**a**) Large-scale model optimization processes; (**b**) Small-scale model optimization processes.

Figure 5 compares the simulative monthly flow and actual monthly flow from 2007 to 2014. The results show that the model is capable of solving the challenges associated with uneven spatial and temporal distribution of water resources. Three observations are evident in Figure 6: (1) The simulative monthly flow is more fluctuant than the actual monthly flow; (2) the simulative monthly flow in the flood season is larger than the actual monthly flow; and (3) the occurrence time of the simulative high flow is more regular than actual high flow. These results are explicable, as more water was needed during the ecologic critical period to generate the characteristic flow regimes. Moreover, the average flow in flood season can be increased to 1400 m^3/s under the premise of meeting the water supply reliability.

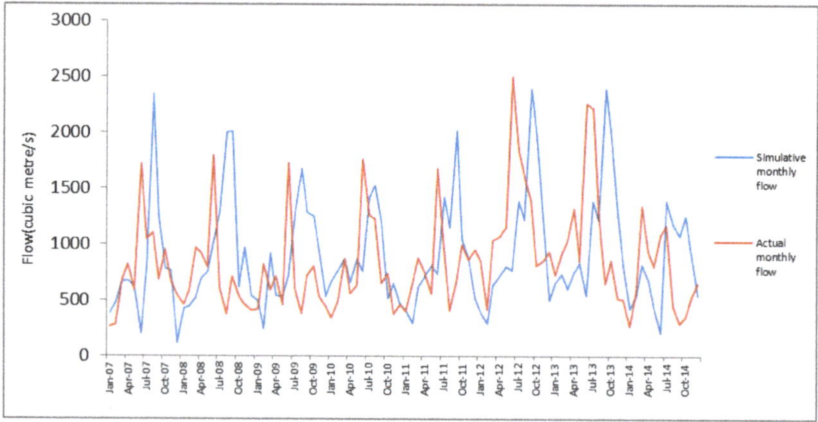

Figure 5. Monthly discharge from Xiaolangdi reservoir.

The 'percent-of-flow' (POF) approach has been widely used in the USA, the European Union, and elsewhere over the past decade [3]. To contrast the above results, the flow was also generated using the POF approach, which sets protection standards by using allowable departures from natural conditions, expressed as percentage alteration. The hydrologic alteration of flow was calculated under the POF approach (Figure 6) and contrasted with the hydrologic alteration of simulative flow (Figure 6). Comparison of Figures 3 and 6 shows that the POF approach is more protective with respect to the magnitude of monthly water conditions—as the Hydrologic alteration of the magnitude of monthly water conditions were reduced from 81% to 65%. Yet, in other groups, particularly in the magnitude and duration of annual extreme conditions, the POF approach does not provide a high degree of protection for natural flow variability. Contrarily, the hydrologic alteration of the simulative flow was reduced in all five groups. The integrated hydrologic alteration was reduced from 86% to 53%. On the one hand, this is because the objective function of the optimization model is to reduce the degree of hydrological alterations. On the other hand, the incoming water in the whole dispatching period of the optimization model is known, and this is not possible in the actual scheduling process. The hydrologic alterations of five groups are 0.60, 0.41, 0.29, 0.67, and 0.67, respectively (see the beginning of this section for corresponding groups). It is worth noting that the simulative flow did not provide a very high degree of protection for the frequency and duration of high pulses, particularly in the high RVA categories. This is explained by the difficulties incurred with providing protection for natural flow variability of different indicators in different RVA categories. For instance, more water was needed to keep the frequency of flood pulses in the high RVA category; and the increased water concentration reduced the unit water value, resulting in a lack of water elsewhere.

From the perspective of hydraulic parameters, the number of days within the appropriate velocity range during the reference series (1958–1977) is 2117, which is 29% of the total number of days. The number of days within the appropriate velocity range during the scheduling period (2007–2014) is 1452, which is 50% of the total number of days. Due to the influence of flow flattening, the number of days suitable for fish has increased. We expected that the duration of the simulated flow within the appropriate velocity is not less than the duration of natural flow. The number of days within the appropriate velocity range in the simulated flow accounted for 46% of the total. And the number of days within the appropriate velocity range under the POF approach accounted for 43% of the total. Obviously, this is an easy goal to achieve.

To solve the problem of uncertainty, we use the autoregressive model to generate nine sets of inflow runoff series. Ten sets of inflow runoff series (Including the actual inflow series from 2007–2014) were used as input for simulation calculation. The final result is the average of 10 sets of calculated results and the variation range of 10 sets of results. The values above are the mean, and the values in brackets are the variation ranges. The quantitative evaluation and biological significance of various indicators under the POF approach and optimization model respectively are shown in Table 3. As it is difficult to simulate decision-makers' decisions, the uncertainty of current scheduling is not shown. The hydrological alterations of the current scheduling are only used as a reference in Table 3. We found that the calculation results of the optimization model performed well in the second, third and fifth categories of indicators. That means the model performed better in terms of vegetation expansion, fluvial topography, natural habitat construction, fish migration, and spawning and life cycle reproduction compared to current scheduling and the POF approach. In terms of uncertainty, the optimization model performs better than the POF approach too. Especially in the second, third and fifth categories of indicators, the performance of the optimization model are far better than the POF approach, and the variation ranges barely overlap. That means the model performed better in addressing uncertain problems compared to current scheduling and the POF approach in terms of the above ecological problems.

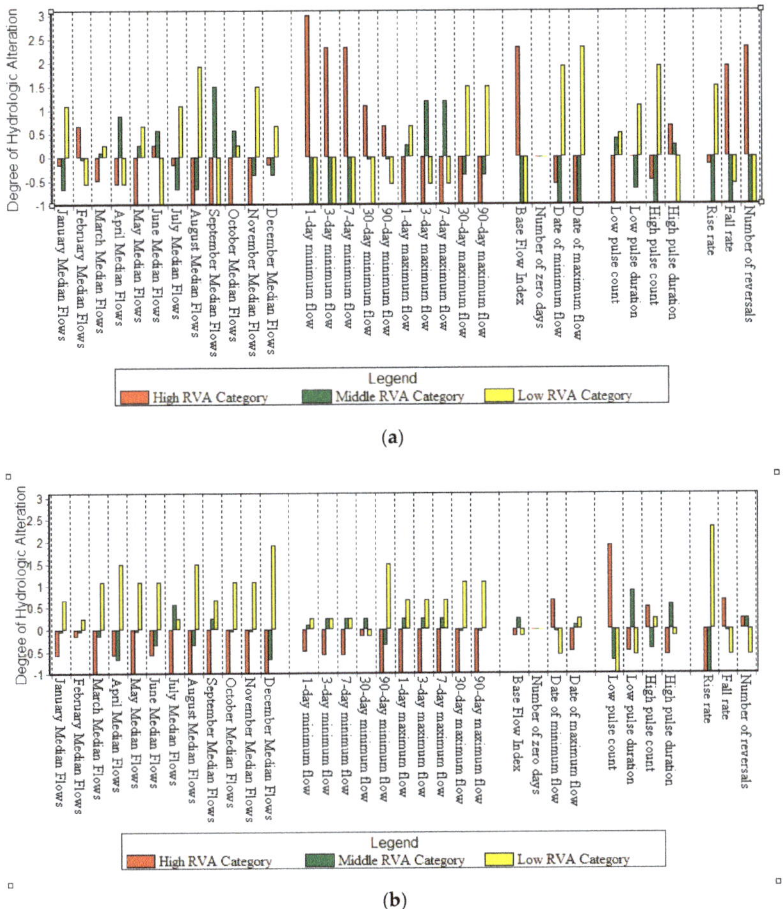

Figure 6. (**a**) Hydrologic alteration of POF approach; (**b**) Hydrologic alteration of simulative flow.

Table 3. Quantitative evaluation of the results.

Index	Current Scheduling	POF Approach	Optimization Model	Ecosystem Influences [38]
Magnitude of monthly discharge conditions	0.81	0.67 (0.62–0.71)	0.60 (0.55–0.63)	To meet the habitat needs of aquatic organisms, the needs of plants for soil moisture content, the water needs of terrestrial organisms with high reliability, the migration needs of carnivores, and the influences of water temperature and oxygen content.
Magnitude and duration of annual extreme discharge conditions	0.72	0.88 (0.80–0.95)	0.41 (0.40–0.42)	To meet the needs of vegetation expansion, river topography and natural habitat construction, nutrient exchange in rivers and flood detention areas, distribution of plant communities in lakes, and ponds and flood detention areas.
Timing of annual extreme discharge conditions	0.99	1.22 (0.88–1.44)	0.27 (0.25–0.29)	To meet the migration of fish spawning, the cycle of life reproduction, biological breeding habitat conditions, and species evolution needs.
Frequency and duration of high/low flow pulses	0.83	0.71 (0.68–0.7)	0.70 (0.69–0.72)	To generate the frequency and magnitude of soil moisture stress for plants.
Rate/frequency of hydrograph changes	0.71	1.12 (0.99–1.20)	0.67 (0.65–0.69)	To meet the drought of plants, the trapping of organics on the island and in the flood detention area, and the drying stress of low-speed organisms.
Suitable ecological velocity	0.29	0.40 (0.38–0.42)	0.46 (0.43–0.49)	To meet the appropriate velocity requirements of fish.

Table 3 shows the quantitative evaluation and biological significance of various indicators under the POF approach and optimization model. The values of the first five categories of indicators are degrees of hydrologic alterations. The value of the last indicator represents the proportion of the number of days in the suitable ecological velocity range to the total number of days.

4. Discussion

Compared to conventional models, the multi-scale coupled ecological dispatching model can obtain the degree of hydrologic alteration of small-scale ecological flow regimes and daily stream flow. At the same time, the results can be manipulated by adjusting the weights afforded to the degree of hydrologic alteration and other targets. In contrast, degrees of hydrologic alteration based on whether or not to consider the characteristic flow requirements and whether or not to introduce the small-scale model are shown in Table 4. When considering the characteristic flow requirements, reservoirs are required to release more water in the ecological critical period. Under these circumstances, the appropriate characteristic flows can be generated by the small-scale model. When the characteristic flow requirements are not taken into consideration, the agricultural water supply dependability increases from 75% to 90%.

Table 4. Contrast of the degree of hydrologic alteration.

Whether Considering Characteristic Flow Requirements	Whether Introducing Small-Scale Model	Hydrologic Alteration	Agricultural Water Supply Reliability	Urban Water Supply Reliability
Yes	Yes	53%	More than 75%	More than 95%
Yes	No	≥53%	More than 75%	More than 95%
No	Yes	87%	More than 90%	More than 95%
No	No	≥87%	More than 90%	More than 95%

Table 4 shows the influence of considering characteristic flow requirements on guaranteed water supply reliability and hydrologic alteration.

Compared with the existing research, the existing model either directly optimizes the scheduling rules [17] or evaluates the existing daily flow process [24]. However, the rule optimization model cannot describe the small scale runoff process in detail, and the runoff evaluation model cannot directly guide reservoir operation. Therefore, a modelling approach which allows scheduling of reservoir operation in such a way that hydrological alteration is reduced was presented in this work. The multi-scale coupled ecological dispatching model provides a framework for the formulation of multi-scale scheduling rules. The more flexible decoupling method can be used by the model to optimize the scheduling rules directly. In further research, the large-scale model can optimize the operation rules, while the small-scale model can optimize the discharge process of reservoirs. This method is suitable for areas where the hydrological regime is seriously affected by dam operation. This model is more suitable for large rivers because of its attention to the degree of hydrological alterations. For other river basins, decision makers can choose the key ecological issues of the river basin, and the weight of the objective function can be formulated according to the key ecological problems and the ecological significance of various indicators. Scheduling rules suitable for this basin can be found by this way.

However, the multi-scale coupled ecological dispatching model established in this study still has some issues. The competitive relationship between the ecological and economical target was coordinated by the weighting method. A more reasonable and flexible multi-objective coordination mechanism should be introduced to the model. Furthermore, considering the impact of hydrological uncertainty, calculation results from the model cannot directly guide the reservoir operation. A multi-scale coupled scheduling rule for the reservoir should be put forward and planned for future studies, which is what we are going to do next.

In general, the multi-scale coupled ecological dispatching model is capable of considering the degree of hydrologic alteration of small-scale ecological flow regimes. The ecological target was regarded as both a water requirement and a difference between the artificial flow and natural flow. Operation rules extracted from results of the coupled ecological dispatching model will include more information and be more reasonable.

5. Conclusions

A multi-scale coupled ecological dispatching model, that considers the degree of hydrologic alteration, was established to satisfy the multi-scale features of ecological water demand. In order to acquire all characteristics of the environmental flows and to avoid the "dimension disaster", the decomposition coordination algorithm was applied to this model. Furthermore, this model is capable of using account steps from months to hours, as needed. Moreover, the general multi-scale model uses results of the superior subsystems only as boundary conditions of the subordinate subsystems; while in our model, multi-scale subsystems are based on the theory of large-scale systems and we consider the coordination of multi-scale objectives.

Simulation results show that the model is able to solve the problem of uneven spatial and temporal distribution of water resources and obtain degrees of hydrologic alteration of characteristic flow regimes. The difficulty of generating characteristic flow regimes is indicated by the results, which will be helpful to the formulation of operation rules. The calculation results indicate that increasing the average flow in flood season to 1400 m^3/s can reduce the degree of hydrologic alteration from 86% to 53% in the downstream Yellow River. Model optimization results were compared with results of the POF approach. The comparison showed that further reduction in hydrologic alteration is possible, but further research is needed to determine which method is more suitable for accomplishing this task.

While the proposed model is overall a success, there are some limitations worth noting. On the model side, a more reasonable and flexible multi-objective coordination mechanism needs to be established. From the mathematics side, a high-efficiency and applicative arithmetic method needs to be developed or improved to extract scheduling rules from artificial daily flow series.

Author Contributions: Conceptualization, T.Z. and Z.D.; methodology, T.Z.; software, T.Z.; investigation, X.G. and Z.H.; resources, X.G. and R.S.; data curation, R.S.; writing original draft preparation, T.Z.; writing—review and editing, T.Z.; W.W. and Z.D.

Funding: This research has been financially supported by the National Key Research and Development Program of China (2018YFC1508200), the Postgraduate Research & Practice Innovation Program of Jiangsu Province (KYCX18—0574) and the Fundamental Research Funds for the Central Universities (2018B603X14).

Conflicts of Interest: The authors declare no conflict of interest. The funders had no role in the design of the study; in the collection, analyses, or interpretation of data; in the writing of the manuscript, or in the decision to publish the results.

References

1. Nilsson, C.; Reidy, C.A.; Dynesius, M.; Revenga, C. Fragmentation and Flow Regulation of the World's Large River Systems. *Science* **2005**, *308*, 405–408. [CrossRef] [PubMed]
2. Symphorian, G.R.; Madamombe, E.; Zaag, P.V.D. Dam operation for environmental water releases; the case of Osborne dam, Save catchment, Zimbabwe. *Phys. Chem. Earth Parts A/B/C* **2003**, *28*, 985–993. [CrossRef]
3. Richter, B.D. Re-thinking environmental flows: From allocations and reserves to sustainability boundaries. *River Res. Appl.* **2009**, *26*, 1052–1063. [CrossRef]
4. Richter, B.; Baumgartner, J.; Wigington, R.; Braun, D. How much water does a river need. *Freshw. Biol.* **1997**, *37*, 231–249. [CrossRef]
5. Hughes, D.A.; Ziervogel, G. The inclusion of operating rules in a daily reservoir simulation model to determine ecological reserve releases for river maintenance. *Water* **1998**, *24*, 293–302.
6. Higgins, J.M.; Brock, W.G. Overview of Reservoir Release Improvements at 20 TVA Dams. *J. Energy Eng.* **1999**, *125*, 1–17. [CrossRef]
7. Richter, B.D.; Warner, A.T.; Meyer, J.L.; Lutz, K. A collaborative and adaptive process for developing environmental flow recommendations. *River Res. Appl.* **2006**, *22*, 297–318. [CrossRef]
8. Poff, N.L.; Allan, J.D.; Bain, M.B.; Karr, J.R.; Prestegaard, K.L.; Richter, B.D.; Stromberg, J.C. The Natural Flow Regime: A Paradigm for River Conservation and Restoration. *Bioscience* **1997**, *47*, 769–784. [CrossRef]
9. Naiman, R.J.; Latterell, J.J.; Pettit, N.E.; Olden, J.D. Flow variability and the biophysical vitality of river systems. *C. R. Géosci.* **2008**, *340*, 629–643. [CrossRef]

10. Jager, H.I.; Smith, B.T. Sustainable reservoir operation: Can we generate hydropower and preserve ecosystem values? *River Res. Appl.* **2010**, *24*, 340–352. [CrossRef]

11. Homa, E.S.; Vogel, R.M.; Smith, M.P.; Apse, C.D.; Huber-Lee, A.; Sieber, J. An Optimization Approach for Balancing Human and Ecological Flow Needs. In Proceedings of the EWRI 2005 World Water and Environmental Resources Congress, ASCE, Anchorage, Alaska, 15–19 May 2005; pp. 1–12.

12. Schlager, E. Rivers for life: Managing water for people and nature. *Ecol. Econ.* **2003**, *55*, 306–307. [CrossRef]

13. Petts, G.E. Instream flow science for sustainable river management. *JAWRA J. Am. Water Resour. Assoc.* **2009**, *45*, 1071–1086. [CrossRef]

14. Richter, B.D.; Davis, M.M.; Apse, C.; Konrad, C. A presumptive standard for environmental flow protection. *River Res. Appl.* **2012**, *28*, 1312–1321. [CrossRef]

15. De Jalón, S.G.; del Tánago, M.G.; de Jalón, D.G. A new approach for assessing natural patterns of flow variability and hydrological alterations: The case of the Spanish rivers. *J. Environ. Manag.* **2019**, *233*, 200–210. [CrossRef]

16. Peñas, F.J.; Barquín, J. Assessment of large-scale patterns of hydrological alteration caused by dams. *J. Hydrol.* **2019**, *572*, 706–718.

17. Suen, J.P.; Eheart, J.W. Reservoir management to balance ecosystem and human needs: Incorporating the paradigm of the ecological flow regime. *Water Resour. Res.* **2006**, *42*, 178–196. [CrossRef]

18. Yin, X.A.; Yang, Z.F. Development of a coupled reservoir operation and water diversion model: Balancing human and environmental flow requirements. *Ecol. Model.* **2011**, *222*, 224–231. [CrossRef]

19. Yin, X.A.; Yang, Z.F.; Petts, G.E. Optimizing Environmental Flows Below Dams. *River Res. Appl.* **2012**, *28*, 703–716. [CrossRef]

20. Steinschneider, S.; Bernstein, A.; Palmer, R.; Polebitski, A. Reservoir Management Optimization for Basin-Wide Ecological Restoration in the Connecticut River. *J. Water Resour. Plan. Manag.* **2014**, *140*, 04014023. [CrossRef]

21. Tsai, W.P.; Chang, F.J.; Chang, L.C.; Herricks, E.E. AI techniques for optimizing multi-objective reservoir operation upon human and riverine ecosystem demands. *J. Hydrol.* **2015**, *530*, 634–644. [CrossRef]

22. İlker, A.R.; Osman, C. A Descriptive Decision Process Model for Hierarchical Management of Interconnected Reservoir Systems. *Water Resour. Res.* **1984**, *20*, 803–811.

23. Olden, J.D.; Poff, N.L. Redundancy and the choice of hydrologic indices for characterizing streamflow regimes. *River Res. Appl.* **2003**, *19*, 101–121. [CrossRef]

24. Yang, T.; Zhang, Q.; Chen, Y.D.; Tao, X.; Xu, C.Y.; Chen, X. A spatial assessment of hydrologic alteration caused by dam construction in the middle and lower Yellow River, China. *Hydrol. Process.* **2008**, *22*, 3829–3843. [CrossRef]

25. Richter, B.D.; Baumgartner, J.V.; Powell, J.; Braun, D.P. A Method for Assessing Hydrologic Alteration within Ecosystems. *Conserv. Biol.* **1996**, *10*, 1163–1174. [CrossRef]

26. Mathews, R.; Richter, B.D. Application of the Indicators of Hydrologic Alteration Software in Environmental Flow Setting. *JAWRA J. Am. Water Res. Assoc.* **2007**, *43*, 1400–1413. [CrossRef]

27. Junk, W.J. The flood pulse concept in river-floodplain systems. *Can. J. Spec. Publ. Fish. Aquat. Sci.* **1989**, *106*, 110–127.

28. Mosley, M.P. Analysis of the effect of changing discharge on channel morphology and instream uses in a Braided River, Ohau River, New Zealand. *Water Resour. Res.* **1982**, *18*, 800–812. [CrossRef]

29. Ubertini, L.; Manciola, P.; Casadei, S. Evaluation of the minimum instream flow of the Tiber River basin. *Environ. Monit. Assess.* **1996**, *41*, 125. [CrossRef]

30. Gippel, C.J.; Stewardson, M.J. Use of wetted perimeter in defining minimum environmental flows. *River Res. Appl.* **1998**, *14*, 53–67. [CrossRef]

31. Changming, Ecological hydraulic radius method for estimating ecological water demand in river channels. *Prog. Nat. Sci.* **2007**, *1*, 42–48.

32. Liu, J. *Advanced Aquatic Biology*; Science Publishing Company: Beijing, China, 1999.

33. Xu, Z. *Theory and Practice of River and Lake Ecological Water Demand, China*; Water&Power Press: Beijing, China, 2005.

34. Haimes, Y.Y. *Hierarchial Analysis of Water Resources Systems: Modeling and Optimization of Large-Scale Systems*; McGraw-Hill International Book Company: New York, NY, USA, 1977.

35. Nocedal, J.; Wright, S.J. *Numerical Optimization*; Springer: Berlin, Gemany, 2006; pp. 29–76.

36. Yellow River Conservancy Commission. *Comprehensive Planning of the Yellow River Basin: 2012–2030*; Yellow River Water Conservancy Press: Zhengzhou, China, 2013; pp. 108–110.
37. Yellow River Conservancy Commission. *Hydrological Yearbook of the People's Republic of China*; China Water Power Press: Zhengzhou, China, 2015.
38. Richter, B.D. A spatial assessment of hydrologic alteration within a river network. *Regul. Rivers Res. Manag.* **1998**, *14*, 329–340. [CrossRef]

Article

SPH Simulation of Interior and Exterior Flow Field Characteristics of Porous Media

Shijie Wu [1], Matteo Rubinato [2] and Qinqin Gui [1,*]

[1] Faculty of Maritime and Transportation, Ningbo University, Ningbo 315211, China; wushijie009@163.com
[2] School of Energy, Construction and Environment & Centre for Agroecology, Water and Resilience, Coventry University, Coventry CV1 5FB, UK; ad2323@coventry.ac.uk
* Correspondence: guiqinqin@nbu.edu.cn; Tel.: +86-182-5878-7809

Received: 16 January 2020; Accepted: 17 March 2020; Published: 24 March 2020

Abstract: At the present time, one of the most relevant challenges in marine and ocean engineering and practice is the development of a mathematical modeling that can accurately replicate the interaction of water waves with porous coastal structures. Over the last 60 years, multiple techniques and solutions have been identified, from linearized solutions based on wave theories and constant friction coefficients to very sophisticated Eulerian or Lagrangian solvers of the Navier-Stokes (NS) equations. In order to explore the flow field interior and exterior of the porous media under different working conditions, the Smooth Particle Hydrodynamics (SPH) numerical simulation method was used to simulate the flow distribution inside and outside a porous media applied to interact with the wave propagation. The flow behavior is described avoiding Euler's description of the interface problem between the Euler mesh and the material selected. Considering the velocity boundary conditions and the cyclical circulation boundary conditions at the junction of the porous media and the water flow, the SPH numerical simulation is used to analyze the flow field characteristics, as well as the longitudinal and vertical velocity distribution of the back vortex flow field and the law of eddy current motion. This study provides innovative insights on the mathematical modelling of the interaction between porous structures and flow propagation. Furthermore, there is a good agreement (within 10%) between the numerical results and the experimental ones collected for scenarios with porosity of 0.349 and 0.475, demonstrating that SPH can simulate the flow patterns of the porous media, the flow through the inner and outer areas of the porous media, and the flow field of the back vortex region. Results obtained and the new mathematical approach used can help to effectively simulate with high-precision the changes along the water depth, for a better design of marine and ocean engineering solutions adopted to protect coastal areas.

Keywords: Smooth Particle Hydrodynamics (SPH); porous media; mathematical model; coastal structure; ocean and engineering

1. Introduction

In ocean and water conservancy projects, pore structures are widely used, such as reservoir dams, artificial reefs, and breakwaters. When water flows through pore structures, factors such as porosity rate and volume of porous media can be the major cause for different flow field fluctuations [1]. Therefore, to investigate the influence of porosity rate and volume of porous media on the flow field, which is crucial to characterize the features of flow field characteristics around the porous media [2], this paper analyzes the internal and external flow field characteristics of typical structures by simulating these physical phenomena.

The commonly used simulation method to study these effects is CFD (Computational Fluid Dynamics). CFD simulation can effectively simulate the flow field characteristics, however multiple complications and lengthy simulations are associated with the establishment of a proper mesh [3,4].

Boundary element simulations can effectively simulate the influence of water flow on pore structure, but these methods only divide the elements on the boundary domain to meet the approximations applied to the functions implemented for the boundary conditions of the control equation. Therefore, compared with the finite element method, the boundary element method has the advantage of fewer elements to consider and a simpler data preparation. However, when the boundary element method is used to solve the nonlinear problem, which corresponds to the integral corresponding to the nonlinear term, more complications are frequently encountered because this kind of integration has strong singularity near single points, making an accurate solution very challenging to be obtained. Therefore, to reduce and avoid these complications, the key properties of the Smooth Particle Hydrodynamics (SPH) method, which is based on the pure Lagrangian numerical simulation method, were adopted for this study. This method can track the trajectory of each particle, and has good advantages in terms of visualization, high precision, and large processing. Many studies have been completed to date to explore the hydrodynamic characteristics of porous media. For example, Hossei Basser et al. [5] used SPH method to characterize the gravity flow of a single fluid in porous media and multi-fluid flows in impermeable layers and porous media, as well as the influence of unidirectional flow on pore structure and particle motion. The results that were obtained confirmed how the SPH numerical simulation method had the characteristics of high accuracy and visualization, which could effectively and accurately simulate the movement of each particle in the flow field inside and outside the pore structure, describing very accurately the characteristics of the flow field inside and outside the pore structure.

Pore flow field simulation can be simulated focusing on two aspects: internal flow field and external flow field, which refer to the specific flow field formed in large-scale ocean engineering, such as artificial reefs, dam break and breakwater, under the condition of laminar and turbulent flows. Boundary conditions are very important and are typically periodic or open. The first category means that the flow particles run from the inlet boundary to the outlet boundary, re-entering the inlet boundary and form a continuous loop. The second category refers to the flow particles moving in from the inlet boundary to the outlet boundary without re-entering from the inlet boundary, because this approximation considers that new flow particles will move in following an uninterrupted process. Under the open boundary conditions, researchers typically simulate the internal and external flow fields, effectively realizing the continuous process of particles motion [6]. However, by using these conditions although the accuracy of the inlet and outlet flows is high, the interaction with the porous media is relatively complex and affected by lower accuracy. Therefore, periodic boundary conditions accompanied with a damping zone were employed in this paper to complete the work mentioned above.

When simulating pore structure models, porosity must be considered, because factors such as pore permeability can affect the flow field around the pore, like for example in processes such as dam break, where the dam body is composed of pore structure. Therefore, under different porosity, the dam break strength is different, generating different impacts in case of failure of the dam taken into consideration. Additionally, when the flow comes in contact with the pore structure, the seepage characteristics of the pore structure can further affect the surrounding flow field [7], so the porosity that affects the pore structure is one of the key factors to be analyzed.

To date, multiple studies have been conducted on the characterization of pore structures. Qinqin Gui et al. [8] studied the coupling law of wave and pore structure by using the incompressible SPH method. Through numerical simulation and experiment, the typical flow around and inside the underwater porous structure was compared, however the analysis of the internal and external flow field of the porous structure was not considered as well as the influence of water flow on the porous structure. Abbas Khayyer et al. [9] proposed a numerical solution for variable porosity, because the purpose of this study was to investigate the effect of waves on structures with different porosity. Vanneste and Troch [10] conducted a detailed two-dimension numerical study on the wave propagation and verified the calculation results in terms of average water level and spatial pore pressure distribution from the incident wave field to the normal wave field. Shao [11] proposed an Incompressible Smooth Particle Hydrodynamics (ISPH) method to simulate the interaction between waves and porous media, and achieved interesting results on the analysis of flows and pore structures: through comparative analysis, the error obtained by Shao [11] was between 10% and 15%, which better reflects the accuracy of numerical simulations when the SPH is applied [8]. To summarize what was found in literature, to date, traditional SPH method is generally used to estimate the influence of waves on the porous media, but seldom for the coupling the effects of water flow and pore separately. Therefore, there is a need for research to be conducted to demonstrate that the SPH method can be adopted to simulate the action process of water flow on porous materials and the consequent effects. To fulfill this gap, SPH is used in this work to simulate coupling effects of water flow and porous media with the periodic circulation flow as a boundary condition. The influence of the flow field on the porous media was analyzed, together with the flow field inside and outside the porous media, the longitudinal and vertical velocity distribution, and the eddy current law of the eddy current field.

2. Smooth Particle Hydrodynamics (SPH) Methodology

The Smooth Particle Hydrodynamics method [12–22] is commonly used to describe a continuous fluid (or solid) as a group of interacting particles, despite each of these particles carries independent physical features such as different mass and dissimilar velocity. This method tends to characterize the continuous fluid by solving the behavior of the entire group of particles to simplify the problem and to determine the mechanical behavior of the entire system. In principle, as long as the number of particles is sufficient, the mechanical process can be accurately described. In practice, these particles are distributed irregularly, but in theory, they all move according to the motion law of the conservation governing equation. Any field of particles can be represented by applying a sufficiently accurate integral obtained by approximating the interpolation between adjacent particles. Considering this aspect, spatial derivatives such as gradient and scatter operators can be similarly estimated using the properties of all the particles in the Navier-Stokes equations, and the kernel approximation is commonly used to approximate the field function.

The procedure applied for this research uses the Smooth Particle Fluid Dynamics (SPH) method, which can be approximated by integral interpolation:

$$A(r) = \int A(r')W(r - r', h)dr' \tag{1}$$

where $A(r)$ represent any function. For numerical simulations, the integral interpolation is calculated as follows:

$$A_a(r) = \sum_b m_b \frac{A_b(r)}{\rho_b} W_{ab} \tag{2}$$

where a and b represent particles; m_b and ρb represent particle mass and density, W_{ab} is a kernel function, r is the distance between the reference particle and the adjacent (m), h is the smoothing length (m). To ensure a linear and angular momentum conservation, the asymmetric pressure gradient terms can be obtained as follows:

$$\left(\frac{1}{\rho}\nabla P\right)_a = \sum_b m_b \left(\frac{P_a}{P_{a^2}} + \frac{P_b}{P_{b^2}}\right)\nabla_a W_{ab} \tag{3}$$

where $\nabla_a W_{ab}$ represents the kernel gradient taken relative to the position of the particle a and P represents the pressure; similarly, the divergence of the vector at a given particle a can be estimated by u:

$$\nabla u_a = \rho_a \sum_b m_b \left(\frac{u_a}{P_{a^2}} + \frac{u_b}{P_{b^2}}\right)\nabla_a W_{ab} \tag{4}$$

The viscous force can be calculated by combining the gradient and the divergence term, as follows:

$$\left(v_0 \nabla^2 u\right)_a = \sum_b m_b \frac{2(v_a + v_b)}{(\rho_a + \rho_b)} \cdot \frac{u_{ab} r_{ab} \cdot \nabla_a W_{ab}}{|r_{ab}|^2 + \eta^2} \tag{5}$$

where v represents the viscosity coefficient. Since pressure is highly sensitive to turbulence, it has been necessary to adopt the Laplace term in the pressure Poisson equation, written as follows [23]:

$$\nabla \cdot \left(\frac{\nabla P}{\rho}\right)_a = \sum_b m_b \frac{8 \cdot P_{ab} r_{ab} \cdot \nabla_a W_{ab}}{(\rho_a + \rho_b)^2 \cdot (|r_{ab}|^2 + \eta^2)} \tag{6}$$

$$P_{ab} = P_a - P_b \quad r_{ab} = r_a - r_b \tag{7}$$

The kernel function uses the cubic spline kernel function developed by Monaghan [24], which is related to the dimension of the field investigated and the distance between the relevant points. The kernel function is displayed below:

$$W(r,h) = \begin{cases} (10 \times (1 - 1.5q^2 + 0.75q^3))/7\pi h^2 & 0 \le q < 1 \\ (10(2-q)^3)/28\pi h^2 & 1 \le q \le 2 \\ 0 & q > 2 \end{cases} \tag{8}$$

where r is the distance between the reference particle and the adjacent one; $q = r/h$ is the relative distance.

3. SPH Model

3.1. Equations for Flow Field Porous Media

Based on the previous studies conducted [19,25], the flow outside of the porous media is typically considered laminar and can be solved by the two-dimensional Navier-Stokes equations. The external flow field of pore structure is as follows:

$$\frac{d\rho}{\rho dt} + \nabla \cdot u_w = 0 \tag{9}$$

$$\frac{du_w}{dt} = -\frac{1}{\rho}\nabla P + g + v\nabla^2 u_w \tag{10}$$

where ρ is the flow particle density; t is the time; u_w is the external flow rate of the pore.

Equation (10) is the governing equation of the external flow field of the porous media. This equation incorporates the influence of the current on the porous media during the movement, and the parameters to calculate the motion of the eddy current.

The internal flow field inside pore structure can be obtained as follows [19,25]:

$$\frac{du_p}{dt} = -\frac{1}{\rho}\nabla P + g + v \cdot \nabla^2 u_p - \frac{C_f n_w^2}{\sqrt{K_P}} u_p |u_p| \tag{11}$$

$$K_p = 1.643 \times 10^{-7}\left[\frac{d_{50}}{d_0}\right]^{1.57}\left[\frac{n_w^3}{(1-n_w)^2}\right] \quad C_f = 100[d_{50}(m) \cdot \left(\frac{n_w}{K_p}\right)^{1/2}]^{-1.5} \tag{12}$$

where u_{p1} is the conveying speed (m/s); n_w is the porosity (/); K_p is the permeability (/); C_f is the nonlinear resistance coefficient (/).

3.2. Numerical Model Solving Process

In the first prediction step, the velocity and the position of the particles are calculated according to the momentum equations, and these equations can be expressed as follows [11]:

$$\Delta u_* = \left(g + v\nabla^2 u_p - \frac{v n_w}{K_P} u - \frac{C_f n_w^2}{\sqrt{K_P}} u|u|\right)_t \Delta t \tag{13}$$

$$u_* = u_t + \Delta u_* \tag{14}$$

$$r_* = r_t + u_* \Delta t \tag{15}$$

where Δu is the particle velocity that changes during the prediction step; Δt is the time increment; u_t and r_t represents the particle velocity and position at time t, respectively. The pressure term is based on the classical pressure Poisson equation that can be expressed as follows [23]:

$$\nabla \cdot \left(\frac{1}{\rho} \nabla P_{t+1}\right) = \frac{\rho_0 - \rho_*}{\rho_0 \Delta t^2} \tag{16}$$

where ρ_0 represents the initial constant particle density; ρ_* represents the central particle density after the prediction step, and P_{t+1} is the pressure of the particles at the $t+1$ time step. In the calibration of the second step, the pressure gradient term is combined with the momentum equation to ensure incompressibility. Pressure can be used to correct particle velocity as follows:

$$\Delta u_{**} = \frac{-1}{\rho_*} \nabla p_{t+1} \Delta t \tag{17}$$

$$u_{t+1} = u_* + \Delta u_{**} \tag{18}$$

$$r_{t+1} = r_t + \frac{u_{t+1} + u_t}{2} \Delta t \tag{19}$$

where u_{t+1} is represents the particle velocity and r_{t+1} represents position at the moment of $t + 1$.

3.3. Boundary Conditions

3.3.1. Free Surface Boundary

Shao and Lo [23] proposed a free surface treatment method where if the density of a particle is 10% lower than the reference density, it can be considered as a surface particle, and then zero pressure can be applied as a known boundary condition [8].

These assumptions are adopted for this study.

3.3.2. Fluid-Structure Coupling Boundary

The interface region is placed between the exchange flow and the porous flow region at the boundary of the pure fluid interface, and the interface line is located at the center of the region. Therefore, the width of the two regions (the fluid portion and the porous portion) at the interface boundary line is twice the distance between particles. By using the proposed SPH interface model, the average value of the interface region is obtained by solving the pressure Poisson equation in each calculation time step using the SPH kernel function. Only the adjacent particles located in the interface area are included in the summation.

At the same time, the interface boundary conditions can automatically satisfy the normal and tangential continuity.

3.3.3. Impermeable/Fixed Solid Wall Boundary

The impermeable/fixed solid wall is maintained along the boundary line. The pressure gradient between particles is adopted and when fixed virtual particles are used, the anti-skid boundary is selected.

3.3.4. Periodic Inflow and Outflow Boundaries Accompanied with a Damping Zone

According to previous studies conducted on pore structure logistics field, this paper adopts the boundary of inflow and outflow as displayed in Figure 1 below.

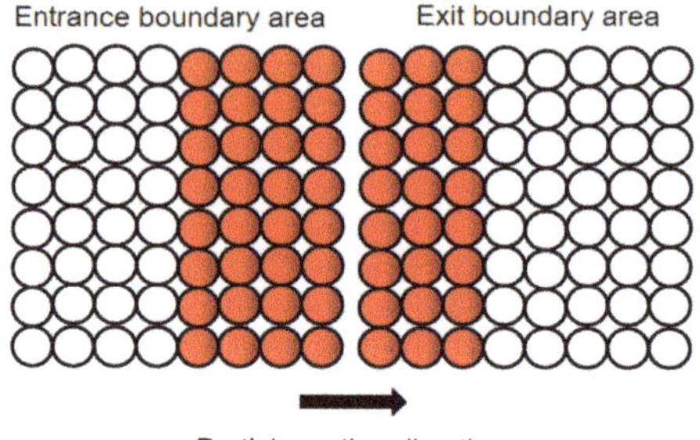

Figure 1. Flow boundary diagram.

Particles leaving one side of the computational domain move directly through the other side, and particles near the open side boundary interact with particles near the complementary open side boundary. The fluid particles near the left end of the channel can also interact with the fluid particles near the right end in the core affected area. In this sense, the upstream and downstream boundaries can actually be considered as overlapping. In the SPH numerical algorithm, we used the particle linked list to search for neighborhoods through a square grid with a side length of 2*H. Therefore, the fluid particles in the first column of the block located near the upstream boundary can be in contact with the particles in the last column of the block located near the downstream boundary, and vice versa, so the exit boundary flow velocity is also consistent with the inlet boundary flow velocity trend. As a preliminary test to verify this hypothesis, we set the calculation area as 3 m, the inlet flow rate as 0.00936 m^3/s, and the particle size D as 0.1 m. In order to further test the performance of our method, three time stages t = 18.0 s, t = 20.0 s, and t = 22.0 s were considered to analyze its inlet and outlet flow rate, respectively. x = 0.0 represents the inlet flow rate distribution curve; x = 60.0 represents the outlet flow rate distribution curve. Analyzing the tests shown in Figure 2, it is possible to notice that the flow distribution of the particles is very similar between inlet (x = 0.0) and outlet (x = 60.0). Furthermore, the velocity ranges from 0 *m/s* to 1.5 *m/s* for each time step displayed proving the periodic boundary conditions of the inlet and outlet.

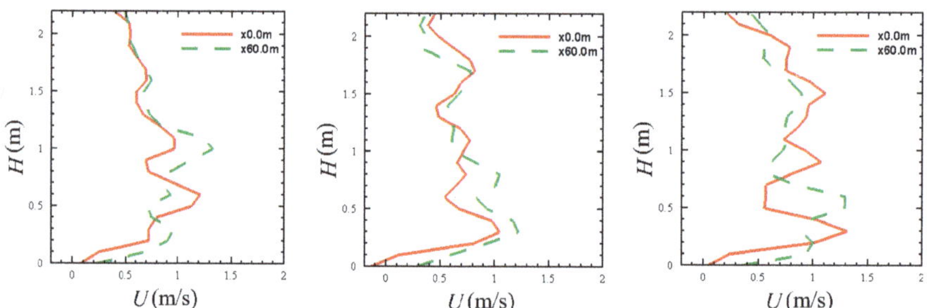

Figure 2. Comparison of flow velocities at inlet and outlet boundary at different times at T = 18.0 s, T = 20.0 s and T = 22.0 s.

It can also be observed from Figure 2 that the inlet and the exit boundary particles are basically consistent in the horizontal direction despite different time steps considered, confirming that the particles at the exit layer can smoothly enter the inlet across the boundary.

However, if the flow is significantly disturbed before exiting the outlet boundary, the inflow condition will inherit these disturbances and affect the downstream flow. In that case, at least a damping zone is needed downstream the inlet boundary and a numerical calibration is needed to clarify the effectiveness of the damping zone [26,27]. For this study, a damping zone is set at the junction of the exit boundary and the entrance boundary.

Figure 3 displays the distribution of the velocity of the flow at the exit boundary along the depth of the water after passing through the damping zone. It can be seen that the speed variations remain relatively stable along the depth of the water at different times (t = 60 s; t = 70 s; t = 80 s; t = 90 s), hence the flow velocity at the outlet remains basically identical (the average flow velocity is stable at about 0.13 *m/s* at all times) as well as the inlet velocity. Under the effect of the periodic boundary, the inlet velocity can be stably maintained at 0.13 *m/s* so that the flow is significantly disturbed before it departures the exit boundary without affecting the downstream flow. This further validates the effectiveness of the damping region.

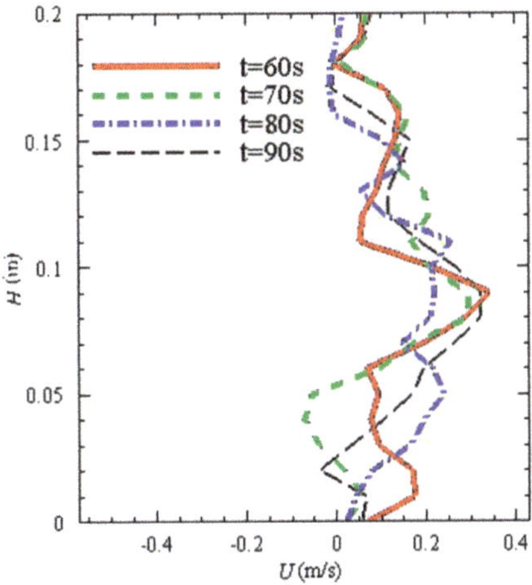

Figure 3. Comparison of flow velocities at inlet and outlet boundary at different times at t = 60.0 s, t = 70.0 s, t = 80.0 s and t = 90 s.

For all these reasons, it is possible to confirm the numerical characteristics of the periodic cyclic boundary of the flow.

4. Model Verification

In order for SPH to work in numerical sink simulations, characterizing the correct flow into the outflow boundary is key to achieve a continuous steady flow cycle. In this study, the flow was simulated using periodic boundary conditions and a damping zone. The position and velocity of the particles were firstly initialized, simulating the state of the current at the beginning of the movement. At the same time, during the interaction between the current and the porous media, the flow velocity was considered to change with the variation of the porosity. According to the experiments in the literature, porosity values of 0.349 and 0.475 [7] are the most commonly used and hence have been considered in this work to verify the model adopted.

The numerical water tank used for verification is shown in Figure 4 below [7]. The numerical flume has a pore structure placed 2.915 m away from the inlet boundary. The numerical model tank has a length of 6.0 m, an inlet flow rate of 0.13 *m/s*, a pore structure long 0.15 m and elevated 0.075 m, a sink channel slope of 0.005, and a viscosity of 1.0×10^{-6} (Pa·s). Finally, the flow field characteristics were simulated at the porosity of 0.349 and 0.475.

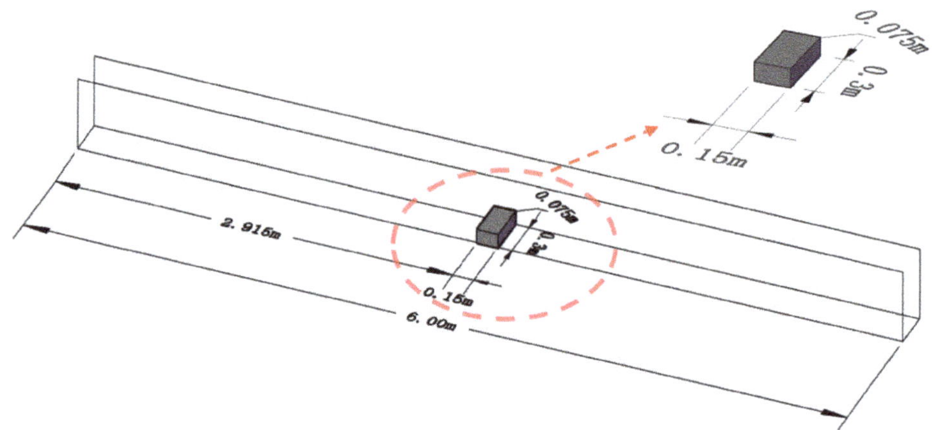

Figure 4. Schematic diagram of the numerical water tank.

The flow fields of velocity vectors for porosity equal to 0.349 (Figure 5a) and 0.475 (Figure 5b) are shown below in Figure 5.

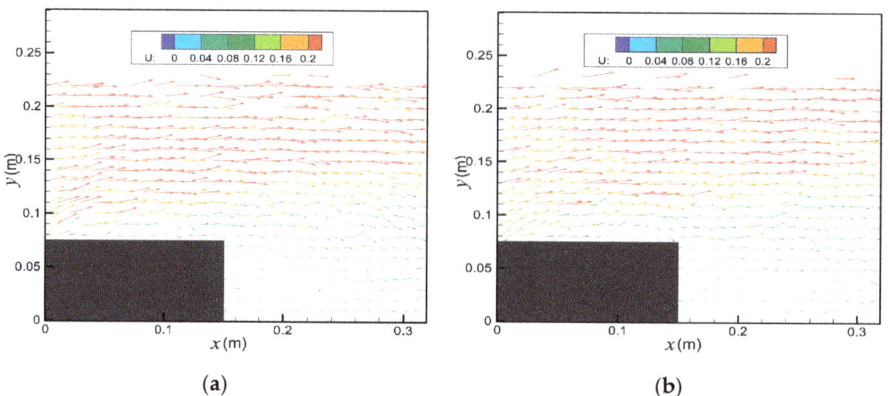

(**a**) (**b**)

Figure 5. (**a**) represents the velocity field using porosity = 0.349; (**b**) represents the velocity field using porosity = 0.475. U: Longitudinal velocity (*m/s*).

It can be seen from Figure 5 that the flow simulation based on the periodic boundary basically simulates the correct natural movement of the water flow. The internal and external flow fields and back vortex flow fields of the porous media are also visible in Figure 4. The flow field characteristics, the flow velocity U, V, and water depth H were processed dimensionless to verify the flow field velocity distribution characteristics of the porous media at 0.475 and 0.349 and results are shown in Figures 6 and 7, respectively.

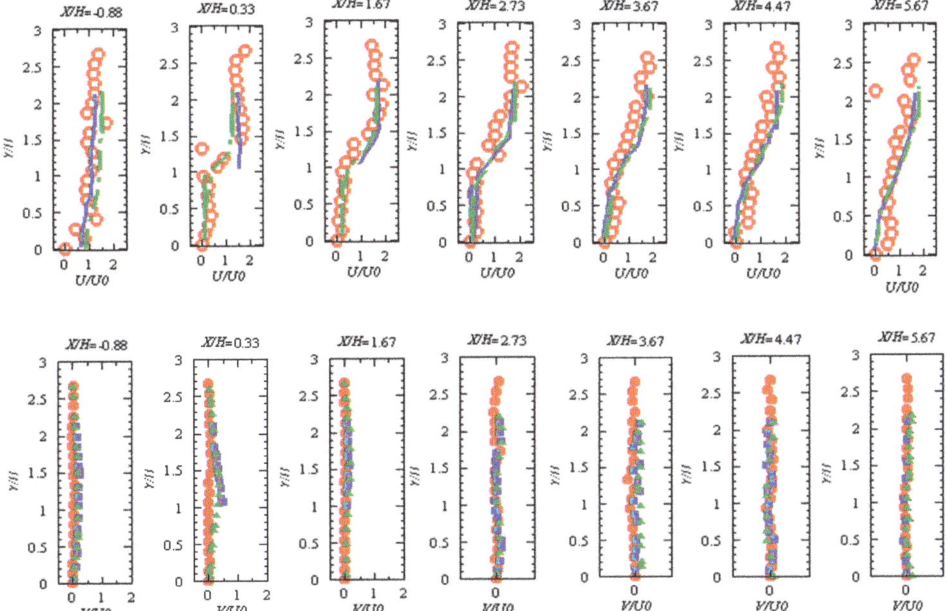

Figure 6. Comparison of longitudinal and vertical velocity for porosity 0.475. Legend: Red open circle: numerical simulation of longitudinal velocity of smooth particle hydrodynamics (SPH); Green dash dot: numerical simulation of longitudinal velocity presented by Chan et al. [7]; Blue solid line: experimental longitudinal velocity values from Chan et al. [7]; Red solid circle: numerical simulation of vertical velocity of SPH; Green solid triangle: numerical simulation of vertical velocity of Chan et al. [7]; Blue solid square: experimental vertical velocity values from Chan et al. [7]. H: water depth (m); U: longitudinal flow rate (*m/s*), V: vertical flow rate (*m/s*), U0: Inlet boundary velocity (*m/s*). X is the distance between the particle and the left boundary of the pore structure, H is the height of the pore structure, and X/H is a dimensionless treatment of the X axis. The pore structure is placed at the origin of this system.

The numerical simulation of the longitudinal and vertical velocity distributions at different locations around the porous media fit well (±5–10%) with the velocity curves measured for the experimental experiments. This confirms that the SPH method is feasible for the simulation of the flow field around the porous media and the analysis of the longitudinal and vertical velocity at a porosity of 0.475.

According to the above model verification displayed in Figure 7, it can also be seen that the SPH method can simulate the flow field inside and around the porous media at a porosity of 0.349, and results are consistent with the experiment of the flow field inside and outside the porous media. The model error is between 10% and 15%.

Observing Figures 6 and 7, which show the average flow direction and vertical velocity components at different locations, it is also possible to notice specific features that are worth to be highlighted. At X/H = 0.33, the velocity gradient is large. In the vertical position, the upstream edge of the structure (X/H = 0.33) is 1 < Y/h < 1.5, and the velocity is small. At the downstream edge of the structure (X/H = 1.67), according to its velocity curve, it can be seen that the permeability of porous media is good, and its reverse velocity is small.

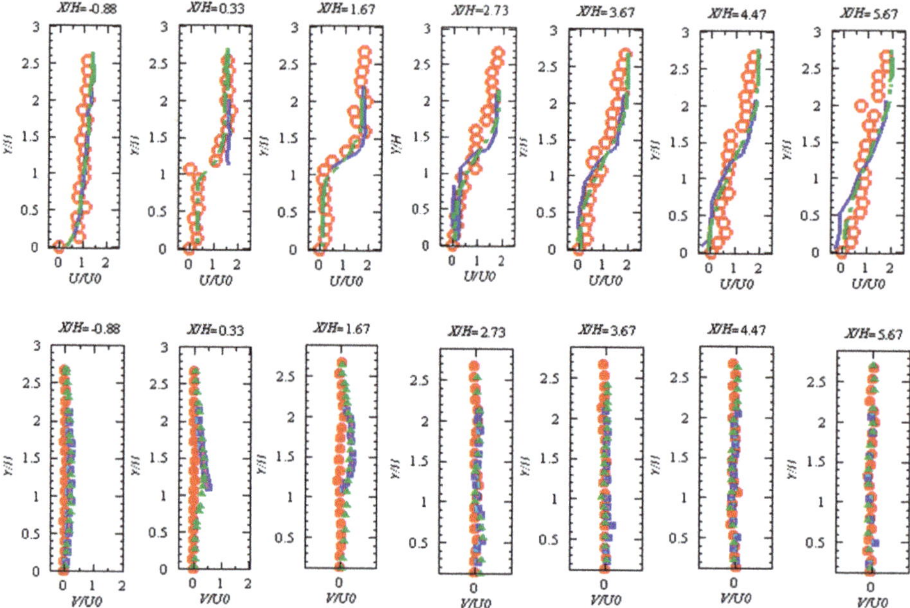

Figure 7. Comparison of longitudinal and vertical velocity for porosity 0.349 Legend: Red open circle: numerical simulation of longitudinal velocity of SPH; Green dash dot: numerical simulation of longitudinal velocity presented by Chan et al. [7]; Blue solid line: experimental longitudinal velocity values from Chan et al. [7]; Red solid circle: numerical simulation of vertical velocity of SPH; Green solid triangle: numerical simulation of vertical velocity of Chan et al. [7]; Blue solid square: experimental vertical velocity values from Chan et al. [7]. H: water depth (m); U: longitudinal flow rate (*m/s*), V: vertical flow rate (*m/s*), U0: Inlet boundary velocity (*m/s*). X is the distance between the particle and the left boundary of the pore structure, H is the height of the pore structure, and X/H is a dimensionless treatment of the X axis. The pore structure is placed at the origin of this system.

X/H = 0.33 indicates that the position is 0.02475 m from the left boundary of the pore structure which is placed at the origin. This location is characterized by the velocity distribution of the flow field when the water flow is coupled with the left boundary of the pore structure. Because this is a periodic boundary and a damping region is added to the flow field, the damping region typically causes some disturbance to the flow field in the initial stage, but the effect of the damping region on the flow field velocity gradually decrease after the particles run for 20 s. It was verified within the model that the deviation has a small effect on the flow field around the pores, and it can be seen that the flow field around the pores is not greatly affected by the damping area.

However, in order to ensure that in the oncoming flow field of the pore structure, the flow simulation of the three parts of the pore flow field and the back vortex field of the pore structure have good convergence, the convergence analysis (Figure 8) was completed.

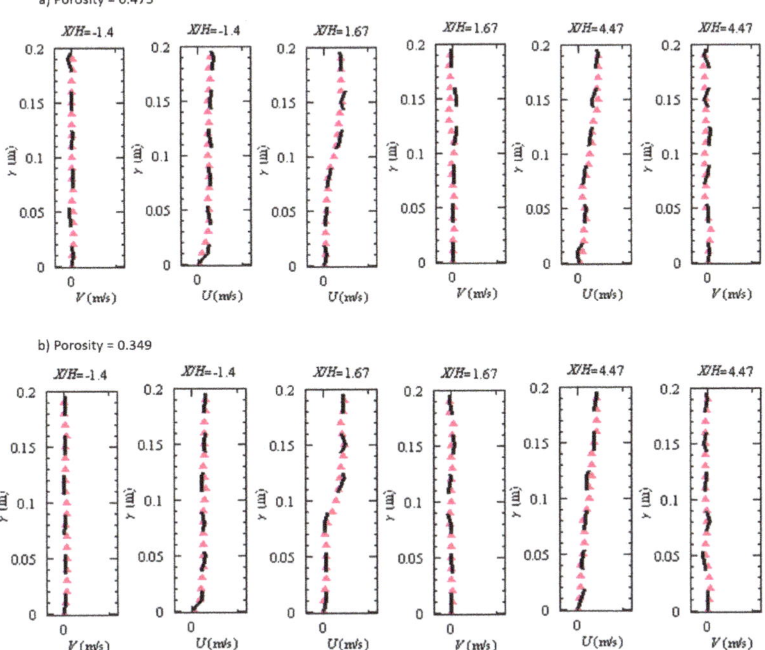

Figure 8. Solid purple triangle D = 0.0104; Dashed line when D = 0.01. X/H = −1.4 indicates the oncoming flow field of the pore structure; X/H = 1.67 indicates that the water flows through the pore structure field; X/H = 4.47 indicates the back vortex field of the pore; Y (m) indicates that the particles of the water flow from the horizontal plane vertical distance; U (*m/s*) represents the longitudinal water flow velocity; V (*m/s*) represents the vertical water flow velocity.

Figure 8 depicts the convergence analysis of the vortex area behind the pore structure and the vortex structure in the oncoming flow area when the porosity is 0.475 and 0.349, respectively. Analyzing Figure 8, it is possible to state that the particles pass through these three areas, and that due to the good convergence and good curve fitting that can be observed, the simulation with different particle sizes still meets the requirements of good convergence.

In practical engineering, the seepage problem of permeable structure will be helpful to the analysis of important engineering structures. It is necessary to study the flow state of fluid in permeable structure. The traditional velocity measurement method is difficult to be used in permeable buildings. Particle image velocimetry (PIV) and particle tracking velocimetry (PTV) are other robust experimental techniques to solve this problem [28–35]. In this study, the Lagrangian particle tracking method is used to measure the velocity distribution along the water depth at different positions.

5. Model Application

According to the model verification previously described in Section 4, it can be confirmed that the numerical simulation method based on SPH simulates appropriately the water flowing on the porous media, thereby it was decided to apply this model to a water tank and a porous media featuring periodic boundaries and a damping zone. The length of the domain is 60 m, at a porosity value is 0.5, particle size is 0.15, pore length, width, and height ranges within 0.6 m and 1.5 m, water depth considered corresponds to 2.4 m, viscosity is 1.0×10^{-6} (Pa·s), structure coordinates are 30.0, 0.0. The schematic diagram of this configuration is displayed in Figure 9.

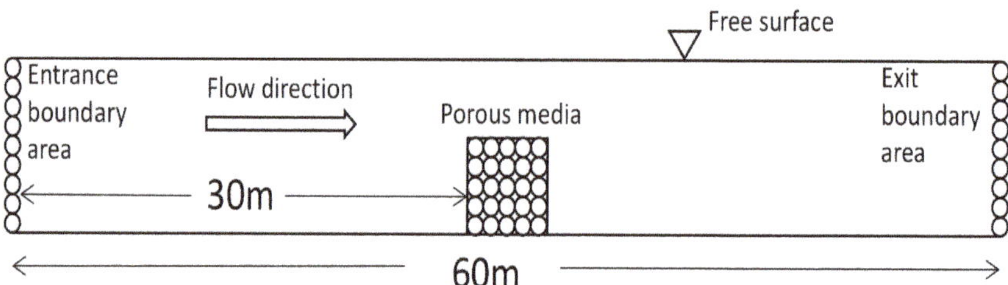

Figure 9. Schematic diagram of numerical water tank.

5.1. Flow Field Velocity Distribution Diagram of Different Volumes of Porous Media

Figure 10 shows the flow field velocity distribution of different volumes of porous media at the porosity of 0.5. According to the flow field displayed below, as the volume of the porous media increases (from top to bottom), the longitudinal flow velocity U of the upper layer of the porous media gradually increases, and the upward-incidence flow region is also expanding, but the enlarged region is not well defined. During the expansion process, the back vortex area expands, and multiple vortex motions are formed. The vortex area increases with the volume of the porous media, and begins to form when the vertical section of the porous media is equal to 1.2 m × 1.2 m. While there is an increase of pore structure volume, the back vortex area expands as well but the eddy strength decreases.

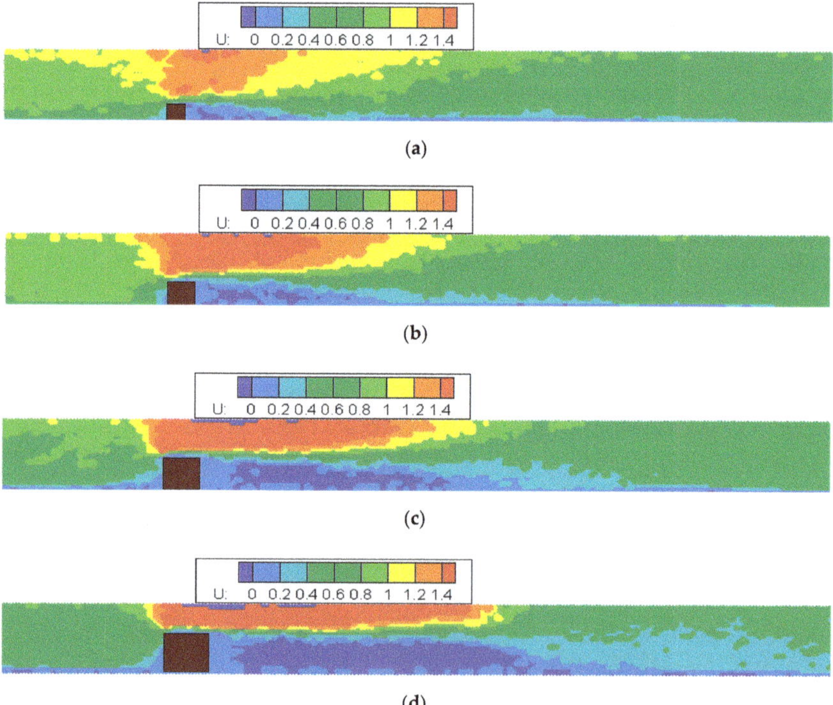

Figure 10. Flow field velocity distribution of different volumes of porous media at a porosity of 0.5. U: Longitudinal velocity (*m/s*). (**a**) Pore structure: 0.6×0.6 (m^2); (**b**) Pore structure: 0.9×0.9 (m^2); (**c**) Pore structure: 1.2×1.2 (m^2); (**d**) Pore structure: 1.5×1.5 (m^2)

5.2. Analysis of Flow Field Inside and Outside the Porous Media

Figure 11 displays the horizontal and vertical velocity of the flow in the upstream area of 0.6 m × 0.6 m pore structure (the area where water flows through the pore structure and the vortex area). In this Figure 11, as well as for Figures 12–14, x = 25–29.0 m represents the horizontal coordinate range of the upstream area, x = 30.1–30.5 m represents the horizontal coordinate range of the flow through the pore structure, and x = 31.0–36.0 m represents the horizontal coordinate range of the vortex area.

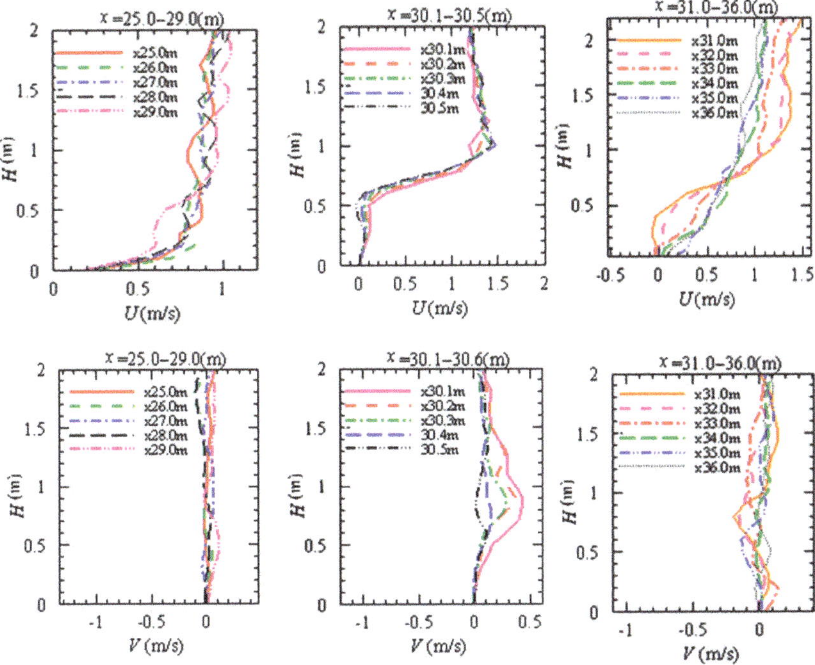

Figure 11. Characteristics of longitudinal and vertical flow fields when the vertical section of the pore is 0.6 m × 0.6 m; solid line: x = 25.0; x = 30.1; x = 31.0; dashed line: x = 26.0; x = 30.2; x = 32.0; dotted line: x = 27.0; x = 30.3; x = 33.0. Long dashed line: x = 28.0; x = 30.4; x = 34.0; double dotted line: x = 29.0; x = 30.5; x = 35.0; dotted line: x = 36.0.

Moving from x = 0.0 m towards the pore structure (located at 30.0 m) when the distance between the water particles and the pore material becomes smaller, the longitudinal velocity *u* on the upper side of the pore material increases, as well as the internal velocity of the pore material. The average velocity of this area tends to be stable at about 1 *m/s*, and the vertical average velocity V ranges between −0.2–0.1 *m/s*.

When x = 30.1–30.5 m, the water flows through the pore area. Increasing x, at the initial part or the pore structure, the longitudinal velocity *u* increases from 0.15 *m/s* to 1.5 *m/s*, showing the trend of short-term reflux. Therefore, it can be considered that the pore structure of 0.6 m × 0.6 m has an obvious energy dissipation effect, while in the upper water area of the pore structure, the flow velocity shows a slow upward trend and tends to be stable at 1.3 *m/s*. In the meantime, the vertical average velocity V ranges between −0.2 and 0.4 *m/s*.

When x = 31.0–36.0 m, this stage is the eddy current area at the back of the pore structure. It can be seen from the change of the back vortex flow field that with the increase of the volume of porous media, there is a turning point at the speed of H (m) = 0.6 m. There are two stages of velocity curves, before and after the turning point, that can be described independently. The first stage represents the velocity curve inside the pore structure, and the second stage corresponds to the velocity curve outside the pore structure. When *H* is less than the height of pore structure, the velocity of the flow particles increases with the increase of distance from the right boundary of pore structure, with an average velocity of 0.5 *m/s*; when the water level *H* is greater than the height of the pore structure, the velocity decreases getting far away from the right boundary of the pore structure, while when x > 30.6 m, and H < 0.6, the velocity of the flow particles on the right boundary of pore structure decreases with the increase of distance from the right boundary of the pore structure. When x = 36.0 m, the flow particle velocity reaches the peak of 1.3 *m/s*. For this scenario, the vertical average velocity V is between −0.2 and 0.1 *m/s*.

Figure 12 displays the horizontal and vertical velocity of the flow in the upstream area of 0.9 m × 0.9 m pore structure (the area where water flows through the pore structure and the vortex area).

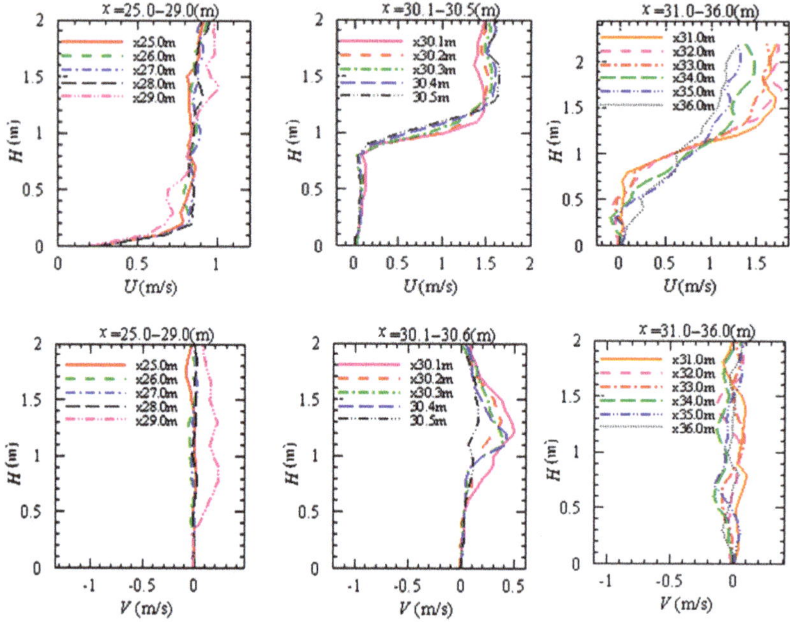

Figure 12. Characteristics of longitudinal and vertical flow fields when the vertical section of the pore is 0.9 m × 0.9 m; solid line: x = 25.0; x = 30.1; x = 31.0; dashed line: x = 26.0; x = 30.2; x = 32.0; dotted line: x = 27.0; x = 30.3; x = 33.0. Long dashed line: x = 28.0; x = 30.4; x = 34.0; double dotted line: x = 29.0; x = 30.5; x = 35.0; dotted line: x = 36.0.

When x = 25–29.0 m, with the increase of x, that is, when the distance between water particles and pore structure becomes smaller, the particle flow velocity in the upstream area of pore structure slightly decreases when the longitudinal velocity u on the upper side of pore structure is 0.6 m higher than the one recorded on the side, while the internal velocity growth of the pore structure is still almost negligible. The pore structure is more stable when the overall average velocity is 0.6 m higher than the side length, and the average up-flow velocity tends to be stable when it is close to 1 *m/s*, while the vertical average velocity V ranges between −0.2 and 0.1 *m/s*. With the increase of x, the longitudinal velocity u on the upper side of the porous material increases slightly, and the decrease observed on the left side is equal to the height of the porous material. The total velocity of the upwelling increases gradually, and tends to be stable near 0.9 *m/s*. The vertical velocity of the upwelling area remains almost stable at 0 *m/s*.

When x = 30.1–30.5 m, water flows through the pore structure area. With the increase of x, the longitudinal velocity *u* of the fluid in the pore structure decreases gradually, but this phenomenon is not significant, and the reflux trend weakens. Therefore, it can be judged that the energy dissipation effect of the pore structure is slightly stronger when the side length is 0.9 m than when the side length is 0.6 m, while in the upper water area of the pore structure, the flow velocity shows a slow upward trend. When the height h is constant, the flow velocity decreases with the increase of x, and the overall flow velocity still increases with the increase of H.

When x = 31.0–36.0 m, an eddy current area at the back of pore structure can be noticed. It can be seen from the change of the back vortex flow field that there is a turning point in the velocity distribution curve at H = 1.5 m. When H is less than the height of porous material, the velocity increases with the increase of x, and the average velocity is 0.25 *m/s*; when h is greater than the height of porous material, the velocity decreases with the increase of x, and reaches 1.7 *m/s* at x = 36.0 m, and the back vortex strength decreases with the increase of the distance from the hole. The vertical velocity is relatively stable, in the range of −0.15 and 0.1 *m/s*.

Figure 13 displays the horizontal and vertical velocity of the flow in the upstream area of 1.2 m × 1.2 m pore structure (the area where water flows through the pore structure and the vortex area). When x = 25–29.0 m, with the increase of x, the particle velocity in the upstream area of the pore structure decreases slightly when the longitudinal velocity u on the upper side of the pore structure is 0.9 m longer than that on the side, and the average velocity of upwelling tends to be stable at the place close to 0.8 *m/s*, and the vertical average velocity V is between −0.15–0.1 *m/s*. When x = 30.1–30.5 m, water flows through the pore area.

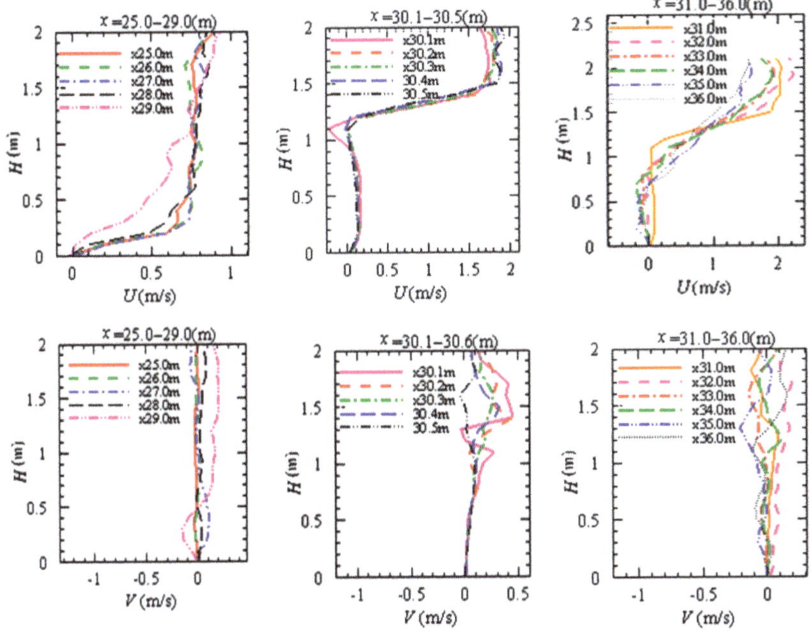

Figure 13. Characteristics of longitudinal and vertical flow fields when the vertical section of the pore is 1.2 m × 1.2 m; solid line: x = 25.0; x = 30.1; x = 31.0; dashed line: x = 26.0; x = 30.2; x = 32.0; dotted line: x = 27.0; x = 30.3; x = 33.0. Long dashed line: x = 28.0; x = 30.4; x = 34.0; double dotted line: x = 29.0; x = 30.5; x = 35.0; dotted line: x = 36.0.

With the increase of x, the longitudinal velocity U of the fluid in the pore structure is gradually stable at 0.1 *m/s*, and the reflux trend is not strong. It can be seen that for 1.2 m × 1.2 m porous material, the energy dissipation effect of porous material is also stronger than before. When x = 31.0–36.0 m, when H is greater than the hole height, the velocity decreases with the increase of x, and the velocity at the right boundary of the pore structure decreases with the increase of the distance from the right boundary of the hole to 2.0 *m/s*. The back vortex intensity also decreases with the increase of x. The vertical velocity is relatively stable, in the range of −0.20 and 0.35 *m/s*.

Finally, Figure 14 displays the horizontal and vertical velocity of the flow in the upstream area of 1.5 m × 1.5 m pore structure (the area where water flows through the pore structure and the vortex area). When x = 25–29.0 m, the overall velocity of the upwelling increases gradually, and tends to be stable (approximately 0.75 *m/s*), while the vertical velocity ranges between −0.2 and 0.2 *m/s*.

When x = 30.1–30.5 m, water flows through the pore area. With the increase of x, the longitudinal velocity *u* of the fluid in the pore structure tends to 0, and the reflux trend is not noticeable. It can be concluded that the hole has obvious energy dissipation effect on the hole of 1.5 m × 1.5 m, while in the upper water area of the hole, the velocity keeps rising slowly. With the increase of distance, the water content of the upper part of the pore structure shows a slowly rising trend, the vertical velocity increases at the same time. When x = 31.0–36.0 m, when h is less than the height of porous material, the flow velocity increases with the increase of the right boundary distance of porous material, and the average flow velocity is approximately 0.02 *m/s*; when h is greater than the height of porous material, the flow velocity decreases with the increase of the right boundary distance of porous material, and the flow velocity at the right boundary of porous material reaches 2.2 *m/s*, and the back vortex strength decreases with the increase of the distance from the hole. The vertical velocity is relatively stable, in the range between −0.35 and 0.35 *m/s*.

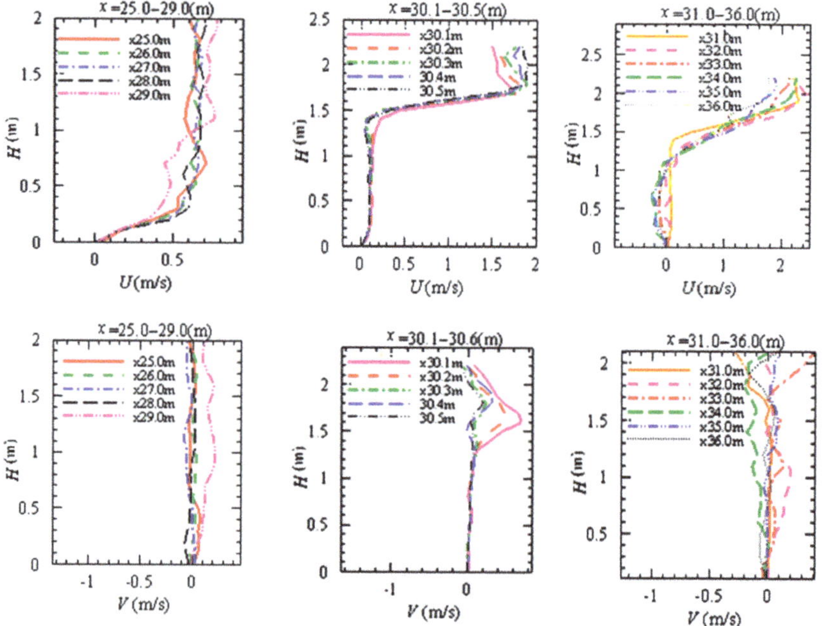

Figure 14. Characteristics of longitudinal and vertical flow fields when the longitudinal section of the pore is 1.5 m × 1.5 m; solid line: x = 25.0; x = 30.1; x = 31.0; dashed line: x = 26.0; x = 30.2; x = 32.0; dotted line: x = 27.0; x = 30.3; x = 33.0. Long dashed line: x = 28.0; x = 30.4; x = 34.0; double dotted line: x = 29.0; x = 30.5; x = 35.0; dotted line: x = 36.0.

In summary, based on the changes of up-flow field tested, it can be stated that by increasing the pore volume, the up-flow velocity typically decreases and the energy dissipation becomes higher, but changes associated with the vertical velocity are not significant with the change of volume for the pore structure. Focusing on the change of the flow field inside and outside the hole it is possible to confirm that the internal velocity of the pore is mainly in the range of 0 and 0.2 *m/s*. With the increase of the porous media' volume, the internal velocity difference of the pore decreases gradually, while the particle velocity directly above the pore structure increases with the increase of the pore volume, and the velocity difference of the upper layer flow also gradually increases.

Furthermore, it can be seen from the change of the back vortex flow field that, with the increase of the volume of porous media, there are inflexion points at H (m) = 0.6, 0.9, 1.2 and 1.5 for the four groups of horizontal velocity, respectively, thus forming multiple velocity curves before and after the inflexion point. When H is less than the height of the hole, the flow velocity increases with the increase of x, however when H is greater than the height of the hole, the flow velocity increases with the increase of x. The intensity of the back vortex also decreases with the increase of x. From the change of vertical velocity, it can be observed that the velocity basically varies from −0.5 *m/s* to 0.5 *m/s*.

5.3. Longitudinal and Vertical Flow Field Distribution under Different Porosity

Porosity is also an important factor influencing the characteristics of the flow field in the porous media [36]. Therefore, the pore size was 1.2 × 1.2 (m), and the oncoming flow and reflux vortices were analyzed for porosity values of 0.5, 0.25 and 0.125, respectively. The velocity distribution and the flow field along the water depth were analyzed for each porosity investigated.

Figure 15 displays the flow field in the upstream area of pore structure under different porosities 0.5, 0.25, and 0.125 for x = 25.0–29.0. As the porosity decreases, it can be seen that the longitudinal velocity curve of the upstream flow tends to be gradually stable.

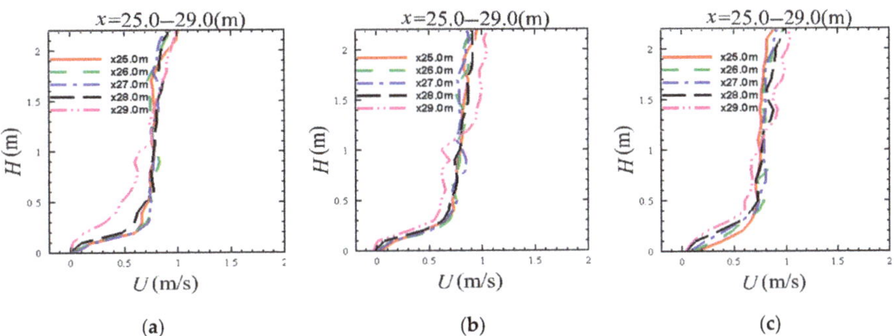

Figure 15. Longitudinal velocity distribution of the flow field in the upstream area of the pore structure when the porosity is 0.5, 0.25 and 0.125 and x = 25.0–29.0 m. (a) n_w = 0.5; (b) n_w = 0.25; (c) n_w = 0.125.

Figure 16 shows the flow field when the water flows through the pore structure under different porosities 0.5, 0.25 and 0.125 for x = 30.1–30.5. The longitudinal velocity difference of the internal flow field of the pore structure increases with the decrease of porosity.

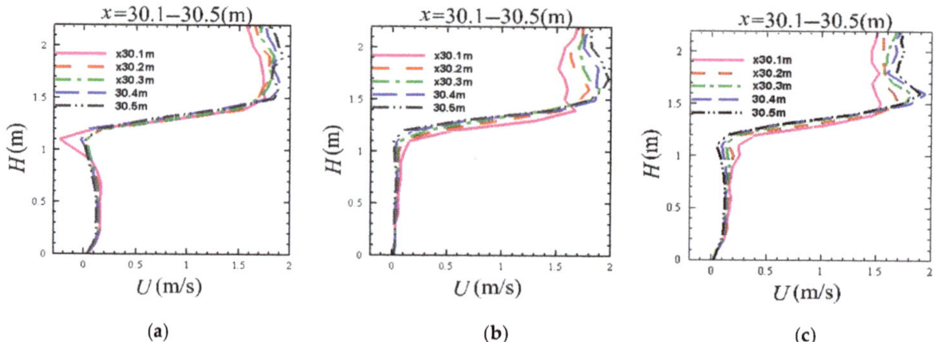

(a) **(b)** **(c)**

Figure 16. Longitudinal velocity distribution of the flow field inside and directly above the pore structure when the porosity is 0.5, 0.25 and 0.125 and x = 30.1–30.5.0 m. (**a**) $n_w = 0.5$; (**b**) $n_w = 0.25$; (**c**) $n_w = 0.125$.

Figure 17 presents the flow field in the back vortex region with different porosities 0.5, 0.25 and 0.125 for x = 31.0–36.0. With the decrease of porosity, the flow rate changes are not significant, but the back vortex region increases slightly, and the flow rate at the inflection point also initially increases but later decreases in combination with the increase of porosity. When the porosity is 0.25, the particle longitudinal velocity at the inflection point is the maximum calculated.

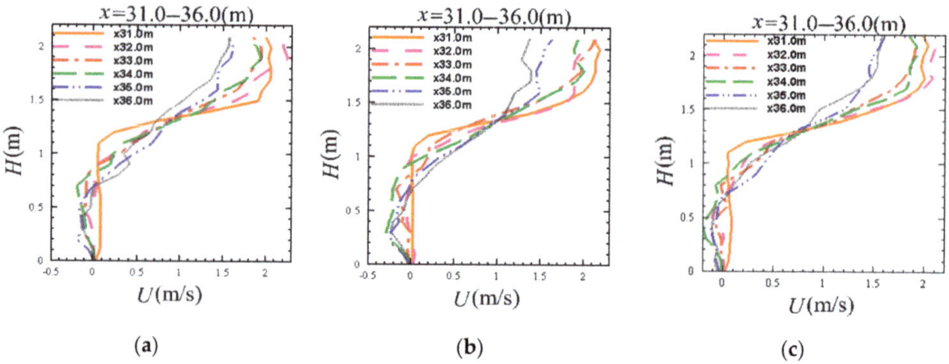

(a) **(b)** **(c)**

Figure 17. Longitudinal velocity distribution in the back vortex flow field when the porosity is 0.5, 0.25, and 0.125 and x = 31.0–36.0 m. (**a**) $n_w = 0.5$; (**b**) $n_w = 0.25$; (**c**) $n_w = 0.125$.

Figure 18 displays the vertical velocity distribution of the flow field in the upstream area of the pore structure under different porosities 0.5, 0.25, and 0.125 for x = 25.0–29.0. With the gradual decrease of porosity, it can be seen that the vertical velocity difference of each position of the upstream flow is gradually reduced.

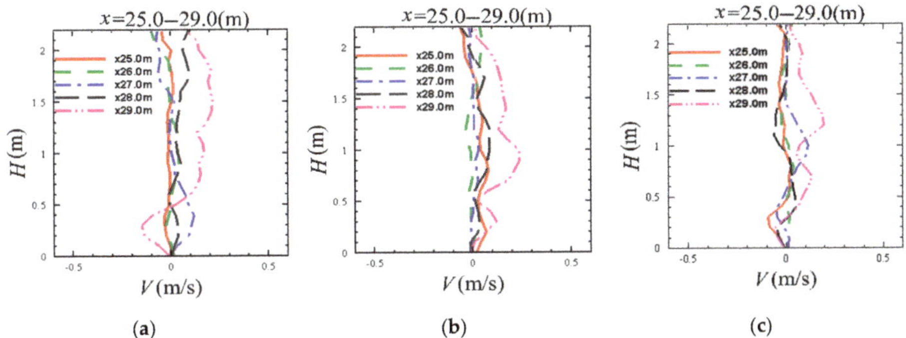

Figure 18. Vertical velocity distribution of the flow field in the upstream area of the pore structure when the porosity is 0.5, 0.25 and 0.125 and x = 25.0–29.0 m. (**a**) $n_w = 0.5$; (**b**) $n_w = 0.25$; (**c**) $n_w = 0.125$.

Figure 19 shows the vertical velocity distribution of the flow field inside and directly above the pore structure under different porosities 0.5, 0.25, and 0.125 for x = 30.1–30.5. With the gradual decrease of porosity, it can be seen that in the fluid solid interface area, the change of horizontal velocity becomes more evident, and the velocity reduces.

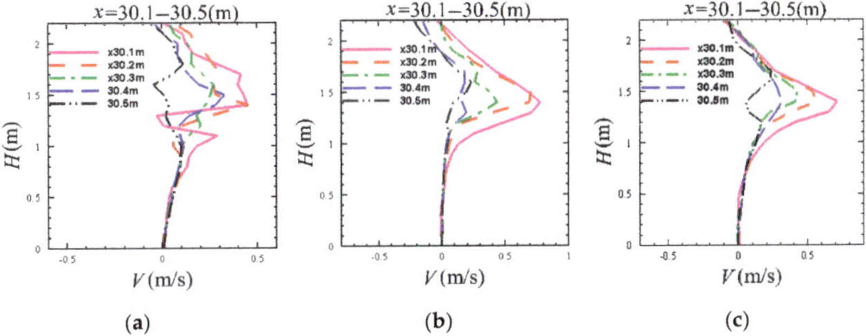

Figure 19. Vertical velocity distribution of the flow field inside and directly above the pore structure when the porosity is 0.5, 0.25 and 0.125 and x = 30.1–30.5 m. (**a**) $n_w = 0.5$; (**b**) $n_w = 0.25$; (**c**) $n_w = 0.125$.

Figure 20 presents the vertical velocity distribution of the back vortex flow field under different porosities 0.5, 0.25 and 0.125 for x = 31.0–36.0. As the porosity decreases gradually, it can be seen that when the height H is constant, the vertical velocity difference at each position is characterized by a gradually increasing trend.

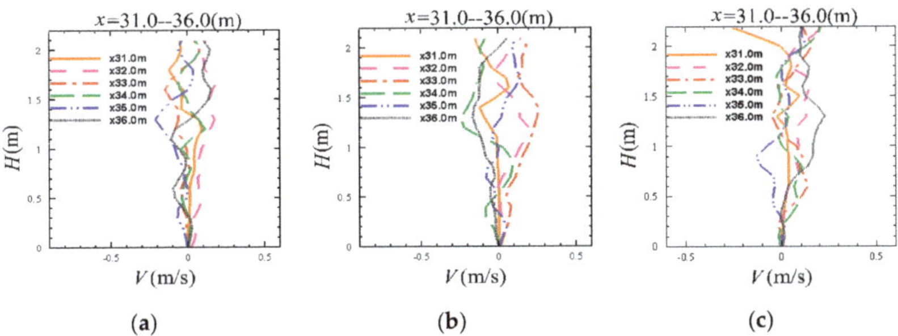

Figure 20. Vertical velocity distribution in the back vortex flow field when the porosity is 0.5, 0.25 and 0.12 and x = 31.0–36.0 m. (**a**) $n_w = 0.5$; (**b**) $n_w = 0.25$; (**c**) $n_w = 0.125$.

In summary, it can be seen from Figures 15–20 that as the porosity increases, the longitudinal flow velocity of the oncoming flow slightly changes, but the overall change is not significant, plus the longitudinal flow velocity of the back vortex decreases. Considering the vertical flow velocity of the upstream area, despite the increase of porosity, this parameter remains within the range between -0.2 *m/s* and 0.2 *m/s*, and the vertical flow velocity through the porous media decreases with the increase of porosity. From the vertical flow velocity of the back vortex, as the porosity increases, the vertical flow velocity tends to decrease slightly overall.

Due to the infiltration of the water, the permeable structure has a significant impact on its discharge capacity [37]. It can be seen from the horizontal and vertical comparison of the flow field in Figure 16 that with the increase of porosity, the flow resistance decreases. The distribution of longitudinal velocity of porous material changes in the upper layer of porous material.

The maximum turbulence intensity near the surface and behind the structure increases proportional to the porosity [38] at the same downstream position.

5.4. Convergence Verification of Pore Logistics Field Model

In order to verify the convergence of the SPH in the numerical model, the flow field characteristics of each section previously investigated were analyzed by applying multiple particle size distributions [8]. Up-flow area, pore flow area, back vortex area, and the horizontal and vertical velocity distribution along the depth were calculated with the particle size D of 0.096 m, 0.01 m, and 0.104 m as shown in Figures 21 and 22 to verify convergence, at the selected positions x = 28.0 m, 30.3 m, 33 m.

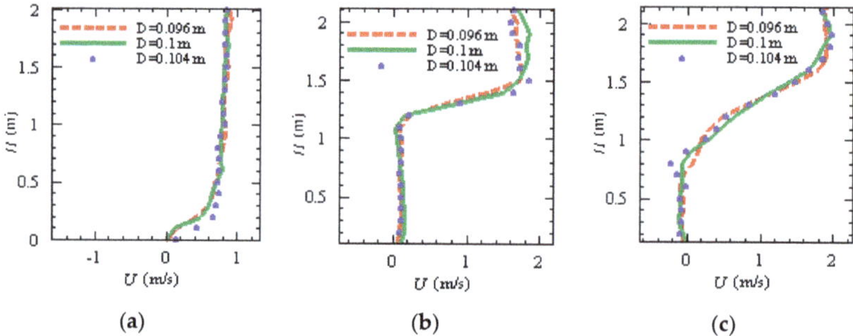

Figure 21. Longitudinal velocity distribution of the inflow field with different particle spacing (**a**), longitudinal velocity distribution of the sea current flowing through the porous media with different particle spacing (**b**), and longitudinal velocity distribution of the rear vortex field with different particle spacing (**c**).

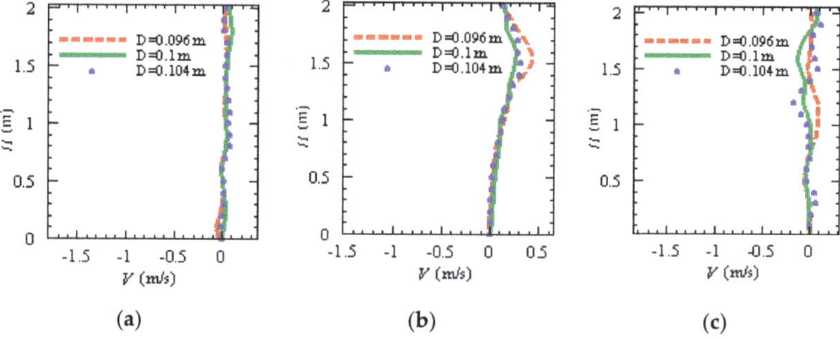

Figure 22. Vertical velocity distribution of the inflow field with different particle spacing (**a**), vertical velocity distribution of the sea current flowing through the porous media with different particle spacing (**b**), and vertical velocity distribution of the rear vortex field with different particle spacing (**c**).

It was previously noticed that when the height of the pores is 1.2 m, the flow direction is opposite to the others when D = 0.096. This elevation corresponds to the upper flow area of the pore structure and justify the presence of turbulence as shown in the graph. However, results displayed in Figures 21 and 22 show that the velocity distributions of the internal and external flow fields formed by the oncoming flow field, the back vortex flow field and the water flow through the porous media are consistent despite different particle spacing and the error range is between 10% and 15%, hence providing an acceptable grade of convergence.

Finally, Figures 23 and 24 show the back vortex field of the pore structure when the porosity is 0.349 and 0.475, respectively.

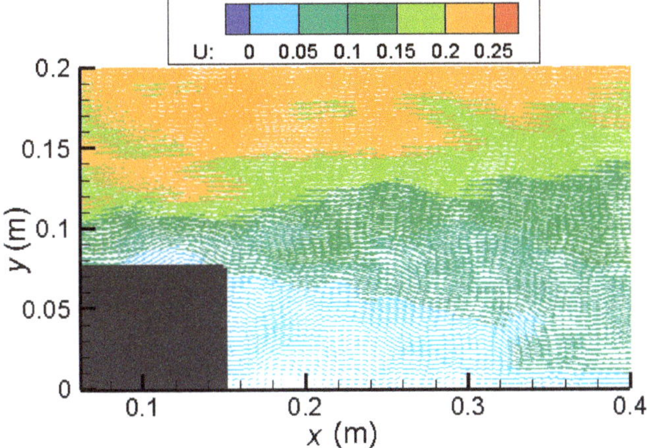

Figure 23. Back vortex field of the pore structure when the porosity is 0.349. U: Longitudinal velocity (*m/s*).

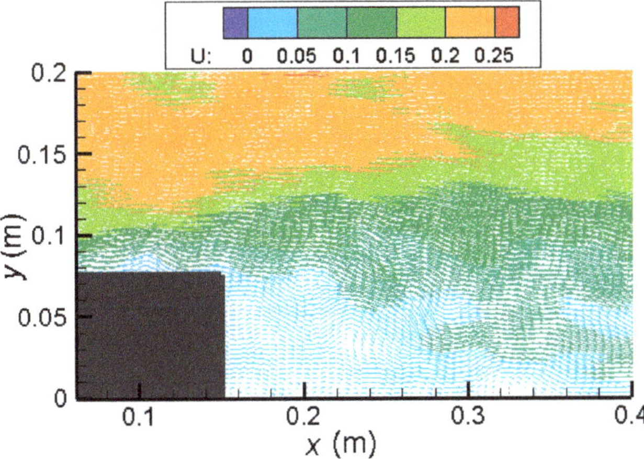

Figure 24. Back vortex field of the pore structure when the porosity is 0.475. U: Longitudinal velocity (*m/s*).

It can be seen from the flow field distribution diagrams in Figures 23 and 24 that the energy in the vicinity of the pore structure presents a dissipative trend and shows different vorticities under different porosities. When the porosity is 0.475, the energy dissipation in the vicinity of the pore structure is weak, the vortex flow pattern is more noticeable and multiple vortex flow patterns are formed on the back of the pore structure. However, when the porosity is 0.349, the energy dissipation near the pore structure is relatively strong, and the intensity of the vortex flow is relatively weak.

To enable academic colleagues to replicate the presented numerical data and have a better estimation of the numerical capabilities of the numerical method, details about the total number of particles used, the particle size, the physical simulation times plus the CPU time and the CPU cores used are displayed and summarized in Table 1 for all the configurations tested (D = 0.0104 m, D = 0.01 m, D = 0.096 m, D = 0.1 m, and D = 0.104 m).

Table 1. Simulation details for each configuration tested.

Particle Size D (m)	Total Number of Particles	Total Physical Simulation Time (s)	CPU Time	CPU Cores
0.0104	13,914	100	27 h 05 min	4
0.01	15,000	100	27 h 14 min	4
0.096	16,225	100	8 h 28 min	20
0.1	15,000	100	8 h 23 min	20
0.104	13,914	100	8 h 18 min	20

6. Conclusions

Nowadays, a variety of man-made structures such as submerged breakwaters, fishing reefs, outfall protections, armor layers, rubble-mound or berm breakwaters, built of gravel or artificial units, can be found in coastal areas worldwide for multiple purposes (e.g., coastal flooding protection, coastal erosion protection). All these structures are characterized by the fact that either some of their layers or their full structure are porous, hence they are permeable to flows induced by waves and currents. Therefore, the hydrodynamics, the stability and performance of these structures are dependent on the characteristics of the waves and their interaction with permeable material and it is crucial to provide mathematical formulations to model these features and the role played by the porous material considering its geometry, its location, and its characteristics (which could influence the wave propagation, diffusion, overtopping, and the consequent turbulence generated).

This paper enabled the calibration and validation of a SPH model utilized to characterize the influence of porous media on flow propagation and outcomes can be summarized as follows:

1. Based on the analysis of the area upstream the porous media, the longitudinal flow velocity of the ascending flow decreases slightly with the increase of the volume of the pore structure;

2. Based on the analysis of the internal and external flow fields of the porous media, the internal flow velocity of the porous media for the configurations tested is in the range of 0–0.2 *m/s*. It was possible to notice that with the increase of the volume of the porous media, the flow velocity of the upper layer of the porous media firstly tends to accelerate and then stabilizes. The longitudinal flow velocity of the upper layer of the flow increases with the increase of water depth, while the vertical velocity fluctuates sharply at the fluid-solid boundary;

3. Based on the back vortex analysis, there is an inflection point in the flow velocity distribution of the upper and lower layers of the porous media. As the distance from the porous media increases, the flow velocity increases, so that the velocity distribution curve of the back vortex flow forms different concavities and convexities before and after the porous media, and as the pore volume gradually increases, a plurality of vortices are formed on the backflow surface. With the rotary motion, the area of the back vortex increases as the volume of the pore increases, but the strength decreases as the volume increases.

4. Overall, with the increase of porosity, the longitudinal flow and back vortex flow slightly decrease in the vertical direction, and the SPH-based model calculation has good convergence (error between experimental and numerical results within 10%).

Author Contributions: Conceptualization, Q.G.; methodology, S.W. and Q.G.; software, S.W. and Q.G.; validation, S.W. and Q.G. and M.R.; formal analysis, S.W. and Q.G.; investigation, S.W., Q.G. and M.R.; resources, S.W. and Q.G.; data curation, S.W., Q.G. and M.R.; writing—original draft preparation, S.W., Q.G. and M.R.; writing—review and editing, S.W., Q.G. and M.R.; visualization, S.W. and Q.G.; supervision, Q.G. and M.R.; project administration, Q.G.; funding acquisition, Q.G. All authors have read and agreed to the published version of the manuscript.

Funding: This research was funded by Zhejiang Province Natural Science Foundation, grant number LQ16E090001, the National Natural Science Foundation of China, grant number 51509134.

Acknowledgments: The author is very grateful to Ningbo University, the National Natural Science Foundation of China and the Natural Science Foundation of Zhejiang Province. All authors are very grateful to Dr Songdong Shao for his constructive comments and continuous inspiration to complete this work.

Conflicts of Interest: The authors declare no conflict of interest.

References

1. Espinoza-Andaluz, M.; Velasco-Galarza, V.; Romero-Vera, A. On hydraulic tortuosity variations due to morphological considerations in 2D porous media by using the Lattice Boltzmann method. *Math. Comput. Simul.* **2020**, *169*, 74–87. [CrossRef]
2. Akbari, H. Modified moving particle method for modeling wave interaction with multi layered porous structures. *Coast. Eng.* **2014**, *89*, 1–19. [CrossRef]
3. Houreh, N.B.; Shokouhmand, H.; Afshari, E. Effect of inserting obstacles in flow field on a membrane humidifier performance for PEMFC application: A CFD model. *Int. J. Hydrog. Energy* **2019**, *44*, 30420–30439. [CrossRef]
4. Mei, T.-L.; Zhang, T.; Candries, M.; Lataire, E.; Zou, Z.-J. Comparative study on ship motions in waves based on two time domain boundary element methods. *Eng. Anal. Bound. Elem.* **2020**, *111*, 9–21. [CrossRef]
5. Basser, H.; Rudman, M.; Daly, E. SPH modelling of multi-fluid lock-exchange over and within porous media. *Adv. Water Resour.* **2017**, *108*, 15–28. [CrossRef]
6. Aganetti, R.; Lamorlette, A.; Thorpe, G.R. The relationship between external and internal flow in a porous body using the penalisation method. *Int. J. Heat Fluid Flow* **2017**, *66*, 185–196. [CrossRef]
7. Chan, H.C.; Leu, J.M.; Lai, C.J. Velocity and turbulence field around permeable structure:Comparisons between laboratory and numerical experiments. *J. Hydraul. Res.* **2007**, *45*, 216–226. [CrossRef]
8. Gui, Q.; Dong, P.; Shao, S.; Chen, Y. Incompressible SPH simulation of wave interaction with porous structure. *Ocean Eng.* **2015**, *110*, 126–139. [CrossRef]
9. Khayyer, A.; Gotoh, H.; Shimizu, Y.; Gotoh, K.; Falahaty, H.; Shao, S. Development of a projection-based SPH method for numerical wave flume with porous media of variable porosity. *Coast. Eng.* **2018**, *140*, 1–22. [CrossRef]
10. Vanneste, D.; Troch, P. 2D numerical simulation of large-scale physical model tests of wave interaction with a rubble-mound breakwater. *Coast. Eng.* **2015**, *103*, 22–41. [CrossRef]
11. Shao, S. Incompressible SPH flow model for wave interactions with porous media. *Coast. Eng.* **2010**, *57*, 304–316. [CrossRef]
12. Gnanasekaran, B.; Liu, G.-R.; Fu, Y.; Wang, G.; Niu, W.; Lin, T. A Smoothed Particle Hydrodynamics (SPH) procedure for simulating cold spray process—A study using particles. *Surf. Coat. Technol.* **2019**, *377*, 124812. [CrossRef]
13. Shadloo, M.S.; Oger, G.; Le Touze, D. Smoothed particle hydrodynamics method for fluid flows, towards industrial appliactions: Motivations, current state and challenges. *Comput. Fluids* **2016**, *136*, 11–34. [CrossRef]
14. Niu, X.; Zhao, J.; Wang, B. Application of smooth particle hydrodynamics (SPH) method in gravity casting shrinkage activity prediction. *Comput. Part. Mech.* **2019**, *6*, 803–810. [CrossRef]
15. Yin, J.P.; Shi, Z.X.; Chen, J.; Chang, B.H.; Yi, J.Y. Smooth particle hydrodynamics-based characteristics of a shaped jet from different materials. *Strength Mater.* **2019**, *51*, 85–94. [CrossRef]
16. Avesani, D.; Dumbser, M.; Chiogna, G.; Bellin, A. An alternative smooth particle hydrodynamics formulation to simulate chemotaxis in porous media. *J. Math. Biol.* **2017**, *74*, 1037–1058. [CrossRef]

17. Eghtesad, A.; Knezevic, M. A new approach to fluid–structure interaction within graphics hardware accelerated smooth particle hydrodynamics considering heterogeneous particle size distribution. *Comput. Part. Mech.* **2018**, *5*, 387–409. [CrossRef]

18. Yang, H.X.; Li, R.; Lin, P.Z.; Wan, H.; Feng, J. Two-phase smooth particle hydrodynamics modeling of air-water interface in aerated flows. *Sci. China Technol. Sci.* **2017**, *60*, 479–490. [CrossRef]

19. Shao, S. Incompressible smoothed particle hydrodynamics simulation of multifluid flows. *Int. J. Numer. Methods Fluids* **2011**, *69*, 11. [CrossRef]

20. Khayyer, A.; Gotoh, H.; Shao, S.D. Corrected incompressible SPH method for accurate water-surface tracking in breaking waves. *Coast. Eng.* **2008**, *55*, 236–250. [CrossRef]

21. Shao, S.; Gotoh, H. Turbulence particle models for tracking free surfaces. *J. Hydraul. Res.* **2010**, *43*, 276–289. [CrossRef]

22. Wang, S.; Shu, A.; Rubinato, M.; Wang, M.; Qin, J. Numerical Simulation of Non-Homogeneous Viscous Debris-Flows based on the Smoothed Particle Hydrodynamics (SPH) Method. *Water* **2019**, *11*, 2314. [CrossRef]

23. Shao, S.; Lo, E.Y.M. Incompressible SPH method for simulating Newtonian and non-Newtonian flows with a free surface. *Adv. Water Resour.* **2003**, *26*, 787–800. [CrossRef]

24. Monaghan, J.J. Smoothed Particle Hydrodynamics. *Annu. Rev. Astron. Astrophys.* **1992**, *30*, 543–574. [CrossRef]

25. Huang, C.-J.; Chang, H.-H.; Hwung, H.-H. Structural permeability effects on the interaction of a solitary wave and a submerged breakwater. *Coast. Eng.* **2003**, *49*, 1–24. [CrossRef]

26. Sun, P.N.; Colagrossi, A.; Marrone, S.; Antuono, M.; Zhang, A.M. Multi-resolution Delta-plus-SPH with tensile instability control: Towards high Reynolds number flows. *Comput. Phys. Commun.* **2018**, *224*, 63–80. [CrossRef]

27. Sun, P.N.; Colagrossi, A.; Le Touze, D.; Zhang, A.M. Extension of the δ-plus-SPH model for simulating vortex-induced-vibration problems. *J. Fluids. Struct.* **2019**, *90*, 19–42. [CrossRef]

28. Rubinato, M.; Martins, R.; Kesserwani, G.; Leandro, J.; Djordjevic, S.; Shucksmith, J. Experimental investigation of the influence of manhole grates on drainage flows in urban flooding conditions. In Proceedings of the 14th IWA/IAHR International Conference on Urban Drainage, Prague, Czech Republic, 10–15 September 2017.

29. Rubinato, M.; Martins, R.; Kesserwani, G.; Leandro, J.; Djordjevic, S.; Shucksmith, J. Experimental calibration and validation of sewer/surface flow exchange equations in steady and unsteady flow conditions. *J. Hydrol.* **2017**, *552*, 421–432. [CrossRef]

30. Rubinato, M.; Lee, S.; Martins, R.; Shucksmith, J. Surface to sewer flow exchange through circular inlets during urban flood conditions. *J. Hydroinform.* **2018**, *20*, 564–576. [CrossRef]

31. Martins, R.; Rubinato, M.; Kesserwani, G.; Leandro, J.; Djordjević, S.; Shucksmith, J.D. On the Characteristics of Velocities Fields in the Vicinity of Manhole Inlet Grates During Flood Events. *Water Resour. Res.* **2018**, *54*, 6408–6422. [CrossRef]

32. Nichols, A.; Rubinato, M. Remote sensing of environmental processes via low-cost 3D free-surface mapping. In Proceedings of the 4th IHAR Europe Congress, Liege, Belgium, 27–29 July 2016.

33. Lopes, P.; Shucksmith, J.; Leandro, J.; de Fernandes Carvalho, R.; Rubinato, M. Velocities profiles and air-entrainment characterization in a scaled circular manhole. In Proceedings of the 13th ICUD, Sawarak, Malaysia, 7–12 September 2014.

34. Rojas Arques, S.; Rubinato, M.; Nichols, A.; Shucksmith, J.D. Cost effective measuring technique to simultaneously quantify 2D velocity fields and depth-averaged solute concentrations in shallow water flows. *Flow Meas. Instrum.* **2018**, *64*, 213–223. [CrossRef]

35. Beg, M.N.A.; Carvalho, R.F.; Tait, S.; Brevis, W.; Rubinato, M.; Schellart, A.; Leandro, J. A comparative study of manhole hydraulics using stereoscopic PIV and different RANS models. *Water Sci. Tech.* **2018**, *2017*, 87–98. [CrossRef]

36. Cavelan, A.; Boussafir, M.; Rozenbaum, O.; Laggoun-Défarge, F. Organic petrography and pore structure characterization of low-mature and gas-mature marine organic-rich mudstones: Insights into porosity controls in gas shale systems. *Mar. Pet. Geol.* **2019**, *103*, 331–350. [CrossRef]

37. Zheng, X.; Ma, Q.; Shao, S.; Khayyer, A. Modelling of Violent Water Wave Propagation and Impact by Incompressible SPH with First-Order Consistent Kernel Interpolation Scheme. *Water* **2017**, *9*, 400. [CrossRef]
38. Lo, E.Y.; Shao, S. Simulation of near-shore solitary wave mechanics by an incompressible SPH method. *Appl. Ocean. Res.* **2002**, *24*, 275–286.

Article

An Experimental Study of Focusing Wave Generation with Improved Wave Amplitude Spectra

Guochun Xu [1],*, Hongbin Hao [2], Qingwei Ma [2] and Qinqin Gui [1]

[1]   Faculty of Maritime and Transportation, Ningbo University, Ningbo 315211, China; guiqinqin@nbu.edu.cn
[2]   College of Shipbuilding Engineering, Harbin Engineering University, Harbin 150001, China;
      haohongbin1989@163.com (H.H.); qw_ma2004@yahoo.co.uk (Q.M.)
*   Correspondence: xuguochun@nbu.edu.cn; Tel.: +86-150-584-37295

Received: 15 October 2019; Accepted: 26 November 2019; Published: 28 November 2019

Abstract: We experimentally investigate the generating results of space-time focusing waves based on two new wave spectra, i.e., the quasi constant wave amplitude spectrum (QCWA) and the quasi constant wave steepness spectrum (QCWS), in which amplitude and steepness for each wave component can be adjusted with fixed wave energy. The wavemaker signal consists of a theoretical wavemaker motion signal and two different auxiliary functions at two ends of the signal. By testing a series of focusing waves in a physical wave tank, we found that with given wave energy, the QCWA spectrum can produce a focusing wave with larger crest elevation and farther focusing location from the wavemaker flap, as compared with the QCWS spectrum. However, both spectra lead to larger focusing wave crests when the wave frequency bandwidth was narrowed down and a positive correlation between the generated relative wave crest elevation and the input wave elevation parameter. The two spectra produce different focusing wave positions for the same wave frequency range. We also found that the focusing time strongly relates to the energy of the highest-frequency wave component of the wave spectrum.

Keywords: focusing waves; wave amplitude spectra; space-time parameter; experimental investigations

---

1. Introduction

Focusing wave is a special water wave, different from regular wave or stochastic wave, with a single large wave crest when it happens [1,2]. On the basis of investigations of the triggering mechanisms of focusing waves [3–6], many causes have been identified, such as the space-time focusing of transient waves, wave–current interaction, geometrical focusing due to the seabed topography, atmospheric forcing, nonlinear self-focusing, etc. If the wave heights of focusing waves exceed two to 2.2 times their significant wave heights, they are generally defined as freak waves or rogue waves [7,8]. Hence, focusing waves in laboratory are often employed to model freak wave events observed in extreme sea state [9,10], in order to better understand the generation process, the mechanisms of those extreme waves, and the hydrodynamic loads on floating or fixed ocean structures in extreme sea environments [11–13].

Over the past few decades, various theoretical models, such as wave energy focusing, wave–wave interaction, wave-air coupling, etc. [14–16], have been established to express the generation mechanisms of focusing waves. Among them, the space-time focusing theory of wave energy and wave modulation instability (also named nonlinear self-focusing) theory are the most widely used and are often applied to numerical simulations of focusing or extreme waves. The former assumes that the focusing wave consists of a number of small harmonic wave components which can be superposed to form the focusing wave. Research based on the space-time focusing theory was initially carried out by the linear wave superposition [17,18] and, then, it was followed by low-order wave–wave interactions [19], and wave directional spread [20]. Recently, based on spatio-temporal focusing of wave energy, focusing waves have been produced in fully nonlinear potential or viscous numerical wave tanks [21–24]. As for the latter, i.e., wave modulation instability, a wave group is designed to be composed of carrier waves and their sideband waves. When the sideband disturbance occurs in a wave travelling process, the wave energy of carrier waves is transferred to their sideband waves, leading to wave energy focusing. Waves generated this way are are also called rogue waves or freak waves. Theoretical study on wave modulation instability dates back to the last sixties with investigations on a Stokes wave train with small perturbation [25]. Then, studies were extended to nonlinear four-wave interactions and random wave group [26,27]. In the meantime, several mathematical equations were derived under different physical assumptions, such as the nonlinear Schrödinger equation, the Davey–Stewartson system, the Korteweg–de Vries equation, and the Kadomtsev–Petviashvili equation. Detailed reviews of these mathematical models can be found in [3,28].

On the basis of the abovementioned generating mechanisms and theoretical models, the focusing waves have been mainly produced by two methods in physical wave flumes. One is to produce focusing waves through space-time focusing of wave energy. To achieve wave energy focusing at a specified position and time, the phases of wave components are modulated as zero or $\pi/2$. This phase modulation method has been adopted in the single wave model [19], the double wave model [29], the triple wave model [30], and the NewWave model [31]. The main difference among those models is the ways to specify the amplitudes of wave components. In the first three models, their amplitudes are assigned by some predefined wave spectrum, such as the constant wave amplitude spectrum (CWA), the constant wave steepness spectrum (CWS), and the random sea wave spectrum or their combined spectra, whereas the wave component amplitudes in the NewWave model are determined by the autocorrelation function of the wave energy density spectrum. In addition to the above phase modulation method, focusing waves can also be spatio-temporally produced by means of the wave dispersion method, i.e., focusing wave components are individually generated and their frequencies linearly vary with the largest one at the start [32]. By contrast, the phase modulation method can produce focusing waves with higher frequency wave components. When the wave dispersion method is adopted to generate focusing waves, the higher frequency waves are severely constrained by the stroke limitation of wavemaker.

The other way to generate focusing waves, in a laboratory, is to employ wave modulation instability. On the basis of this mechanism, Li et al. [33] experimentally observed focusing wave occurrence in a random wave group by adjusting the wave steepness parameter and the Benjamin–Feir Index (BFI). By assigning carrier wave amplitude and steepness, Chabchoub et al. [34] generated super focusing waves in their physical wave flume. Moreover, several breather solutions of the nonlinear Schrödinger equation, such as Kuznetsov–Ma solution, Akhmediev breather, and the Peregrine breather solution, are sometimes adopted to experimentally simulate extreme waves in a physical wave tank [35,36]. Apart from above methods, recently a phase-amplitude iteration scheme based on space-time focusing of wave energy has been developed to generate tailored focusing wave [37,38]. Nevertheless, the investigation from Deng et al. [39] indicated that the above phase-amplitude iteration scheme may be less effective for focusing waves with severe phase coupling. The phase-amplitude iteration method probably can be improved by directly adjusting wave component amplitude or steepness.

In order to better understand focusing waves generated under various experimental conditions, several experimental investigations have also been carried out. The testing results from Liu et al. [40] showed that some additional high-order wave components are produced, and the amplitudes of these extra components increase when focusing wave elevation becomes large. The frequency parameter of focusing wave was experimentally examined by Ma et al. [41] who demonstrated that focusing waves with a wider frequency range transfer more wave energy to their high-frequency components. The local wave steepness of focusing waves with three wave amplitude spectra types, (i.e., the linear wave steepness spectrum, the CWA spectrum, and the CWS spectrum) were examined by Wu and Yao [42] in a physical wave-current flume. Their results demonstrate that focusing waves generated by the linear steepness wave spectrum have larger local wave steepness. In addition, focusing waves with the CWA spectrum were investigated by Baldock et al. [19] who observed that the focusing position, time, and wave crest elevation increase along with a larger amplitude parameter. However, despite all the wave amplitude spectra used to generate focusing waves in published researches, their wave component amplitudes or steepness are not adjustable under assigned wave energy. Indeed, in the recently developed phase-amplitude iteration scheme [37] used for focusing wave generation, the steepness of wave components are fixed once the wave component amplitudes are determined. This means that the wave nonlinearity effect arising from phase coupling among wave components cannot be better controlled in focusing wave generation. Thus, more flexible spectra with adjustable amplitude and steepness of the components should be explored in the laboratory, in order to better represent real focusing waves or freak waves with strong nonlinearity [43].

In this study, two new wave spectra, i.e., quasi constant wave amplitude spectrum (QCWA) and quasi constant wave steepness spectrum (QCWS), are developed by modifying the conventional CWA and CWS spectra [44] used to produce focusing waves. In comparison to previous wave spectra, the advantage of the new wave spectra is that the amplitudes and steepness of wave components can be adjusted by adjusting the water depth of wave tank. On the basis of the two new wave spectra, two-dimensional nonbreaking focusing waves are spatio-temporally generated in a physical wave tank and the parameters of the generated waves are investigated under different component groups.

2. Generating Principle of Focusing Waves in a Physical Wave Tank

2.1. Experimental Setup

The experiment on focusing wave generation was carried out in the towing wave tank at Harbin Engineering University. The experimental sketch is shown in Figure 1. The wave tank has a length of 108 m, a width of 7 m, and a depth of 3.5 m. The hydraulic flap-type wavemaker is installed at one side of the wave tank and generates regular waves and random waves, with the maximum wave height of 0.4 m and the wave period ranging from 0.4 s to 4.0 s. At the other side, an absorbing shore is arranged to reduce the reflective waves from the end wall of the wave tank.

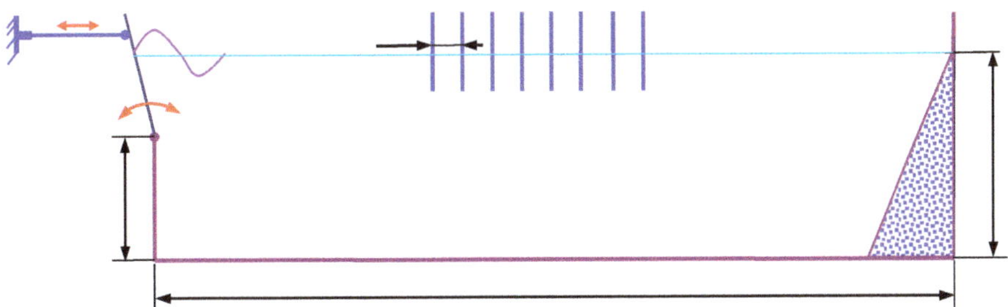

Figure 1. Experimental arrangement sketch.

To record the free surface elevations, 24 wave gauges were installed with an interval of 0.4 m along the length of the wave tank. When the space parameter of focusing waves was set as 30 m, 40 m, and 50 m, the gauge array could be shifted with the first wave gauge (the far left one in Figure 1) installed at 21.905 m, 31.495 m, and 38.905 m from the flap, respectively. The hydraulic piston movement was also monitored by optic equipment (including markers and the matching cameras) to validate the accuracy of output signals.

The experimental procedure is summarized in Figure 2. To generate the desired focusing waves in the physical wave tank, it is essential to have appropriate and practical wave-making signals. The theory and method used in this research for signal generation are described in the following Sections 2.2 and 2.3.

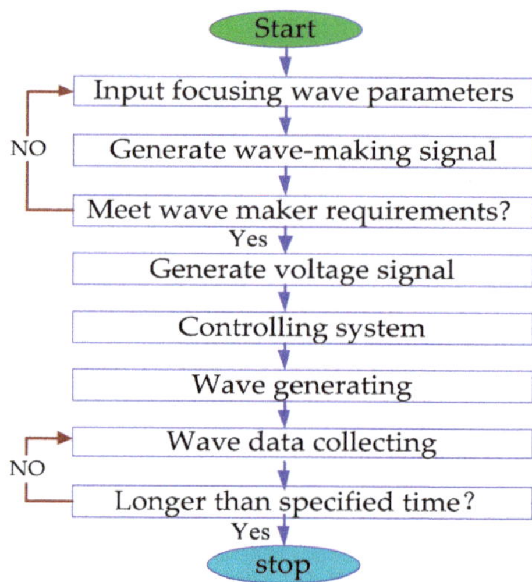

Figure 2. Flow chart of focusing wave generation.

2.2. Theoretical Wave-Making Signal

Focusing waves are generated by the spatio-temporal focusing principle of wave energy. The origin of the coordinate system is defined at the intersection point of the flap and the still water surface. The z-axis is vertically upward, and the x-axis is horizontally towards the wave travelling direction. The free surface of the focusing wave in the defined coordinate system can be expressed by:

$$\eta = \sum_{n=1}^{N} a_n \cos[k_n(x - x_f) - \omega_n(t - t_f)] \tag{1}$$

where x_f and t_f are focusing location and time and ω_n is the frequency of each wave component, which linearly increases from ω_1 to ω_N, as expressed by Equation (2). In this work, a_n is newly developed to be a variable which is determined by wave spectra, i.e., the quasi constant wave amplitude spectrum and the quasi constant wave steepness spectrum. They can be expressed by Equations (3) and (4), respectively.

$$\omega_n = \omega_1 + (n-1) \times \frac{\omega_N - \omega_1}{N-1} \tag{2}$$

$$a_n = \frac{A_f}{N} \frac{\cosh(k_n h)}{\sinh(k_n h)} \tag{3}$$

$$a_n = \frac{A_f}{k_n \sum_{n=1}^{N} 1/k_n} \frac{\cosh(k_n h)}{\sinh(k_n h)} \tag{4}$$

From Equations (3) and (4), it can be seen that $\cosh(k_n h)/\sinh(k_n h)$ approaches one when the water depth tends to be infinite. Thus, the previous CWA and CWS spectra used to generate focusing waves are two special expressions of the QCWA and QCWS spectra at infinite water depth.

The wave component amplitudes of the QCWA spectrum and wave steepness of the QCWS spectrum at different water depths are demonstrated in Figure 3a,b, respectively. Figure 3a also shows that the wave component amplitude of the QCWA spectrum approaches that of the CWA spectrum when water depth increases. For a certain water depth, the wave component amplitude gradually becomes a constant as the wave frequency becomes higher. This is because these high-frequency wave components have short wave lengths and can be considered as deep-water waves. Therefore, Equation (3) expresses the CWA spectrum for high frequency range, i.e., $a_n = A_f/N$. The similar correlation could also be seen between the QCWS spectrum and the CWS spectrum in Figure 3b. According to QCWA (or QCWS) spectrum, for the finite water depth, the crest elevation of the generated focusing wave (defined as $A_f{}^0$) is equal to $\sum a_n$ (calculated by the linear wave theory), which is larger than A_f resulting from the CWA (or CWS) spectrum. According to the focusing wave-free surface elevation expressed by Equation (1), the hydraulic piston movement to drive wavemaker flap yields

$$s(t) = \sum_{n=1}^{N} \frac{a_n}{F_n} \cos[\omega_n t + (k_n x_f - \omega_n t_f)] \tag{5}$$

where,

$$F_n = \frac{4\omega_n{}^2}{gH_1} \frac{\cosh(k_n h)[\cosh(k_n(h - H_0)) - \cosh(k_n h) + k_n h \sinh(k_n h)]}{k_n^2(2k_n h + \sinh(2k_n h))} \tag{6}$$

and $H_1 = 2.3$ m for the wavemaker of this experiment. The displacement signal calculated by Equation (5) is the theoretical wave-making signal to generate focusing waves.

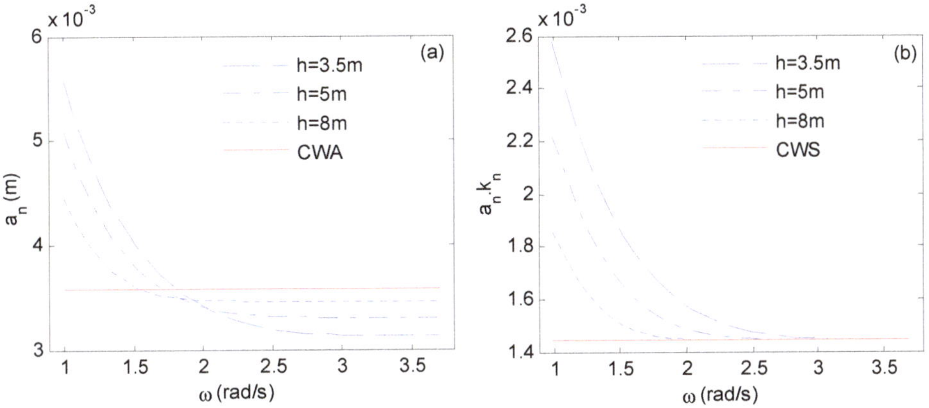

Figure 3. Two wave spectra changing with water depth: (**a**) The quasi constant wave amplitude QCWA spectrum and (**b**) the quasi constant wave steepness QCWS spectrum.

2.3. Implementing Wave-Making Signal

The theoretical wave-making signal cannot be directly employed to generate focusing waves, because of nonzero displacement at initial and terminating instants. Practically, the wavemaker flap needs to slowly start from its static status, and it has to gradually terminate at static status after the wave generation is completed. Thus, the theoretical wave-making signal needs to be processed to match the actual motion of the flap. An example of the theoretical signal of a QCWA focusing wave is illustrated by the red line in Figure 4 with $A_f = 0.1$ m, $x_f = 50$ m, $t_f = 25$ s, $N = 32$, and $\omega_n = 1.336$ rad/s~2.695 rad/s, and the wave component amplitude is computed by Equation (3).

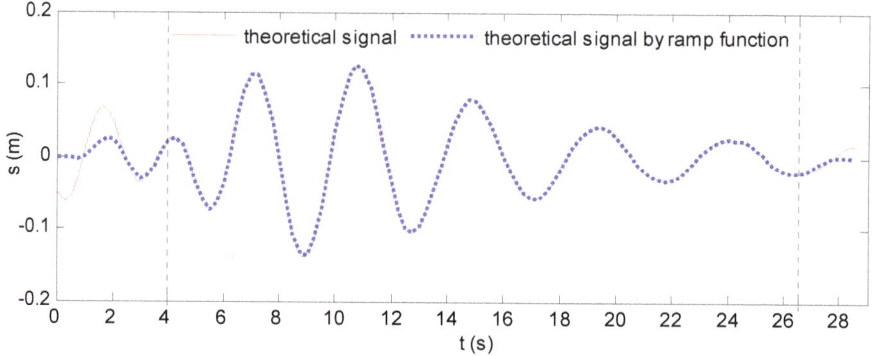

Figure 4. Theoretical signal and processed signal by ramp function.

From the theoretical wave-making signal, it is seen that the initial displacement is smaller than zero and the terminating value is larger than zero. This is generally remedied by multiplying the theoretical signal by a ramp function, as shown in Figure 4. It is observed that the flap gradually starts from the static position and stops at the initial static position, matching with the actual motion requirements of wavemaker.

However, by comparing the rump function modified and the original wave-making signals in Figure 4, we observe that the theoretical movement signal has been distorted by the ramp function at the beginning and terminating stages. Considering the strong transient characteristics of focusing waves, the ramp function method may not be suitable for processing the theoretical wave-making signal. In this study, a new scheme is proposed to generate the practicable wave-making signal of focusing waves through inserting a small section auxiliary signal expressed as

$$s_0(t) = \left(\sum_{j=1}^{m} \mu_j(t)\chi_j \right) \times \left(\sin^{\gamma}\left(\frac{\pi}{2t_0}t \right) \right)(0 \leq t \leq t_0) \tag{7}$$

where, the base function μ_j is written as $[1, t, t_2]$ for $m = 3$ and the coefficient matrix χ could be solved from Equation (8). For matching with the initially static state of the flap, the parameter γ should be larger than one. In this study, $\gamma = 2$, $t_0 = 1.5$ s.

$$\begin{cases} s_0(t_0) = s(0) \\ s'_0(t_0) = s'(0) \\ s''_0(t_0) = s''(0) \end{cases} \tag{8}$$

To make the wavemaker flap return to its initial position at the terminating stage, the theoretical stroke signal is extended, and the extension is processed by a ramp down function. The newly proposed scheme is applied to the above focusing wave example in Figure 4, and the modified wave-making signal result is illustrated in Figure 5.

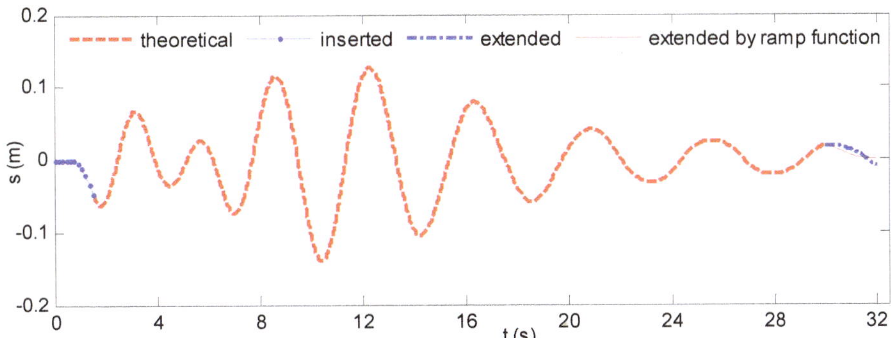

Figure 5. Wave-making signal generated by the new scheme.

It is observed that the wave-making signal processed by this research meets the actually wave-making requirements by extensions at two ends of the theoretical signal. This method will be adopted in the following focusing wave generation.

3. Experimental Results and Discussions

All the tested focusing wave cases in this research are listed in the Appendix A. In order to examine the accuracy of the wavemaker movement controlling system, the hydraulic piston motion of wavemaker is captured and is compared with the input stroke signal. The motion signal comparison of one focusing wave case (i.e., A101f61X50t36 in Appendix A) is plotted in Figure 6, showing that the tested signal has good agreement with the input signal. The maximum relative error is less than 2%.

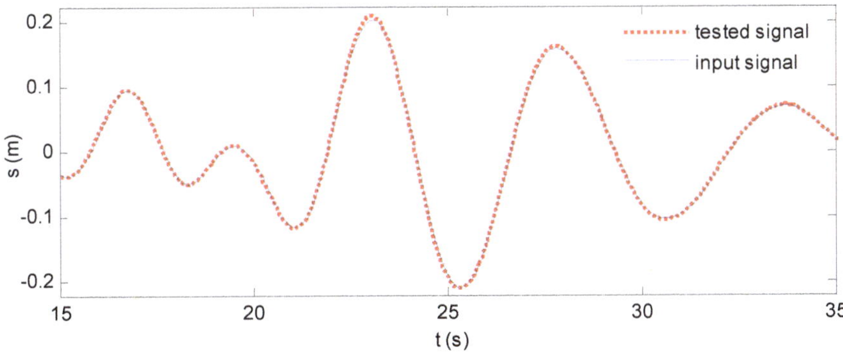

Figure 6. Comparison of input and tested stroke signals.

The repeatability of the whole experiment system also has been examined by each wave gauge recording result in the wave tank. All focusing wave cases are repeatedly tested twice and the percentage error of the maximum wave crest elevation (or the minimum wave trough) at each tested position is estimated by

$$E_r^{+(-)} = \left[\frac{\eta_1^{+(-)} - \eta_2^{+(-)}}{\eta_1^{+(-)}} \right] \times 100\% \qquad (9)$$

The relative percentage errors of one focusing wave case (i.e., A101f61X50t36 in Appendix A) are illustrated in Figure 7a,b, showing the maximum errors of the tested wave crests and wave troughs are about 4% and 3%, respectively. The rest of the experimental focusing wave cases are also validated by repeating tests and their maximums of E_r^+ and E_r^- are both less than 5% [45].

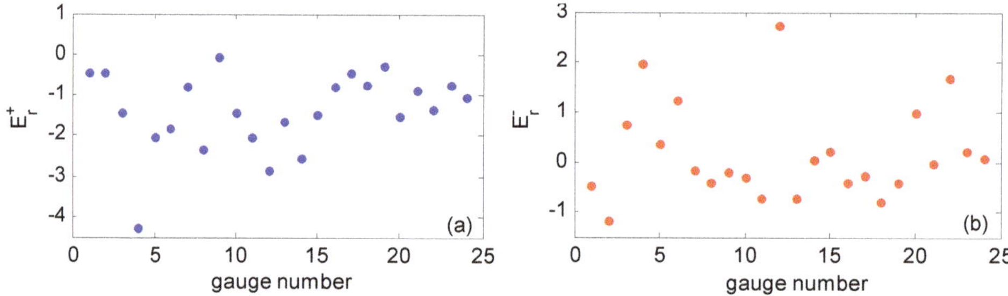

Figure 7. Experimental repeatability error: (**a**) Relative error of focusing wave crests and (**b**) relative error of focusing wave troughs.

Moreover, the wave modulation instability of the experimental focusing wave cases in this research has also been examined depending on the Benjamin–Feir Index (BFI). The BFI is evaluated by half-frequency width at half maximum of the wave spectrum [46,47]. The corresponding results of all experimental focusing wave cases are summarized in Table A1. We found that the BFIs of the focusing waves are all less than one, which indicates the wave focusing in this research is dominated by the space-time focusing of wave energy, rather than the modulation instability. In the following sections, the effects of the spectrum types and focusing wave parameters will be investigated according to the tested focusing waves. It should be noted that the focusing wave in this study is defined as the wave with the maximum wave crest elevation.

3.1. Wavefree Surface Evolution for QCWA and QCWS Focusing Waves

To compare the evolving processes of the focusing waves generated by QCWA and QCWS spectra, a focusing wave case with parameters of $A_f = 0.1$ m, $t_f = 36$ s, $x_f = 50$ m, $\omega_n = 0.997$ rad/s~3.696 rad/s, and N = 32 was tested. The wave amplitude spectra and wave steepness spectra are compared in Figure 8a,b, respectively. It is seen that, the QCWA spectrum provides larger wave amplitudes and steepness for the high-frequency wave components than the QCWS spectrum. This implies the focusing wave generated by the QCWA spectrum will have stronger nonlinearity.

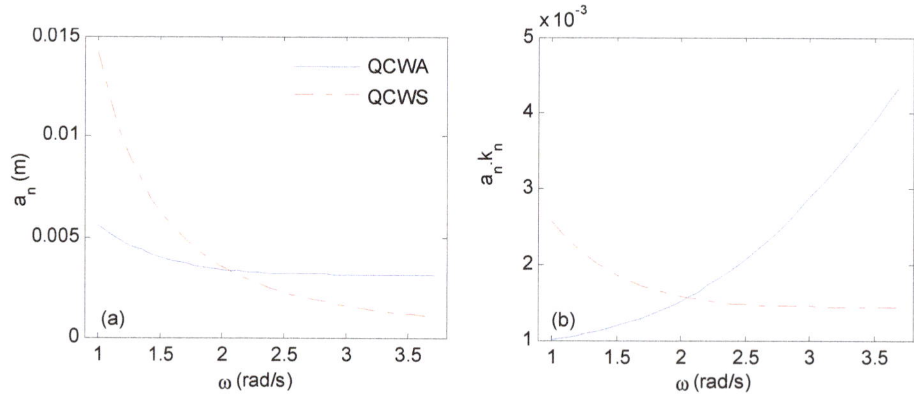

Figure 8. Comparison of two wave spectra: (**a**) Wave amplitude spectrum and (**b**) wave steepness spectrum.

The wave-free surface time histories of focusing waves based on the above two spectra at different positions are plotted in Figure 9a,b showing that the two focusing waves both start from a deep wave trough, followed by the largest wave crest to form the focusing wave. The deep trough then arises and disappears again to complete the wave energy focusing and diffusing [48]. The focusing waves based on the two wave spectra both appear at 45.305 m from the flap, being smaller than the assigned position of 50 m. The shift of the focusing point towards upstream was also observed in the study of the CWA focusing waves [49], while the opposite phenomenon was found in investigations on the CWA and CWS focusing waves by Li and Liu [50]. Hennig and Schmittner [51] pointed out that the shift of focusing point is due to the wave group celerity alteration in focusing wave generation.

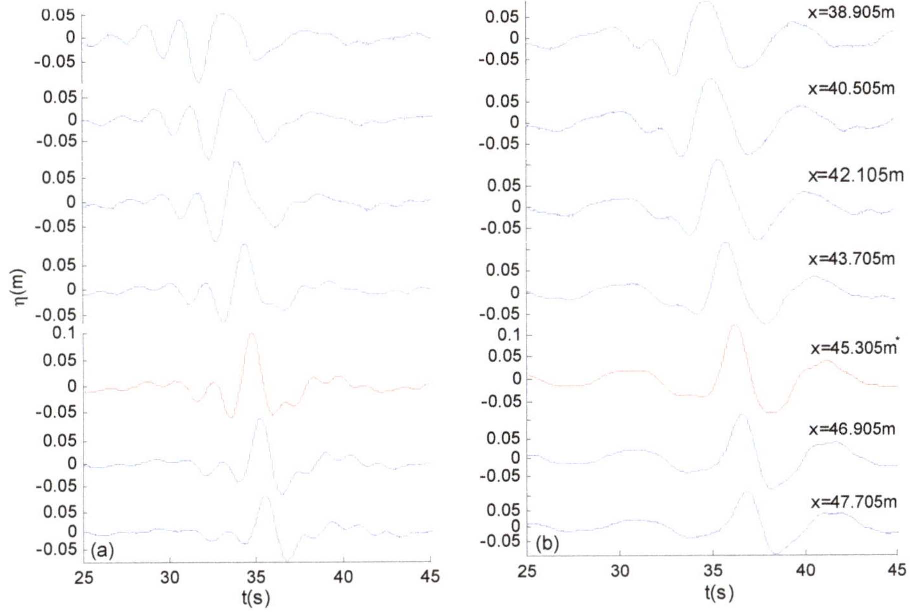

Figure 9. Evolution of two focusing waves: (**a**) The QCWA focusing wave and (**b**) the QCWS focusing wave.

Relative to the focusing wave generating instant in their time history curves (the wave-free surface curves of the wave-free plotted by red line), the two focusing wave-free surfaces are both asymmetry, which implies the focusing wave generation shown in Figure 9 is not only a linear superposition process. Comparing time history curves at each testing position in Figure 9, we observe that the QCWA focusing wave-free surface is steeper and deeper, whereas the QCWS focusing wave-free surface shape is flatter and shallower. This is mainly introduced by their wave amplitude spectra differences, as shown in Figure 8. The QCWS focusing wave has more wave energy in low frequency range which determines the overall shape of focusing waves, while the QCWA focusing wave distributes more wave energy in high frequencies which affects the local wave-free surface.

To further analyze focusing wave results from the aspect of the spectrum, the wave component amplitude spectra are calculated via a fast Fourier transform (FFT) analysis based on recording results of wave gauges at different positions. Because of the disturbance of reflective waves, only a limited time length of wave surface recording is available in FFT. It should be noted that the resolution of wave spectra calculated by FFT is lower than the input wave spectra. From the FFT results shown in Figures 10 and 11, it is clearly seen that the extra lower-frequency (<0.997 rad/s) and higher-frequency harmonics (>3.696 rad/s) are generated in generating processes of focusing waves. The higher-frequency components are mainly the second-order harmonics (3.696 rad/s~7 rad/s). These higher-frequency harmonics of the QCWA spectrum continually become larger and the wave components in input frequency range reduce as the wave group approaches the focus position. Nevertheless, one opposite result is found when the wave group passes the focus position. For the wave spectra variation of the QCWS focusing wave shown in Figure 11, it is observed that the apparent variation of wave component amplitudes occurs within the input frequency range (i.e., the wave energy is redistributed in input wave frequency).

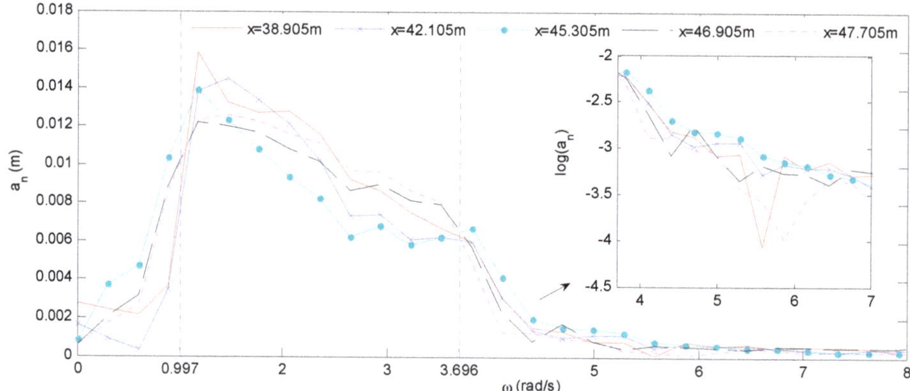

Figure 10. Wave amplitude spectra variation of the QCWA focusing wave at different positions.

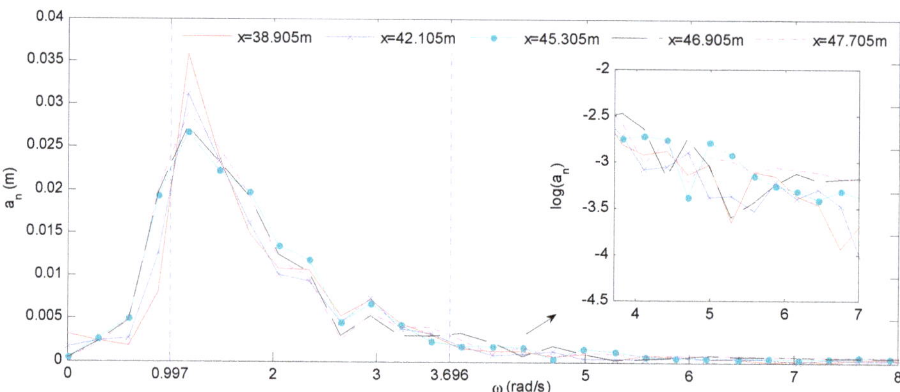

Figure 11. Wave amplitude spectra variation of the QCWS focusing wave at different positions.

Moreover, for qualitatively identifying the nonlinear effects on focusing wave generation, the local Ursell number at each testing location is calculated by:

$$U_r = \frac{\eta_{max}\lambda^2}{h^3} \tag{10}$$

The Ursell numbers at different positions are plotted in Figure 12a,b respectively for the QCWA focusing wave and the QCWS focusing wave. From Figure 12a, it is seen that the Ursell number arrives at the maximum as the wave group approaches the focus position, which implies that the wave nonlinearity plays an important role in the QCWA focusing wave generation. This is also observed through the extra higher-order harmonics' variation at different positions, as shown in Figure 10. However, from Figure 12b, one can observe that the Ursell number of the QCWS focusing wave keeps rising although the wave group has passed the focus position. It is deduced that the QCWS focusing wave generation is probably dominated by wave components in input frequency range. Additionally, the extra higher-order harmonics have smaller amplitudes in wave spectra of Figure 11 which could be considered as evidence of this deduction.

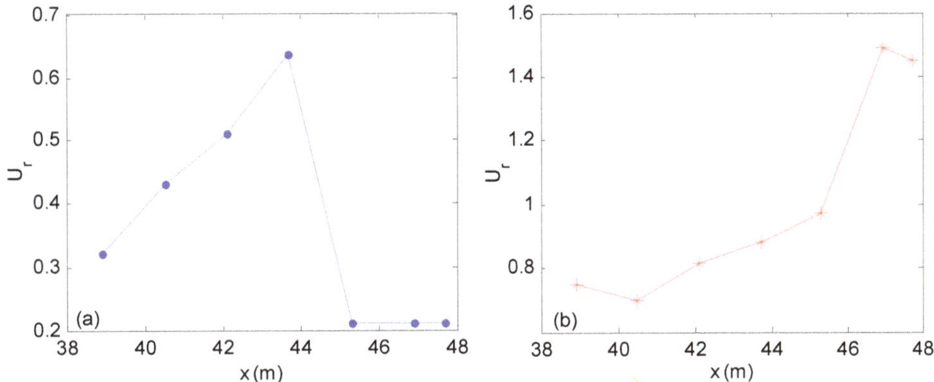

Figure 12. Ursell number changing at different space positions for two generations of focusing waves: (**a**) The QCWA focusing wave and (**b**) the QCWS focusing wave.

3.2. Focusing Wave Generation under Different a_n and ω_n Parameters

The QCWA and QCWS focusing waves with different frequency ranges are generated in the physical wave tank. The required maximum displacements (i.e., S_{max}) of hydraulic piston are summarized in Table 1. For wave groups 1-1 and 2-a, their maximum displacements of hydraulic piston both increase with their lower bonds of wave frequencies being extended, whereas, when the upper bonds of the frequency decrease, for wave groups 2-a, 3-b and 4-c, the corresponding maximum displacements also increase. These comparisons indicate that focusing waves with more low-frequency wave components require wider wavemaker stroke range.

Table 1. Required maximum displacements of hydraulic piston.

No	a_n	ω_n (rad/s)	S_{max} (m)	$S^{QCWS}_{max}/S^{QCWA}_{max}$
1-1	QCWA	0.997~3.696	0.202	1.834
	QCWS		0.371	
2-a	QCWA	1.336~3.696	0.127	1.772
	QCWS		0.225	
3-b	QCWA	1.336~3.142	0.164	1.518
	QCWS		0.249	
4-c	QCWA	1.336~2.094	0.387	1.062
	QCWS		0.411	
$A_f = 0.1$ m; $x_f = 50$ m; $t_f = 36$ s; $N = 32$				

By comparing the maximum displacements of the QCWA and QCWS focusing waves in each group, we find the QCWS focusing wave requires wider movement range of the wavemaker and this difference becomes more obvious when the frequency range gets wider. For the example of the focusing wave group 1-1 in Table 1, the maximum displacement of the QCWS focusing wave is almost twice that of the QCWA focusing wave. Therefore, the QCWA spectrum is more effective for the generation of focusing waves containing a wider frequency range from the aspect of the wavemaker stroke.

The generated focusing wave crest elevations of the last three experimental groups (i.e., groups 2-a, 3-b, 4-c, in Table 1) are plotted in Figure 13a. The wave crest elevations from the two spectra both raise when their frequency upper bounds are decreased. A similar relationship between the frequency bandwidth and the focusing wave crest elevation is also found in experimental results of the CWA focusing waves [19] and in numerical investigations [24]. We infer that the wave crest elevation is significantly affected by the wave frequency bandwidth but slightly affected by the type of wave amplitude spectrum, as shown in Figure 13a.

Figure 13b compares focusing positions of the QCWA and QCWS spectra. We observed that both waves focus closer to the flap, with the frequency bandwidth narrower. At the same time, the above results also demonstrate that the focusing position changes with the corresponding central frequencies. Li and Liu [50] presented an opposite correlation between focusing position and frequency bandwidth for the focusing waves based on CWA and CWS spectra, but the correlation between the focusing position and the central frequency in their work is similar to that illustrated in Figure 13b. The comparison of the two tested spectra in this study, showed that the focusing position of the QCWA focusing wave is a bit farther from the flap. Therefore, if a focusing wave with a larger wave crest elevation and a farther location from wavemaker is required, the QCWA spectrum is better than the QCWS. In addition, the QCWA focusing wave can also include a wider frequency range for a specific wavemaker with the fixed stroke range, according to Table 1. The QCWA focusing waves will be further investigated in the following sections.

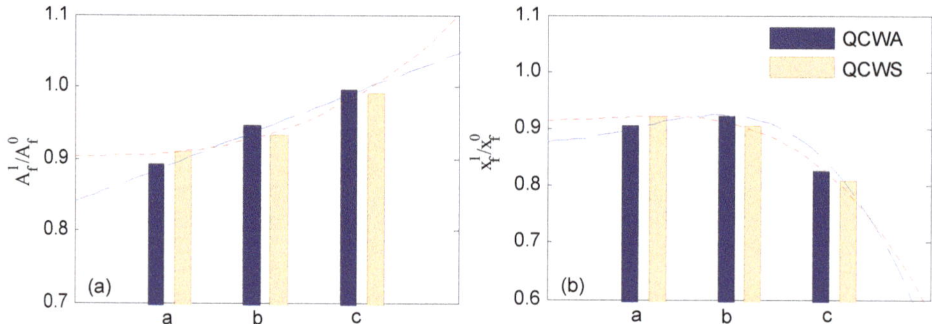

Figure 13. Comparison of generating results of the QCWA and QCWS focusing waves: (**a**) Focusing wave crest elevation and (**b**) focusing wave position (a: ω_n = 1.336 rad/s~3.696 rad/s, b: ω_n = 1.336 rad/s~3.142 rad/s, and c: ω_n = 1.336 rad/s~2.094 rad/s).

3.3. Focusing Wave Generation under Different A_f and x_f Parameters

In order to investigate the effect of A_f and x_f on the generating results, focusing wave groups with two different wave frequency ranges are tested in a physical wave tank. In the first experiment group, ω_n = 0.997 rad/s~3.696 rad/s and A_f is chosen as 0.1 m and 0.14 m. For the second group of focusing waves, ω_n is varied from 1.336 rad/s to 3.142 rad/s with three wave crest elevations, i.e., A_f = 0.1 m, 0.14 m, and 0.20 m. All these focusing waves are tested at two different focusing locations, i.e., x_f = 30 m and x_f = 50 m. The focusing time parameter t_f and wave component number N are assigned as 36 s and 32.

Figure 14a,b illustrates tested focusing wave crest elevations under different A_f and x_f parameters. The generated relative wave crest elevations increase nonlinearly with A_f being larger at both specified focusing locations, i.e., the increasing ratio is not a constant. The same conclusion can be also drawn from the comparison of the CWS and CWA focusing waves by Li and Liu [50] and in experimental investigations on the CWA focusing waves by Baldock et al. [19]. The wave crest elevation increase is generally attributed to the generated nonlinear wave components (i.e., extra higher-frequency components) produced in focusing wave generation. By maintaining A_f and varying x_f parameter from 30 m to 50 m, it is seen that the tested focusing wave crest elevations show a slight decrease in Figure 14a,b. This may be because the wave energy does not fully focus at the assigned location or a small portion of wave energy is dissipated in focusing wave propagation.

Figure 14c,d demonstrates the actual positions of focusing waves with two wave frequency ranges. As for focusing wave cases with ω_n = 0.997 rad/s~3.696 rad/s, the larger wave elevation A_f results in a forward moving of the focusing point. But the focusing wave group with the other frequency range shows an inversed changing trend. Despite the irregular focusing position variation existing, the overall shift of the focusing position is moderate and the waves approximately focus at $0.9 \times x_f$ in the physical wave tank.

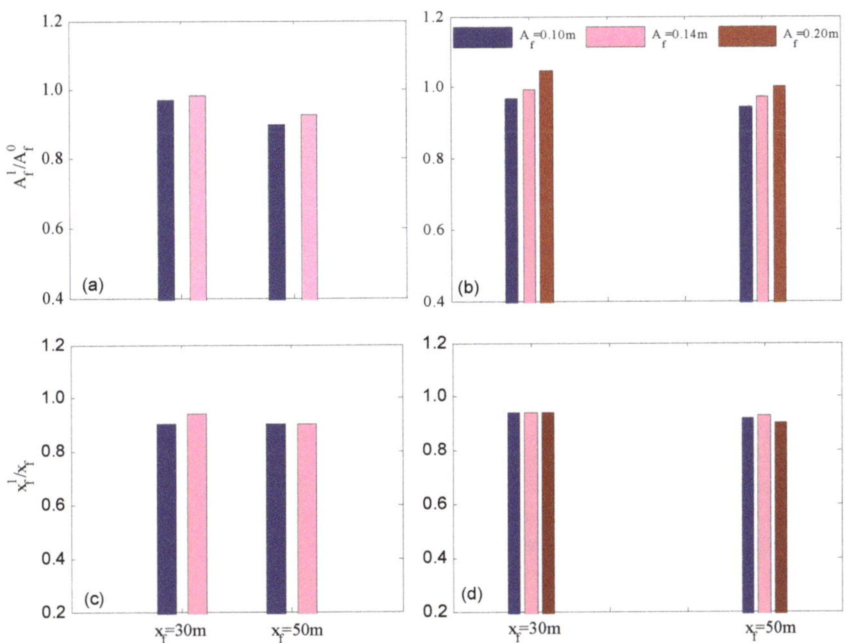

Figure 14. Generated results of focusing waves under different A_f and x_f parameters: (**a**) Focusing wave crest elevations for ω_n = 0.997 rad/s~3.696 rad/s, (**b**) focusing wave crest elevations for ω_n = 1.336 rad/s~3.142 rad/s, (**c**) focusing wave positions for ω_n = 0.997 rad/s~3.696 rad/s, and (**d**) focusing wave positions for ω_n = 1.336 rad/s~3.142 rad/s.

3.4. Focusing Wave Generation under Different x_f and t_f Parameters

A series of focusing wave cases are tested against various focusing position and time parameters with a fixed amplitude of A_f = 0.1 m. The total number of wave components is 32 and the wave component frequency ω_n is varied from 1.336 rad/s to 3.696 rad/s. The focusing time parameter is, respectively, set as 25 s, 30 s, and 36 s at each focusing position of x_f = 30 m, 40 m, and 50 m.

The measured crest elevations of focusing waves are illustrated in Figure 15a. It is found that when x_f = 30 m the wave crest elevations are almost the same for three focusing times and they are about 10% higher than the expected wave crest elevations (i.e., A_f^0). As the focusing position parameter is increased to 40 m, the generated wave crest elevations with t_f being 30 s and 36 s basically stay the same as those tested at $x_f = 30$ m. However, the wave crest elevation with t_f being 25 s shows a significant drop. When the focusing position is further increased to 50 m, the wave crest elevations highly depend on the focusing time parameter and they significantly increase from about 0.6 to 1.0 as the focusing time parameter ranges from 25 s to 36 s. Comparing the wave crest elevations under three different x_f parameters, it is found that the wave crest elevations with the focusing time t_f being 25 s and 30 s both show a decreasing trend as the focusing position is changed from 30 m to 50 m. Nevertheless, when the focusing time is increased to 36 s, the generated focusing wave crest elevations are largely maintained as the focusing position parameter varies. We deduce that only when the focusing time parameter t_f exceeds a threshold, i.e., about 36 s for this testing case, the expected wave crest elevation can be easily achieved in the physical wave tank.

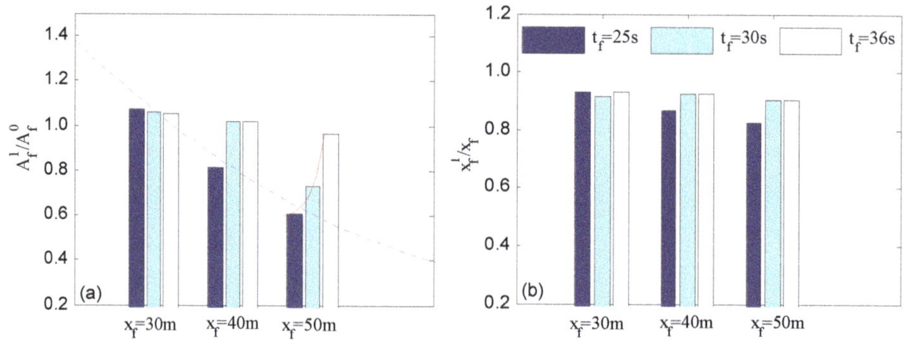

Figure 15. Focusing wave generating results under different focusing position and time parameters: (a) Focusing wave crest elevations and (b) focusing wave positions.

Figure 15b compares the actual focusing positions against theoretical focusing positions under different focusing times. We observed that experimental focusing waves appear in the position of about $0.9 \times x_f$ from the flap, as the focusing time t_f is 30 s or 36 s. When the focusing time decreases to 25 s, the ratio of actual and assigned positions is below one and it becomes smaller as the given x_f increases from 30 m to 50 m. Thus, according to above analysis, a sufficient time length t_f for focusing wave generation is suggested to achieve smaller deviation of the actual focusing position.

To find the relationship between the focusing position, x_f, and the focusing time, t_f, from the wave energy aspect, the required minimum focusing time t_{cr}, i.e., the time for the highest-frequency wave component energy propagating to the specified focusing position, is calculated by the linear wave theory. As for x_f being specified to be 30 m, 40 m, and 50 m, the calculated t_{cr} are 22.59 s, 30.12 s, and 37.65 s, respectively. According to the comparison between the t_f parameters of the above focusing wave cases and their corresponding t_{cr} results, we deduced that the smaller wave crest elevation and the forward shift of focusing positions, as seen in Figure 15, result from the wave energy not being completely focused at the specified positions. Therefore, for focusing wave generation in a physical wave tank, the assigned time parameter t_f should be longer than t_{cr} as predicted by linear wave theory. It should be noted that for focusing wave cases with large crest elevation the time parameter, t_{cr}, predicted by the linear wave theory may not be applicable due to strong wave nonlinearity.

4. Conclusions

In this paper, two new wave amplitude spectra (i.e., the QCWA and QCWS spectra) used for focusing wave generation in a wave tank have been formulated and experimentally tested. With fixed wave energy, the amplitude and steepness of wave components in these two wave spectra are adjustable. Focusing waves have been generated in the physical wave tank by modifying the wave generator signal through adding two extensions at the beginning and terminating stages. A series of focusing waves based on the two newly-proposed wave amplitude spectra are tested in physical wave tank, demonstrating that by comparing with the QCWS spectrum the focusing waves generated based on the QCWA spectrum can achieve larger wave crest elevation, and are focused further downstream away from the wavemaker. Additionally, to generate a focusing wave with the same wave parameters, the QCWA spectrum requires a smaller wavemaker movement range. The spectral analysis results show that the wave nonlinearity plays an important role in the QCWA focusing wave generation, whereas the redistribution of wave energy in input frequency range significantly affects the focusing wave generated by the QCWS spectrum. It is also found that two spectra have similar effects on wave crest alternation through adjusting wave frequency range and wave crest elevation parameter, whereas actual focusing positions are highly dependent on the type of the wave spectrum.

Experimental investigations on space-time parameters have also been carried out. We found that the space parameter, $x_{f,}$, and the time parameter, $t_{f,}$, jointly affect the generated focusing wave crest elevation and focusing position. The focusing time parameter is constrained by a critical value, which could be predicted by linear wave theory for linear or weakly nonlinear focusing waves. On the basis of this research, future work should be carried out on strongly nonlinear focusing wave generation using these two newly formulated spectra and investigating the effects of the space-time parameters on focusing waves with large-crest elevation.

Author Contributions: Investigation, G.X.; validation, H.H.; supervision, Q.M.; writing—original draft; writing—Review and Editing, Q.G.

Funding: The National Natural Science Foundation of China (grant No. 51739001 and No. 51909125), the Zhejiang Provincial Natural Science Foundation of China (grant No. LY17A020002), and the Ningbo University Science Foundation (grant No. XYL19030).

Acknowledgments: The authors want to thank Fenglai Li and Shizhou You, working in towing wave tank of the Harbin Engineering University. They provided help and support for the focusing wave experiments.

Conflicts of Interest: The authors declare no conflicts of interest.

Nomenclature

H	water depth	g	gravitational acceleration
k_n	wave number	$s_0(t)$	the inserted auxiliary signal
t_0	time length of $s_0(t)$	a_n	wave component amplitude
μ_j	base function of $s_0(t)$	η	wave-free surface elevation
t_f	focusing time parameter	$x_f{}^1$	the tested focusing position
N	wave component number	x_f	focusing position parameter
χ	the coefficient matrix of $s_0(t)$	ω_N	the maximum wave frequency
t_{cr}	the minimum focusing time	$S(t)$	Hydraulic piston stroke signal
ω_n	wave component frequency	γ	the controlling parameter of $s_0(t)$
ω_1	the minimum wave frequency	A_f	focusing wave elevation parameter
$A_f{}^1$	the tested focusing wave crest elevation		
$E_r{}^{+(-)}$	the percentage error between $\eta_1{}^{+(-)}$ and $\eta_2{}^{+(-)}$		
$s'_0(t_0)/s''_0(t_0)$	the first- and second-order derivatives of $s_0(t)$		
η_{max}	local maximum wave crest of tested focusing wave		
$s(0)/s'(0)/s''(0)$	initial displacement, velocity, and acceleration of hydraulic piston		
$A_f{}^0$	focusing wave crest elevation predicted by linear wave theory		
H_0	distance from hydraulic piston to wavemaker flap rotating shaft		
H_1	distance from still water surface to wavemaker flap rotating shaft		
F_n	transfer function between wave component and hydraulic polar stroke		
$\eta_1{}^{+(-)}/\eta_2{}^{+(-)}$	the first and second tested focusing wave elevation (+, maximum; −, minimum)		
λ	local wavelength computed by local wave period of tested focusing wave		
CWS	constant wave steepness	QCWS	quasi constant wave steepness
CWA	constant wave amplitude	QCWA	quasi constant wave amplitude

Appendix A

Focusing wave cases involved in this research are summarized in Table A1, in which the related parameters and wave amplitude spectrum types of each focusing wave case are listed in detail. For parameters in Table A1, their definitions can be found in the Nomenclature section. From Table A1, it is found that the BFI value of each focusing wave case is quite small, which indicates that the nonlinear self-focusing of wave components almost has little influence on focusing wave generation of this research. The focusing wave generation of this research is dominated by the space-time focusing of wave energy.

Table A1. Focusing wave cases in a physical wave tank.

No	Name	Inputting Parameters of Focusing Waves							BFI
		A_f (m)	$A_f{}^0$ (m)	t_f (s)	T_f (s)	ω_n (rad/s)	X_f (m)		
1	A101f61X30t36	0.1	0.1145	36	1.7~6.3	0.997~3.696	30	QCWA	0.0057
2	A101f41X30t36	0.1	0.1078	36	1.7~4.7	1.336~3.696	30	QCWA	0.00891
3	A101f42X30t36	0.1	0.1099	36	2.0~4.7	1.336~3.142	30	QCWA	0.00898
4	A101f41X30t25	0.1	0.1078	25	1.7~4.7	1.336~3.696	30	QCWA	0.00891
5	A101f41X30t30	0.1	0.1078	30	1.7~4.7	1.336~3.696	30	QCWA	0.00891
6	A141f61X30t36	0.14	0.1603	36	1.7~6.3	0.997~3.696	30	QCWA	0.0080
7	A141f42X30t36	0.14	0.1539	36	2.0~4.7	1.336~3.142	30	QCWA	0.01257
8	A201f42X30t40	0.2	0.2199	36	2.0~4.7	1.336~3.142	30	QCWA	0.01796
9	A101f41X40t36	0.1	0.1078	36	1.7~4.7	1.336~3.696	40	QCWA	0.00891
10	A101f41X40t25	0.1	0.1078	25	1.7~4.7	1.336~3.696	40	QCWA	0.00891
11	A101f41X40t30	0.1	0.1078	30	1.7~4.7	1.336~3.696	40	QCWA	0.00891
12	A101f61X50t36	0.1	0.1145	36	1.7~6.3	0.997~3.696	50	QCWA	0.0057
13	A101f41X50t36	0.1	0.1078	36	1.7~4.7	1.336~3.696	50	QCWA	0.00891
14	A101f42X50t36	0.1	0.1099	36	2.0~4.7	1.336~3.142	50	QCWA	0.00898
15	A101f43X50t36	0.1	0.1201	36	3.0~4.7	1.336~2.094	50	QCWA	0.01160
16	A101f41X50t25	0.1	0.1078	25	1.7~4.7	1.336~3.696	50	QCWA	0.00891
17	A101f41X50t30	0.1	0.1078	30	1.7~4.7	1.336~3.696	50	QCWA	0.00891
22	A141f61X50t36	0.14	0.1603	36	1.7~6.3	0.997~3.696	50	QCWA	0.0080
23	A141f42X50t36	0.14	0.1539	36	2.0~4.7	1.336~3.142	50	QCWA	0.01257
24	A201f42X50t36	0.2	0.2199	36	2.0~4.7	1.336~3.142	50	QCWA	0.01797
18	A101f61X50t36c	0.1	0.1270	36	1.7~6.3	0.997~3.696	50	QCWS	0.0140
19	A101f41X50t36c	0.1	0.1132	36	1.7~4.7	1.336~3.696	50	QCWS	0.0148
20	A101f42X50t36c	0.1	0.1145	36	2.0~4.7	1.336~3.142	50	QCWS	0.0141
21	A101f43X50t36c	0.1	0.1219	36	3.0~4.7	1.336~2.094	50	QCWS	0.0158

References

1. Adcock, T.; Taylor, P. Focusing of unidirectional wave groups on deep water: An approximate nonlinear Schrödinger equation-based model. *Proc. R. Soc. A* **2009**, *465*, 3083–3120. [CrossRef]
2. Schmittner, C.; Brouwer, J.; Henning, J. Application of focusing wave groups in model testing practice. In Proceedings of the 33rd International Conference on Ocean, Offshore and Arctic Engineering, San Francisco, CA, USA, 8–13 June 2014.
3. Kharif, C.; Pelinovsky, E. Physical mechanisms of the rogue wave phenomenon. *Eur. J. Mech. B Fluids* **2003**, *22*, 603–634. [CrossRef]

4. Manolidis, M.; Orzech, M.; Simeonov, J. Rogue Wave Formation in Adverse Ocean Current Gradients. *J. Mar. Sci. Eng.* **2019**, *7*, 26. [CrossRef]

5. Steer, N.; McAllister, L.; Borthwick, G.; Van den Bremer, S. Experimental Observation of Modulational Instability in Crossing Surface Gravity Wavetrains. *Fluids* **2019**, *4*, 105. [CrossRef]

6. Yan, S.; Ma, Q. Numerical simulation of interaction between wind and 2D freak waves. *Eur. J. Mech. B Fluids* **2010**, *29*, 18–31. [CrossRef]

7. Mori, N.; Liu, P.C.; Yasuda, T. Analysis of freak wave measurements in the Sea of Japan. *Ocean Eng.* **2002**, *29*, 1399–1414. [CrossRef]

8. Kokorina, A.; Slunyaev, A. Lifetimes of Rogue Wave Events in Direct Numerical Simulations of Deep-Water Irregular Sea Waves. *Fluids* **2019**, *4*, 70. [CrossRef]

9. Cui, C.; Zhang, N.; Yu, Y.; Li, J. Numerical Study on the Effects of Uneven Bottom Topography on Freak Waves. *Ocean Eng.* **2012**, *54*, 132–141. [CrossRef]

10. Clauss, G.; Stempinski, F.; Stuck, R. On Modelling Kinematics of Steep Irregular Seaway and Freak Waves. In Proceedings of the 27th International Conference on Offshore Mechanics and Arctic Engineering, Berlin, Germany, 15–20 June 2008.

11. Pang, H.; An, Z. Wavelet transform analysis of freak waves and the ringing response of vertical cylinder in a numerical wave tank. In Proceedings of the 19th International Offshore and Polar Engineering Conference, Osaka, Japan, 26 June 2009.

12. Fonseca, N.; Soares, C.; Pascoal, R. Global loads on a FPSO induced by a set of freak waves. *J. Offshore Mech. Arct. Eng. Trans. ASME* **2009**, *131*, 011103. [CrossRef]

13. Yan, B.; Luo, M.; Bai, W. An Experimental and Numerical Study of Plunging Wave Impact on a Box-Shape Structure. *Mar. Struct.* **2019**, *66*, 272–287. [CrossRef]

14. Hu, Z.; Tang, W.; Xue, H. A Probability-Based Superposition Model of Freak Wave Simulation. *Appl. Ocean Res.* **2014**, *47*, 284–290. [CrossRef]

15. Yan, S.; Ma, Q. Improved Model for Air Pressure Due to Wind On 2D Freak Waves in Finite Depth. *Eur. J. Mech. B Fluids* **2011**, *30*, 1–11. [CrossRef]

16. Dyachenko, A.; Zakharov, V. On the Formation of Freak Waves on the Surface of Deep Water. *JETP Lett.* **2009**, *88*, 307–311. [CrossRef]

17. Rapp, R.; Melville, W. Laboratory measurements of deep water breaking waves. *Philos. Trans. R. Soc. A Math. Phys. Eng. Sci.* **1990**, *A331*, 735–800. [CrossRef]

18. Pei, Y.; Zhang, N.; Zhang, Y. Efficient Generation of Freak Waves in Laboratory. *China Ocean Eng.* **2007**, *3*, 515–523.

19. Baldock, T.; Swan, C.; Taylor, P. A laboratory study of nonlinear surface waves on water. *Philos. Trans. R. Soc. A Math. Phys. Eng. Sci.* **1996**, *354*, 649–676.

20. Li, J.; Wang, Z.; Liu, S. Experimental Study of Interactions between Multi-Directional Focused Wave and Vertical Circular Cylinder, Part I: Wave Run-Up. *Coast Eng.* **2012**, *64*, 151–160. [CrossRef]

21. Ma, Q. Numerical Generation of Freak Waves Using MLPG_R and QALE-FEM Methods. *CMES-Comp. Model. Eng. Sci.* **2007**, *18*, 223–234.

22. Ning, D.; Zang, J.; Liu, S.; Eatock, T.; Teng, B.; Taylor, P. Free-Surface Evolution and Wave Kinematics for Nonlinear Uni-Directional Focused Wave Groups. *Ocean Eng.* **2009**, *36*, 1226–1243. [CrossRef]

23. Fochesato, C.; Grilli, S.; Dias, F. Numerical Modeling of Extreme Rogue Waves Generated by Directional Energy Focusing. *Wave Motion* **2007**, *44*, 395–416. [CrossRef]

24. Ai, C.; Ding, W.; Jin, S. A General Boundary-Fitted 3D Non-Hydrostatic Model for Nonlinear Focusing Wave Groups. *Ocean Eng.* **2014**, *89*, 134–145. [CrossRef]

25. Benjamin, T.; Feir, J. The disintegration of wave trains on deep water Part 1—Theory. *J. Fluid Mech.* **1967**, *27*, 417–430. [CrossRef]

26. Janssen, P. Nonlinear Four-Wave Interactions and Freak Waves. *J. Phys. Oceanogr.* **2003**, *33*, 863–884. [CrossRef]

27. Onorato, M.; Osborne, A.; Serio, M.; Cavaleri, L.; Brandini, C.; Stansberg, C. Extreme Waves, Modulational Instability and Second Order Theory: Wave Flume Experiments on Irregular Waves. *Eur. J. Mech. B Fluids* **2006**, *25*, 586–601. [CrossRef]

28. Manzetti, S. Mathematical Modeling of Rogue Waves: A Survey of Recent and Emerging Mathematical Methods and Solutions. *Axioms* **2018**, *7*, 42. [CrossRef]

29. Deng, Y.; Yang, J.; Li, X.; Xiao, L. Experimental and numerical investigation on kinematics of freak waves. In Proceedings of the 25th International Ocean and Polar Engineering Conference, Kona, Big Island, HI, USA, 21–26 June 2015.

30. Deng, Y.; Yang, J.; Zhao, W.; Xiao, L.; Li, X. An efficient focusing model of focusing wave generation considering wave reflection effects. *Ocean Eng.* **2015**, *105*, 125–135. [CrossRef]

31. Whittaker, C.; Fitzgerald, C.; Raby, A.; Taylor, P.; Borthwick, A. Extreme coastal responses using focused wave groups: Overtopping and horizontal forces exerted on an inclined seawall. *Coast Eng.* **2018**, *140*, 292–305. [CrossRef]

32. Touboul, J.; Giovanangeli, J.; Kharif, C.; Pelinovsky, E. Freak waves under the action of wind: Experiments and simulations. *Eur. J. Mech. B Fluids* **2006**, *25*, 662–676. [CrossRef]

33. Li, J.; Yang, J.; Liu, S.; Ji, X. Wave Groupiness Analysis of the Process of 2D Freak Wave Generation in Random Wave Trains. *Ocean Eng.* **2015**, *104*, 480–488. [CrossRef]

34. Chabchoub, A.; Hoffmann, N.; Onorato, M.; Akhmediev, N. Super rogue waves: Observation of a higher-order breather in water waves. *Phys. Rev. X* **2012**, *2*, 11015. [CrossRef]

35. Chabchoub, A.; Peri, R.; Hoffmann, N. Dynamics of Unstable Stokes Waves: A Numerical and Experimental Study. In Proceedings of the 33rd International Conference on Ocean, Offshore and Arctic Engineering, San Francisco, CA, USA, 8–13 June 2014.

36. Onorato, M.; Residori, S.; Bortolozzo, U.; Montina, A.; Arecchi, F. Rogue Waves and their Generating Mechanisms in Different Physical Contexts. *Phys. Rep.* **2013**, *528*, 47–89. [CrossRef]

37. Clauss, G.; Schmittner, C.; Klein, M. Generation of rogue waves with predefined steepness. In Proceedings of the 25th International Conference on Offshore Mechanics and Arctic Engineering, Hamburg, Germany, 4–9 June 2006.

38. Fernández, H.; Sriram, V.; Schimmels, S.; Oumeraci, H. Extreme wave generation using self correcting method—Revisited. *Coast Eng.* **2014**, *93*, 15–31.

39. Deng, Y.; Yang, J.; Tian, X.; Li, X.; Xiao, L. An experimental study on deterministic freak waves: Generation, propagation and local energy. *Ocean Eng.* **2016**, *118*, 83–92. [CrossRef]

40. Liu, S.; Sun, Y.; Li, J.; Zang, J. Experimental Study on 2-D Focusing Wave Run-up on A Vertical Cylinder. *China Ocean Eng.* **2010**, *24*, 499–512.

41. Ma, Y.; Dong, G.; Liu, S.; Zang, J.; Li, J.; Sun, Y. Laboratory study of unidirectional focusing waves in intermediate depth water. *J. Eng. Mech.* **2010**, *136*, 78–90. [CrossRef]

42. Wu, C.; Yao, A. Laboratory measurements of limiting freak waves on currents. *J. Geophys. Res.* **2004**, *109*, C12002. [CrossRef]

43. Clauss, G.; Klein, M. Experimental investigation on the vertical bending moment in extreme sea states for different hulls. *Ocean Eng.* **2016**, *119*, 181–192. [CrossRef]

44. Liang, S.; Zhang, Y.; Sun, Z.; Chang, Y. Laboratory study on the evolution of waves parameters due to wave breaking in deep water. *Wave Motion* **2017**, *68*, 31–42. [CrossRef]

45. Xu, G. Research on Generation of Freak Wave and Its Effect on Truss SPAR Motions. Ph.D. Thesis, Harbin Engineering University, Harbin, China, 2016.

46. Gramstad, O. Modulational Instability in JONSWAP Sea States Using the Alber Equation. In Proceedings of the 36th International Conference on Ocean, Offshore and Arctic Engineering, Trondheim, Norway, 25–30 June 2017.

47. Li, J.; Li, P.; Liu, S. Observations of Freak Waves in Random Wave Field in 2D Experimental Wave Flume. *China Ocean Eng.* **2013**, *5*, 659–670. [CrossRef]

48. Cui, C.; Zhang, N. Research on the time-Frequency energy structure of freak wave generation and evolution. In Proceedings of the 30th International Conference on Ocean, Offshore and Arctic Engineering, Rotterdam, The Netherlands, 19–24 June 2011.

49. Yan, S.; Ma, Q. Numerical study on significance of wind action on 2-D freak waves with different parameters. *J. Mar. Sci. Tech. TAIW* **2012**, *20*, 9–17.

50. Li, J.; Liu, S. Focused Wave Properties Based on A High Order Spectral Method with A Non-Periodic Boundary. *China Ocean Eng.* **2015**, *29*, 1–16. [CrossRef]
51. Hennig, J.; Schmittner, C. Experimental variation of focusing wave groups for the investigation of their predictability. In Proceedings of the 28th International Conference on Ocean, Offshore and Arctic Engineering, Honolulu, HI, USA, 31 May–5 June 2009.

Article

Numerical Investigation of Vortex Induced Vibration for Submerged Floating Tunnel under Different Reynolds Numbers

Ruijia Jin [1,2], Mingming Liu [3,*], Baolei Geng [1], Xin Jin [3], Huaqing Zhang [1] and Yong Liu [2]

[1]   Tianjin Research Institute for Water Transport Engineering, M.O.T., Tianjin 300456, China;
      ruijia_jin@163.com (R.J.); stonegeng@163.com (B.G.); tjzhq1@163.com (H.Z.)
[2]   College of Engineering, Ocean University of China, Qingdao 266100, China; liuyong@ouc.edu.cn
[3]   College of Energy, Chengdu University of Technology, Chengdu 610059, China; jinx@cdut.edu.cn
*   Correspondence: liumingming19@cdut.edu.cn

Received: 9 October 2019; Accepted: 3 January 2020; Published: 7 January 2020

Abstract: A 2D numerical model was established to investigate vortex induced vibration (VIV) for submerged floating tunnel (SFT) by solving incompressible viscous Reynolds average Navier-Stokes equations in the frame of Abitrary Lagrangian Eulerian (ALE). The numerical model was closed by solving SST k-ω turbulence model. The present numerical model was firstly validated by comparing with published experimental data, and the comparison shows that good achievement is obtained. Then, the numerical model is used to investigate VIV for SFT under current. In the simulation, the SFT was allowed to oscillate in cross flow direction only under the constraint of spring and damping. The force coefficients and motion of SFT were obtained under different reduced velocity. Further research showed that Reynolds number has not only a great influence on the vibration amplitude and 'lock-in' region, but also on the force coefficients on of the SFT. A large Reynolds number results in a relatively small 'lock-in' region and force coefficient.

Keywords: Navier-Stokes equation; SST k-ω turbulence model; vortex-induced vibration (VIV); Arbitrary Lagrangian Eulerian (ALE) method; finite element method (FEM)

---

1. Introduction

Submerged floating tunnel (SFT) is a new type of traffic structure that crosses the strait, bay and lake. It is usually suspended more than 30 m below the water. The SFT has a large internal space, which is sufficient to meet the requirements of roads and even railways. For some fjords with harsh natural conditions, due to environmental conditions and technical constraints, traditional spanning methods (such as: cross-sea bridges, immersed tunnels) are not feasible, and SFT offers the possibility of crossing. Since the SFT is always in a deep-water depth, whose location is more the half wave length of normal wave, the normal wave has less influence on it. In addition, due to severe natural environment in the fjord, the flow velocity is usually fast, which has a greater influence on the SFT.

In view of the coupled analysis of flow and SFT, many scholars have done the relevant researches. Mai [1] considered the fluid-structure interaction effect, and studied the effects of surface velocity, tunnel section form and support form on the dynamic response of the SFT. It is found that the surface velocity would significantly affect the response displacement of the SFT, but it did not affect the stress distribution along the axial direction. Wang [2] analysed the variation law of the load on the SFT structure under the lateral lift force of the flow with the submerged depth, water depth, flow velocity and section size. Long [3] studied the dynamic response of SFT in different Buoyancy Weight Ratios (BWRs), and proposed the optimal range of BWRs for SFT under flow loading. The empirical lift formula based on Morison's formula was used in the above studies when discussing the flow load, but the effect of SFT on the flow field was not considered in detail. In order to find the optimal section of SFT, Luo et al. [4,5] compared the flow field distribution and force acting on the fixed SFT with different cross-section forms by large eddy numerical simulation. It was found that the ear-shaped (Figure 1) SFT structure had smaller lift coefficient and drag coefficient, which was the most reasonable cross-section shape, followed by circle, ellipse, hexagon and rectangular.

Figure 1. A sketch of ear-shape SFT structure.

Since the SFT is suspended in the sea via the mooring system, it will move under the flow action. The periodically varying lift makes the SFT with elastically support vibrate perpendicular to the inlet flow direction, that is, 'Vortex-Induced Vibration' (VIV). When the vortex shedding frequency is close to the natural frequency of the structure, the phenomenon of resonance or lock-in occurs, which reflects the complex interaction between the fluid and the structure. Therefore, when frequency lock-in occurs in the SFT under the flow action, the fatigue damage of the structure will be significantly increased, which will have a negative impact on the safety of the project. Many experimental researches and numerical simulations on the VIV problem were carried out, such as Morse and Williamson [6], Govardhan and Williamson [7], Yan et al. [8], Luo et al. [9], Zheng et al. [10]. VIV experiment of rigid cylinder with elastically support under wind load was successfully studied by Feng [11]. Williamson and Khalak [12,13] and Govardhan et al. [14] performed VIV experiments on rigid columns with low mass ratio and elastic support in the wave tank, which became the verification test for many subsequent numerical simulations. Lu and Dalton [15] numerically studied the cylindrical VIV problem with cross-flow motion in the case of Reynolds number Re = 13,000, and the model used large eddy simulation to close the turbulence equation. Dong and Karniadakis [16] used direct numerical simulation (DNS) to study the force vibration of cylinders with a Reynolds number of 10,000. For the vibration analysis of SFT, Ge et al. [17] and Kang et al. [18] applied the von der Pol equation to simulate the VIV of the SFT in flow, and studied the influence of the spacing of the anchor chain on the vibration amplitude of the tunnel. Su and Sun [19] utilised the wake oscillation model to simulate the VIV of the SFT. It was found that the vortex-induced vibration resonance occurred and the axial stress of the structure increased significantly. Chen et al. [20] proposed a simplified theoretical model for vibration analysis of the coupled SFT tube-cable system under wave and current.

In general, the empirical formulas are employed in most research about the VIV of SFT, and the study based on computational fluid dynamic (CFD) are limited on a specific Reynolds number. For this reason, the investigation of VIV based on CFD in a serious of Reynolds number under a specific engineering background should be proceeded here. The vibration phenomenon of SFT under different Reynolds number is different because the size of the structure and flow velocity vary greatly in different situations. Based on the above background, this study aims to analyse the coupled motion of flow and SFT in the range of Reynolds number from 1000 to 100,000, and study the influence of different Reynolds numbers on the vibration of SFT as well as force coefficient under the flow action, so as to provide reference value for practical engineering.

2. Numerical Model

For underwater SFT, the flow around the structure is usually turbulent due to the large scale of the structure itself and the relatively fast flow velocity. At present, the simulation of turbulence can be approximated by direct numerical simulation (DNS) or by using a suitable turbulence model, and the turbulence model is used to simulate the problem in this paper. At the same time, since the length of the SFT structure is much larger than the section size, we can simulate this problem as a two-dimensional (2D) flow-structure interaction problem approximately. Although it is well known that the flow is indeed three-dimensional (3D) effect for large Reynolds number, however, 3D simulation cost a lot of computational resources. Therefore, a 2D numerical model was adopted in this study. Although 2D numerical model will overpredict the numerical results, it still can reveal the relationship between reduced velocity, vibration amplitude and force coefficient. In addition, many other researchers have also adopted 2D numerical model to solve similar problems (Lu and Dalton [15], Dong and Karniadakis [16]).

2.1. Governing Equation and Turbulence Model

The two-dimensional incompressible Reynolds-Averaged Navier-Stokes equations are adopted to describe the turbulence flow of incompressible viscous fluid. The governing equations in the Arbitrary Lagrangian-Eulerian (ALE) frame can be written as (Liu et al. [21])

$$\frac{\partial u_i}{\partial t} + (u_j - u_j^m)\frac{\partial u_i}{\partial x_j} = -\frac{1}{\rho}\frac{\partial p}{\partial x_i} + \frac{\partial}{\partial x_j}[2\nu S_{ij} - \overline{u_i' u_j'}] \tag{1}$$

$$\frac{\partial u_i}{\partial x_i} = 0 \tag{2}$$

where $x_1 = x$, $x_2 = y$ are the horizontal and vertical coordinates, respectively, u_i is the fluid velocity in the x_i-direction, t is the time, u_j^m is the velocity of moving grid in the x_j-direction, p is the pressure, ρ is the fluid density, ν is the kinematic viscosity of the fluid, S_{ij} is the mean strain rate tensor with $S_{ij} = (\partial u_i/\partial x_j + \partial u_j/\partial x_i)/2$. The Reynolds stress term in Equation (1) reads can be expressed as

$$\overline{u_i' u_j'} = \nu_t(\partial u_i/\partial x_j + \partial u_j/\partial x_i) + \frac{2}{3}k\delta_{ij} \tag{3}$$

where ν_t is the turbulent eddy viscosity, k is the turbulence kinetic energy and δ_{ij} is the Kronecker operator.

In order to close the governing equations, the Shear Stress Transport (SST) k-ω turbulence model (Menter [22]; Menter et al. [23]) is adopted. The parameters in the equation have been widely accepted and successfully applied, which has shown good performance in simulating the boundary layer flows with significant adverse pressure gradient. The governing equation of the SST k-ω turbulence model can be written as follows:

$$\frac{\partial k}{\partial t} + (u_j - u_j^m)\frac{\partial k}{\partial x_j} = \frac{\partial}{\partial x_j}\left[(\nu + \sigma_k \nu_t)\frac{\partial k}{\partial x_j}\right] + P_k - \beta^* \omega k \tag{4}$$

$$\frac{\partial \omega}{\partial t} + (u_j - u_j^m)\frac{\partial \omega}{\partial x_j} = \frac{\partial}{\partial x_j}\left[(\nu + \sigma_\omega \nu_t)\frac{\partial \omega}{\partial x_j}\right] + \alpha S^2 - \beta \omega^2 + 2(1 - F_1)\sigma_{\omega 2}\frac{1}{\omega}\frac{\partial k}{\partial x_j}\frac{\partial \omega}{\partial x_j} \tag{5}$$

where P_k is the production of turbulent kinetic energy and the related parameters in Equations (4) and (5) are calculated as follows

$$\nu_t = \frac{\alpha_1 k}{\max(\alpha_1 \omega, \Omega F_2)}$$

$$P_k = \min\left[\nu_t \frac{\partial u_j}{\partial x_i}\left(\frac{\partial u_j}{\partial x_i} + \frac{\partial u_i}{\partial x_j}\right), 10\beta^* k\omega\right]$$

$$F_1 = \tanh\left\{\left\{\min\left[\max\left(\frac{\sqrt{k}}{\beta * \omega y*}, \frac{500\nu}{y*^2 \omega}\right), \frac{4\rho\sigma_{\omega 2}k}{D_{k\omega}y*^2}\right]\right\}^4\right\}$$

where Ω is the absolute value of vorticity, y^* is the distance to the nearest solid wall, and the parameters F_2 and $D_{k\omega}$ are

$$D_{k\omega} = \max\left(2\sigma_{\omega 2}\rho\frac{1}{\omega}\frac{\partial k}{\partial x_j}\frac{\partial \omega}{\partial x_j}, 10^{-10}\right), F_2 = \tanh\left\{\left[\max\left(\frac{2\sqrt{k}}{\beta * \omega y*}, \frac{500\upsilon}{y*^2\,\omega}\right)\right]^2\right\}$$

By using the blending function F_1, the following parameters can be calculated, i.e.,

$$\sigma_k = F_1\sigma_{k1} + (1 - F_1)\sigma_{k2}; \sigma_\omega = F_1\sigma_{\omega 1} + (1 - F_1)\sigma_{\omega 2}$$

$$\alpha = F_1\alpha_1 + (1 - F_1)\alpha_2; \beta = F_1\beta_1 + (1 - F_1)\beta_2$$

The model constant in the SST k-ω model are listed in Table 1.

Table 1. Parameters of SST k-ω turbulent model.

β^*	α_1	β_1	σ_{k1}	$\sigma_{\omega 1}$	α_2	β_2	σ_{k2}	$\sigma_{\omega 2}$
0.09	5/9	3/40	0.85	0.5	0.44	0.0828	1.0	0.856

When the flow field and the pressure field are obtained, the fluid force acting on the structure can be obtained by integrating the surface pressure and the viscous shear force over the body surface. The dimensionless drag coefficient C_D and the lift coefficient C_L are respectively

$$C_D = -\int_0^{2\pi} p\cos\theta d\theta - \frac{1}{\mathrm{Re}}\int_0^{2\pi}\left(\frac{\partial\upsilon}{\partial x} - \frac{\partial u}{\partial y}\right)\sin\theta d\theta \tag{6}$$

$$C_L = -\int_0^{2\pi} p\sin\theta d\theta + \frac{1}{\mathrm{Re}}\int_0^{2\pi}\left(\frac{\partial\upsilon}{\partial x} - \frac{\partial u}{\partial y}\right)\sin\theta d\theta \tag{7}$$

2.2. Motion Response of SFT

For the vibration problem of the SFT under the flow action, since it is necessary to ensure the anchor cable is always in the elastic range, the whole system can be simplified as a mass-damping-spring system. In this paper, only the vibration response of the SFT in cross flow direction is considered, and its motion equation can be expressed as follows.

$$m\ddot{y} + c\dot{y} + ky = F_y \tag{8}$$

where m, c, and k are the mass, damping and stiffness of SFT, respectively. F_y is the fluid force of SFT in cross flow direction which is determined by the flow equation.

Utilising the relationship of structural dynamics, $c/m = 4\pi\xi f_n$, $k/m = (2\pi f_n)^2$, the ratio of the mass of the SFT to the mass of water discharged from SFT is defined $m* = 4m/\pi\rho D^2$. Therefore, the expression can be obtained as follow

$$\ddot{y} + 4\pi\xi f_n\dot{y} + (2\pi f_n)^2 y = \frac{4F}{\pi\rho D^2 m*} \tag{9}$$

where ξ is the damping ratio of structure, f_n is the natural frequency and m^* is mass ratio.

The following dimensionless relationship can be defined further

$$\ddot{Y} = \frac{\ddot{y}D}{U^2}, \dot{Y} = \frac{\dot{y}}{U}, Y = \frac{y}{D}, F_n = \frac{f_n D}{U} \tag{10}$$

At the same time according to the definition of lift coefficient, $F_y = 0.5\rho U^2 DC_L$, Equation (9) can be transformed in dimensionless form

$$\ddot{Y} + 4\pi\xi F_n\dot{Y} + (2\pi F_n)^2 Y = \frac{2C_L}{\pi m*} \tag{11}$$

By introducing the definition of the reduced velocity $U_r = U/f_nD$, the motion equation of the SFT can also be expressed as a dimensionless equation in the form of reduced velocity.

$$\ddot{Y} + \frac{4\pi\xi}{U_r}\dot{Y} + \left(\frac{2\pi}{U_r}\right)^2 Y = \frac{2C_L}{\pi m*} \tag{12}$$

The lift coefficient at the right end of the above formula has been given before, and the dynamic response of the SFT can be calculated by the above formula.

2.3. Calculation Model and Boundary Conditions

The calculation model and boundary conditions are shown in Figure 2. Let the origin of the coordinate be at the initial centre of the cylinder, and the dimensionless cylinder diameter $D = 1$. The dimensionless velocity $u = 1$, $v = 0$ are set up at inlet. Symmetrical boundary conditions are applied at side wall $\partial u/\partial y = 0$, $v = 0$. The outlet velocity boundary condition is $\partial u_i/\partial t + c\partial u_i/\partial x_i = 0$, where c is local average flow velocity. non-slip boundary conditions applied to the cylindrical surface $u = dx/dt$, $v = dy/dt$. In the calculation, in order to ensure the first layer of the grids is in the viscous boundary layer, the distance between the surface of the circular cylinder and the first layer of the grids is less than 0.05% D. In addition, the choice of boundary layer can be employed by the method of Palm et al. [24]. In the calculation, the pressure at outlet $p = 0$ and the pressure boundary condition $\partial p/\partial \mathbf{n} = 0$ is applied at other boundary conditions, $\mathbf{n}$ is unit normal vector pointing out the fluid domain. At initial time, the velocity and pressure in the fluid domain are set up to zero, i.e., the initial velocity field satisfies the continuous equation.

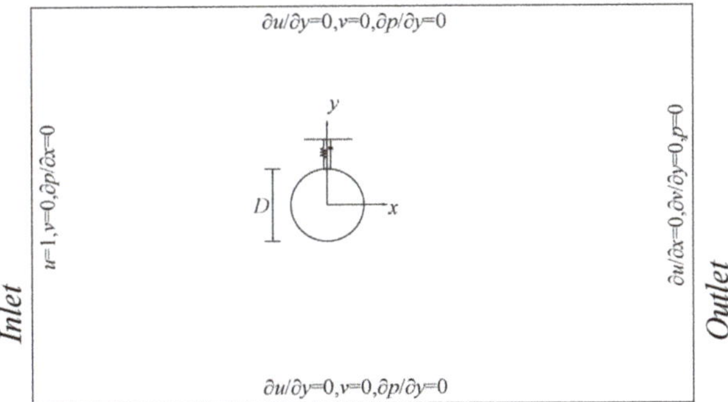

Figure 2. Sketch definition of computational domain and boundary conditions.

3. Numerical Dispersion and Grid Update

The convection-diffusion equation is solved by streamline upwind/petrov-galerkin finite element method in this paper (Brooks and Hughes [25]), and this method has been applied in the solution of impressive flow problem successfully (Mochida and Murakami [26], Kim et al. [27], Guilmineau and Queutey [28]). The distribution method is utilised in the time integration of the momentum equation. First, the pressure term is neglected, and the intermediate velocity of convection and diffusion term is considered; then the pressure equation is solved to calculate the pressure of the next time step; finally, the pressure gradient term is considered to correct the flow field. Streamline upwind method is used to predict the velocity.

The Newmark-β method is used to solve the motion equation of the structure. Given the displacement, velocity and acceleration at an initial moment, the time step Δt, the parameters β and γ are selected, and then the equivalent stiffness is formed. The effective load at time $t + \Delta t$ is obtained, and the displacement at the time $t + \Delta t$ can be solved. For time advance, according to CFL (Courant-Friedrichs-Lewy) conditions, the following dynamic time steps are adopted:

$$\Delta t = C_s \min\left(\sqrt{S_c} / |u_e| \right) \tag{13}$$

where S_c is the mesh area, u_e is the flow velocity at grid centre, min indicates the minimum in the computational domain, C_s is the safe coefficient, $C_s = 0.2$. Due to the reciprocating motion of the SFT under the flow action, the dynamic grid method based on ALE is used to simulate the fluid-structure coupling problem. In this paper, the mesh in the computational domain is assumed to be an elastic one, as shown in Figure 3, in order to achieve the purpose of adapting to the movement of the grid boundary nodes and internal nodes. The motion and deformation of the mesh can be obtained by solving the governing equation of linear elastodynamics (Johnson and Tezduyar [29]). The mesh updating method make the displacement of the mesh nodes more uniform and improve the stability of numerical calculation. In addition, the possibility of mesh distortion can be reduced by controlling the elastic modulus of the computing element. Specifically, the balance length of the mesh is equal to the length of the mesh itself at the initial moment. When the two ends of the joint move relative to each other, the mesh will be stretched or compressed, correspondingly. The mesh still satisfies Hooke's law, so the total force vector of any node i is

$$\mathbf{F}_i = \sum_{j=1}^{v_i} \alpha_{ij}\left(\delta_j - \delta_i\right) \tag{14}$$

where $\mathbf{F}_i$ is the total force vector on node i, α_{ij} is the mesh stiffness between the nodes, v_i represents the number of nodes connected to node i, $j = [i, v_i]$, δ_i and δ_j are displacement vector of node i and j. In order to avoid collisions of mesh nodes, the following expression is usually used to calculate the mesh stiffness

$$\alpha_{ij} = \frac{1}{\sqrt{\left(x_j - x_i\right)^2 + \left(y_j - y_i\right)^2}} \tag{15}$$

where x_i, x_j are the position vectors of nodes i and j, that is, the value of α_{ij} is considered to be the reciprocal of the side length. This mesh updating method has also been widely used in the VIV research (Tang et al. [30], Lu et al. [31]).

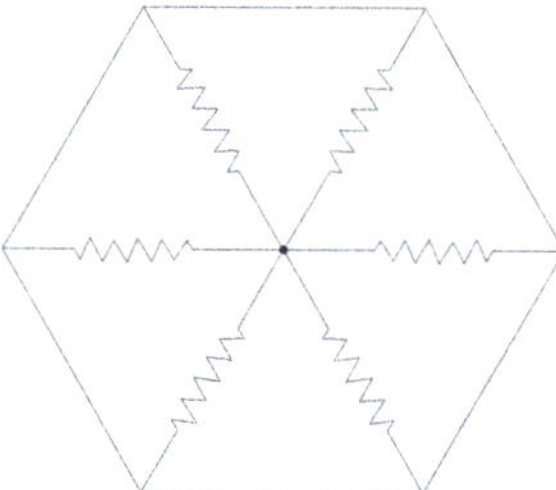

Figure 3. Illustration of the elastic mesh.

4. Model Validation

In order to obtain reliable numerical results, this paper firstly uses the free vibration cylinder problem under Re = 30,000 and $U_r = 6.00$ as an example to verify the mesh convergence of the numerical model. Four different meshes are considered in Table 2. From Table 2, it can be seen that the numerical results under the four meshes are very close, indicating that the numerical results have converged under the current grid density. Considering the computational efficiency, the latter numerical calculation takes mesh 3 (Figure 4) as the benchmark. In the table, Y_{max} denotes the maximum vibration amplitude of the cylinder, Dis_{min} denotes the minimum distance between the circular cylinder and the first layer gird, C_D^M is the mean drag coefficient, C_L^{RMS} is root mean square (RMS) of lift coefficient.

Table 2. Comparisons of numerical results with different mesh solutions.

	Boundary Division	**Element**	**Nodes**	Dis_{min}/D	Y_{max}/D	C_D^M	C_L^{RMS}	f_n
Mesh 1	80	18,200	18,490	0.0045	0.7166	1.295	0.148	0.173
Mesh 2	120	21,900	22,300	0.0045	0.7284	1.325	0.187	0.172
Mesh 3	160	25,600	25,970	0.0040	0.7292	1.332	0.214	0.172
Mesh 4	200	29,300	29,710	0.0038	0.7293	1.335	0.224	0.172

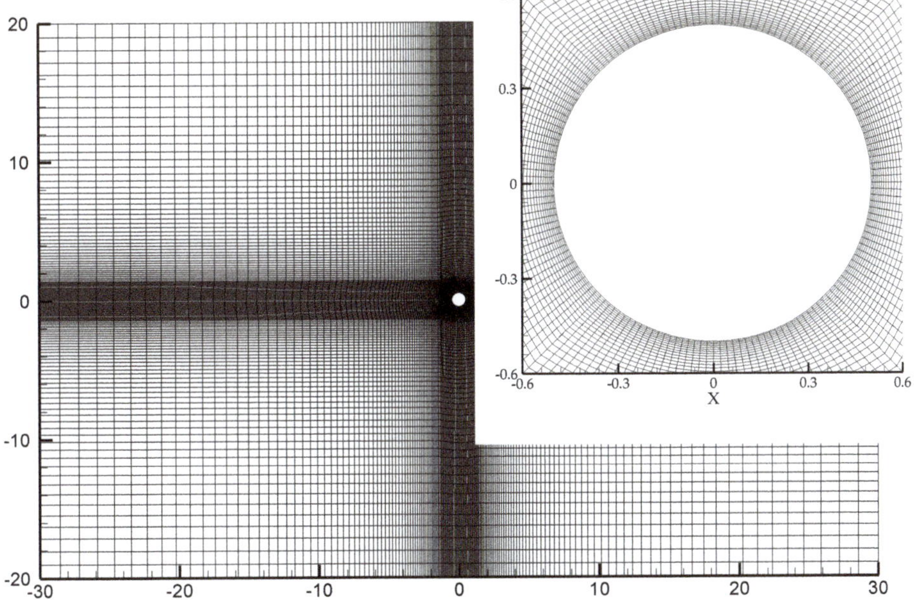

Figure 4. Sketch of mesh 3 in the model validation.

In order to further verify the reliability of the numerical model in this paper, the coupling analysis of uniform flow and fixed cylinder is verified firstly. The diameter of cylinder is $D = 1.0$, the Reynolds number Re = 10,000. The calculation results containing the mean drag coefficient C_D^M, amplitude of lift coefficient C_L^A and strouhal number S_t are compared with experiment results conducted by Gopalkrishnan [32] and the numerical results of Dong and Karmiadakis [16], Zhao et al. [33], Song [34], which are shown in Table 3.

Table 3. Calculation results comparisons of fixed single cylinder.

	Gopalkrishnan [32]	**Dong and Karmiadakis [16]**	**Zhao et al. [33]**	**Song [34]**	**Present**
C_D^M	1.186	1.143	1.078	1.288	1.262
C_L^A	0.384	0.448	0.667	0.638	0.668
S_t	0.193	0.203	0.211	0.166	0.202

From the comparisons in Table 3, the calculation results of C_D^M, C_L^A and S_t are almost consistent with those of other scholars, which proves the accuracy of the numerical model in the case of high Reynolds number. The problem of flow around a fixed cylinder is actually only a unilateral hydrodynamic calculation problem. It does not involve the motion response calculation of the circular cylinder itself, so it is not a true fluid-structure coupling problem. The vibration of cylinder with elastic support under the flow action involves fluid-structure coupling problem. There have been many experimental results on the VIV of rigid cylinders with elastic supports under the flow action, such as Khalak and Williamson [12], which have done systematic experimental studies. In order to facilitate comparison with the experimental results, the same calculation parameters as in the tests of Khalak and Williamson [12] were used. The Reynolds number Re = 12,000, the mass ratio m^* = 2.4, and the mass damping ratio $m^*\xi$ = 0.013, and the maximum dimensionless vibration amplitudes versus reduced velocity are calculated in the paper.

From the comparison results in Figure 5, it can be seen that the maximum dimensionless vibration amplitude is close to 1 and the 'lock-in' region is from U_r = 4.0~10.0. In addition, the upper branch and lower branch can also be described in the numerical model clearly as shown in the experiment. Therefore, the results in the numerical model agree well with the experimental results of Khalak and Williamson [12]. It is illustrated that the model established in this paper can be used to investigate the fluid-structure coupling problem with high Reynolds number.

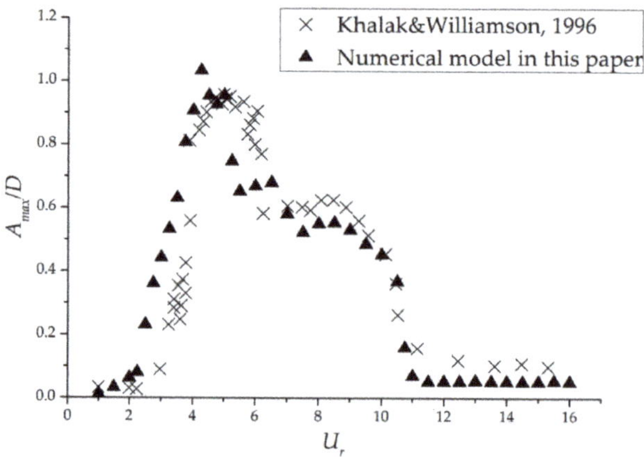

Figure 5. Calculation result comparisons of cylinder with spring and damping.

5. Example Analysis

Based on the above numerical model, this paper calculates the motion of the SFT under different constraint stiffness and different Reynolds numbers. The Reynolds number is calculated from 1000 to 100,000. Firstly, the time history curves of the cross-flow direction under the conditions of Reynolds number 50,000, mass ratio m^* = 2.5, damp ξ = 0.007 and reduced velocity U_r = 2.0, 5.0 and 12.0, respectively, are introduced, as shown in Figure 6. Then, the Fast Fourier Transform (FFT) is utilised in the lift force coefficient time history, and the result is displayed in Figure 7.

From Figure 5, it can be found that when reduced velocity are 2.0 and 12.0, the vibration amplitude in cross flow direction of the SFT is small, and when the reduced velocity is 5.0, the vibration amplitude is large. Subsequently, the flow field is analysed in the case of larger and smaller vibration amplitudes, as shown in Figures 8 and 9.

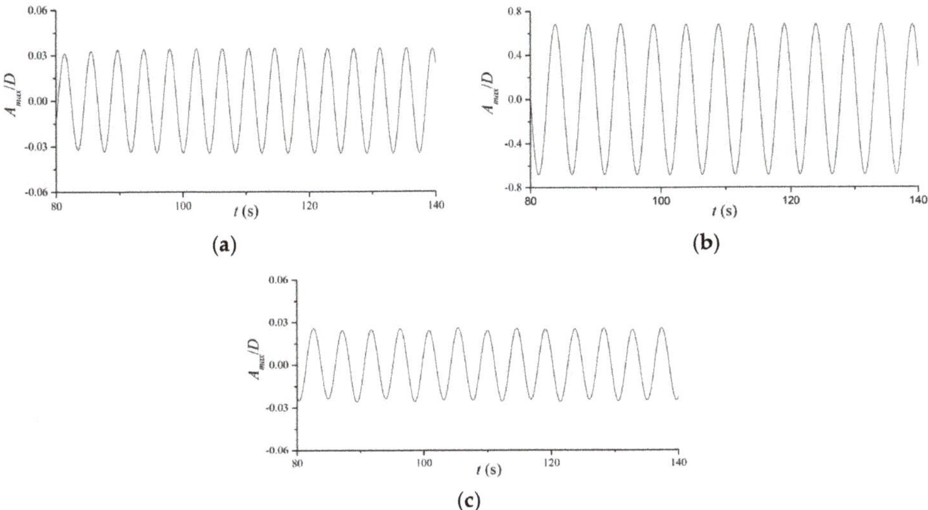

Figure 6. Time histories of submerged floating tunnel (SFT) motion under different reduced velocity: (a) $U_r = 2$; (b) $U_r = 5$; (c) $U_r = 12$ (Re = 50,000, $m^* = 2.5$, $\xi = 0.007$).

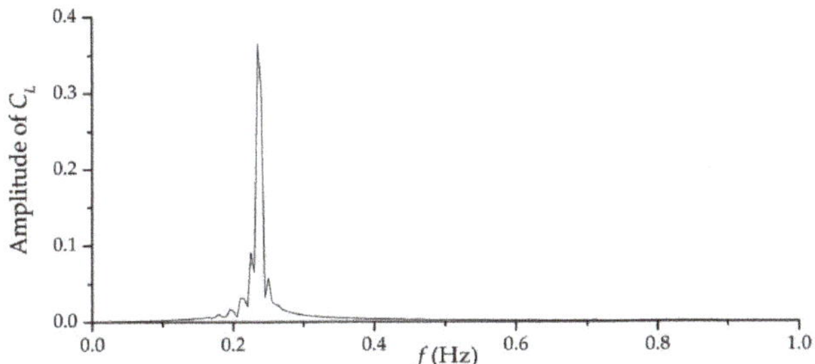

Figure 7. Fourier transform of SFT lift force coefficient when reduced velocity $U_r = 2.0$ (Re = 50,000, $m^* = 2.5$, $\xi = 0.007$).

From Figure 6, it can be seen that the vortex shedding frequency is about 0.23 Hz. According to the definition of $U_r = U/f_n D$, the natural frequencies (f_n) of SFT in the case here are 0.5 Hz, 0.2 Hz and 0.083 Hz, respectively. Therefore, it can be seen that when the reduced velocity of the SFT are 2.0 and 12.0, the vortex shedding frequency is far away from the natural frequency, and the VIV of the SFT is not obvious. The wake pattern is shown in Figure 9, and the wake shape is also regular in one vibration period. However, when the reduced velocity is 5.0 by adjusting the spring stiffness, the frequency of vortex shedding is close to the natural frequency. Under the action of flow lift force, a large VIV phenomenon occurs in the SFT. The vibration amplitude is larger, even reaching 0.7 times of the outer diameter of the SFT, whose wake pattern is shown in Figure 8. It can be seen that the wake has a long 'tail' after being separated, and the wake shape is irregular. Next, we compare the coupling effect of the flow and the SFT versus different reduced velocity under this Reynolds number, and the maximum dimensionless vibration amplitude of the SFT has been statistically obtained, as shown in Figure 10.

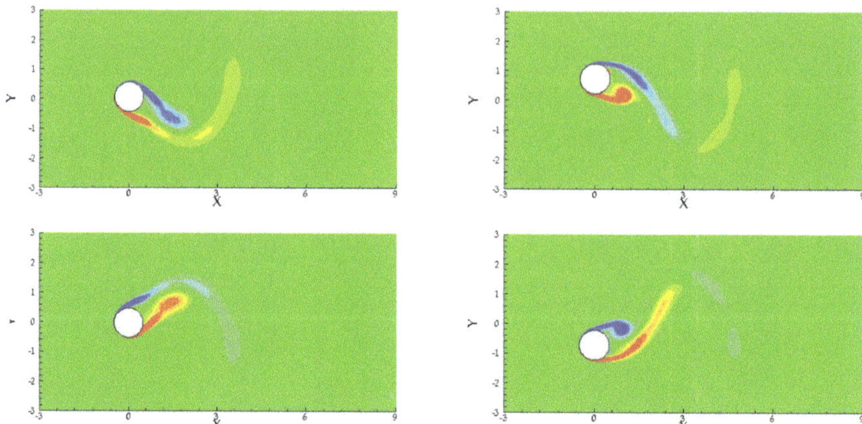

Figure 8. Vibration mode and wake vortex shedding mode of SFT in one cycle when reduced velocity $U_r = 5.0$ (Re = 50,000, $m^* = 2.5$, $\xi = 0.007$).

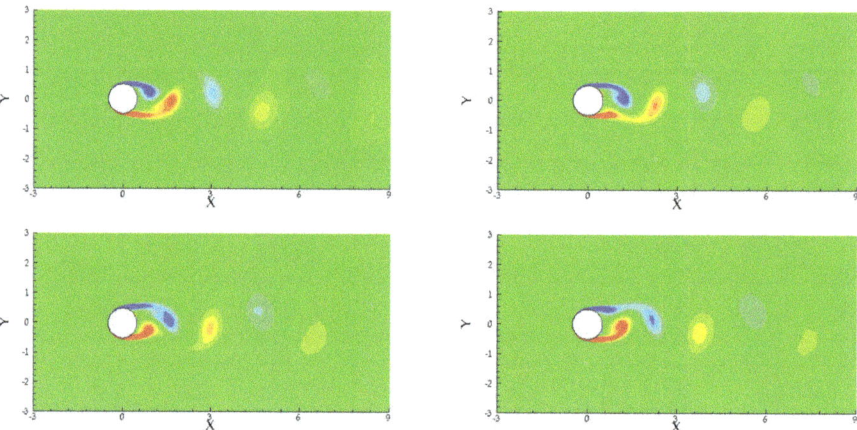

Figure 9. Vibration mode and wake vortex shedding mode of SFT in one cycle when reduced velocity $U_r = 2.0$ (Re = 50,000, $m^* = 2.5$, $\xi = 0.007$).

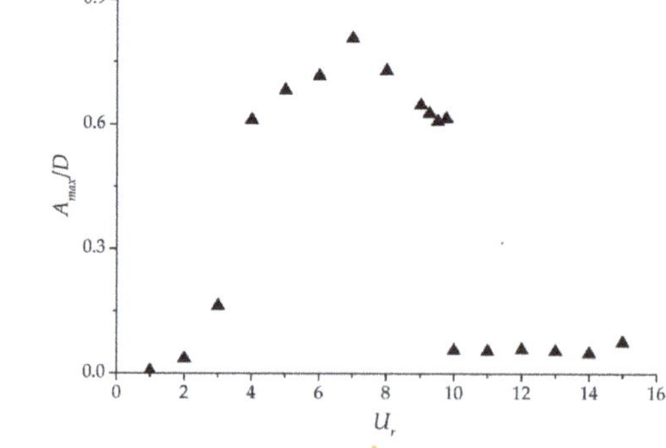

Figure 10. Vibration amplitude of SFT versus reduced velocity under Re = 50,000, $m^* = 2.5$ and $\xi = 0.007$.

It can be seen from the calculation results that when the reduced velocity is from 4.0 to 10.0, the structure is 'locked' under the flow action, while at other reduced velocity, the vibration amplitude of the structure is small. Then the VIV of the SFT under different Reynolds numbers are compared, and the calculation results are shown in Figure 11. From the results of VIV of SFT under different Reynolds numbers, it can be seen that Reynolds number has not only a great influence on the vibration amplitude, but also on the 'lock-in' region. In general, the lower the Reynolds number is, the larger the amplitude is out of the 'lock-in' region. The minimum vibration amplitude in the 'lock-in' region is about $0.4D$, while the maximum vibration amplitude in the 'lock-in' region can reach to $0.8D$. It can also be seen that larger Reynolds number leads to narrow 'lock-in' region in Figure 10. Therefore, when the size of the SFT is small or the inlet flow velocity is slow, VIV 'lock-in' phenomenon is more likely to occur for SFT because of lower Reynolds number.

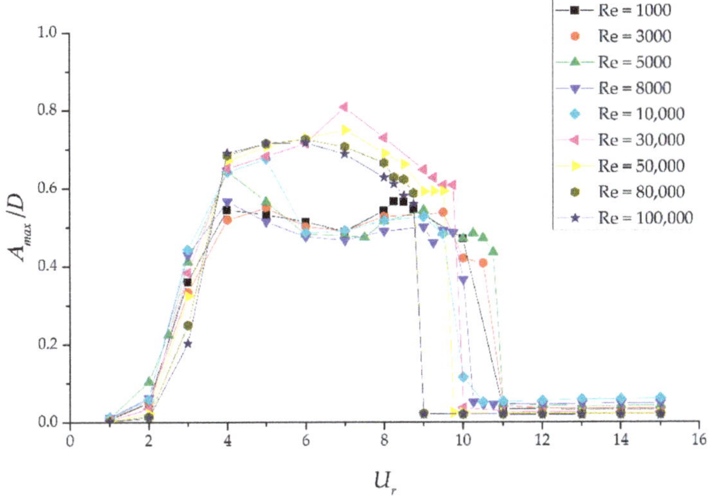

Figure 11. Vibration amplitude of SFT versus natural frequency in different Reynolds numbers.

Figure 12 is the force coefficient on SFT versus reduced velocity at different Reynolds numbers. It can be seen that reduced velocity has a greater influence on the mean drag coefficient and RMS of lift coefficient. When the Reynolds number is low, the mean drag coefficient and RMS of lift coefficient of the SFT are relatively large from 1000 to 10,000. As the Reynolds number increases, the mean drag coefficient and RMS of lift coefficient become smaller. Therefore, when the size of the SFT is small or the inlet flow velocity action on the structure is slow, the force coefficient is large, and when the size is large or the flow velocity is fast, the mean drag coefficient and lift force coefficient of the structure are small.

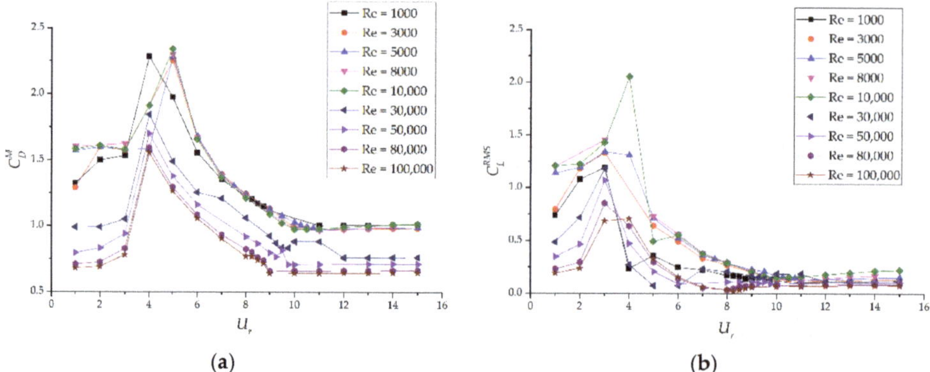

Figure 12. Non-dimensional force coefficient on SFT versus natural frequency at different Reynolds numbers: (**a**) mean drag coefficient; (**b**) RMS of lift coefficient.

6. Conclusions

Based on the FEM solution of incompressible viscous Reynolds average Navier-Stokes equations, combining the frame of Abitrary Lagrangian Eulerian, through accurate computational fluid dynamic numerical simulation, the vortex-induced vibration problems of submerged floating tunnel with different Reynolds numbers are studied. Main conclusions are as follows:

Firstly, the analysis of uniform flow and fixed single cylinder proves the accuracy of the model in the case of high Reynolds number. Then, by simulating the vortex-induced vibration of a cylinder at high Reynolds number and comparing with other scholars' experimental results, it is proved that the model established in this paper can be used to study the fluid-structure coupling problem at high Reynolds number.

Through the research of vortex-induced vibration of SFT under the flow action, the force coefficient and motion of the SFT versus different reduced velocity at different Reynolds numbers are analysed. The results show that the Reynolds number has not only a great influence on the vibration amplitude and 'lock-in' region, but also on the force coefficient on the SFT. When the Reynolds number is low, the 'lock-in' region, the mean drag coefficient and RMS of lift coefficient of the SFT are relatively large. As the Reynolds number increases, the 'lock-in' region, the mean drag coefficient and RMS of lift coefficient become smaller. Therefore, when the size of the SFT is small or the flow velocity action on the structure is slow, the force coefficient and 'lock-in' region are relatively large, while when the size is large or the flow velocity is fast, the force coefficient and 'lock-in' region are relatively small.

Author Contributions: Formal analysis, X.J.; data curation, B.G.; writing—original draft preparation, R.J.; writing—review and editing, M.L. and Y.L.; funding acquisition, H.Z. All authors have read and agreed to the published version of the manuscript.

Funding: The authors would like to acknowledge the support from State Key Laboratory of Coastal and Offshore Engineering Projects Program (Grant No. LP1912), National Natural Science Foundation of China (Grant No. 51809133) and China Postdoctoral Science Foundation (Grant No. 2019M652479).

Conflicts of Interest: The authors declare no conflict of interest.

References

1. Mai, J.T. The Study of Responses of a Submerged Floating Tunnel Subjected to the Wave and Current. Ph.D. Thesis, Southwest Jiaotong University, Chengdu, China, 2005.
2. Wang, G.D. Numerical Analysis and Experiment Research of Submerged Floating Tunnel Subjected to the Wave and Current Effects. Ph.D. Thesis, Southwest Jiaotong University, Chengdu, China, 2008.
3. Long, X. Dynamic Response of Submerged Floating Tunnel with Different Buoyancy-Weight Ratios under Wave and Current Loads. Master's Thesis, Chinese Academy of Sciences, Beijing, China, 2009.

4. Luo, G.; Zhou, X.J.; Li, D.F.; Zhang, C. Analysis on characteristics of flow passing submerged floating tunnels of different sections. *J. China Railw. Soc.* **2013**, *35*, 115–120.

5. Luo, G.; Zhou, X.J.; Zhang, C.; Shen, Q. Numerical simulation analysis on rational cross-section of submerged floating tunnel. *Railw. Eng.* **2012**, *12*, 32–36.

6. Morse, T.L.; Williamson, C.H.K. Prediction of vortex-induced vibration response by employing controlled motion. *J. Fluid Mech.* **2009**, *634*, 5–39. [CrossRef]

7. Govardhan, R.; Williamson, C.H.K. Critical mass in vortex-induced vibration of a cylinder. *Eur. J. Mech. B Fluids* **2004**, *23*, 17–27. [CrossRef]

8. Yan, B.; Luo, M.; Bai, W. An experimental and numerical study of plunging wave impact on a box-shape structure. *Mar. Struct.* **2019**, *66*, 272–287. [CrossRef]

9. Luo, M.; Reeve, D.E.; Shao, S.D.; Karunarathna, H.; Lin, P.; Cai, H. Consistent Particle Method simulation of solitary wave impinging on and overtopping a seawall. *Eng. Anal. Bound. Elem.* **2019**, *103*, 160–171. [CrossRef]

10. Zheng, X.G.; Chen, R.D.; Luo, M.; Kazemi, E.; Liu, X.N. Dynamic hydraulic jump and retrograde sedimentation in an open channel induced by sediment supply: Experimental study and SPH simulation. *J. Mt. Sci.* **2019**, *16*, 1913–1927. [CrossRef]

11. Feng, C.C. The Measurement of Vortex Induced Effected in Flow Past Stationary and Oscillating Circular and d-Section Cylinder. Ph.D. Thesis, University of British Columbia, Vancouver, SC, Canada, 1968.

12. Khalak, A.; Williamson, C.H.K. Dynamics of a hydroelastic cylinder with very low mass and damping. *J. Fluids Struct.* **1996**, *10*, 455–472. [CrossRef]

13. Khalak, A.; Williamson, C.H.K. Investigation of relative effects of mass and damping in vortex-induced vibration of a circular cylinder. *J. Wind Eng. Ind. Aerodyn.* **1997**, *69*, 341–350. [CrossRef]

14. Govardhan, R.; Williamson, C.H.K. Modes of vortex formation and frequency response of a freely vibrating cylinder. *J. Fluid Mech.* **2000**, *420*, 85–130. [CrossRef]

15. Lu, X.Y.; Dalton, C. Calculation of the timing of vortex formation from an oscillating cylinder. *J. Fluids Struct.* **1996**, *10*, 527–541. [CrossRef]

16. Dong, S.; Karniadakis, G.E. DNS of flow past a stationary and oscillating cylinder at Re = 10000. *J. Fluids Struct.* **2005**, *20*, 519–531. [CrossRef]

17. Ge, F.; Long, X.; Wang, L.; Hong, Y. Flow-induced vibrations of long circular cylinders modeled by coupled nonlinear oscillators. *Sci. China Ser. G Phys. Mech. Astron.* **2009**, *52*, 1086–1093. [CrossRef]

18. Kang, L.; Ge, F.; Hong, Y.S. A numerical study on responses of submerged floating structures undergoing vortex-induced vibration and seismic excitation. *Procedia Eng.* **2016**, *166*, 91–98. [CrossRef]

19. Su, Z.B.; Sun, S.N. Vortex-Induced Dynamic Response of Submerged Floating Tunnel Tether Based on Wake Oscillator Model. *Adv. Mater. Res.* **2014**, *919*, 1262–1265. [CrossRef]

20. Chen, Z.; Xiang, Y.Q.; Lin, H.; Yang, Y. Coupled vibration analysis of submerged floating tunnel system in wave and current. *Appl. Sci.* **2018**, *8*, 1311. [CrossRef]

21. Liu, M.M.; Lu, L.; Teng, B.; Zhao, M.; Tang, G.Q. Numerical modeling of local scour and forces for submarine pipeline under surface waves. *Coast. Eng.* **2016**, *116*, 275–288. [CrossRef]

22. Menter, F.R. Two-equation eddy-viscosity turbulence models for engineering applications. *AIAA J.* **1994**, *32*, 1598–1605. [CrossRef]

23. Menter, F.R.; Kuntz, M.; Langtry, R. Ten years of industrial experience with the SST turbulence model. *Turbul. Heat Mass Transf.* **2003**, *4*, 625–632.

24. Palm, J.; Eskilsson, C.; Paredes, G.M.; Bergdahl, L. Coupled mooring analysis for floating wave energy converters using CFD: Formulation and validation. *Int. J. Mar. Energy* **2016**, *16*, 83–99. [CrossRef]

25. Brooks, A.N.; Hughes, T.J.R. Streamline Upwind/Petrov-Galerkin formulations for convection dominated flows with particular emphasis on the incompressible Navier-Stokes equations. *Comput. Methods Appl. Mech. Eng.* **1982**, *32*, 199–259. [CrossRef]

26. Johnson, A.A.; Tezduyar, T.E. Mesh update strategies in parallel finite element computations of flow problems with moving boundaries and interfaces. *Comput. Methods Appl. Mech. Eng.* **1994**, *119*, 73–94. [CrossRef]

27. Mochida, A.; Murakami, S. On turbulent vortex shedding flow past 2D square cylinder predicted by CFD. *J. Wind Eng. Ind. Aerodyn.* **1995**, *54–55*, 191–211.

28. Kim, D.H.; Yang, K.S.; Senda, M. Large eddy simulation of turbulent flow past a square cylinder confined in a channel. *Comput. Fluids* **2004**, *33*, 81–96. [CrossRef]

29. Guilmineau, E.; Queutey, P. Numerical simulation of vortex-induced vibration of a circular cylinder with low mass-damping in a turbulent flow. *J. Fluids Struct.* **2004**, *19*, 449–466. [CrossRef]

30. Tang, G.Q.; Lu, L.; Teng, B.; Liu, M.M. Numerical simulation of vortex-induced vibration with three-step finite element method and Arbitrary Lagrangian-Eulerian formulation. *Adv. Mech. Eng.* **2013**, *5*, 1–14. [CrossRef]

31. Lu, L.; Guo, X.L.; Tang, G.Q.; Liu, M.M.; Chen, C.Q.; Xie, Z.H. Numerical investigation of flow-induced rotary oscillation of circular cylinder with rigid splitter plate. *Phys. Fluids* **2016**, *28*, 093604. [CrossRef]

32. Gopalkrishnan, R. Vortex-Induced Forces on Oscillating Bluff Cylinders. Ph.D. Thesis, Massachusetts Institute of Technology, Cambridge, MA, USA, 1993.

33. Zhao, M.; Cheng, L.; Teng, B.; Dong, G. Hydrodynamic forces on dual cylinders of different diameters in steady currents. *J. Fluids Struct.* **2007**, *23*, 59–83. [CrossRef]

34. Song, J.N. Experimental Investigation and Numerical Simulation by a Discrete Vortex Method on VIV of Marine Risers. Ph.D. Thesis, Dalian University of Technology, Dalian, China, 2012.

Article

Aeroelastic Performance Analysis of Wind Turbine in the Wake with a New Elastic Actuator Line Model

Ziying Yu [1], Zhenhong Hu [1,*], Xing Zheng [1], Qingwei Ma [1,2] and Hongbin Hao [1]

[1] College of Shipbuilding Engineering, Harbin Engineering University, Harbin 150001, China; djwnfbnx@foxmail.com (Z.Y.); zhengxing@hrbeu.edu.cn (X.Z.); q.ma@city.ac.uk (Q.M.); haohongbin1989@163.com (H.H.)

[2] Schools of Engineering and Mathematical Sciences, City, University of London, London EC1V 0HB, UK

[*] Correspondence: huzhenhong@hrbeu.edu.cn; Tel.: +86-451-8256-8147

Received: 8 March 2020; Accepted: 21 April 2020; Published: 26 April 2020

Abstract: The scale of a wind turbine is getting larger with the development of wind energy recently. Therefore, the effect of the wind turbine blades deformation on its performances and lifespan has become obvious. In order to solve this research rapidly, a new elastic actuator line model (EALM) is proposed in this study, which is based on turbinesFoam in OpenFOAM (Open Source Field Operation and Manipulation, a free, open source computational fluid dynamics (CFD) software package released by the OpenFOAM Foundation, which was incorporated as a company limited by guarantee in England and Wales). The model combines the actuator line model (ALM) and a beam solver, which is used in the wind turbine blade design. The aeroelastic performances of the NREL (National Renewable Energy Laboratory) 5 MW wind turbine like power, thrust, and blade tip displacement are investigated. These results are compared with some research to prove the new model. Additionally, the influence caused by blade deflections on the aerodynamic performance is discussed. It is demonstrated that the tower shadow effect becomes more obvious and causes the power and thrust to get a bit lower and unsteady. Finally, this variety is analyzed in the wake of upstream wind turbine and it is found that the influence on the performance and wake flow field of downstream wind turbine becomes more serious.

Keywords: elastic actuator line model; OpenFOAM; NREL 5 MW wind turbine; aeroelastic performance

---

1. Introduction

With the improvement of wind power technology and the demand of high-power generation, the target of wind turbine design turns to large scale and offshore [1–7]. In 2009, the National Renewable Energy Laboratory (NREL) in America defined a 5 MW reference wind turbine for offshore system, in which the rotor diameter is 126 m [8]. Technical University of Denmark described a 10 MW reference wind turbine whose rotor diameter is 178.3 m [9] in 2013. As the blade of wind turbine gets longer, it becomes less stiff, more susceptible, and easily deformable, which will lead to increased fatigue damage and reduced production. When simulating the aeroelastic performances of a floating offshore wind turbine or a wind farm, there are many challenges to solve.

There are mainly three methods to study the aerodynamic performances of wind turbines. The first one is the Blade Element Momentum (BEM) theory, which combines the blade element theory and momentum theory. It has high efficiency and is widely used in the industrial application, but the information of flow field is not considered. The second one is the Computational Fluid Dynamics (CFD) method, which calculates the velocity and pressure fields by solving the Navier-Stokes equations. It can obtain quite accurate results and is gradually used in recent years, e.g., Tran et al. [10] analyzed the loading of 5 MW offshore wind turbine, Miao et al. [11] investigated the wake characteristics of double wind turbines, Liu et al. [12] established a fully coupled model for simulating floating offshore wind turbine. Nevertheless, this research usually takes consuming time to run on a computer, because of the refinement mesh in the wake region and motion of the platform. The last one is Vortex lattice method, which can achieve the velocity distribution behind the rotor and calculate faster than CFD method. In this method, the vortex core model, using the empirical equations, is considered to avoid the unrealistic result and the dissipation of the wake structure. Compared with BEM theory, it pays more attention to the wake region rather than performance on the blade (both of them are considered in the CFD method). However, it may gain inaccurate results in far wake area and is seldom applied [13].

The structural dynamic characteristics are studied mainly by three methods, modal approach, Finite Element Method (FEM), and multi-body dynamics method. In the modal approach, the description of deformation can be made as a linear superimposition by some physical realistic modes. This method can reduce the number of DOF (Degree Of Freedom) and compute very fast [14]. FEM divides the structure into finite elements, in which a shape function is used to approximate the deformation [15]. However, its number of DOF is much bigger than that of former method and it will cost more computational time. In the multi-body dynamic method, every rigid part is connected by springs and hinges. This method is cheaper than FEM, but more expensive than the modal approach [14].

The aeroelastic performances of wind turbine contain the aerodynamic properties and structural responses. Multi combination methods of aerodynamic performances and structural dynamic characteristics method can solve this complex problem and have their own advantages and disadvantages. Fatigue, Aerodynamics, Structures, and Turbulence (FAST) is developed by the National Renewable Energy Laboratory (NREL), which combined the BEM and modal approach to simulate the aeroelastic performances. In a recent version, the multi-body dynamics method is also added into FAST. Ferede et al. [16] gave a framework for aeroelastic wind turbine blade analysis with BEM and multi-body dynamics method. Li et al. [17] coupled CFD and multi-body dynamics method by overset technology predict the aerodynamic performances. Ageze et al. [18] and Heinz et al. [19] used CFD and FEM to solve the fluid–structure interaction (FSI) in a wind turbine simulation.

Since BEM cannot predict wake behavior and CFD is a high computational cost method, the actuator line method (ALM), which combines advantages of BEM and CFD, was introduced by Shen and Sørensen in 2002 [20]. In ALM, the blade is replaced by a series of aerodynamic loading and this loading which acts as the body forces is added into the source item of Navier-Stokes equations. Thus, ALM is very effective and can be widely used in wind farm simulation [21,22]. Moreover, Meng et al. [23] proposed the elastic actuator line (EAL) model firstly, in which ALM and a finite difference structural model are used to analyze the wake-induced fatigue. Ma et al. [13] combined ALM and FEM to discuss the wake characteristics. The aim of the current research is developing a rapid technique to predict the aeroelastic performances. Therefore, a beam solver, which is used in the wind turbine blade design [24], is combined with ALM to accomplish this task.

In this paper, the concept of ALM is reviewed in brief, and the technique about how to add the structural solver method into turbinesFoam in OpenFOAM [25] is given. The NREL 5 MW baseline wind turbine is studied under the uniform inlet wind speed of 8 m/s and 11.4 m/s. The aerodynamic properties, including power and thrust, and the tip displacement are calculated and compared with related research and NREL's reference data, which can validate the new elastic actuator line model (EALM). Then, the difference of the aerodynamic performances with and without aeroelastic is discussed. Moreover, this variety is also studied in the wake of upstream wind turbine. It is to be observed that the foundation of the wind turbine is fixed in this research, and the floating one combined with structural dynamic characteristics of wind turbine blade is too complex and requires more validation which needs further study.

2. Theoretical Model

2.1. Actuator Line Model

In the actuator line model, the blades of wind turbine are replaced by lines with the aerodynamic force distributed on it, as shown in Figure 1. The lines are discretized into actuator sections, in which there are different airfoils, chords and structure twists. The loading in each section, which is usually named as the aerodynamic force, can be calculated by Equation (1).

$$f = (L, D) = \frac{1}{2}\rho U^2 c(C_l \vec{e_L} + C_d \vec{e_D}) \tag{1}$$

where f is the aerodynamic force; L and D are lift and drag, respectively; ρ is the air density, and $\rho = 1.225$ kg/m^3 in this paper; U is the relative velocity of the blade element in actuator sections; c is the chord length; C_l and C_d are the coefficients of lift and drag, which can be looked up from airfoil data tables derived from physical experiment; $\vec{e_L}$ and $\vec{e_D}$ represent the unit direction vectors of L and D.

Besides, the aerodynamic loads are concentrated on the aerodynamic center of the blade element, which are located in 0.25 chord length. To describe the influence of wind turbine blade on the flow field, the aerodynamic loads should be dispersed on the grid point. There are many distributional ways, and three-dimensional Gaussian distribution η_ε is used in this paper, whose function is shown in Equation (2). This equation is a smooth function and can avoid numerical singularity.

$$\eta_\varepsilon(d) = \frac{1}{\varepsilon^3 \pi^{3/2}} \exp[-(\frac{d}{\varepsilon})^2] \tag{2}$$

where ε is a constant to adjust the strength of the distribution formula, and it is two times of local grid scale in this study according to Shives [26]; and d is the distance between the force point and the grid point. Therefore, the forces f_ε on the nearby mesh can be calculated by Equation (3) and added into the momentum equation to solve the flow field as the body forces.

$$f_\varepsilon = f \times \eta_\varepsilon(d) \tag{3}$$

$$\frac{\partial V}{\partial t} + V \cdot \nabla V = -\frac{1}{\rho}\nabla p + \upsilon \nabla^2 V + f_\varepsilon \tag{4}$$

where V and p are the wind speed and pressure in the flow; υ is the kinematic viscosity, and it is set as 1.5×10^{-5} m^2/s ; f_ε is the source item calculated through Equation (3).

Moreover, the Prandtl–Glauert tip loss model is used to the consideration that the velocity is zero at the tip. The function of tip loss model is shown as Equation (5).

$$F_{tip} = \frac{2}{\pi}\arccos[\exp(-\frac{B(R-r)}{2r\sin\phi})] \tag{5}$$

where F_{tip} is the aerodynamic corrected force at the tip; B is the blades number; R is the radius of the blades; r is the distance between the root of blades and the action point of aerodynamic force; and ϕ is the angle between U and the rotor plane [27]. Tower and hub effect are also considered in this research by similar technology, and their airfoils are cylinder.

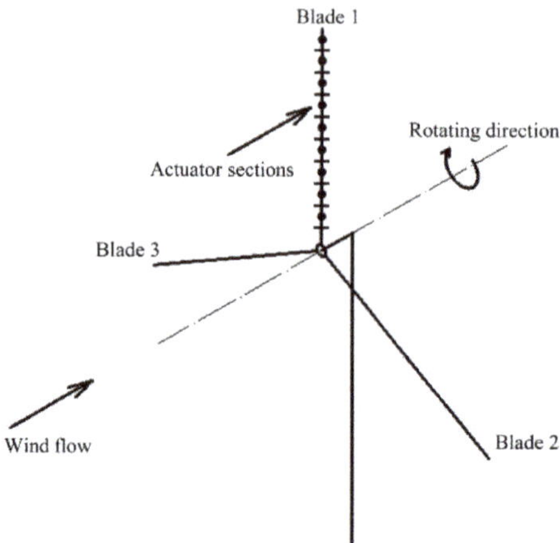

Figure 1. Blades represented by actuator line and discretized into actuator sections.

2.2. Rotating Beam Solver

The blades of wind turbine are considered as rotating variable cross-section projecting beams in this study. In this section, an approach of solving these beams in actual design is introduced [24]. Although this method is faulty and not suitable for structural solution, it still could catch some phenomena in vibration.

Figure 2 shows the diagram of the blade deflection, in which the blades have been substituted by actuator lines. In the picture, direction 0 is out of the rotating plane and called flapwise. Similarly, direction 1 is in the rotating plane and called edgewise. In this research, the subscript 0 means the component of the physical quantity in 0 direction and the subscript 1 represents component of the corresponding parameter in 1 direction. The diagrammatic sketch of the forces on the blade, which is considered as a cantilever beam, is given in Figure 3. According to the Newton's second law of a microelement, the sheer stresses T_0, T_1 and the bending moments M_0, M_1 of each blade element can be calculated from Equation (6) to Equation (9).

$$\frac{dT_0}{dx} = -p_0(x) - m(x)g\sin\theta + m(x)\ddot{y}_0(x) \tag{6}$$

$$\frac{dT_1}{dx} = -p_1(x) - m(x)g\cos\theta + m(x)\ddot{y}_1(x) \tag{7}$$

$$\frac{dM_1}{dx} = T_0 + N_0 \tag{8}$$

$$\frac{dM_0}{dx} = -T_1 + N_1 \tag{9}$$

where $p(x)$ is the aerodynamic force distribution; $m(x)$ is the blade mass distribution; g is the gravitational acceleration, and it is set as 9.8 m/s^2 in this paper; θ is the shaft tilt angle, and it is 5 degree here; $\ddot{y}$ is the displacement y to the second derivative of time, and it can be calculated by prediction and correction of modified Euler method, given from Equation (10) to Equation (13); N_0, N_1 are the component of centrifugal force in 0 direction and 1 direction.

$$\widetilde{y}_{n+1} = 2y_n - y_{n-1} + \ddot{y}_n dtdt \tag{10}$$

$$\widetilde{\ddot{y}}_{n+1} = h\left(\widetilde{y}_{n+1}, t_{n+1}\right) \tag{11}$$

$$y_{n+1} = 2y_n - y_{n-1} + 0.5(\ddot{y}_n + \widetilde{\ddot{y}}_{n+1})dtdt \tag{12}$$

$$\ddot{y}_{n+1} = h(y_{n+1}, t_{n+1}) \tag{13}$$

where the subscript $n + 1$ represents at time t_{n+1}; the subscript n represents at time t_n; $\widetilde{y}$ is predicted value; function h can be calculated from transposition of structure equations, as shown in Equation (15).

$$M(x)\ddot{y} + K(x)y = F(x, t) \tag{14}$$

$$\ddot{y} = \frac{F(x, t)}{M(x)} - \frac{K(x)y}{M(x)} = g(y, t) \tag{15}$$

where Equation (14) is the structure equations without damping; $M(x)$ is mass of blade; $K(x)$ is the stiffness of blade; $F(x)$ is the external loading.

According to Figure 4, the bending moments M_0 and M_1 can be transformed into principal axes direction by Equation (16) and Equation (17).

$$M_{11} = M_1 \cos\beta - M_0 \sin\beta \tag{16}$$

$$M_{22} = M_1 \sin\beta - M_0 \cos\beta \tag{17}$$

where M_{11} and M_{22} are the bending moments on the first principal axis and the second principal axis; β is the structure twist angel.

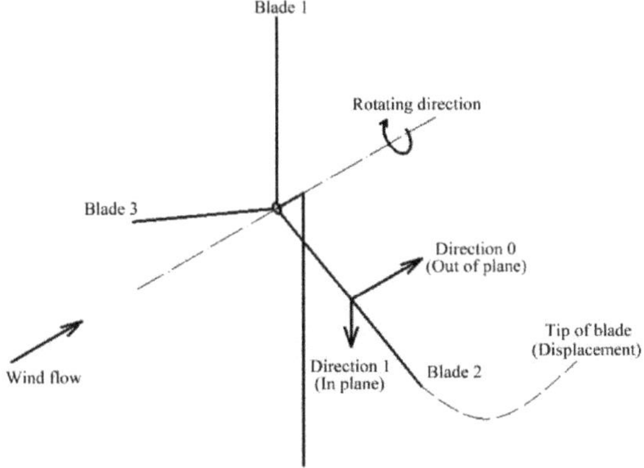

Figure 2. Coordinate system of the blade deflection.

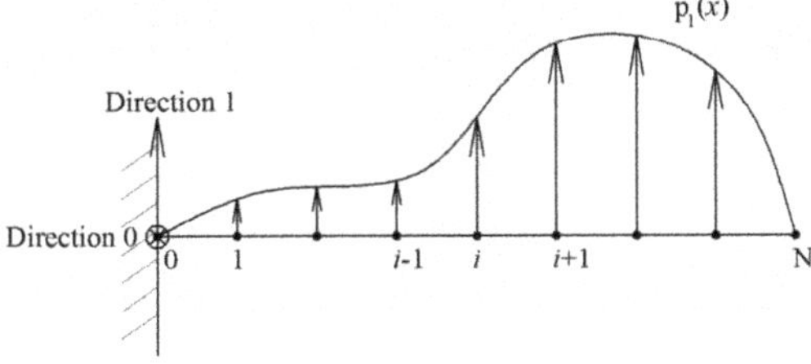

Figure 3. The diagrammatic sketch of the forces on the cantilever beam.

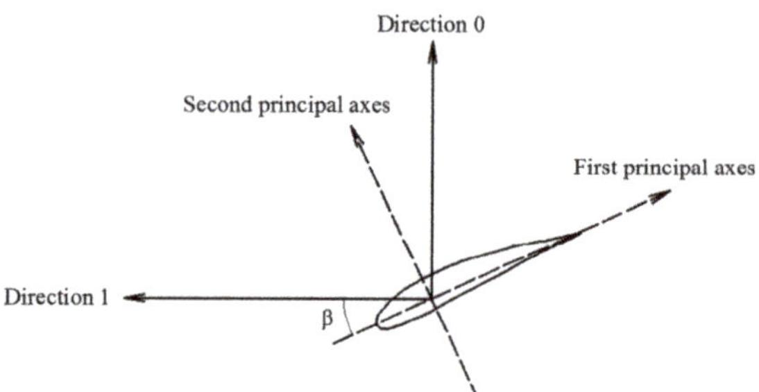

Figure 4. The blade element in the local coordinate system.

Based on the beam theory, the curvature of principal axes κ_{11} and κ_{22} can be obtained from Equation (18) and Equation (19).

$$\kappa_{11} = \frac{M_{11}}{EI_1} \tag{18}$$

$$\kappa_{22} = \frac{M_{22}}{EI_2} \tag{19}$$

where EI is the stiffness of the blade element.

Equation (18) and Equation (19) are converted back to direction 0 and direction 1 through the formulas as following:

$$\kappa_0 = -\kappa_{11} \sin\beta + \kappa_{22} \cos\beta \tag{20}$$

$$\kappa_1 = \kappa_{11} \cos\beta + \kappa_{22} \sin\beta \tag{21}$$

Therefore, the angular deformation θ and deflection y can be calculated by:

$$\frac{d\theta_1}{dx} = \kappa_1 \tag{22}$$

$$\frac{d\theta_0}{dx} = \kappa_0 \tag{23}$$

$$\frac{dy_0}{dx} = -\theta_1 \tag{24}$$

$$\frac{dy_1}{dx} = \theta_0 \tag{25}$$

The root of the blade is clamped, so the boundary condition at root is:

$$\theta_0 = 0, \theta_1 = 0, y_0 = 0, y_1 = 0 \tag{26}$$

The tip is free end, so the boundary condition at tip is:

$$T_0 = 0, T_1 = 0, M_0 = 0, M_1 = 0 \tag{27}$$

2.3. New Elastic Actuator Line Model

In this section, the ALM is improved into a new elastic actuator line model (EALM) based on turbinesFoam library, which is developed by Bachant et al. [25]. This library uses ALM to simulate wind and marine hydrokinetic turbines in OpenFOAM. Its interpolation, Gaussian projection, and vector rotation functions are all adapted from NREL's Simulator for Off/Onshore Wind Farm Applications (SOWFA).

In EALM, the body forces are calculated by traditional ALM and the blade deflection is computed by rotating beam solver, which is defined in Section 2.2. The computation process of EALM is given in

Figure 5, in which the difference between EALM and ALM is marked by dashed rectangle. The part of structure solver is added into the actuatorLineSource class, and the actuator point is changed in the actuatorLineElement class. From Figure 5, it can be found that this combination of ALM and structure model is one-way coupling. Compared with the research by Meng et al. [23], the part of aerodynamics solver is similar, but the way of dealing with structural solver is different (see Section 2.2). The advantages of this model are that the EALM allows large time step when the position of actuator line changes every time and computes long terms of wind turbine working. This technology will be further improved and used in the simulation of the floating offshore wind turbine in the sea, which needs more simulation time to keep the floating foundation stability under waves.

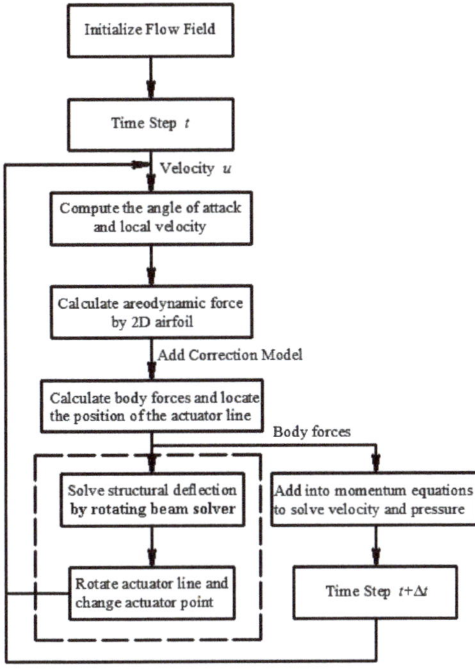

Figure 5. The computational flow chart of new elastic actuator line model.

2.4. Computational Wind Turbine Model

In this study, the aerodynamic performances, which are power and thrust, and the structural responses, which are represented mainly by the blade tip displacement, will be compared with different research using varieties methods to validate EALM. According to the theory above, all these three physical quantities are obtained from the aerodynamic forces along the blade. Different cases use varieties method to achieve these forces. Only if the blade properties, including the aerodynamic and structural properties, and the computational condition such as wind speed are all the same, these three quantities could make equivalent comparisons. Thus, the computational condition will be given next. The model used in this research is NREL 5 MW wind turbine, and its gross properties are listed in Table 1. The details of the blade structural and aerodynamic properties can be referenced in Jonkman's technical report [8]. To be exact, the shaft tilt angle is considered in the structural part, which will make the external force of blade increased because of the gravity component.

Table 1. Properties of NREL 5 MW baseline wind turbine (NREL: National Renewable Energy Laboratory).

Properties	Content
Rotor orientation	Upwind
Rotor configuration	Three blades
Rotor diameter	126 m
Hub diameter	3 m
Hub height	90 m
Shaft tilt angle	5°
Rotor mass	110,000 kg
Rated power	5 MW
Rated wind speed	11.4 m/s
Rated rotor speed	12.1 rpm

The simulation physical domains used in this paper are the same as Yu et al. [28]. Their three-dimensional sketches are shown in Figures 6 and 7. In these pictures, A is the outer mesh region and B is the refined mesh region, where the grid is refined by 4 levels. The mesh refined ratio is 0.5 between each level. The mesh independence test has been completed in that research and 1.5 m is chosen as the minimum size of grid. Besides, it is confirmed that the wind turbine gets a better working status at the wind speed of 8 m/s and its corresponding tip speed ratio is 7.55. As a consequence of this model, the cases studied in present research are mainly in two situations, 8 m/s and 11.4 m/s. To analyze the aeroelastic performance in the wake, double NREL 5 MW wind turbines set in a line are studied, and its simulation domain is shown in Figure 7. The main contents of this research are focused on the downstream wind turbine.

Compared to the work of Yu et al. [28], the parameter settings are all the same expect that the LES (Large Eddy Simulation) model is used instead of RANS (Reynolds Average Navier-Stokes). That is because the LES model will get more accurate results about the vorticities than RANS and the influence of the wake flows to elastic blade is studied in this research. In addition, the wind turbine blade deflection solver is added in ALM, named EALM.

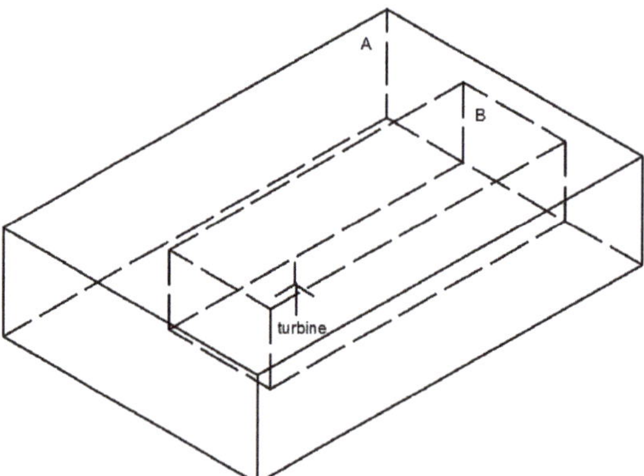

Figure 6. The simulation domain of single wind turbine.

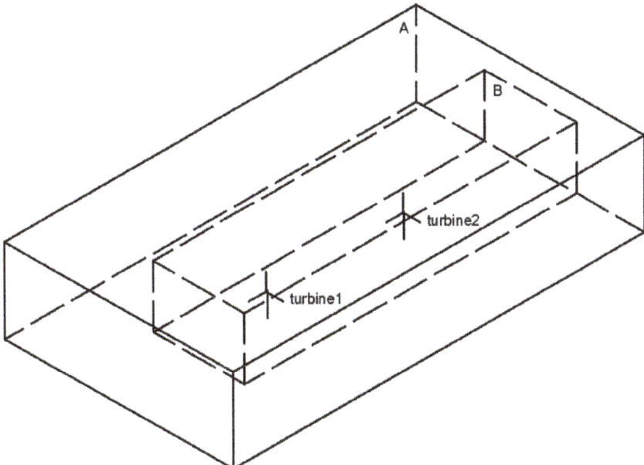

Figure 7. The simulation domain of double wind turbines.

3. Verification and Analysis

In this section, the mesh independence and uncertainty analysis of actuator point number are given first. Then, the aerodynamic performance and the structural responses are compared to validate EALM. The aerodynamic performance contains the power and the thrust. The structural responses are represented mainly by the blade tip displacement.

3.1. Mesh Independence and Uncertainty Analysis

Four levels of grids are tested to prove the mesh independence under the rated wind speed, and the non-dimension coefficients of power and thrust are compared in Table 2. They are defined as power coefficient C_p and thrust coefficient C_t as below:

$$C_p = \frac{P}{0.5\rho V^3 \pi R^2} \tag{28}$$

$$C_t = \frac{T}{0.5\rho V^3 \pi R^2} \tag{29}$$

where P is the mechanical power; ρ is the air density; V is the wind speed in the flow; T is the thrust on the blades. It shows that the results drop slow when the grid levels up especially from level 3 to level 4. Therefore, the level 3 mesh is used in the following research.

Table 2. Mesh independence test of power and thrust coefficient.

Grid Level	Number of Cell	Size of Wake Region Cell (m)	C_p	C_t
1	0.38 M	2.5	0.4678	0.8665
2	0.75 M	2.0	0.4642	0.8599
3	2.90 M	1.5	0.4634	0.8584
4	5.96 M	1.0	0.4627	0.8572

According to Shives' research [26], the number of actuator point has the rule that the maximum of the distance between the adjacent point should not more than the size of blade region. Different actuator point numbers are also tested in Table 3 and Figure 8. Besides, the mesh size and computational condition are kept as constant during this test. To improve the accuracy the uncertainty coefficient U_ξ is calculated as follow steps.

(1) Calculate the difference of aerodynamic coefficients between neighbor level, which are defined as $\varepsilon_{\zeta 21}$ and $\varepsilon_{\zeta 32}$;

(2) The convergence ratio R_ξ can be computed by Equation (30).

$$R_\xi = \frac{\varepsilon_{\zeta 32}}{\varepsilon_{\zeta 21}} \tag{30}$$

(3) The power coefficient convergence ratio is 0.125 and the thrust coefficient convergence ratio is 0.167. They are both located in the interval which greater than 0 and less than 1. So, the Richardson extrapolation method is used to get the order of accuracy p_ξ and the estimated value of error $\delta_{\mathrm{Re}\xi 1}$ in Equation (31) and Equation (32).

$$p_\xi = \frac{\ln(\varepsilon_{21}/\varepsilon_{32})}{\ln(r_\xi)} \tag{31}$$

$$\delta^*{}_{\mathrm{Re}\xi 1} = \frac{\varepsilon_{\xi 32}}{r_\xi^{p_\xi} - 1} \tag{32}$$

where r_ξ is refinement ratio of actuator point number.

(4) The uncertainty coefficient U_ξ can be compute by Equation (33).

$$U_\xi = \left(\left| C_\xi \right| + \left| 1 - C_\xi \right| \right) \left| \delta^*{}_{\mathrm{Re}\xi 1} \right| \tag{33}$$

where C_ξ is correction factor and defined by Equation (34).

$$C_\xi = \frac{r_\xi^{p_\xi} - 1}{r_\xi^2 - 1} \tag{34}$$

Following the calculation steps above, the uncertainty coefficient of C_p and C_t can be achieved. They are 0.129% D and 0.126% D, where D is the reference data from the NREL technical report [8]. Because both of them are less than 1% D, that proves the results are credible. Through the verification above, the level 2 actuator point number will be adopted, and it will obtain reliable results.

Table 3. Relationship between actuator point number and aerodynamic coefficients.

Level	Actuator Point Number	C_p	C_t
1	54	0.4650	0.8608
2	72	0.4634	0.8584
3	90	0.4632	0.8580

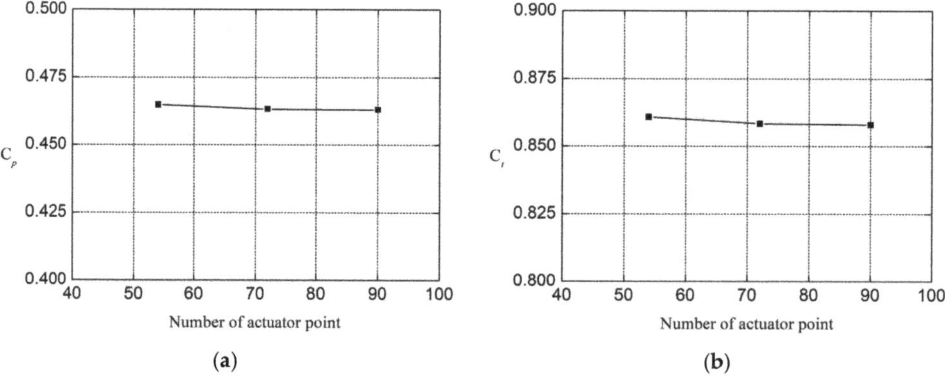

Figure 8. Relationship between actuator point number and aerodynamic coefficient, (a) power coefficient, and (b) thrust coefficient.

In addition, one of this method's advantages is that in the part of the aerodynamic solver it will cost less computational resources than CFD. The BEM theory is not considered here because the wake flow cannot be achieved in this way. The comparison of the computational cost between ALM and CFD method are shown in Table 4. It can be concluded that ALM uses less grids and computes faster than the CFD method under similar accuracy.

Table 4. Comparison of the computational cost between actuator line method (ALM) and computational fluid dynamics (CFD) method.

Method	Software	Number of Cell	Simulated Time	Power
ALM	OpenFOAM	1.49 M	9.2 h	5.036 MW
CFD	Star CCM+	5.01 M	82.0 h	5.050 MW

ALM: Actuator Line Method; CFD: Computational Fluid Dynamics; OpenFOAM: Open Source Field Operation and Manipulation, a free, open source CFD software package released by the OpenFOAM Foundation, which was incorporated as a company limited by guarantee in England and Wales; Star CCM+: Siemens Digital Industries Software, 5800 Granite Parkway, Suite 600, Plano, TX, USA.

3.2. Comparison of the Power and the Thrust

In this part, the results of aerodynamic performance are compared with 8 cases, in which 7 cases are the previous research and 1 case is the present one. Case 1 is the simulation of EALM and it is the present study. Case 2 is from the official technical report even though some data are given by NREL's FAST code [8]. In this report, the blade mode is derived by the mode's program and then passes a best-fit polynomial to get the equivalent polynomial representations of the mode shapes needed by FAST. Case 3 is simulated by HAWC2 (Horizontal Axis Wind turbine simulation Code 2nd generation) [29], which is an in-house nonlinear aeroelastic model developed by Technical University of Denmark. BEM is used as aerodynamic model and the multi-body dynamics method (MBD) is its structural model, in which each body is a linear Timoshenko beam element. The data of Case 4 are calculated by Li et al. [17] using CFD and MBD. This approach uses a dynamic overset CFD code for aerodynamics and MBD code for motion responses. The coupled way is done by exchanging the information between the fluid and structure solver in explicit form. The results of Case 5 are given by Jeong et al. [30] with BEM. In their research, the FSI (Fluid–Structure interaction) applies a strong coupling method which is using a first order implicit-explicit coupling scheme. Case 6 is set up in the research of Ponta et al. [31] by BEM and dynamic rotor deformation model, where the effects of rotor deformation are incorporates in the computation of aerodynamic loads. Yu et al. [32,33] combined CFD and CSD (Computational Structural Dynamics) in Case 7. The coupling methodology in this research is made by adopting the delta-airload loose-coupling technology. The last Case 8 is the results of the actuator line finite-element beam method (ALFBM) by Ma et al. [13]. The process of this research is three parts: the CFD solver given by OpenFOAM, the aerodynamic solver calculated by ALM, and the structural solver by finite-element beam method. The coupled steps are reading the velocity in the flow field from $t-\Delta t$ and adding it to source term, which is computed by aerodynamic solver and structural solver. To clearly compare the varieties, the detailed information about the solved method on the aerodynamic performance and the structural responses in different cases are given in Table 5. In this research, BEM theory cannot describe the detailed flow field, and the CFD method may cost too much time to calculate. Therefore, ALM is chosen to resolve aerodynamic performance. Compared to ALFBM, this research concentrates on the variation of the aeroelastic characteristics in wake flow caused by upstream wind turbine.

Table 5. Detailed information about the solved method on the aerodynamic performance and the structural responses in different cases.

Case Number	Aerodynamic Method	Elastic Dynamics Method	Studying Contents
1	ALM	Rotating beam solver	Variation in wake flow
2	BEM	Modal approach	Blade response and aerodynamics
3	BEM	MBD	Blade response and aerodynamics
4	CFD	MBD	Influence of wind turbulence
5	BEM	Geometric nonlinearity beam model	Optimal yaw and pitch angle
6	BEM	Dynamic rotor deformation model	Rotor structure deformation
7	CFD	FEM-based CSD beam solver	Yaw and wind shear
8	ALM	Finite-element beam method	Wake behavior

ALM: Actuator Line Method; BEM: Blade-Element Momentum; MBD: Multi-Body Dynamics; CFD: Computational Fluid Dynamics; CSD: Computational Structural Dynamics; FEM: Finite Element Method.

Table 6 lists the results of 8 cases with velocities of 8 m/s and 11.4 m/s, and their mean output power and thrust differences are described in a histogram to visualize disparity of these cases, as shown in Figures 9 and 10. According to the results, the NREL's reported results are chosen as the reference data. It can be seen that all the results of power are higher than the reference data, except Case 6. All the results of thrust are lower than the FAST result. The big difference of thrust results are found from Case 5 to Case 8. The main reason for this is that the tip loss correction is not considered, and it has a great influence on aerodynamic performance. Case 4 has unsatisfied power result and there is no data in Case 3 under the rated situation. Besides, Case 1 has the most approximate result to the reference data among all studies. Its power difference is less than 50 kW and thrust difference is not more than 50 kN.

Table 6. Comparison of the power and the thrust with different research.

Case Number	8 m/s		11.4 m/s	
	Power (MW)	Thrust (kN)	Power (MW)	Thrust (kN)
1	1.891	432	5.024	772
2	1.856	466	5.000	817
3	1.928	391	No data	No data
4	1.865	389	5.407	759
5	2.242	410	5.249	645
6	1.817	340	5.130	659
7	2.000	362	5.334	671
8	1.924	384	5.350	663

Case 1: Elastic Actuator Line Model (present study); Case 2: NREL's Fatigue, Aerodynamics, Structures, and Turbulence (FAST) code; Case 3: Horizontal Axis Wind turbine simulation Code 2nd generation results; Case 4: research by Li et al. [17]; Case 5: research by Jeong et al. [30]; Case 6: research by Ponta et al. [31]; Case 7: research by Yu et al. [32,33]; Case 8: research by Ma et al. [13].

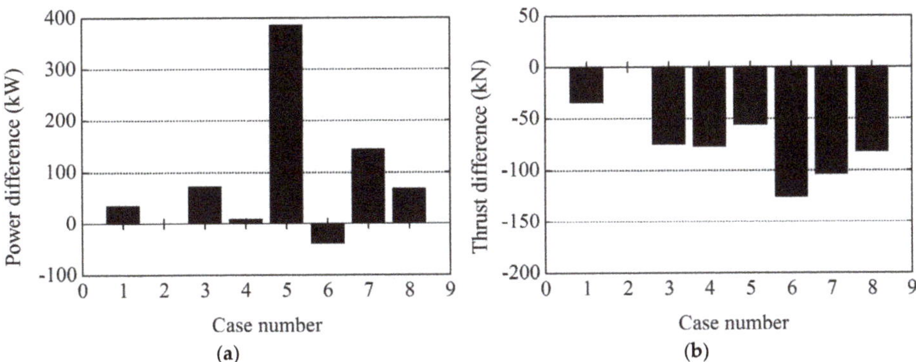

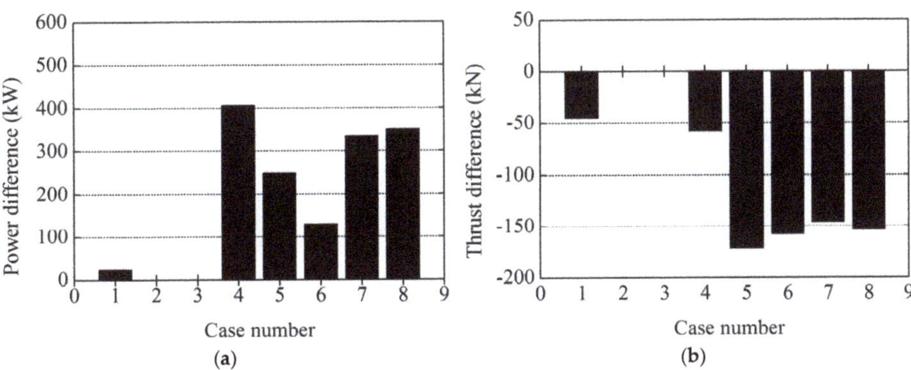

Figure 9. Comparison of different cases results in 8 m/s: (**a**) power; and (**b**) thrust Case 1: Elastic Actuator Line Model (present study); Case 2: NREL's FAST code; Case 3: Horizontal Axis Wind turbine simulation Code 2nd generation results; Case 4: research by Li et al. [17]; Case 5: research by Jeong et al. [30]; Case 6: research by Ponta et al. [31]; Case 7: research by Yu et al. [32,33]; Case 8: research by Ma et al. [13].

Figure 10. Comparison of different cases results in 11.4 m/s: (**a**) power; and (**b**) thrust Case 1: Elastic Actuator Line Model (present study); Case 2: NREL's FAST code; Case 3: Horizontal Axis Wind turbine simulation Code 2nd generation results; Case4: research by Li et al. [17]; Case 5: research by Jeong et al. [30]; Case 6: research by Ponta et al. [31]; Case 7: research by Yu et al. [32,33]; Case 8: research by Ma et al. [13].

To further examine the accuracy of EALM under different wind speed conditions, power and thrust are computed and compared with Case 2, which are shown in Figure 11. Besides, their relative errors are given in Table 7. The errors are smaller when the wind speed is close to the rated speed, which is 11.4 m/s, than those when the wind speed is below 8 m/s. This phenomenon is caused by the coefficients of lift and drag C_l and C_d in Equation (1). These two parameters are referenced from airfoil data tables and these tables are achieved by physical experiment which is given by NREL official report [8]. In that report, the airfoil data is only obtained under nearly rated wind speed. Therefore, the lift and drag of a blade element can get more accurate if the wind speed closed to 11.4 m/s. On the contrary, the result of the force in the blade element will not be satisfactory if the wind speed is far away from 11.4 m/s. In case of application that the airfoil data corresponds to the wind speed, the result will be perfect, which will be improved in future research. From results of Picture 11, the predict results show good agreements with NREL's reference data. In addition, the differences between NREL and EALM are getting small with the inlet wind velocity increasing whether the output power or thrust. Especially in the rated situation, the relative errors of both power and thrust are less than 5%.

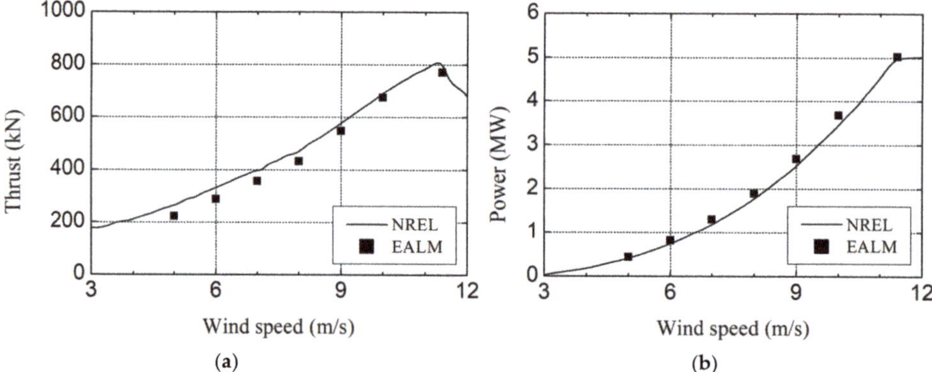

Figure 11. Comparisons of the aerodynamic performance results between elastic actuator line model (EALM) and NREL's FAST code at different wind speeds: (**a**) thrust; and (**b**) power.

Table 7. Comparisons of the power and the thrust relative error between elastic actuator line model (EALM) and NREL's FAST code at different wind speeds.

Wind Speed (m/s)	Power (MW)	Error	Thrust (kN)	Error
5.0	0.4355	2.828%	222.6153	15.965%
6.0	0.8194	7.080%	287.8339	13.253%
7.0	1.2986	6.099%	356.5589	10.233%
8.0	1.8906	3.856%	432.3593	7.543%
9.0	2.6880	3.564%	546.8937	5.374%
10.0	3.6799	3.405%	674.6184	2.392%
11.4	5.0241	0.482%	772.1718	3.948%

According to the validation above, it could be concluded that EALM can predict the power and thrust accurately. This conclusion can be predictable because the aerodynamic calculations of EALM are based on blade element theory. In this theory, the lift and drag of blade element are achieved from two-dimensional airfoil data, which is obtained from physical experiments.

3.3. Comparison of the Tip Displacement

In this section, the structural responses of blade tip are compared in 5 cases. Case 1 is the present research and obtained by EALM. Case 2 is from the official technical report and its data are given by NREL's FAST code [8]. Case 3 is the results of CFD and multi-body dynamics method by Li et al. [17]. Case 4 is set up by Yu et al. [32,33], in which CFD and CSD (Computational Structural Dynamics) are combined to calculate the coupled problem. The last Case 5 is using the actuator line finite-element beam method (ALFBM) by Ma et al. [13]. Detailed information in different cases can refer to Table 5.

Similarly, Table 8 lists the tip displacement results of 5 cases with 8 m/s and 11.4 m/s, and their differences are provided in the form of histograms, which are shown in Figures 12 and 13. Additionally, the NREL's reported results are chosen as the reference data. In these pictures, 0 direction means out of the rotor plane and 1 direction is in the rotor plane. The results of Case 3 and Case 5 are larger than those in Case 2, while the results of Case 4 are smaller. At the same time, the tip displacement of Case 1 is the nearest with the reference data in 0 direction and the farthest away from that reference data in 1 direction among 5 cases. It indicates that the tip displacement calculated by the beam solver, used in actual design, is similar to the modal approach, used to achieve structural dynamic characteristics, in 0 direction. Besides, the means used in this research could only predict approximate results in 1 direction. Even so, this beam solver can be still utilized to provide structural responses because of little deflection in edgewise.

Table 8. Comparison of tip deflection with different research.

Case Number	8 m/s		11.4 m/s	
	Out of Plane Tip Displacement (m)	In Plane Tip Displacement (m)	Out of Plane Tip Displacement (m)	In Plane Tip Displacement (m)
1	3.159	−0.418	5.560	−0.762
2	3.220	−0.350	5.550	−0.592
3	3.592	−0.345	6.379	−0.579
4	2.958	No data	4.851	−0.624
5	3.675	−0.298	6.212	−0.584

Case 1: Elastic Actuator Line Model (present study); Case 2: NREL's FAST code; Case 3: research by Li et al. [17]; Case 4: research by Yu et al. [32,33]; Case 5: research by Ma et al. [13].

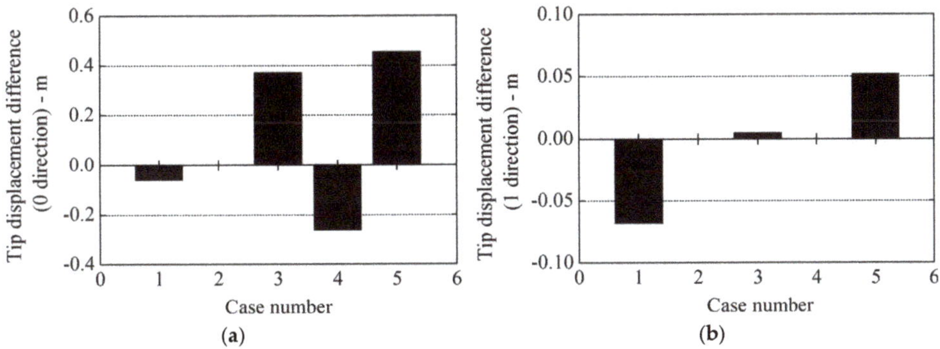

Figure 12. Comparison of different cases tip displacement results in 8 m/s: (**a**) 0 direction; and (**b**) 1 direction Case 1: Elastic Actuator Line Model (present study); Case 2: NREL's FAST code; Case 3: research by Li et al. [17]; Case 4: research by Yu et al. [32,33]; Case 5: research by Ma et al. [13].

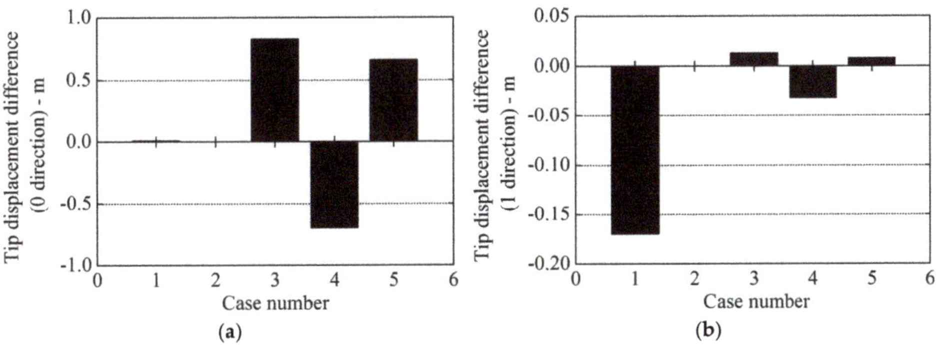

Figure 13. Comparison of different tip displacement results in 11.4 m/s: (**a**) 0 direction; and (**b**) 1 direction Case 1: Elastic Actuator Line Model (present study); Case 2: NREL's FAST code; Case 3: research by Li et al. [17]; Case 4: research by Yu et al. [32,33]; Case 5: research by Ma et al. [13].

The tip displacement and its relative error in varied wind speeds are compared with Case 2 in Figure 14 and Table 9. The predict tendency of the tip displacement in 0 direction is almost the same as the reference data from Figure 13a, and its relative error is less than 5%. That indicates once again that the result of beam solver used in this study is approximate to that of the modal approach in 0 direction. Although the predicted value is not satisfied in high wind velocity in 1 direction, the relative differences are still not more than 15% except in the velocity of 10 m/s.

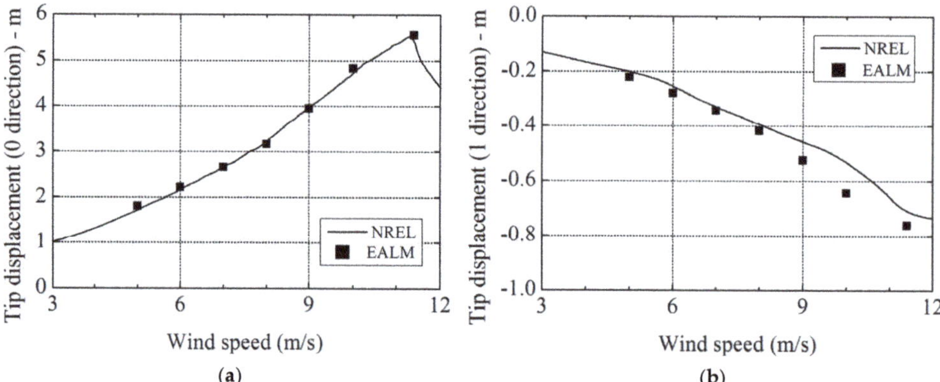

Figure 14. Comparison of the tip displacement results between elastic actuator line model (EALM) and NREL's FAST code at different wind speeds: (**a**) in 0 direction; and (**b**) in 1 direction.

Table 9. Comparison of the tip displacement relative errors between elastic actuator line model (EALM) and NREL's FAST code at different wind speeds.

Wind Speed (m/s)	Out of Plane Tip Displacement (m)	Error	In Plane Tip Displacement (m)	Error
5.0	1.8092	5.340%	−0.2215	−10.536%
6.0	2.2194	2.598%	−0.2793	−8.828%
7.0	2.6592	0.345%	−0.3437	−3.692%
8.0	3.1594	2.090%	−0.4177	−4.895%
9.0	3.9490	0.596%	−0.5247	−14.135%
10.0	4.8308	2.069%	−0.6441	−20.776%
11.4	5.5591	2.545%	−0.7616	−3.225%

To represent the exactitude of blade tip deformation in different positions when considering the rotating motion, the azimuthal variations of tip displacement are compared with FAST code in Figures 15 and 16. In addition, the Pearson simplified correlation coefficient r is introduced to describe the fit degree of two curves. It is a measure of the linear correlation between two variables in statistics. It has a value from −1 to +1, where 1 is total positive linear correlation, 0 is no linear correlation, and −1 is total negative linear correlation. Moreover, the larger the absolute value of the correlation coefficient is, the stronger proximity of two variables becomes. In this study, the coefficient r is rearranged in the formulas like Equation (35).

$$r = \frac{n \sum_{i=1}^{n} x_i y_i - \sum_{i=1}^{n} x_i \sum_{i=1}^{n} y_i}{\sqrt{n \sum_{i=1}^{n} x_i^2 - \left(\sum_{i=1}^{n} x_i \right)^2} \sqrt{n \sum_{i=1}^{n} y_i^2 - \left(\sum_{i=1}^{n} y_i \right)^2}} \tag{35}$$

where n is sample size; x_i and y_i are the individual sample points indexed with i, and they represent the data from FAST and EALM in this research.

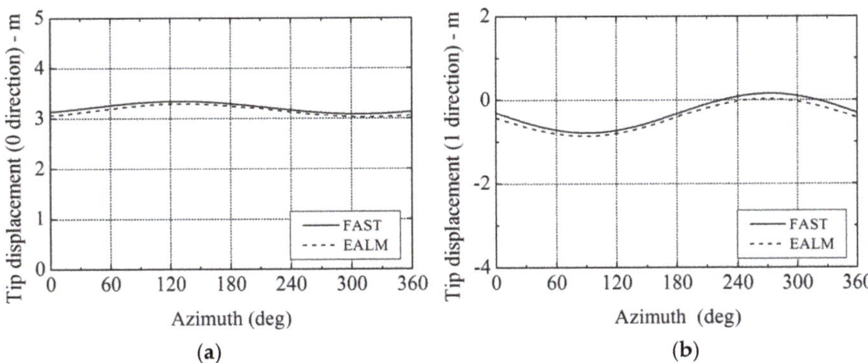

Figure 15. Comparison of azimuthal variations of blade tip deformations between elastic actuator line model (EALM) and NREL's FAST code at 8 m/s: (**a**) in 0 direction; and (**b**) in 1 direction.

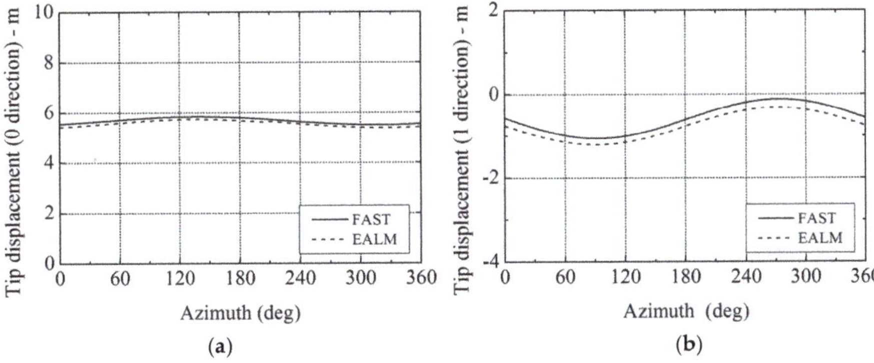

Figure 16. Comparison of azimuthal variations of blade tip deformations between elastic actuator line model (EALM) and NREL's FAST code at 11.4 m/s: (**a**) in 0 direction; and (**b**) in 1 direction.

From the Figure 15, the results agree well with each other. Moreover, it can be seen that all the Pearson correlation coefficients are not less than 99% from Table 10, which means the curves of two methods have strong relativity. Consequently, it could be concluded that EALM can calculate reliable structural responses.

Table 10. Comparison of blade tip displacement Pearson simplified correlation coefficient between elastic actuator line model (EALM) and NREL's FAST code.

Wind Speed (m/s)	Direction	R
8.0	0	99.18%
	1	99.91%
11.4	0	99.40%
	1	99.97%

4. Aeroelastic Performance Analysis

4.1. Influence on Aerodynamic Performance

This section will discuss the influence of elastic blade on the aerodynamic performance. Figures 17 and 18 depict the power and thrust azimuthal history with traditional actuator line model (ALM) and elastic actuator line model (EALM) at 8 m/s and 11.4 m/s. The ALM results are achieved by hiding the part of rotating beam solver in EALM, and the parameters setting are all the same. The curves fluctuate every 60 degrees, which is caused by the tower shadow effect. When the structural deformation is considered, the mean power reduces about 11.38 kW in 8 m/s and 11.43 kW in 11.4 m/s. Besides, the averaged thrust decreases approximately 1.08 kN in 8 m/s and 0.35 kN in 11.4 m/s. It indicates that the influence of elasticity on mean power and thrust is small in the rated condition. As the blade passes through tower, the reduction of output power turns into 15.33 kW in 8 m/s and 31.45 kW in 11.4 m/s. The decrement of thrust changes into 1.32 kN in 8 m/s and 0.96 kN in 11.4 m/s. The distance between tower and blade gets closer due to the blade deformation. Hence, the tower shadow effect becomes more serious and the decrement of power and thrust is larger. Because the blade displacement in high wind speed is large, the tower effect is the most serious in the rated wind speed and its reduction is more than that in 8 m/s.

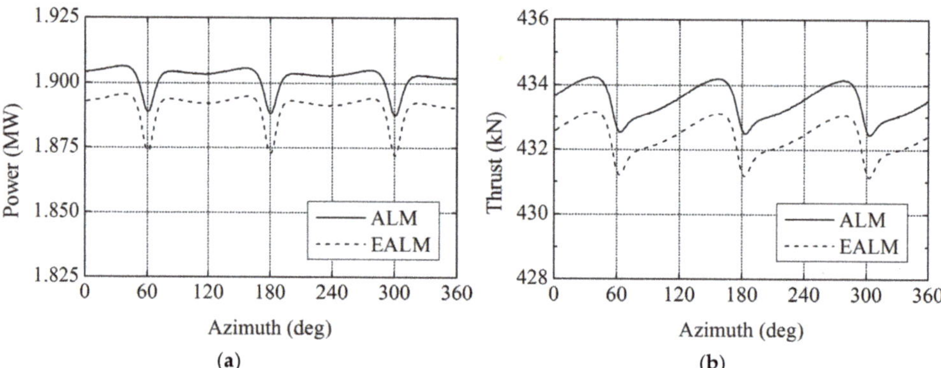

Figure 17. Comparisons of the aerodynamic performance results between traditional actuator line model (ALM) and elastic actuator line model (EALM) at 8 m/s: (**a**) power; and (**b**) thrust.

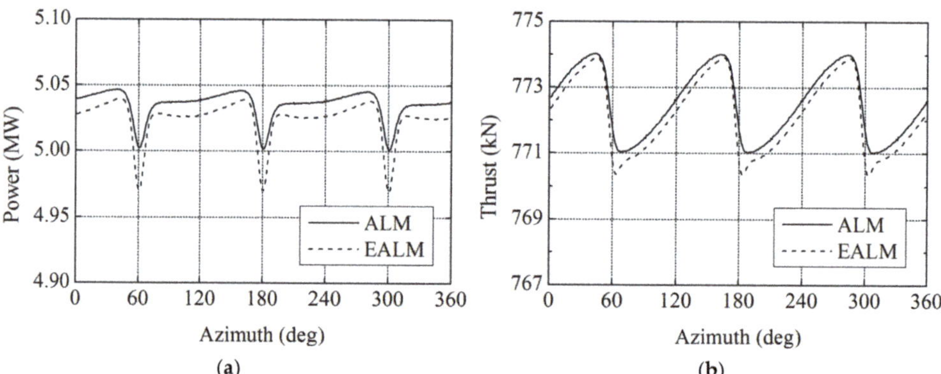

Figure 18. Comparisons of the aerodynamic performance results between traditional actuator line model (ALM) and elastic actuator line model (EALM) at 11.4 m/s: (**a**) power; and (**b**) thrust.

The stabilization of the normal and tangential force is present by standard deviation, as shown in Figure 19. The standard deviation s is defined as follows:

$$s = \sqrt{s^2} = \sqrt{\dfrac{\sum\limits_{i=1}^{n}(x_i - \bar{x})^2}{n-1}} \tag{36}$$

where n is sample size; x_i is the individual sample points indexed with I; and $\bar{x}$ is expectation, which is the mean value of sample. From the figure, the standard deviation of the loads on blade increases a little when the elasticity of the blades is considered.

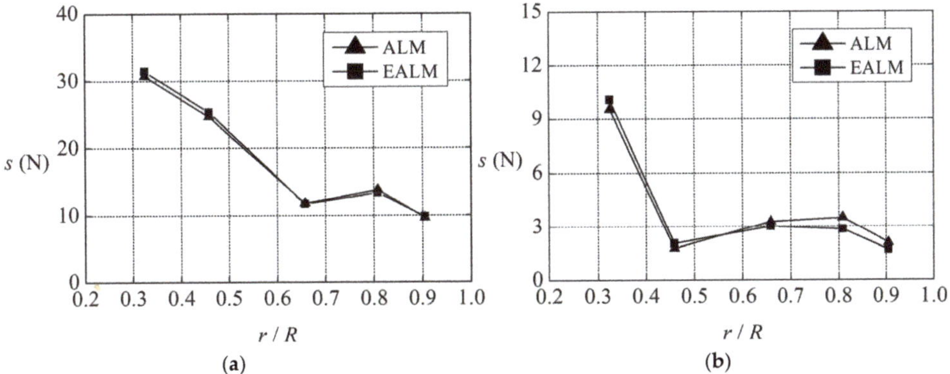

Figure 19. Stabilization of forces on blade: (**a**) normal force; and (**b**) tangential force.

Figure 20 shows the wake structure comparison with EALM and ALM at 8 m/s. The wake vorticities are described by Q criterion, in which Q is calculated by the second invariant of the velocity gradient tensor and are stained by velocity field. The tip and hub vorticities are clear to be seen behind the rotor. The velocity in the internal surface of vorticities is smaller than the outside one. That is because the wind turbine extract momentum from the incoming airflow passing through it, and the wind energy is turned into mechanical energy, which causes the wind velocity after the rotor decreased.

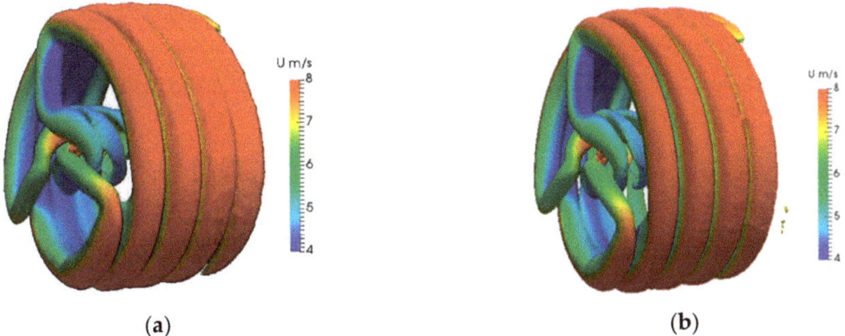

Figure 20. The wake structures in terms of Q field at 8 m/s: (**a**) by EALM; and (**b**) by ALM.

The blade deflection is abbreviated and described in Figure 21. The deformation is too small to be discovered in 1 direction from left side view. Besides, it focuses on blade tip in 0 direction from front and vertical side view.

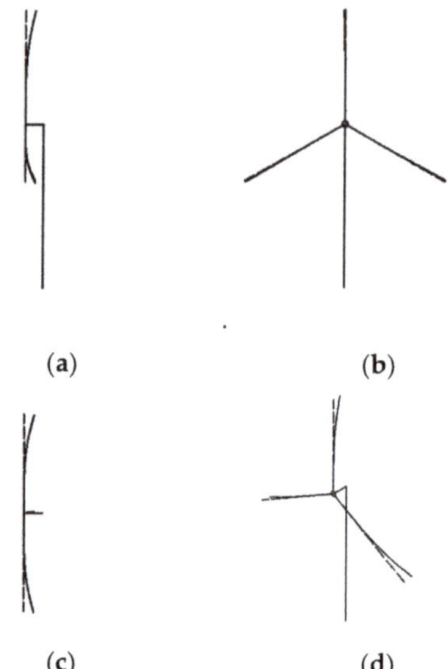

(a)   (b)

(c)   (d)

Figure 21. The abbreviated sketches of blade deflection: (**a**) front side view; (**b**) left side view; (**c**) vertical side view; and (**d**) southwest isometric side view.

4.2. Wake Flows Analysis

To analyze the influence of elastic blade in the wake flows, an array of two NREL 5 MW wind turbines is studied (see Figure 7) with EALM. The results are extracted after the upstream wind turbine rotating 40 revolutions, and the corresponding simulation time is around 260 s. According to research [28], the rotor speed of downstream wind turbine is set as 8.11 rpm to maximize its output power.

The comparison of mean output power among EALM, ALM, and Jha et al. [34] is presented in Table 11, in which the power generated by downstream wind turbine reduces a lot and only about 40% of upstream wind turbine because of the wake flow effects. In addition, the elastic blade makes the mean power of downstream wind turbine decrease about 14.6 kW, which is bigger than 11.38 kW of upstream wind turbine. Furthermore, the power and thrust azimuthal history of downstream wind turbine are depicted in Figure 22. When the blade passes through tower, the output power reduces 24.83 kW and the thrust decreases 7.01 kN. In contrast with the reduction of 8 m/s in Section 4.1, which are 15.33 kW and 1.32 kN, respectively, the tower shadow effect becomes more serious. This phenomenon is caused by the instability of the flow filed around the blade. The velocity and turbulence intensity in the wake of upstream wind turbine are changed into unstable. In the flow field of the downstream wind turbine blade, the variety of blade deflection with time will further disturb them. That will impact the relative velocity of the blade element. Thus, it can be concluded that the decrement of power and thrust caused by elastic blade are getting larger in the wake flows according to the comparisons above.

Table 11. Comparisons of mean output power with Jha et al. [34].

Research	Mean Power of Upstream Wind Turbine (MW)	Difference with Jha et al. [34]	Mean Power of Downstream Wind Turbine (MW)	Difference with Jha et al. [34]
EALM	2.0231	5.28%	0.7943	8.11%
ALM	2.0345	5.87%	0.8089	6.42%
Jha et al. [34]	1.9217	-	0.8644	-

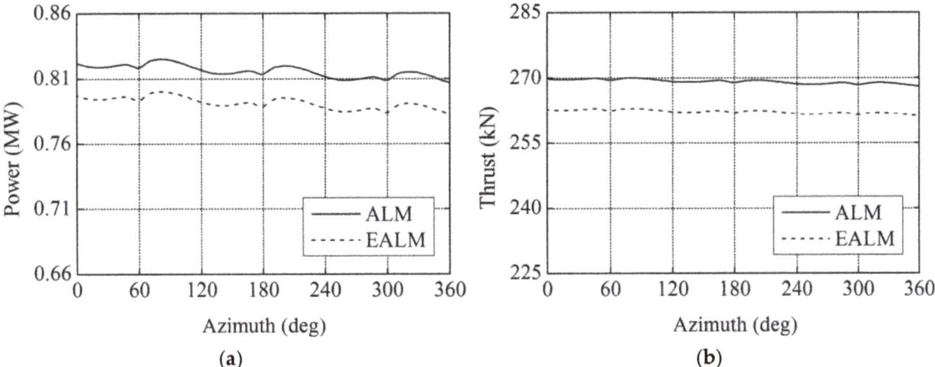

Figure 22. Comparison of downstream wind turbine aerodynamic performance results between Table 8. m/s: (**a**) power; and (**b**) thrust.

The stabilization of the normal and tangential force on downstream wind turbine blade are compared again in Figure 23. In contrast with Figure 19, the standard deviation of the loads on blade in the wake caused by upstream wind turbine getting larger when the elasticity of the blades is considered. It indicates that the influence of elastic blade becomes more serious when it is in the wake of upstream wind turbine. That will aggravate the fatigue loads and should be paid much more attention during corresponding studies.

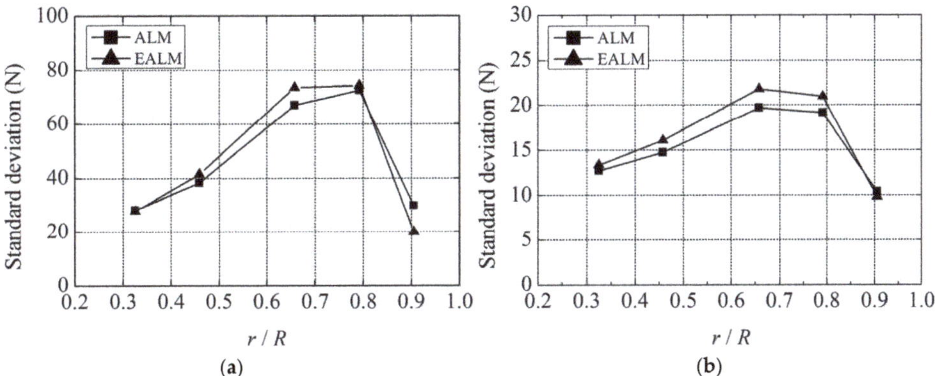

Figure 23. Stabilization of forces on downstream wind turbine blades: (**a**) normal force; and (**b**) tangential force.

Figure 24 illustrates the process of double wind turbines array vorticity development. The wake expands as expected and the tip vorticity turns into continuum gradually in Figure 24a. Then the vorticity breakdown and the smaller-scale turbulence appears with distance increasing from upstream wind turbine (referring to Figure 24b). When the downstream wind turbine meets with this complicated wake, where the turbulence levels raise because of upstream wind turbine wakes and the momentum has not recovery completely, its tip vorticities break down earlier. That makes the wakes behind downstream wind turbine more complex, which can be seen in Figure 24c.

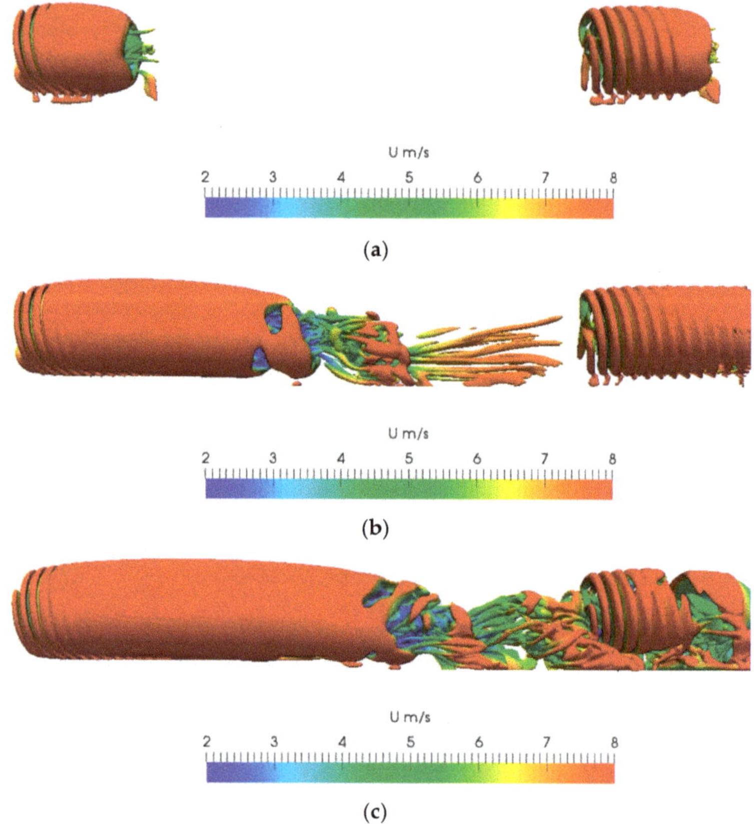

Figure 24. The wake structure development of double wind turbine aligned: (**a**) initial status without disturb; (**b**) developing status; and (**c**) interference status.

Figure 25 shows the development of wake velocity field correspondingly. The phenomenon of velocity deficit is easy to observe behind the rotor. Before the downstream wind turbine disturbed from the wakes generated by upstream wind turbine, the velocity field is stable relatively. As the interference status appears, the distribution of velocity field in nearby regions of downstream wind turbine is in a muddle condition. That will increase the fatigue loads on blade of downstream wind turbine.

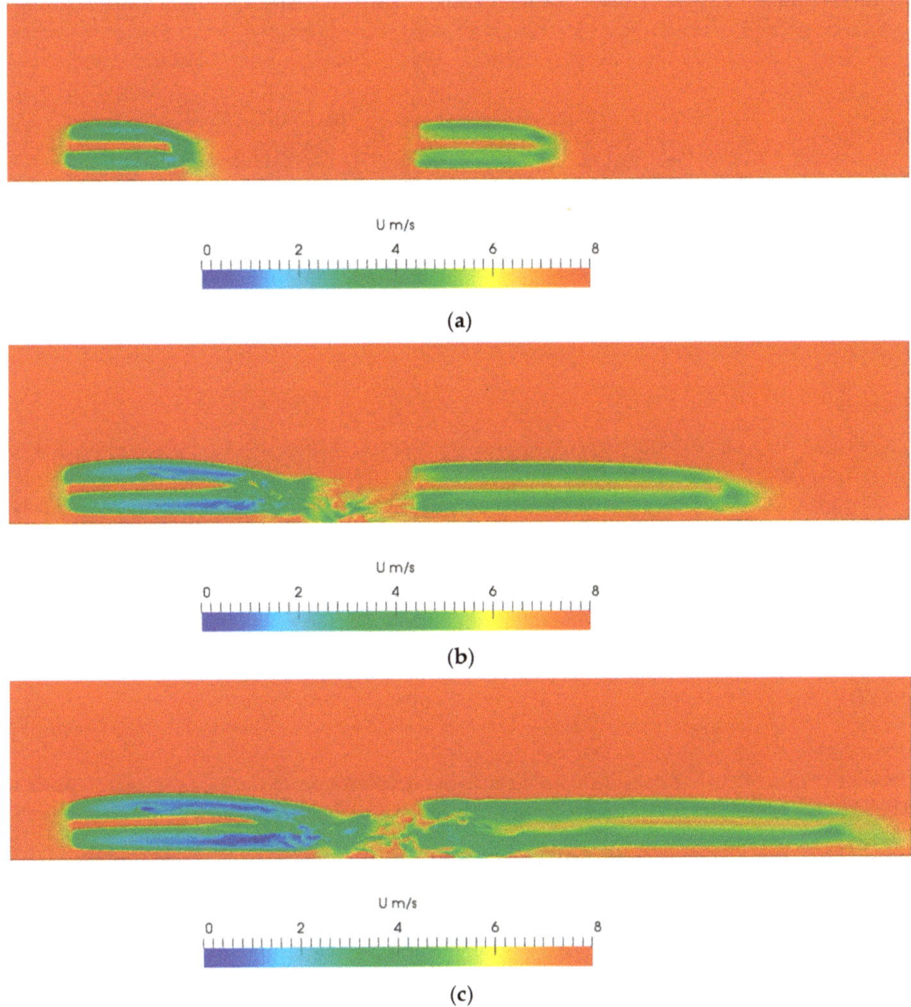

Figure 25. The wake velocity field development of double wind turbine aligned: (**a**) initial status without disturb; (**b**) developing status; and (**c**) interference status.

5. Conclusions and Discussions

A new elastic actuator line model, which combines the traditional actuator line model and a beam solver used in the blade design, is proposed to study the aeroelastic performance of wind turbine efficiently. The validation is set up through comparing the result of the power, the thrust, and the tip deflection with different research. It is found that this new model can obtain acceptable prediction, including aeroelastic performance and wake fields. Besides, the tip displacement in 0 direction is the nearest with the reference data from official technical report given by NREL's FAST code among these studies. Furthermore, the elasticity of the blades can aggravate tower shadow effect and has the possibility of wind turbine instability. The fatigue loads on the blades also become more serious. When the wake effect generated by upstream wind turbine is considered, the impact above will get more obvious.

However, this new elastic actuator line model is incomplete and needs to be improved. For example, the modal shape and natural frequency results are not yet considered. The purpose of this study is to propose an approach for investigating the aeroelastic performance rapidly. Further study is still required to refine the solver of structural dynamic characteristics and wave interactions [35].

Author Contributions: Z.Y. made the computations, conducted data analysis; Z.H. made the comparison of calculation results; X.Z. gave the suggested ideas and did the proof reading; Q.M. guided the whole project; H.H. gave the data analysis of double wind turbine wake flows. All authors have read and agreed to the published version of the manuscript.

Funding: This research received no external funding.

Acknowledgments: This research work was supported by the National Natural Science Foundation of China (Nos. 51879051; 51739001; 51639004). The fourth author also thanks the Chang Jiang Visiting Chair Professorship scheme of the Chinese Ministry of Education, hosted by HEU.

Conflicts of Interest: The authors declare no conflict of interest.

References

1. Agarwal, P.; Manuel, L. Simulation of offshore wind turbine response for long-term extreme load prediction. *Eng. Struct.* **2009**, *31*, 2236–2246. [CrossRef]
2. Ghassempour, M.; Failla, G.; Arena, F. Vibration mitigation in offshore wind turbines via tuned mass damper. *Eng. Struct.* **2019**, *183*, 610–636. [CrossRef]
3. Kim, D.H.; Lee, S.G.; Lee, I.K. Seismic fragility analysis of 5MWoffshore wind turbine. *Renew. Energy* **2014**, *65*, 250–256. [CrossRef]
4. Santangelo, F.; Failla, G.; Santini, A.; Arena, F. Time-domain uncoupled analyses for seismic assessment of land-based wind turbines. *Eng. Struct.* **2016**, *123*, 275–299. [CrossRef]
5. Rendon, E.A.; Manuel, L. Long-term loads for a monopile-supported offshore wind turbine. *Wind Energy* **2012**, *17*, 209–223. [CrossRef]
6. Alati, N.; Failla, G.; Arena, F. Seismic analysis of offshore wind turbines on bottom-fixed support structures. *Philos. Trans. R. Soc. A* **2015**, *373*, 20140086. [CrossRef]
7. Dong, W.; Moan, T.; Gao, Z. Long-term fatigue analysis of multi-planar tubular joints for jacket-type offshore wind turbine in time domain. *Eng. Struct.* **2011**, *33*, 2002–2014. [CrossRef]
8. Jonkman, J.; Butterfield, S.; Musial, W.; Scott, G. *Definition of a 5-MW Reference Wind Turbine for Offshore System Development*; National Renewable Energy Laboratory technical report; National Renewable Energy Laboratory (NREL): Golden, CO, USA, 2009.
9. Bak, C.; Zahle, F.; Bitshe, R.; Kim, T.; Yde, A.; Henriksen, L.C.; Natarajan, A.; Hansen, M. *Description of the DTU 10 MW Reference Wind Turbine*; Technical University of Denmark Wind Energy report; DTU Wind Energy: Frederiksborgvej, Denmark, 2013.
10. Tran, T.T.; Ryu, G.J.; Kim, Y.H.; Kim, D.H. CFD-Based Design Load Analysis of 5MW Offshore Wind Turbine. In Proceedings of the 9th International Conference on Mathematical Problems in Engineering, Aerospace and Sciences, Vienna, Austria, 10–14 July 2012; Volume 1493, pp. 533–545.
11. Miao, W.P.; Li, C.; Pavesi, G.; Yang, J.; Xie, X.Y. Investigation of wake characteristics of a yawed HAWT and its impacts on the inline downstream wind turbine using unsteady CFD. *J. Wind Eng. Ind. Aerodyn.* **2017**, *167*, 60–71. [CrossRef]
12. Liu, Y.C.; Xiao, Q.; Incecik, A.; Peyrard, C.; Wan, D.C. Establishing a fully coupled CFD analysis tool for floating offshore wind turbines. *Renew. Energy* **2017**, *112*, 280–301. [CrossRef]
13. Ma, Z.; Zeng, P.; Lei, L.P. Analysis of the coupled aeroelastic wake behavior of wind turbine. *J. Fluids Struct.* **2019**, *84*, 466–484. [CrossRef]
14. Zhang, P.T.; Huang, S.H. Review of aeroelasticity for wind turbine: Current status, research focus and future perspectives. *Front. Energy* **2011**, *5*, 419–434. [CrossRef]
15. Zienkiewicz, O.C.; Taylor, R.L. The finite element method. *McGraw-Hill Lond.* **1977**, *3*, 39–84.
16. Ferede, E.; Abdalla, M.M.; van Bussel, G.J.W. Isogeometric based framework for aeroelastic wind turbine blade analysis. *Wind Energy* **2017**, *20*, 193–210. [CrossRef]
17. Li, Y.; Castro, A.M.; Sinokrot, T.; Prescott, W.; Carrica, P.M. Coupled multi-body dynamic and CFD for wind turbine simulation explicit wind turbulence. *Renew. Energy* **2015**, *76*, 338–361. [CrossRef]

18. Ageze, M.B.; Hu, Y.; Wu, H.C. Comparative Study on Uni- and Bi-Directional Fluid Structure Coupling of Wind Turbine Blades. *Energies* **2017**, *10*, 1499. [CrossRef]

19. Heinz, J.C.; Sørensen, N.N.; Zahle, F. Fluid-structure interaction computations for geometrically resolved rotor simulations using cfd. *Wind Energy* **2016**, *19*, 2205–2221. [CrossRef]

20. Sørensen, J.N.; Shen, W.Z. Numerical modeling of wind turbine wakes. *J. Fluids Eng.* **2002**, *124*, 393–399. [CrossRef]

21. Fleming, P.; Gebraad, M.O.; Lee, S.; van Wingerden, J.W.; Johnson, K.; Churchfield, M.; Michalakes, J.; Spalart, P.; Moriarty, P. Simulation comparison of wake mitigation control strategies for a two-turbine case. *Wind Energy* **2015**, *18*, 2135–2143. [CrossRef]

22. Na, J.S.; Koo, E.; Munoz-Esparza, D.; Jin, E.K.; Linn, R.; Lee, J.S. Turbulent kinetics of a large wind farm and their impact in the neutral boundary layer. *Energy* **2016**, *95*, 79–80. [CrossRef]

23. Meng, H.; Lien, F.-S.; Li, L. Elastic actuator line modelling for wake-induced fatigue analysis of horizontal axis wind turbine blade. *Renew. Energy* **2018**, *116*, 423–437. [CrossRef]

24. Hansen, M.O.L. *Aerodynamics of Wind Turbines*; Earthscan: London, UK; Sterling, AV, USA, 2008.

25. TurbinesFoam. 2014. Available online: https://github.com/turbinesFoam (accessed on 16 January 2019).

26. Shives, M.; Crawford, C. Mesh and load distribution requirement for actuator line CFD simulations. *Wind Energy* **2013**, *16*, 1183–1196. [CrossRef]

27. Shen, W.Z.; Mikkelsen, R.; Sørensen, J.N. Tip Loss Corrections for Wind Turbine Computations. *Wind Energy* **2005**, *8*, 457–475. [CrossRef]

28. Yu, Z.Y.; Zheng, X.; Ma, Q.W. Study on Actuator Line Modeling of Two NREL 5-MW Wind Turbine Wakes. *Appl. Sci.* **2018**, *8*, 434. [CrossRef]

29. Ferziger, J.H.; Peric, M. *Computational Methods for Fluid Dynamics*; Springer Science & Business: Berlin, Germany, 2012.

30. Jeong, M.; Cha, M.; Kim, S.; Lee, L. Numerical investigation of optimal yaw misalignment and collective pitch angle for load imbalance reduction of rigid and flexible hawt blades under sheared wind inflow. *Energy* **2015**, *84*, 518–532. [CrossRef]

31. Ponta, F.L.; Otero, A.D.; Lago, L.I.; Rajan, A. Effect of rotor deformation in wind-turbine performance: Dynamic rotor deformation blade element momentum model (DRD-BEM). *Renew. Energy* **2016**, *92*, 157–170. [CrossRef]

32. Yu, D.O.; Kwon, O.J. A coupled CFD-CSD method for predicting HAWT rotor blade performance. In Proceedings of the 51st AIAA Aerospace Science Meeting Including the New Horizons Forum and Aerospace Exposition, Grapevine, TX, USA, 7–10 January 2013; p. 911.

33. Yu, D.O.; Kwon, O.J. Time-accurate aeroelastic simulations of a wind turbine in yaw and shear using a coupled CFD-CSD method. In Proceedings of the Science of Making Torque from Wind 2014 (TORQUE 2014), Copenhagen, Denmark, 18–20 June 2014; p. 524.

34. Choi, N.J.; Nam, S.H.; Jeong, J.H.; Kim, K.C. Numerical study on the horizontal axis turbines arrangement in a wind farm: Effect of separation distance on the turbine aerodynamic power output. *J. Wind Eng. Ind. Aerodyn.* **2013**, *117*, 11–17. [CrossRef]

35. Zheng, X.; Shao, S.D.; Khayyer, A.; Duan, W.Y.; Ma, Q.W.; Liao, K.P. Corrected first-order derivative ISPH in water wave simulations. *Coast. Eng. Res.* **2017**, *59*, 1750010. [CrossRef]

MDPI
St. Alban-Anlage 66
4052 Basel
Switzerland
Tel. +41 61 683 77 34
Fax +41 61 302 89 18
www.mdpi.com

Water Editorial Office
E-mail: water@mdpi.com
www.mdpi.com/journal/water